ALLE ZEIT WACH 1842

Paul G. J. Maquet

Biomechanics of the Knee

With Application to the Pathogenesis and the Surgical Treatment of Osteoarthritis

2nd Edition, Expanded and Revised

With 243 Figures

Springer-Verlag
Berlin Heidelberg NewYork Tokyo 1984

Docteur Paul G.J. Maquet
25, Thier Bosset, B-4070 Aywaille

ISBN-13: 978-3-642-61733-1 e-ISBN-13: 978-3-642-61731-7
DOI: 10.1007/978-3-642-61731-7

Library of Congress Cataloging in Publication Data

Maquet, Paul G.J., 1928 –
Biomechanics of the knee.

Translation of: Biomécanique du genou.
Bibliography: p.
Includes index.
1. Knee. 2. Knee-Surgery. 3. Osteoarthritis-Surgery. 4. Human mechanics. I. Title.
RD561.M3613 1983 617'.582 83-14833

Softcover reprint of the hardcover 2nd edition 1984

Typesetting, printing, and bookbinding: Universitätsdruckerei H. Stürtz AG, Würzburg.
2124/3130-543210

Ce n'eſt autre choſe practique
Sinon l'effect de Theorique.

Sience ſans experience
N'apporte pas grande aſſeurance.

AMBROISE PARE', DE LA VAL CONSEILLER ET PREMIER CHIRVRGIEN du Roy.

CANONS ET REIGLES Chirurgiques de l'Autheur.

"Les Oeuvres d'Ambroise Paré, conseiller, et premier chirurgien du Roy, corrigées et augmentées par lui-même, peu avant son décès. A Paris, chez Barthelemy Macé, au mont St-Hilaire, à l'Ecu de Bretagne. 1607".

Foreword

Pathological conditions affecting the hip and knee joints occupy a particular place amongst the important orthopaedic entities affecting the extremities. On the one hand they are relatively frequent and on the other they mean for the patient limitation of his ability to walk, because of their considerable detrimental effects.

A purposeful basic treatment of these joint diseases (and here osteoarthritis takes pride of place) is only possible if it stems from a reliable biomechanical analysis of the normal and pathological stressing of the joint in question. Whilst the situation in the hip can be considered to be fundamentally clarified, a comprehensive representation of the knee is still lacking, particularly when taking into account the latest knowledge of biomechanics. Recently our concepts of the kinematics of the knee have been completely changed, but the clinically important question of articular stressing remains unanswered.

Dr. Maquet has carried out pioneer work in this field for some years in adapting, by analogy, to the knee joint principles already accepted for the hip joint. Since the knee is not a ball and socket joint, a complicated problem arises for which new thoughts are necessary. The results of the numerous operations carried out by Dr. Maquet according to the biomechanical considerations demonstrate that his thinking is fundamentally correct. Above all, it is here again proven (as earlier in the case of the hip) that healing of osteoarthritis depends decisively on reducing and evenly distributing joint pressure.

In the present book Dr. Maquet proceeds from the earlier static analyses for his evaluation of the kinetic stressing, by taking advantage mainly of the very accurate research of O. Fischer about human gait. The result is a survey of the stressing of the knee joint and until now nothing comparable has existed. Moreover, it is shown in an impressive way how judicious surgical procedures based on biomechanics can cause healing even in severe osteoarthritis of the knee. This happens, above all, without implanting endoprostheses which are certainly still more questionable for the knee than they already are for the hip.

May this book be widely disseminated and may much further discussion of this burning problem be stimulated!

Aachen, Summer 1976 F. Pauwels

Preface to the First Edition

Pauwels in 1950, relying on an experience of over 15 years, showed that the clinical and radiological signs of osteoarthritis of the hip could be made to disappear by a proper surgical approach which permitted astonishing regeneration of the diseased joint. His interventions were based on a profound knowledge of hip mechanics and biomechanics in general. The aim of these procedures was to diminish as much as possible the articular pressure to render this pressure supportable by the diseased tissues. These procedures decreased the pressure by reducing the load supported by the hip and increasing the weight-bearing surface of the joint.

The laws of biomechanics enunciated by Pauwels (1950, 1951, 1958, 1959, 1960, 1964, 1965a, b, 1968, 1973a, b) and applied by him to the hip, elbow, and shoulder have general application. It seemed logical, therefore, in osteoarthritis of the knee to apply these rules which had permitted Pauwels to obtain spectacular results in patients suffering from osteoarthritis of the hip. This required a knowledge of the mechanics of the normal knee and of the arthritic knee which we have not been able to find in the literature. Certain movements of the knee have already been studied. However, the influence of the mechanical factors on osteoarthritis of the knee has never been clearly explained. Orthopaedic surgeons have empirically corrected varus deformities of the knee by a valgus osteotomy and valgus ones by a varus osteotomy in either the lower part of the femur or upper part of the tibia. They have obtained inconsistent results. From their experience they have concluded that one has to achieve overcorrection of the femoro-tibial angle in the coronal plane in varus deformities and reasonably precise correction for valgus deformities. These recommendations, deduced from empirical experience, are barely sufficient for one who wishes to base treatment on a sound theoretical foundation and obtain consistently good results. The essential aim of our labours is to furnish this theoretical foundation, rationally justifying the choice of one form of intervention over another, thereby improving the treatment of osteoarthritis of the knee.

Aywaille, Summer 1976 PAUL G.J. MAQUET

Preface to the Second Edition

Since publication of the first edition further research has improved our knowledge of the mechanics of the knee, helping us especially to define more precisely the appropriate approaches to the different types of osteoarthritis. In this edition more emphasis has been put on surgical treatment. Tibial osteotomy remains the method of choice for treating medial osteoarthritis with a varus knee, whereas femoral osteotomy appears the best way of dealing with lateral osteoarthritis with a valgus deformity. Planning and surgical procedures are described more completely. The postoperative results are not only illustrated, but also more thoroughly analysed. These results support our views on biomechanics and osteoarthritis of the knee.

Aywaille, Summer 1983 Paul G.J. Maquet

Acknowledgments

Professor F. Pauwels taught me the basic principles of biomechanics and spent innumerable hours discussing my clinical cases and shaping my way of thinking in orthopaedics. I am also deeply indebted to the many other individuals who made this work possible and first of all to my wife, Josette, for her patient understanding and for her many hours of work in preparing this manuscript.

My good friend, Dr. P. de Marchin, initially germinated the idea of this research. I would never have carried it out without the constant advice and encouragement of Professor J. Lecomte. Professor B. Kummer was most helpful in freely discussing and constructively criticizing the work.

Professor A. Pirard and his assistants, Mrs. G. Pelzer and Mr. F. de Lamotte, not only provided a solution to the difficult mathematical problems but also allowed me to use their laboratory with the help of their technician, Mr. C. Nihard, for the photoelastic studies. Professor E. Betz has kindly furnished me most of the anatomical specimens I needed. Mr. J. Simonet built the apparatus to measure the loads exerted on the specimens. Dr. A. Van de Berg X-rayed them in his private office. Dr. M. Vercauteren supplied the gait movie. After Dr. N. Matsumoto and Dr. T. Yamaguchi had each spent more than a year with me, they carried out arthroscopies of the knee before and after surgery, with Y. Fujisawa, K. Masuhara, N. Mii, H. Fuyihara and S. Shomi. They kindly granted me permission to reproduce some of their pictures in this book. Dr. F. Burny and Mrs. M. Donkerwolcke, using the computer terminal of the Hôpital Erasme at the Université Libre de Bruxelles and their knowledge of statistical analysis, enabled me to write the chapter dealing with the results. Associate Professor E.L. Radin helped me edit the first edition. Mr. R. Furlong F.R.C.S. offered his time, his patience and his accurate knowledge of the English language to prepare the final version of the text for both first and second editions.

Contents

Permission to reproduce

Figures 13, 14, 18, 20, 21, 22, 23, 24, 26, 67, 75, 76, 81, 106, 139 (Maquet, Simonet, and de Marchin, 1967) has been granted by courtesy of the *Revue de Chirurgie Orthopédique et réparatrice de l'Appareil Moteur;*

Figures 11, 15, 65, 74, 82, 107, 109, 116, 236, 242 (Maquet, 1969) by the *Société Internationale de chirurgie Orthopédique et Traumatologique;*

Figures 16, 17, 29, 30, 143, 152, 183a and 184b (Maquet, 1972) and 51, 66, 113, 114, 117 (Maquet, Pelzer, and de Lamotte, 1975) by the *Acta Orthopaedica Belgica;*

Figures 59, 62, 63 (Maquet, Van de Berg, and Simonet, 1975) by the *Journal of Bone and Joint Surgery;*

Figure 54 (Maquet, 1976) by the *Clinical Orthopaedics and Related Research;*

Figures 245, 246 by Fujisawa et al. and by *Igaku Shoin Ltd;*

Figures 155, 156 by Shiomi et al. and by the *Japanese Arthroscopy Association.*

Chapter I. Aims and Limitations of the Work

Biomechanics encompasses a study of:

1. the mechanical stresses to which living tissues are subjected under physiological and pathological conditions;

2. the biological response of the tissues to these mechanical stresses and to their modifications;

3. the possibility of surgically changing the stresses in the living tissues to achieve a therapeutic effect.

This monograph will apply the discipline of biomechanics to the knee.

The forces supported by the normal knee will be analysed successively in standing symmetrically on both feet, in standing on one foot and during gait. These forces are transmitted from the femur to the tibia across joint surfaces the size of which will be measured. They provoke compressive stresses in the joint which will be defined.

There normally exists a physiological balance between the mechanical stress and the resistance of the articular tissues. This equilibrium can be disturbed by several factors: either the resistance of the articular tissues can be lowered by metabolic causes with the mechanical stress remaining normal or the mechanical stress can become abnormally great due to a mechanical disturbance while the integrity of the tissue remains normal. Disturbance of the physiological equilibrium produces reactions in the tissues, leading to osteoarthritis (Müller, 1929; Pauwels, 1973).

If the origin of the imbalance is metabolic and diminishes the tissue resistance, osteoarthritis will initially affect the whole knee. If it is mechanical, the osteoarthritis can primarily affect the medial or lateral part of the femoro-tibial joint. Degeneration can also be localized to the patello-femoral joint.

The causes of mechanically induced degeneration of the knee must be understood. We shall study what can modify the forces exerted on the knee and the mechanical consequences of these changes. The biological phenomena which lower the resistance of the tissues to mechanical stress will not be considered as this is more in the province of metabolic pathology.

After having determined the forces acting on the normal and abnormal knee, as well as the articular joint stresses in both the physiological and pathological states, we shall discuss how to influence the latter by surgery in order to achieve a therapeutic effect. We shall propose several operative procedures which reduce the mechanical stress in the knee. Original techniques will be described. The results of this surgery will be presented to illustrate and substantiate the theoretical analysis and biomechanical principles which are the basis of the treatment.

Methods which replace the whole or parts of the joint by metallic or plastic implants have certain indications but do not use the potential for regeneration possessed by living tissue. By strict definition they are not a biomechanical treatment and will not be considered in this presentation.

Chapter II. Review of the Literature

Several writers have been interested in the mechanics of the knee. Bouillet and Van Gaver (1961) and Debrunner and Seewald (1964) have studied its statics in the coronal plane; Shinno (1961, 1962, 1968a, b), Reilly and Martens (1972) and Bandi (1972) in the sagittal plane and Knese (1955) in both planes. Others have tried to clarify particular points such as the distribution of forces (Rabischong et al., 1970; Engin and Korde, 1974; Kostuik et al., 1975; Izadpanah and Keönch-Fraknóy, 1977), the surface of the femoro-tibial weight-bearing areas (Kettelkamp and Jacobs, 1972; Walker and Hajek, 1972; Walker and Erkman, 1975) or the contact surfaces of the patello-femoral joint (Goodfellow et al., 1976; Townsend et al., 1977). More detailed works have analysed the static and dynamic forces exerted on the knee during gait (Morrison, 1968, 1970; Paul, 1965, 1966, 1967, 1968; Harrington, 1974). Finally, Menschik (1974, 1975), developing and complementing the work of Husson (1974), gives a logical explanation of the kinematics of the knee. We shall review and analyse these papers.

Bouillet and Van Gaver (1961) write that the knee supports only the body weight and that each tibial plateau supports half of it. In other words, during gait the whole body weight would act at the centre of gravity of the weight-bearing surfaces of the knee.[1]

However, the authors observed that in standing on one foot the vertical from the centre of gravity of the body is medial to the knee. Balance could only be maintained by a "contraction of the gluteus medius" or "a translation of the body toward the loaded side." "These two mechanisms complement each other during gait. Because of them the line of action of the body is brought as close to the knee as possible but is never brought directly over the axis of the leg. As a consequence the normal knee is always stressed in varum."

These concepts of Bouillet and Van Gaver would seem to be based on an incomplete analysis. First, the knee does not support the weight of the underlying leg and foot. Consequently, the centre of gravity of the rest of the body is not located in the middle of the body but to the side opposite to the loaded knee. To bring back this centre of gravity above the knee would, at the very least, produce severe limping.

Second, the gluteus medius counterbalances the partial body weight at hip level and in so doing prevents tilting of the pelvis. It cannot prevent the femur from tilting into adduction on the tibia because it is not inserted into the tibia. Another force is thus necessary to balance the partial body weight.

Finally, observation of X-rays shows that the normal knee is not stressed in varum but regularly supports a well-centred load. We shall discuss this.

Bouillet and Van Gaver show three drawings in the sagittal plane (see their Figs. 37, 38, and 39) to explain the action of the patella. These drawings are too inexact and cannot be used to analyse the forces exerted on the knee in the sagittal plane.

Debrunner and Seewald (1964) study the loaded knee in the coronal plane. They analyse the successive positions of the gait. But the projection on one plane of the forces exerted on the joint can only give a partial idea of what happens during gait. Actually, the projection on the coronal plane of the distance between the knee and the line of action of the force exerted by the mass of the body changes very little during the single support period of gait. On the contrary, the projection of this distance on the sagittal plane varies continuously and consider-

[1] The centre of gravity is used here to define the functional centre of the joint which lies between the medial and lateral articulating compartments.

ably. That is why the forces must be analysed in space, or at least their projection on two rectangular planes.

Debrunner and Seewald claim that the action of the body on the ground and the reaction of the ground to the body weight are not exerted along the same line but are parallel. This would, according to the authors, produce a bending moment which would reduce the load on the knee.

If action of the body and reaction of the ground are not exerted along the same direction, they give a torque. No equilibrium is possible. The subject falls or rotates. Since there is equilibrium in the standing position as well as at every instant during gait, no torque is produced. Actually, during gait, as shown by Fischer (1899, 1900), the line of action of the force exerted by the body mass is not vertical but oblique. It results from gravity and from dynamic forces. It passes through the centre of gravity of the body and intersects the supporting surface of the stance foot. Action of the body force and reaction of the ground are exerted along this line but in opposite directions. They balance each other.

Shinno (1961, 1962, 1968a, b) analyses the relations between femur, tibia, patella and menisci during the movements of the knee. He takes into account the body weight and the force developed by muscles to keep equilibrium in several static positions from complete extension to squatting. They are studied in the sagittal plane. But there exists a continual confusion between forces and moments in Shinno's text and figures. Shinno uses the weight of the body minus both lower limbs for the gravity force. The weight of the thighs which acts on the knees is neglected. Morever, in squatting on one foot the weight of the opposite lower leg and foot must be added. The author ignores them.

It is also erroneous to calculate the force acting on the patella by multiplying the force due to the mass of the body by the length of contact between the patella and the femoral condyles. ("Oppressing power (P) against the patella in each flexed angle of the knee is calculated multiplying the pressing power (p) by the length of contact (*l*) between patella and femoral articular surfaces.") Multiplying a force by a length does not give a force. The force developed by the quadriceps muscle to balance the body weight does not depend only on the angle formed by the femoral diaphysis and the tibia, as indicated by Shinno, but on the moment of the partial body weight and on the lever arms of the quadriceps and of the patella tendon. Shinno neglects these lever arms. His conclusions are obviously based on an inadequate analysis. This is confirmed by the fact that, for Shinno, the force compressing the patella against the condyles is not normal to the weight-bearing surfaces (see Figs. 8 and 9, 1961, of the author). Mechanically this is unconceivable in conditions of equilibrium since friction in a human joint is practically negligible.

Reilly and Martens (1972) claim experimentally to determine the force developed by the quadriceps and the force pressing the patella against the femur. In their calculation of the force exerted by the patella tendon to extend the lower leg they use the equation of moments, giving as origin of the lever arms a fixed point between the femur and the tibia ("the point of application of the knee joint reaction force, is taken to be constant throughout the angles of flexion considered"). They then estimate that the force exerted by the patella tendon is equal to that developed by the quadriceps muscle ("the patellar tendon force can be assumed to be equal in magnitude to the quadriceps force"). In reality, the origin of the lever arm of the forces causing flexion and extension of the lower leg on the thigh corresponds to the axis of flexion of the knee. This axis changes its position continuously during movement (Fick, 1910), modifying the length of the lever arms of the forces, the direction of which also varies. Furthermore, the force developed by the quadriceps muscle and that transmitted by the patella tendon differ because their lever arms are not equal and vary differently during movement as a result of the shape of the knee-cap. It must be noticed that the lever arm with which the patella tendon extends the lower leg is not the same as that with which it acts on the knee-cap simply because the femoro-tibial joint and the patello-femoral joint each have their own centre of rotation at any phase of movement. The lever arm with which the patella tendon extends the lower leg stems from the centre of curvature of the femoro-tibial weight-bearing surfaces. That with which the patella tendon acts on the knee-cap has the same origin as that of the quadriceps tendon, at the centre of curvature of the patello-femoral force-transmitting surface. Failure to recognize these elementary facts led Reilly and Martens to erroneous conclusions.

Bandi (1972) in his study of patello-femoral osteoarthritis, also confuses the lever arm with which the patella tendon acts on the tibia and that with which it acts on the patella. He also attributes to the force exerted by the patella tendon the same magnitude as to that developed by the quadriceps muscle ("$M_1 = M_2$"). The quantitative results of the work of Bandi cannot, therefore, be more accurate than those of Reilly and Martens.

Rabischong et al. (1970) measure the distribution of forces at the level of the femoral condyles in static loading, by compressing femora and tibiae separately along their long axis. But the conditions of the experiment do not correspond to the physiological situation. This is also suggested by their results. The authors observe an overloading of the lateral condyle of the femur on one hand and of the medial plateau of the tibia on the other hand. But mechanically the force transmitted by the femoral condyle must equal the force on its corresponding tibial plateau. In fact Rabischong et al. have done experiments in resistance of materials and in transmission of forces on isolated bones without taking their actual function into consideration. No practical conclusion can be drawn from this work as far as physiology of the knee is concerned. As written by the authors themselves, the muscular and ligamentous forces completely change the data of the problem such as it has been summarily considered. We shall see it further.

Engin and Korde (1974) claim "to determine the resultants of the pressure distribution on both lateral and medial plateaux in normal and abnormal knee joint configurations." For these authors, "it is sufficient to consider one position of the lower limb; that of standing was chosen... The problem of determining the muscle forces acting on the knee can now be neglected." This assumption seems preposterous. Try to stand on one leg without any action of your muscles! The authors take a cadaver knee, strip from it the muscles, patella tendon and patella, preserving the collateral and cruciate ligaments. They equip the femur and the tibia with strain-gauges, 4 inches on either side of the joint, and load their preparation in different configurations. "The abnormal configurations were successively obtained by rotating the tibial end holder to the desired angle. This rotation of the tibial end in varus or valgus direction results in a separation of either the medial or the lateral condyles." The authors thus deal not with osteocartilaginous deformity typical of osteoarthritis, but rather with a rotation of the tibia about the centre of curvature of the medial condyle or about that of the lateral condyle of the femur. This means that in these experimental conditions the line of action of the force transmitted from the femur to the tibia must pass either through the centre of curvature of the medial condyle of the femur or through that of the lateral condyle. This does not seem generally to be the case in osteoarthritis, as we shall see.

Kostuik et al. (1975) attempt to determine the displacement of the centre of the weight-bearing surfaces of the knee resulting from a varus or a valgus deformity. In order to create the deformity in a cadaver knee, they simply tilt the femur on the tibia medially or laterally, as do Engin and Korde. Their results are thus open to the same criticism.

Izadpanah and Keönch-Fraknóy (1977a, b) also measure the forces transmitted through the condyles when a bare femur is compressed longitudinally in a machine by two parallel surfaces. As reported by Kummer (1977), their anatomical specimen is forced into a position such that the two condyles are in contact with one surface and the other surface is tangential to one point of the femoral head. The line of action of the load must pass through this point of contact of the femoral head and between the contacts of the condyles, otherwise the specimen would tilt. In the drawing published by the authors, the line of action of the load passes through the lateral condyle. An appropriate lateral displacement of the distal fragment, after a transverse supracondylar osteotomy, shifts the line of action of the load just into the intercondylar notch and ensures a symmetrical distribution of the load between both condyles. This is true only in the conditions of the experiment, which has not much in common with reality. In fact, the knee is stressed by forces due to the partial body weight and to muscles. These forces differ completely from the load applied to the femur in the experiment, from which no practical conclusion can be drawn.

Furthermore, Izadpanah and Keönch-Fraknóy complement their research with photoelastic models. Besides the fact that these plexiglass models are loaded in an incorrect way, as is the anatomical specimen, they display isoclinics which the authors call stress-lines ("Spannungslinien") and from which they deduce the quantitative distribution of the stresses. From all the isoclinics successively obtained by turning the polarizer and analyser step by step from 0° to 90°, a network of trajectories (isostatics) can be drawn. These trajectories indicate the direction of the flow of stresses but not the quantitative distribution of the stresses. Isochromatics are required if the quantitative distribution of the stresses is to be shown (Kummer, 1977).

Kettelkamp and Jacobs (1972) used X-rays to determine the contact surfaces between the femur and the tibia for every 5° from full extension to 35° flexion. Their joint transmitted a load from 3 to 8 kg. But in fact the knee transmits a much greater load, probably in the hundreds of kilograms, in a normal individual (Morrison, 1968, 1970). The weight-bearing articular surfaces are probably larger with such a load than in the conditions of Kettelkamp and Jacobs' experiment. The figures published by Braune and Fischer (1891) show the femorotibial contact to be completely different when the joint is or is not compressed (see their Tables IX and X).

Walker and Hajek (1972) determined the weight-bearing surfaces of the knee by moulding the loaded joint with polymethylmethacrylate after excision of the menisci since "Previous trials using polymethylmethacrylate and silicone casting materials showed that the menisci prevented a coherent cast shape from being obtained. Careful meniscectomy was therefore performed." The mould surrounds the surfaces of direct contact between the femur and the tibia. But from the fibro-cartilaginous nature of the menisci it can be surmised that these are subjected to a compression in the direction of the long axis of the tibia and to stretching in the plane of the tibial plateaux. Squeezed between the femur and the tibia they are an obvious part of the weight-bearing surfaces of the knee which are artificially changed by their excision. Consequently, the results of Walker and Hajek can only be applied to a knee without menisci and cannot be related to the physiology of the normal joint.

More recently, Walker and Erkman (1975) measured the weight-bearing surfaces of the knee with the menisci intact, under loads from 0 to 150 kg. Their results are then very close to ours and roughly confirm them (Maquet et al., 1975).

Goodfellow et al. (1976) delineate the patello-femoral weight-bearing surface by a method of pigmentary impregnation. To this end, the tibia is fixed vertically in a vice. The flexed knee is loaded by hanging 3–10 kg from the upper extremity of the femur. Further flexion is prevented by the quadriceps tendon. According to these authors, the true weight-bearing surfaces, under a physiological load, probably keep the same location but may well be larger.

The same comment can be made on the research of Townsend et al. (1977).

Knese (1955) determined the muscular forces balancing the body weight. He used a skeleton of the lower limb, the tibia of which was rigidly fixed to a large support. Equilibrium was attained by metal springs which replaced the muscles. Such an experimental set-up is somewhat unphysiological. In any position of the living subject the line of action of the force provoked by the body mass must intersect the support, during stance the loaded foot. Otherwise there would be no equilibrium and the subject would fall. But in the experiments of Knese the line of action of the body weight does not necessarily pass through the loaded foot. It can intersect the artificial support far from the foot. This completely modifies the system of forces. The results of Knese are only acceptable in the conditions of his experiments and cannot be generalized to the living knee.

In our opinion, the most coherent analyses of the forces exerted on the knee have been done by Morrison (1968, 1970) and by Paul (1965, 1966, 1967, 1969). These authors have measured the forces exerted on the ground by walking individuals on a force plate (or dynamometer). Theirs was a non-deformable plate sustained by four columns equipped with strain-gauges which measure three moments and three forces, or six parameters, as designed by Cunningham and Brown.

The analysis of such a force plate has been done at the Faculté des Sciences appliquées of Liège[2]. It shows that the sensitivity of the plate is insufficient even to get two parameters. Moreover, the interactions between the columns modify the results and create artefacts.

The engineering faculty of Liège has replaced the columns by circular rings in a first model and by octogonal rings in a second. This artifice considerably improves the sensitivity of the device. But the problem of interactions persists. Practically, these are unavoidable. There are more when there are more components to be measured. One can reasonably count on interactions of 10–20%, more pronounced for the smallest efforts. Paul (1969) estimates a 20% margin of error for his results. If all the causes of inaccuracy which affect it are added, the error can attain 60%, according to the author. This method cannot be relied on for accurate determination.

Paul, as well as Morrison, analyses the displacements of light spots fixed to certain places on the lower limb, in two movie films simultaneously exposed. They first calculate the acceleration of the spots by applying a formula for smoothing the curve. But Paul (1969) observes: "This procedure was adopted to obtain the accelerations, although Felkel has indicated the preferability of graphical smoothing of data followed by check integrations." Knowing the accelerations in three rectangular planes, they calculate the forces exerted by the lower leg and foot and their moments. For this calculation, Paul uses the data of Braune and Fischer ("Using Braune and Fischer's coefficients $C_1 - C_4$"[3]) although in his introduction he criticizes the accuracy of these data ("The accuracy of these figures cannot, therefore, be high."). According to his own words, the results of Paul must be considered inexact.

[2] J. Simonet, Personal communication.

[3] After careful study of the work of Braune and Fischer we could not find the "coefficients" cited by Paul.

Having approximated the forces exerted on the ground by the whole body in movement and calculated the forces exerted by the lower leg and foot, Paul can only approximate the forces exerted on the knee.

Morrison also accepts rough simplifications, ("The joint structures and function as defined in mathematical terms involved a degree of mechanical simplification.") He considers the axis of flexion of the knee as fixed. But, as has been shown by Fick (1910) this axis moves considerably during gait. It changes for each phase of the single support period of gait. Moreover, Morrison accepts a contact line between the femoral condyles and the tibial plateaux and neglects the antero-posterior displacements of this line with movement. ("Anterior-posterior displacements of the line of contact from the Z_s axis due to the rolling of the femoral condyles on the tibial condyles in extension were neglected.") But the femoral condyles are supported by the tibial plateaux through weight-bearing surfaces variable in dimension and in position during the single support period of gait.

Moreover, in calculating the muscular forces, Morrison (1968) deliberately neglects the action of tensor fasciae latae, gluteus maximus, and popliteus ("Tensor fasciae latae and gluteus maximus, by tightening the iliotibial tract, resist adduction of the knee and popliteus which unlocks the knee joint at the beginning of flexion, do not fall naturally in any of the three groups and were omitted from the analysis"). These muscles certainly intervene a great deal in the equilibrium of the knee, as shown by Blaimont et al. (1971) in their electromyographic study of the tensor fasciae latae in walking. Consequently, Morrison cannot quantitatively estimate the forces exerted on the knee.

Harrington (1974) also uses a force plate to measure "the resultant force between ground and foot" in normal subjects and in patients with pathological knee joints waiting for an osteotomy or total replacement of the knee. He then calculates the force transmitted across the knee. He finds that "the centre of joint pressure tends to be located in the medial joint compartment for normal knees", that "the greater proportion of total joint load is transmitted by the medial joint compartment in normal individuals" and that there is "a normal tendency to load the medial side." The author also claims that "the bearing area of the lateral joint compartment is smaller than that of the medial," which is consistent with a larger load on the medial compartment of the knee. However, our experiments have shown little difference between the weight-bearing area of the medial compartment and that of the lateral. In these conditions, if more load is transmitted across the medial plateau than across the lateral, it is difficult to explain why the subchondral scleroses underlining the two plateaux are symmetrical in a normal knee. This image, which anybody can observe in an X-ray, points rather to a line of action of the joint reaction force passing through the centre of gravity of the weight-bearing surfaces of the joint, in the centre of the knee. Actually, if the calculations of Harrington are carried out again, assuming such a central position of the femoro-tibial force, they lead to data which are very close to ours, although our approach is completely different.

Tansen (1976) calculates the forces acting on the knee during gait using a force plate and a spatial linkage. He concludes that the load exerted on the medial plateau is much greater than that on the lateral plateau. As mentioned above, this is at odds with the X-ray picture of a normal knee.

Husson (1973a, b, 1974) and Menschik (1974a, b, 1975) logically demonstrate the complicated kinematics of the knee, which they consider essentially as a "four-bar chain with crossed links." But they do not analyse the mechanical stressing of the joint. Menschik's first contribution (1974a) deals with the cruciate ligaments considered as parts of the four-bar linkage and confirms the previous work of Husson (1973a, b). In his second contribution (1975), Menschik attempts mechanically to explain the screw home mechanism. His explanation is far from convincing. In his third contribution (1974b), he considers the projection of the cruciate and collateral ligaments of the knee on a sagittal plane and asserts that all the ligaments intersect at the instant centre of rotation of the joint in any position of the latter, acting like the spokes of a wheel. With A. Van de Berg, we marked the insertions of the ligaments on cadaver knees which we then X-rayed. This simple experiment failed to confirm Menschik's attractive theory.

In summary, from a critical analysis it is clear that the published works on the mechanics of the knee are either incomplete or based on misinterpretations of the mechanical conditions. Some neglect important anatomical data, or, more frequently, are admitted oversimplifications.

No complete analysis of the mechanical stress in the normal and pathological knee has yet been published. Our researches will attempt to fill this gap within practical limitations.

Chapter III. Methods

In order to determine the mechanical stress in the knee we have used standard techniques: mathematical analysis, arthrography of anatomical specimens, analysis of photoelastic models and clinical and radiological examination of patients. Computerized axial tomography and arthroscopy were also employed, but on a minor scale.

The results attained by these theoretical, experimental, and practical methods complement each other.

In this chapter we shall describe the general principles of the applied techniques. These will be discussed in detail later in the work.

I. Mathematical Analysis

It is not possible at the present time directly to measure either the forces exerted on the knee or the stresses they provoke. Theoretically the stresses could be registered through strain-gauges placed in the joint or in its immediate vicinity. The insertion and presence of such devices would necessarily influence movement and modify articular function. A prosthesis replacing the knee and equipped with proper strain-gauges would allow stress measurements, but such a prosthesis completely changes the function of the joint and, consequently, the gait. The data obtained in this way would not correspond to the physiological state.

The weight of the part of the body carried by the loaded knee can actually be determined. But the knee also supports dynamic forces due to acceleration and deceleration of this part of the body. In order to measure these accelerations directly, an accelerometer could be used which would move with the centre of gravity of the partial mass of the body (S_7 according to our designation, see page 24). But because of the relative displacements of the limbs, the head and the trunk, the partial centre of gravity S_7 moves continuously and does not correspond to a stable anatomically fixed spot. The accelerometer thus should ideally move not only with the body as it is displaced through the space but also inside the moving body. This is obviously impossible. Consequently, it is not possible directly to measure the accelerations of S_7.

Because of these limitations, the most accurate method to determine the forces acting on the knee seems to be calculation based on precise observations taking into account the anatomy, the stance, and the gait of a normal individual. All the basic data are furnished by the fundamental works of Braune (1889, 1891, 1895) and Fischer (1889, 1891, 1895, 1899, 1900, 1901, 1903, 1904).

The calculation is essentially based on trigonometry and analytical geometry, as used by engineers to define the forces acting on a body moving in space.

II. Experiments on Anatomical Specimens

In order to calculate the stresses in the joint, it is necessary to determine both the forces involved and the articular weight-bearing surfaces. They change during movement and this change must be accounted for.

Kettelkamp and Jacobs (1972) have injected opaque substance into the knees of cadavers. They exerted a compressive force of 3–8 kg on the joint which was X-rayed revealing a picture of the tibial plateaux in which the contact areas were radiolucent and surrounded by the opaque

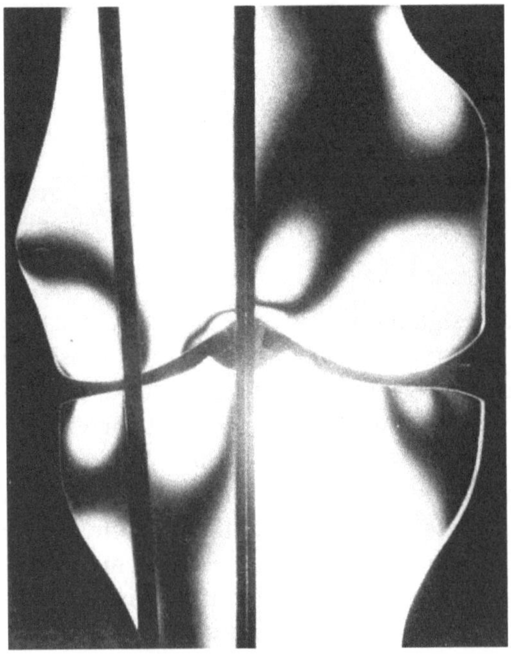

Fig. 1. An isoclinic (in dark). From the isoclinics the isostatics or trajectories can be drawn

III. Photoelastic Models

The distribution of stresses in the knee depends, among other factors, on the form of the articular surfaces. These show no simple geometrical outline. Theoretically calculating the distribution of the contact and internal stresses is thus not only difficult but impossible. An experimental method is then necessary and an optical process, based on photoelasticity, seems the best.

This allows the observation of the direction of the significant stresses as well as of their magnitude at every point, in a single plane. A homogeneous and isotropic model, representing a cross section of the object, can be loaded appropriately. A complete picture of the state of stresses in all the areas of the cross section appears in polarized light. The technique can be applied to a bone such as the femur or the tibia.

substance. They repeated the experiment in from two to four positions from complete extension to 35° of flexion. We have modified their method in order to be able to exert a known load of the same order of magnitude as the physiological load on a knee injected with barium sulphate. The modifications of the technique of Kettelkamp and Jacobs enabled us to show the femoro-tibial weight-bearing surfaces radiographically in several positions from complete extension to complete flexion of the knee.

The experimental details will be further described (page 62).

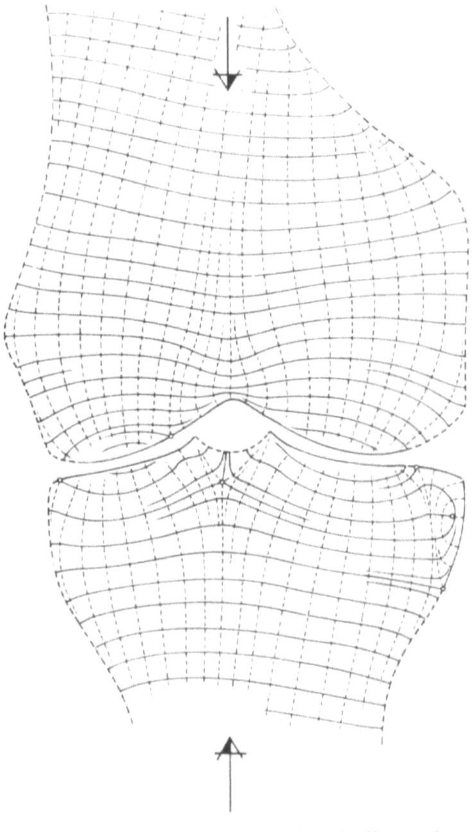

Fig. 2. Pattern of isostatics. They indicate the direction of the principal stresses

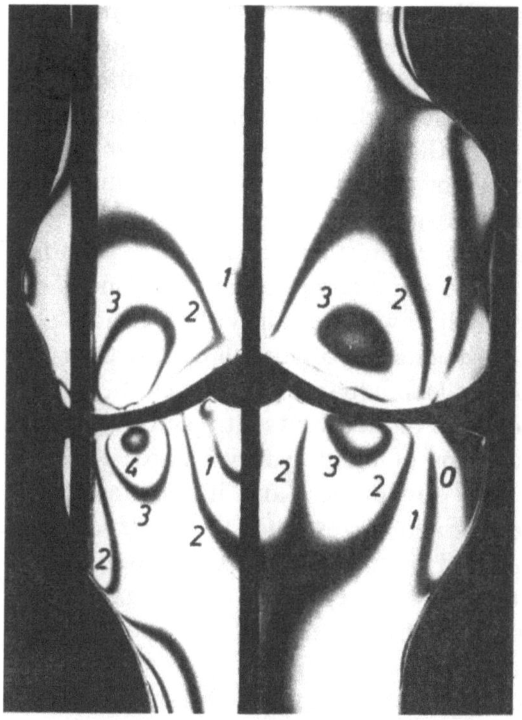

Fig. 3. Isochromatics or curves of equal differences between the principal stresses

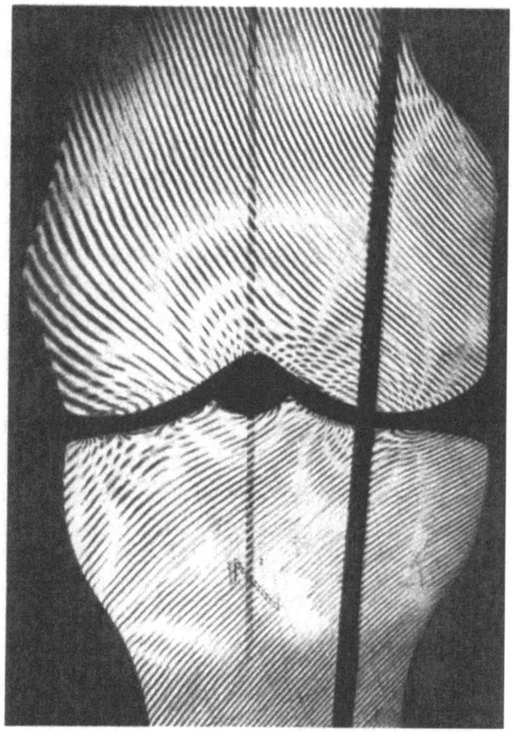

Fig. 4. Pattern of isopachics or curves of equal sums of the principal stresses

A. Theoretical Basis

Photoelasticity uses the accidental double refraction which appears in every monorefringent, translucent material when loaded and consequently subjected to internal stresses. Glass has this property to some degree and gelatine more so. Plexiglass, araldite, bakelite, dekorite and lucite are most often used to make the models. The model when loaded and observed in white polarized light shows continuous, black and coloured lines. The black lines are called isoclinics (Fig. 1). From them the isostatics (Fig. 2) can be constructed indicating the flux of forces. The coloured lines, isochromatics (Fig. 3), indicate the relative magnitude of the stresses.

The basis of the optical phenomenon has been summarized by Kummer (1956, 1959), relying on the work of Föppl and Mönch (1959). Recently the photoelastic technique has been considerably improved and it is now possible to solve certain problems by observing a third arrangement called pattern of isopachics (Fig. 4). Combined isochromatics and isopachics give the absolute magnitude of the stresses. The complete photoelastic effect has been described by Pirard (1960).

B. Historical

Gebhardt (1911) was the first to use photoelasticity in biology. He examined a celluloid model of an epiphysial cartilage and drew the lines he considered to be the trajectories of compression due to loading. But, as Pauwels (1960) has shown, the conditions of the experiment of Gebhardt did not correspond to the physiological stress of a joint.

Gebhardt actually loaded successively several points of the articular surface of his model. He then reproduced in the same drawing the trajectories of compression obtained for several loaded points. In this way he obtained bundles of lines which intersect at acute angles. However, because they do not intersect at right angles, they obviously do not form a true trajectorial pattern. By loading the whole free surface of his model he would have obtained a completely different picture.

Milch (1940) published pictures of a photoelastic model of the upper extremity of the normal, pathological and surgically modified femur. These pictures showed isochromatics covered by isoclinics. The author considered these lines as trajectories of stresses and compared them directly with X-rays.

Pauwels (1940, 1946, 1950, 1954, 1965b, 1980) clearly applied the photoelastic method to functional anatomy. He studied the distribution and the direction of the stresses in long bones and particularly in the upper extremity of the femur.

C. Application and Limitation of the Photoelastic Technique

A plane model corresponding to a plane mechanical object can be rigorously analysed by photoelastimetry, which gives the mechanical state of the object – qualitative as well as quantitative – in an accurate way. Going from the model to the object is quite legitimate since the solution assumes isotropic materials. Only the question of scales (size and force) intervenes and is solvable.

Replacing the bone by a plane cross section roughly through its diameter gives an accurate solution only for a body with symmetry of rotation, which is not the case in bone. Nevertheless the pictures have reasonable qualitative value and give a good quantitative approximation.

The experiments in photoelasticity show the direction and the magnitude of the stresses. These can be used in functional anatomy in order to recognize and study the functional structure and to determine relatively the quantitative distribution of the stresses at several levels of the skeleton.

Since the model is plane and its material homogeneous and isotropic, the experiment will give information about stresses appearing in a cross section of a homogeneous and isotropic body stressed by a force acting in the plane of the cross section.

The only question which can be answered by such an experiment is: does the direction of the elements of an organ (for instance the trabeculae in cancellous bone) correspond to the direction of the trajectories of stresses in a cross section of the model with the same outline as the studied organ?

There exists another possible use of photoelasticity. In some complicated cases one is not able to analyse with certainty the physiological stress of the studied organ. One must then rely on more or less valuable hypotheses. If, in such instances, the model shows a trajectorial picture in conditions of loading which agree with the theoretical possibilities, and if the trajectorial picture attained corresponds to the structure of the analysed organ, one can conclude that the experimentally chosen stress corresponds probably to the physiological stress and that the studied organ has a trajectorial structure. Most biological structures are complicated and of irregular outline. On the other hand the trajectorial pattern of the models is far from simple. It is unlikely that a similarity between them is due to chance. The results of the photoelastic study must thus be considered as an important indication.

IV. Clinical and Radiological Material

Only the observation of patients with osteoarthritis of the knee and also of results after surgery based on well-defined biomechanical principles can confirm the value of the theoretical study of the knee. That is why our work relies on clinical and radiological examination of normal and of osteoarthritic knees, on a critical analysis of the preoperative state and on the results attained by surgery. From this point of view we have studied several hundreds normal and osteoarthritic knees.

The clinical symptoms (pain, resistance to fatigue and function) as well as objective signs (range of movement, laxity or stability) must be considered. The radiological picture indicates truly and objectively the mechanical stress and

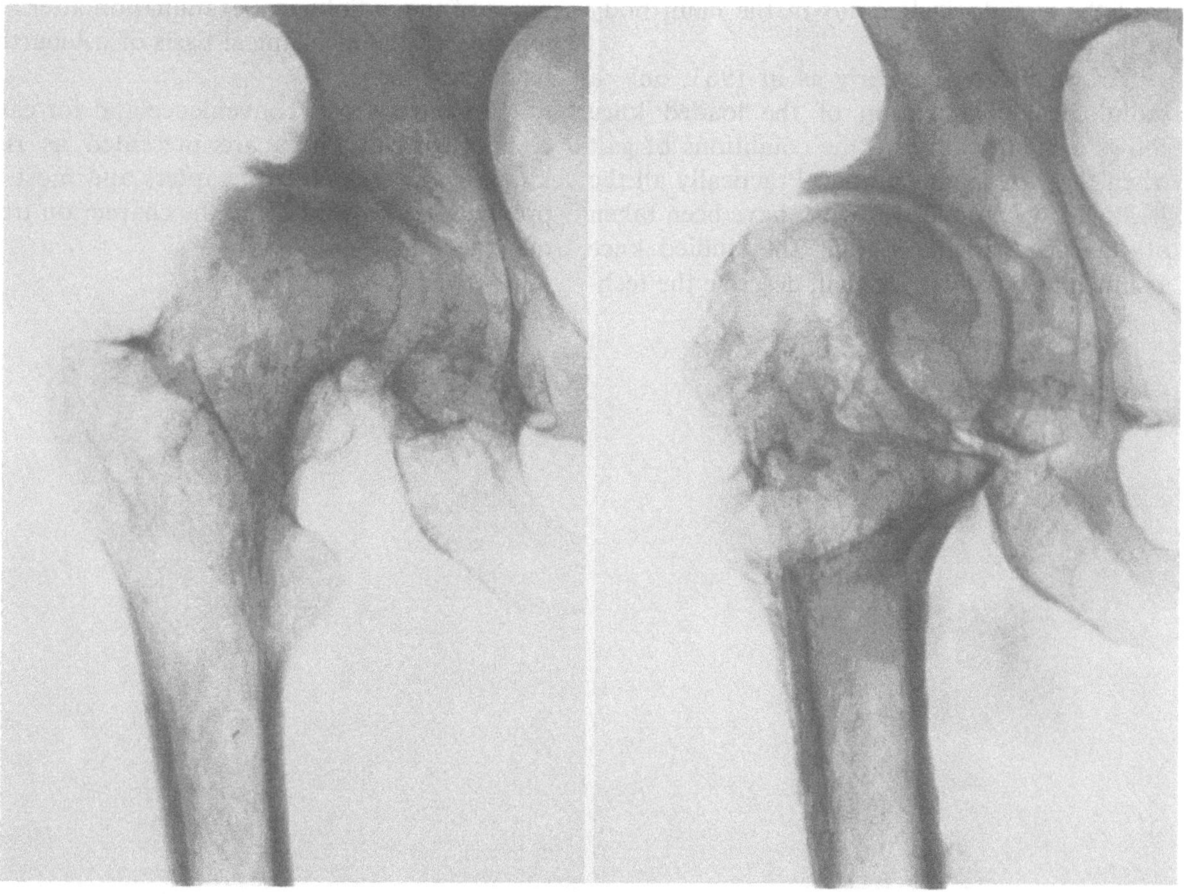

a b

Fig. 5a and b. 57-year-old patient before (a) and 14 years after (b) a valgus intertrochanteric osteotomy which has increased the articular weight-bearing area. Before surgery (a) a dense triangle indicates abnormally high joint pressure concentrated at the rim of the acetabulum. After surgery (b) the triangle is replaced by a subchondral sclerosis of even width throughout, indicating a better distribution of the articular pressure over larger weight-bearing surfaces

consequently is of decisive importance for our analysis. As Pauwels (1950, 1973a, b) has shown, between certain limits of stresses, resorption and apposition of bone balance each other. An increase of the stresses causes an augmentation of bone formation; bone resorption follows a reduction of the stresses. The quantity of bone tissue in the skeleton is therefore proportional to the magnitude of the stresses. The outline of dense bone in the roof of the acetabulum illustrates Pauwels' law. This shape corresponds to the form of the stress diagram. Subchondral dense bone appears as a thin ribbon of equal thickness in a normal hip, becoming thicker in its centre when the joint cartilage no longer distributes the pressure properly. It has the outline of a laterally situated triangle in osteoarthritis with subluxation, of a medially placed triangle near the deep part of the acetabulum in protrusio.

After surgery, appropriate varus or valgus osteotomy of the hip (Pauwels I or Pauwels II operation) which reduces the articular compressive stresses and distributes them over larger weight-bearing surfaces, the dense triangle disappears and is replaced by normal looking density. That is what happened in the patient in Figure 5, who was subjected to surgery at 57 years of age and was reviewed 14 years later (personal case).

The same phenomenon can be observed in the knee. The outline of the subchondral bone densities allows conclusions to be drawn as to the distribution of the stresses in the joint and as to their relative magnitude. The subchondral scleroses make it possible to foresee the development of osteoarthritis before the joint space is narrowed, to follow the evolution of the disease and to appreciate the results of surgery objectively. For these reasons we have systematically used the radiological results in the main body of the work.

As we insisted as early as in 1963, only a radiological examination of the loaded knee shows a picture close to the conditions of gait when the stress is maximum. Practically all the A.-P. X-rays used in this work have been taken of standing patients loading the studied knee as much as possible. We shall describe the technique of the radiological examination after having studied the mechanical basis of osteoarthritis of the knee.

For purposes of convenience and for easier comparison all knees are presented as right knees in the theoretical chapters and most are presented as right knees in the chapter on treatment.

The reader who does not wish to be confronted with mathematics will find on pages 15–21 the definitions of force and stress necessary for understanding. He can then turn to page 110, where he will find a suggestion on how to "read" the stresses in the X-rays.

Chapter IV. Mechanics of the Knee

In this chapter the forces exerted on the knee in different conditions are first calculated. The articular surfaces transmitting the supported load will then be measured. When the load and the weight-bearing surfaces are known, the magnitude of the compressive stresses in the joint can be deduced. These calculations will be preceded by an introductory discussion of load and stress.

I. Load and Mechanical Stresses

Before studying the mechanics of the knee we shall analyse simple theoretical examples of loaded materials. The first of the series is borrowed from Pauwels (1965).

A. Concept of Load and Stresses. Rigid Models

A column of homogeneous material supports a central load of 100 kg (Fig. 6a). The load is an external force which is exerted on the column. By compression it distorts the small particles of the material of which the column is composed. It causes internal compressive stresses and strains (deformations) in the column. These stresses and strains *D* counterbalance the external force, the load. They are the result of the action of the external force on the material of the column.

In the diagram, the small arrows indicate the magnitude and the distribution of the compressive stresses over a horizontal cross section of the column. An external force pulling the column would tend to distract the small elements which compose it, thereby creating tensile stresses.

We express the stresses in kg/cm², i.e. in unit of weight per unit of surface. In this case, their value is 10 kg/cm².

The second column (Fig. 6b) is identical with the first but supports a load five times greater (500 kg). The compressive stresses are increased proportionally to the load. Their value is 50 kg/cm².

The third column (Fig. 6c) supports, as does the first, a load of 100 kg, but it is thinner in diameter. The area of its horizontal cross section is half the size of that of the first. Here the load of 100 kg causes compressive stresses of 20 kg/cm². Thus, the magnitude of the compressive stresses in the column is inversely pro-

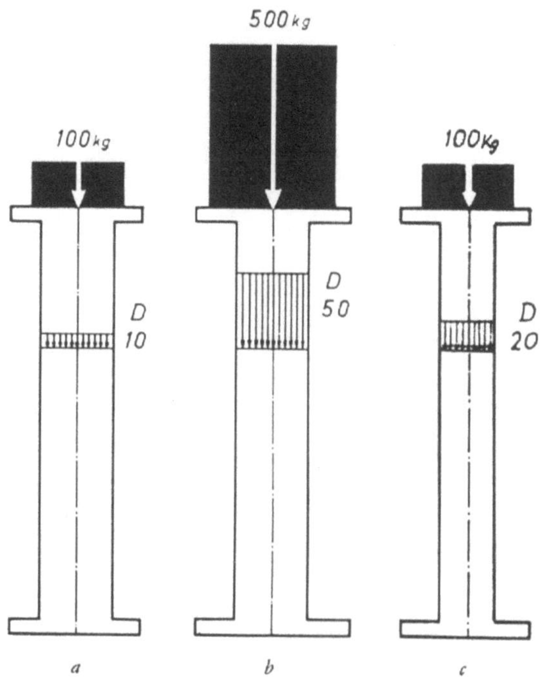

Fig. 6a–c. Stresses in columns supporting a centred load. *D*: compression. (From Pauwels, 1965a)

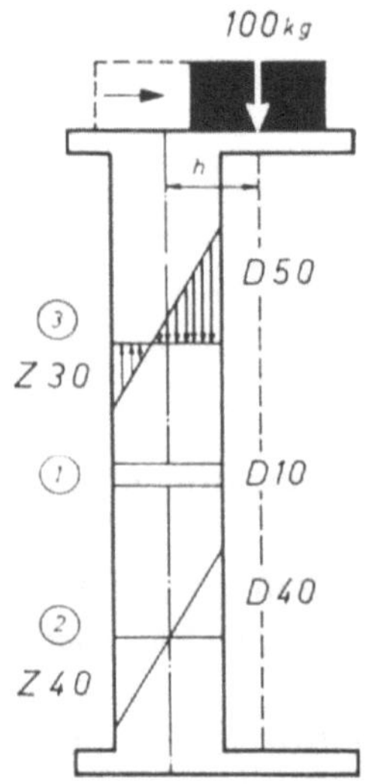

Fig. 7. Stresses in an eccentrically loaded column. *D*: compression. *Z*: tension. Diagram ①: pure compressive stresses. Diagram ②: bending stresses. Diagram ③: diagram resulting from the summation of the stresses of ① and ②. (From Pauwels, 1965a)

portional to the surface of the cross section of the column and directly proportional to the load.

The column (Fig. 7) supports a load of 100 kg but this load has been displaced to the right. In this position it tends to bend the column. The right part of the column is compressed and the left distracted. The small displacement of the load considerably increases the stresses in the column. They attain 50 kg/cm^2 as in the column supporting a central load of 500 kg (Fig. 6b).

An eccentrically placed load generates a higher overall magnitude of stress because two types of stresses are evoked:

1. pure compressive stresses *D*. They are the same as in the first column which supports an axial load (Fig. 7, ①).

2. bending stresses, since the eccentric load tends to bend the column to the right. It produces compressive stresses *D* in the right part of the column, tensile stresses *Z* in the left. Compressive and tensile stresses are maximum at the periphery. They decrease to zero in the middle of the column and are represented in diagram ②.

Both these types of stresses are algebraically added, tensile stresses being positive and compressive stresses negative. Their summation is represented in diagramm ③. The material of the column must be able to withstand the maximum stresses, which amount to 50 kg/cm^2 in the model. The maximum stresses determine the resistance required from the column and the risk of breakage.

If the load is displaced further to the right (Fig. 8a), the pure compressive stresses remain the same, 10 kg/cm^2. The bending tendency of the

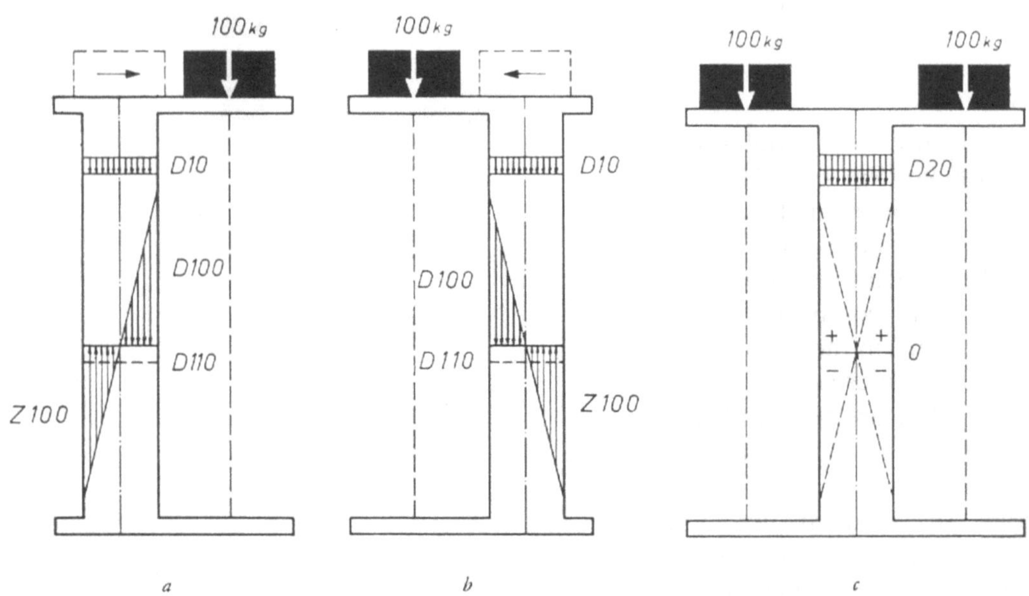

Fig. 8a–c. Eccentrically loaded columns (a and b). Despite a greater load, the stresses are decreased in column (c). (From Pauwels, 1965a)

column, however, is increased as are the compressive and tensile stresses. The column must now resist maximum stresses of 110 kg/cm^2.

The bending stresses are much greater than the pure compressive stresses. They increase rapidly when the load is displaced eccentrically. The pure compressive stresses, on the other hand, remain unchanged under the same load. The bending stresses therefore determine the magnitude of the stress in the column and constitute a danger of breakage.

The column (Fig. 8a) supports an eccentric load of 100 kg on the right side. The other column (Fig. 8b) supports the same eccentric load on the left at the same distance. The pure compressive stresses and the bending stresses are of the same magnitude as in column (Fig. 8a). The compressive stresses generated by the bending tendency are here located on the left and the tensile stresses on the right. If the column (Fig. 8c) supports both loads of 100 kg, compressive stresses and tensile stresses due to bending cancel each other. The pure compressive stresses are added. The total compressive stresses, 20 kg/cm^2, are nevertheless much smaller than the maximum stresses, 110 kg/cm^2, generated by one eccentric load. Despite the increase of the load, the mechanical stress in the column is diminished. It is reduced to one-fifth in spite of the load being doubled.

If the line of action of a force R exerted on a column is inclined at an angle α to the axis of the column, the compressive force R can be resolved into two components (Fig. 9). One, $D = R \cdot \sin \alpha$, is parallel to the axis of the column, and the other, $S = R \cdot \cos \alpha$, is at right angles to the axis of the column. The force D provokes pure compression. The force S is called a shearing force and tends to make the small elementary particles constituting the material of the column slide horizontally on each other. Force S provokes shearing stresses in the column proportional to its magnitude.

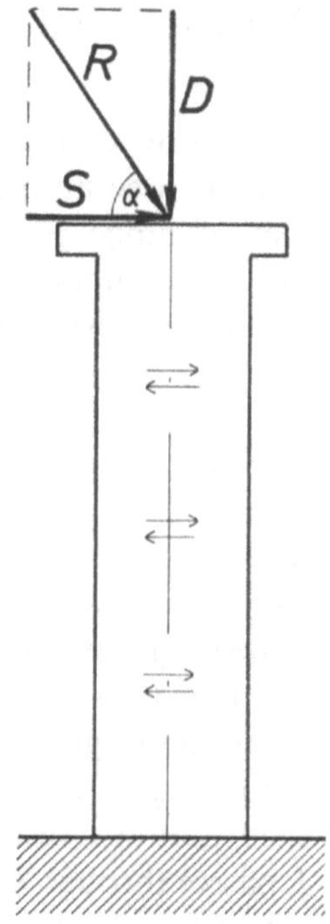

Fig. 9. Force R acting obliquely on the column can be resolved into a compressive component $D = R \sin \alpha$ and a shearing component $S = R \cos \alpha$

A compressive force acting obliquely on the column thus evokes pure compressive stresses, bending stresses and shearing stresses in the column.

B. Articulated Models

Let us now consider an articulated model: two columns one above the other, with a negligible friction hinge joint between them. The lower end of the upper column is convex, the upper end of the lower column concave.

1. Forces

If a weight P is well centred on the upper column, a balance is maintained (Fig. 10a). If the same weight is off-centre, the upper column tilts and falls (Fig. 10b). Balance can be restored by a counterweight L on the other side of the column (Fig. 10c). The counterweight must be such that the resultant R of the forces P and L passes through the axis of rotation of the joint between the columns. R is the vectorial sum of forces P and L and represents the load exerted on the column.

The counterweight L can be of different magnitude than the weight P but their moments relative to the axis of rotation must be equal and opposite in direction. The moment of a force is the product of the force and the distance at which it works

$$P \cdot a = L \cdot b.$$

If the counterweight L is further away from the axis of rotation than the weight P (Fig. 11a), it must be smaller than the weight P. The load R is consequently reduced. If the counterweight L is closer to the axis of rotation than the weight P (Fig. 11b), it must be greater than the weight P. The load R is consequently increased.

Inversely, if the weight P is closer to the axis of rotation (Fig. 11c) the counterweight must be smaller and the load R is decreased. If the weight P is further away from the axis of rotation (Fig. 11d), it must be counterbalanced by a larger counterweight L and the load R is increased.

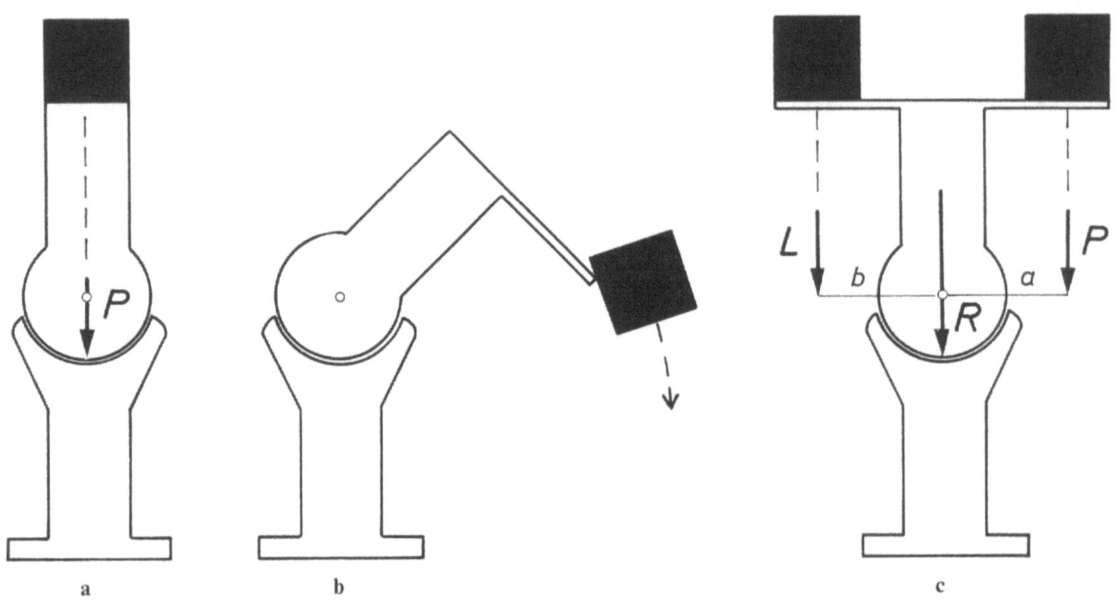

Fig. 10. (a) Load P is centred on an articulated model used for the next figures through 13. (b) An eccentric load tilts. (c) A counterweight L balances the weight P. The column supports the vectorial sum R of forces P and L

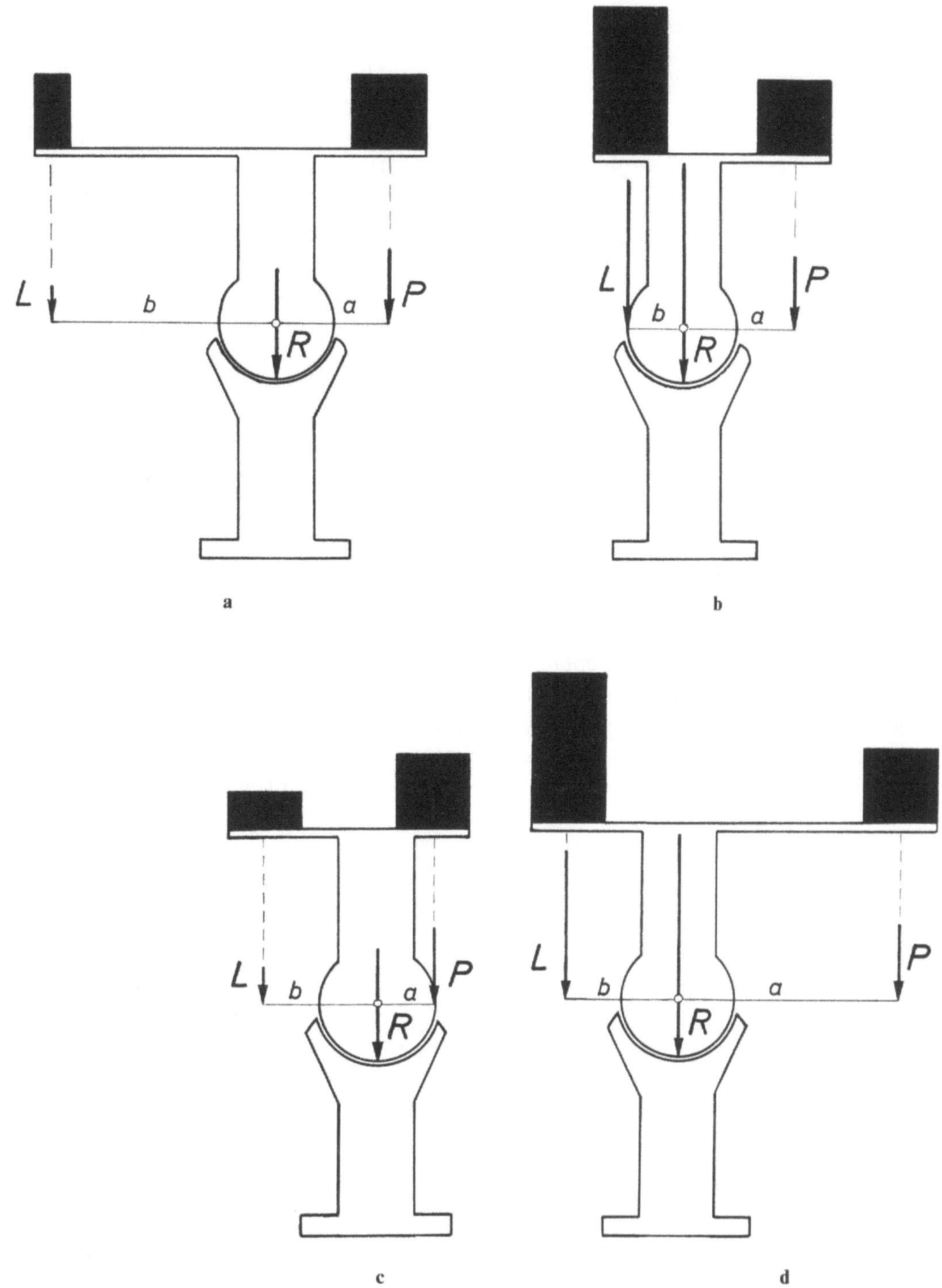

Fig. 11a and b. The magnitude of the counterweight L depends on the distance b from its line of action to the axis of the joint. c and d. The magnitude of the counterweight L depends on the distance a from the line of action of P to the axis of the joint

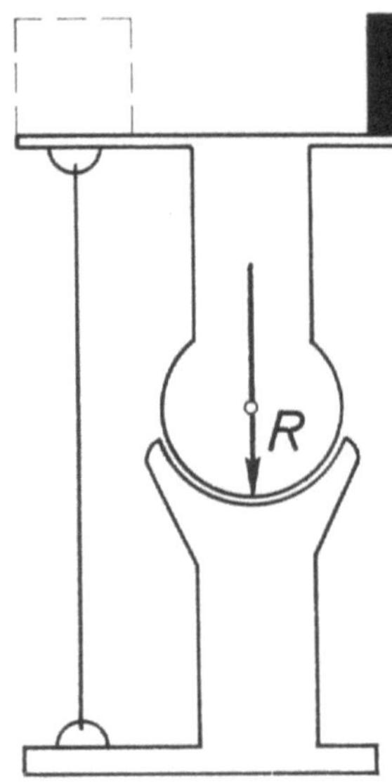

Fig. 12. A stay or tension band can replace the counterweight

A tight rope or chain can replace the counterweight (Fig. 12). It acts as a stay or a tension band. To maintain balance the stay must be stretched to exert the same force as the counterweight it replaces.

2. Contact Stresses

Under the effect of the load *R* (Fig. 13) the contact stresses in the joint can only be compressive stresses σ_D. A joint without friction can only transmit a force *R* the line of action of which passes through the centre of rotation of the joint and which is distributed over the contact surface. These contact compressions constitute the mechanical stress in the joint. They are proportional to the magnitude of the load *R* and roughly inversely proportional to the surface of the weight-bearing areas.

The diagram of distribution of the contact stresses (Fig. 13a) is cup-shaped. The maximum is in the centre and the stresses diminish toward the periphery. The shape of the cup is influenced by the radius of the cylinders or of the spheres forming the articulation and by a coefficient of elasticity characteristic for each material.

If the line of action of *R* is not centrally located, as it is in Figure 13a, but is more and more off-centre, as indicated in Figure 13b and c, the diagram σ_D changes. The maximum of the stresses moves toward the periphery of the contact surface in the same direction as the force

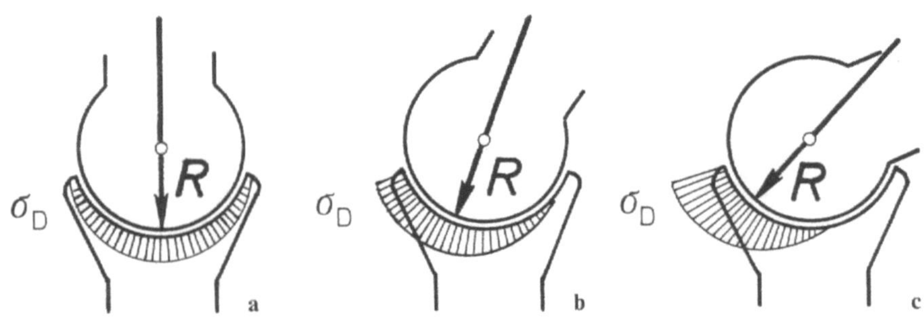

Fig. 13a–c. The stress distribution depends on the line of action of load *R*. (Redrawn from Pauwels, 1963)

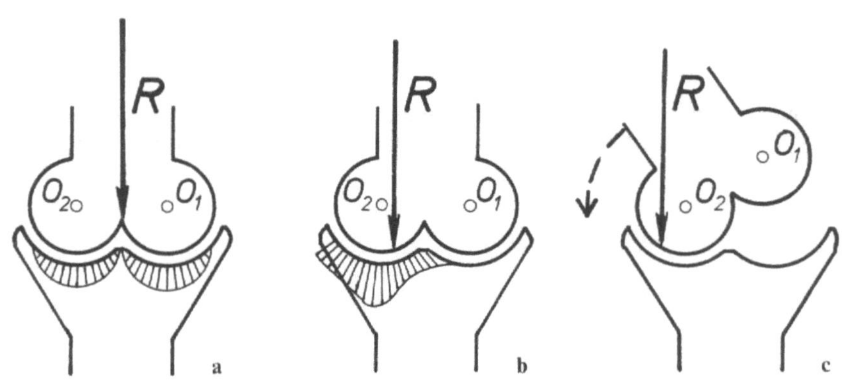

Fig. 14. (a) Centred load. (b) Load off-centre but between the centres of curvature O_1 and O_2. (c) Load outside the centre of curvature O_2. The balance is disrupted

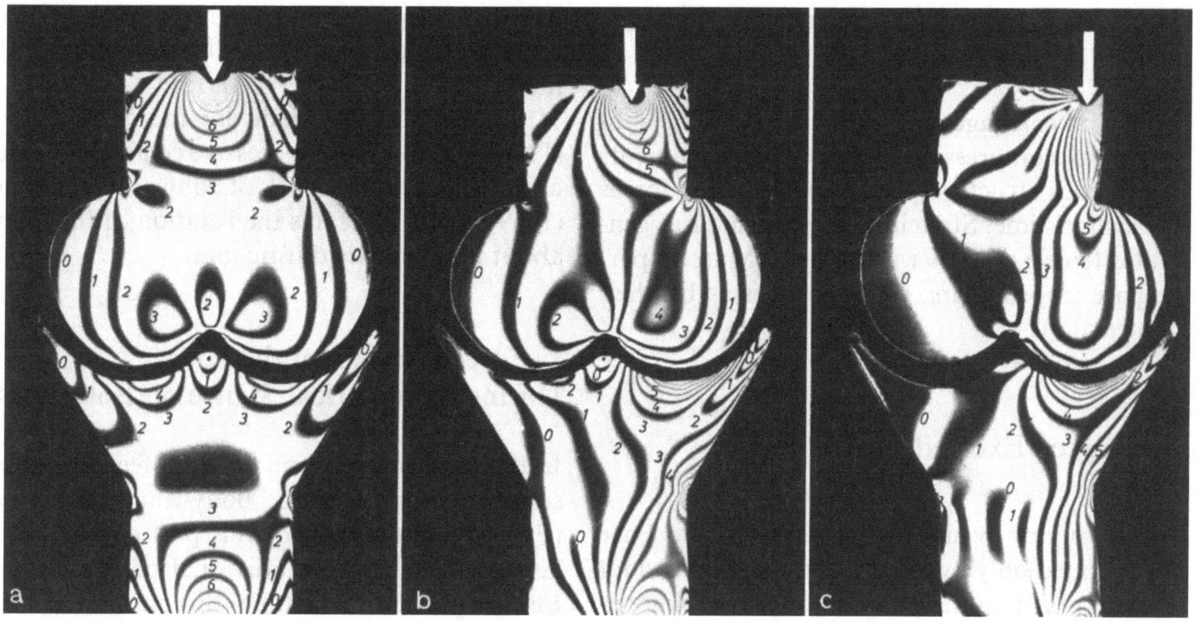

Fig. 15a–c. Photoelastic models. The white arrow represents the load. (a) Centred load. (b) Load off-centre to the right. (c) Load above the centre of curvature of the right half of the joint

and it increases more and more the further the force is displaced.

Now join the two bodies with two cylindrical or spherical surfaces, symmetrically located. The contact surfaces are separated by a non-weight-bearing space as indicated in Figure 14. There are now two centres of curvature, O_1 and O_2. If the force R acts centrally as in the case of Figure 14a, the diagram of the contact stresses presents two symmetrical flat cups, as shown. If the force R is displaced laterally toward O_2, for instance (Fig. 14b), with its line of action between O_1 and O_2, the equilibrium persists. But the contact stresses are obviously no longer symmetrical. The joint compressions σ_D are maximum on the side toward which R has been displaced and minimum on the other side. The maximum of σ_D is considerably increased.

The line of action of R passing through O_2 is a limiting position of equilibrium. Very high stresses are produced when R is in this position and is supported by only one of the weight-bearing surfaces. If R acts outside O_2 (Fig. 14c) the equilibrium is lost. The upper part tilts on the lower about O_2, as shown in the picture. The same phenomenon happens when R is displaced beyond O_1.

Photoelastic models illustrate the distribution of the contact stresses (Fig. 15). In a, the model supports a well-centred load, represented by the white arrow. Under the action of this load coloured lines appear, called isochromatics, equal in number in the right and the left part. From these lines it is possible to draw the diagram of both identical flat cups and the maximum of σ_D. In b, the load acts off-centre. Its displacement to the right has diminished the number of isochromatics in the left part (order 2) and increased the isochromatics in the right one (order 7). The compressive stresses can thus be measured. In c, the limit position is nearly reached. The isochromatics attain order 8 in the right half of the model and only order 1 in the left half.

II. Mechanical Stress in the Knee

Bone and cartilage of a joint are in general subjected to compression. Periarticular ligaments are fibrous structures able to resist great forces for a short time. Stretching these structures generates tensile stresses within them. Muscles participate in producing all these stresses by their contractions.

A. Forces Exerted on the Knee

In the supine position, the stress in the knee is generated only by muscular forces and cannot be determined in the state of our present knowledge.

During standing, the knee supports a part of the body weight. When the line of action of this part does not pass through the knee, muscular forces must intervene to keep balance. The forces due to the partial body weight can be precisely calculated if one knows the body weight, the weight of its several parts, the position of their centre of gravity and, during walking, the displacements of all the parts of the body.

The muscular forces can be determined mathematically with acceptable accuracy by expressing the dynamic equilibrium of the different parts of the body. These muscular forces must intervene automatically to ensure balance.

First, the mechanical stresses in the knee will be studied in symmetrical standing on both legs, second, when standing on one foot and finally during the single support period of gait.

Following the example of Pauwels (1935), our analysis is based on the data provided by Braune and Fischer in their work on the centres of gravity (1889) and on gait (1895, 1899, 1900, 1901, 1903, 1904). Working on frozen cadavers, Braune and Fischer first determined the centres of gravity of the living body and of its various parts. They located them in a system of co-ordinates, x, y, z, in three rectangular planes: a vertical coronal plane zy, a vertical sagittal plane zx, and a horizontal plane xy. From photographs of walking individuals they were able accurately to analyse in this system of co-ordinates the displacements of the total and partial centres of gravity and the displacements of the joints.

In addition, we shall use the data of Fick (1910) regarding the successive centres of flexion of the knee and the work of Johnston and Smidt (1969) which indicates the rotation of the femur about its long axis during gait.

1. Force Exerted on the Knee During Symmetrical Stance on Both Legs

In the standing position on both feet, the knees support the part of the body above the knees (Fig. 16). The weight of this part can be calculated easily. For the individual of Braune and Fischer (weight: 58.700 kg) this part of the body weighed 50.420 kg (85.6% of body weight).

One can assume that this weight is concentrated at its centre of gravity S_3. This is located at the level of the third lumbar vertebra. In the coronal plane the load S_3 is supported by the pelvis acting as a transverse beam which transmits it to the ground through the thigh and lower leg bones, hips, knees, and ankles. Projected on a coronal plane, the centres of these three joints are on the same straight line passing through the ground support.

In the sagittal plane the centre of gravity S_3 also lies on or near the vertical line passing through the centre of rotation of the hip, the flexion centre of the knee and the centre of the ankle joint. In such conditions the muscular force necessary to maintain this equilibrium – which has little inherent stability – is theoretically negligible.

If the support is symmetrical, the load in S_3 is evenly distributed between both knees. Its direction is vertical.

Using the figures of Braune and Fischer, we calculated that each knee bears a vertical load of 25.21 kg for an individual of 58.700 kg. This is approximately 43% of body weight.

Fig. 16. Standing with symmetrical support on both feet. S_3: centre of gravity of the part of the body supported by the knees. The central arrow represents the weight of this part of the body. The lateral arrows indicate the force supported by each knee

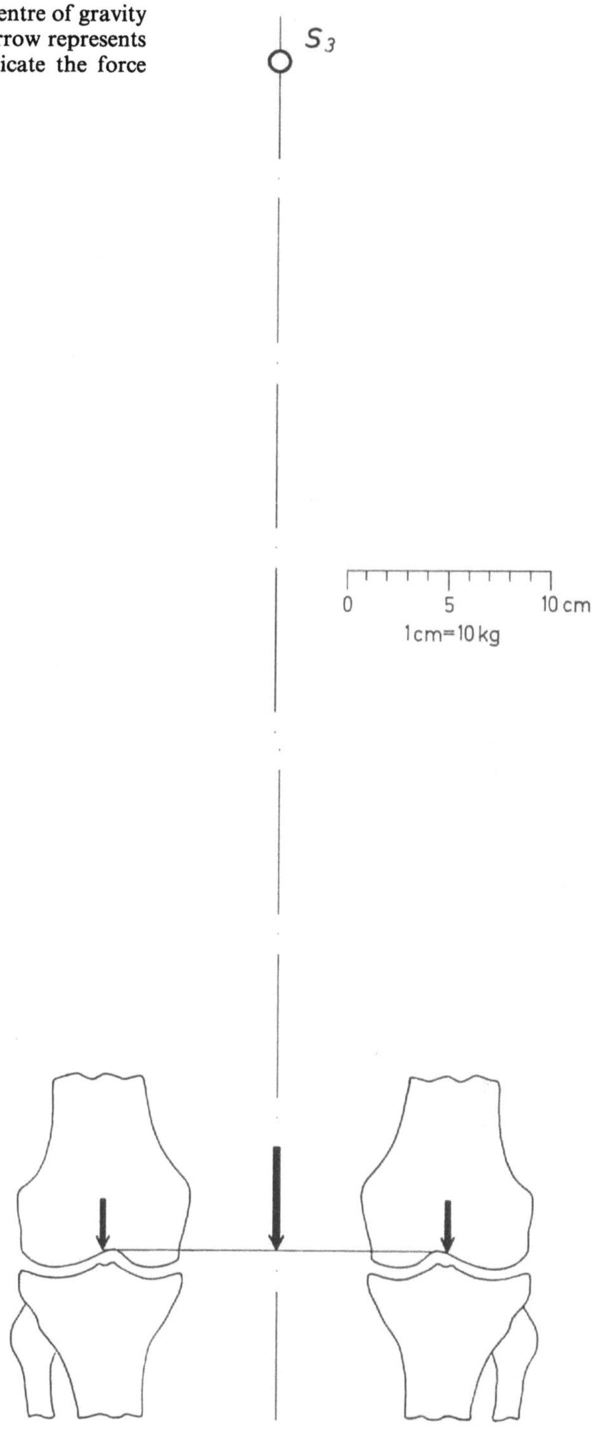

2. Forces Exerted on the Knee in Standing on One Leg

When the subject stands on one foot, the loaded knee supports the head, the trunk, the upper limbs, the loaded thigh and the opposite lower limb. The mass of this part of the body can be considered concentrated at the centre of gravity S_7 (Fig. 17). S_7 is distinct from the centre of gravity S_6 of the whole body. The weight of the part of the body supported by the loaded knee will be called *P*. It can easily be calculated by adding the weights of its constituent parts. For the 58.7 kg subject of Braune and Fischer this would be 54.56 kg, which is about 93% of the body weight.

For didactic reasons, the action of the partial weight *P* on the knee will be analysed first in a coronal plane, then in a sagittal plane. In each of these planes the forces involved correspond to the projection of the forces acting in space, on the respective planes.

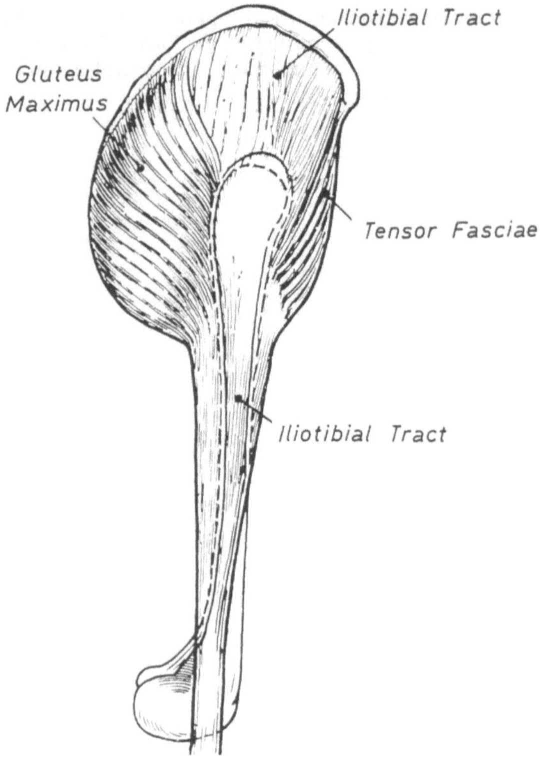

Fig. 18. The pelvic deltoid or lateral muscular tension band. (Redrawn from Henry, 1959)

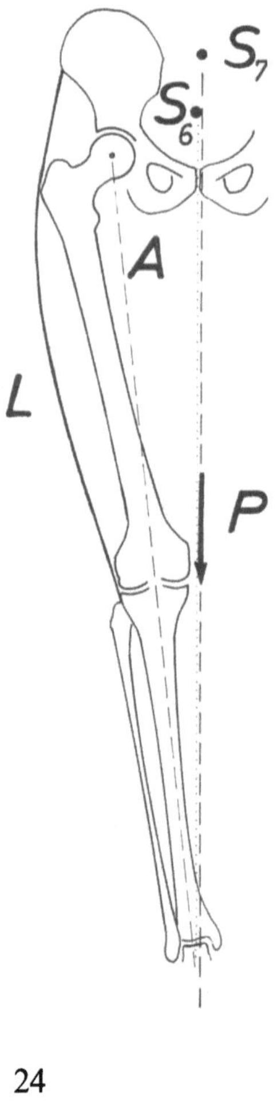

Fig. 17. Standing on one foot. S_6: centre of gravity of the body. S_7: centre of gravity of the part of the body supported by the knee. *P*: weight of this part of the body. *L*: muscular tension band balancing *P*. *A*: so-called mechanical axis of the limb

a) Coronal Plane

The weight *P* is not exerted axially but rather medially to the normal knee. It must, therefore, be balanced by a lateral force *L* which prevents tilting of the femur on the tibia (Fig. 17). In standing on one foot, the lateral tension band *L* is essentially constituted by the "pelvic deltoid": the gluteus maximus, the tensor fasciae latae and the ilio-tibial band (Fig. 18). The pelvic deltoid is also involved in the equilibrium of the hip since it spans both hip and knee. Its tension is determined by the conditions of balance at both joints.

Thus, the knee supports forces *P* and *L* (Fig. 19). For reasons of balance, their vectorial sum, the resultant *R*, must be exerted between the centres of curvature O_1 of the medial condyle and O_2 of the lateral condyle. The magnitude and line of action of the resultant *R* can be calculated for a normal knee.

The subject of Braune and Fischer standing on one foot is drawn in the system of co-ordinates. His centre of gravity S_6 lies above the supporting foot (Fig. 17). Its co-ordinates are $y_{S_6} = 0$; $z_{S_6} = 93$ cm.

The partial body weight P is exerted along the vertical drawn from its centre of gravity S_7. The co-ordinates of S_7 can be calculated from the data of Braune and Fischer introduced in the equations

$$P_6 \cdot y_{S_6} - P_J \cdot y_J = P \cdot y_{S_7}, \quad y_{S_7} = \frac{P_6 \cdot y_{S_6} - P_J \cdot y_J}{P},$$

$$P_6 \cdot z_{S_6} - P_J \cdot z_j = P \cdot z_{S_7}, \quad z_{S_7} = \frac{P_6 \cdot z_{S_6} - P_J \cdot z_J}{P}$$

where P_6 = total body weight.

P = weight of the body minus the supporting leg and foot, or weight exerted on the right knee.

y_{S_7} = co-ordinate y of partial centre of gravity S_7.

z_{S_7} = co-ordinate z of partial centre of gravity S_7.

P_J = weight of supporting leg below the knee.

y_J = co-ordinate y of centre of gravity of supporting leg below the knee.

z_J = co-ordinate z of centre of gravity of supporting leg below the knee.

$y_{S_7} = -0.30$ cm. $\qquad z_{S_7} = 98.27$ cm.

For the individual of Braune and Fischer, standing on one foot, the partial centre of gravity S_7 thus lies 0.30 cm left and 5.27 cm above the centre of gravity S_6 of the whole body.

The weight P is exerted along the vertical through S_7. It acts eccentrically on the knee with a lever arm a, which is the distance between the vertical drawn from S_7 and a central point G on the axis of flexion of the knee (Fig. 19). We shall justify the choice of the point G as origin of the lever arm (page 71).

The line of action of the muscular force L is deduced from anatomical studies. The normal drawn from the point G (on the axis of flexion of the knee) to the line of action of force L represents the lever arm b of L.

The magnitude of L is easily found:

$$L = P \cdot \frac{a}{b}.$$

The construction of the parallelogram of forces shows the magnitude and the direction of the resultant force R graphically. These can also be calculated:

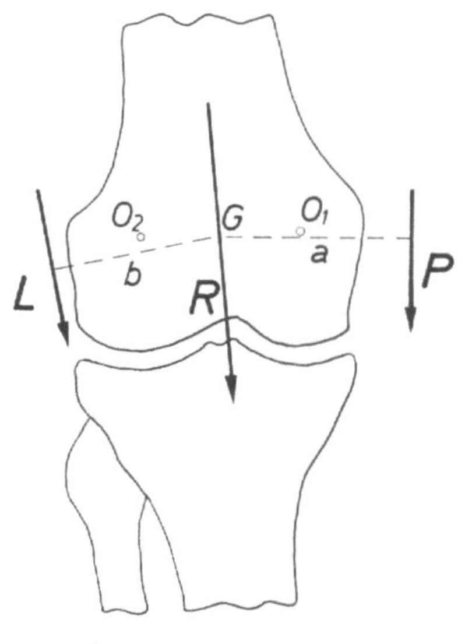

Fig. 19. Loaded knee projected on the coronal plane. P: weight of the part of the body supported by the knee. L: lateral muscular tension band. a: lever arm of P. b: lever arm of L. R: resultant of P and L. O_1: centre of curvature of the medial condyle. O_2: centre of curvature of the lateral condyle. G: central point on the axis of flexion of the knee

$$R = \sqrt{P^2 + L^2 + 2P \cdot L \cdot \cos\psi}.$$

Angle ψ is the acute angle formed by the lines of action of forces P and L.

The direction of resultant R is given by the equation:

$$\sin(\widehat{P \cdot R}) = \frac{L}{R} \sin\psi$$

$(\widehat{P \cdot R})$ is the angle formed by the lines of action of force P and resultant R.

When resultant R is exerted through the centre of gravity of the weight-bearing surfaces of the femoro-tibial joint, its magnitude is 126.824 kg in the studied individual. It is inclined at 5° from the vertical.

When standing on one foot, with equal distribution of compression on the weight-bearing surfaces, the loaded knee of an individual of 58.7 kg thus supports a load of 126.8 kg. This is a little more than twice the body weight. This force is inclined at about 5° from the vertical.

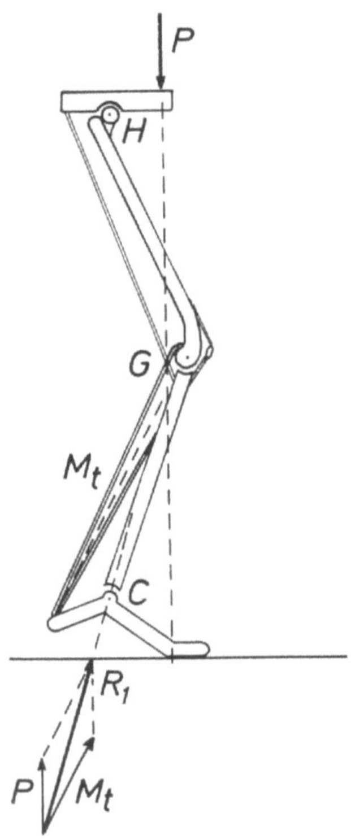

Fig. 20

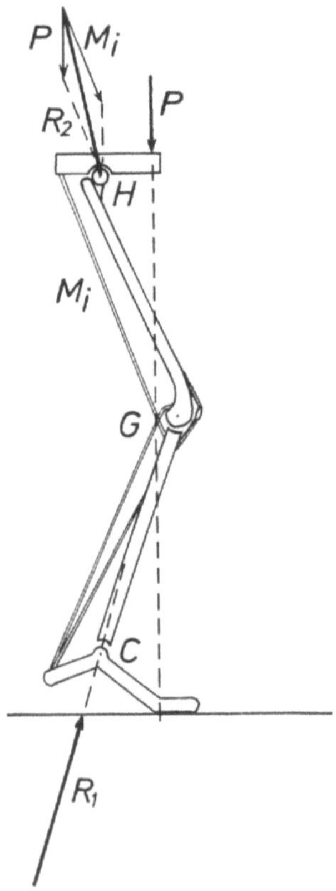

Fig. 21

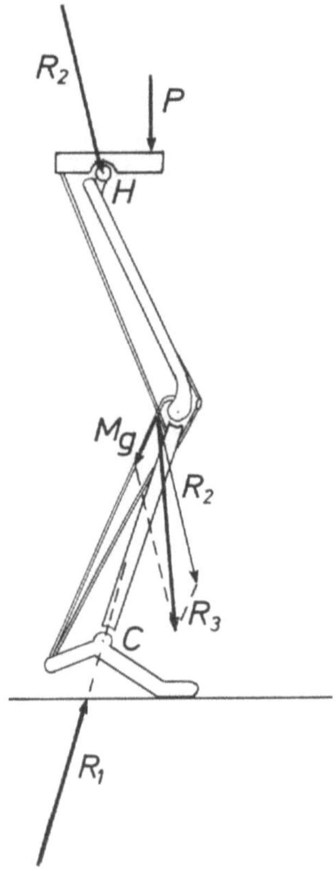

Fig. 22

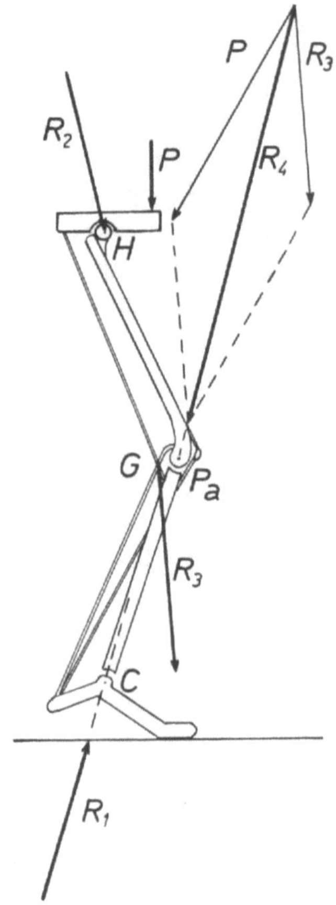

Fig. 23

b) Sagittal Plane

For a schematic analysis in the sagittal plane we choose a slightly flexed position of the joints of the lower limb (Kummer, 1962). This position provides for a better concept of the interplay of the forces during their action on the loaded joints. Both the centre of gravity of the body and the partial centre of gravity S_7 lie on the same vertical line. This intersects the support, i.e. the forefoot in the position represented in Figure 20.

The weight P would dorsiflex the foot on the leg, were it not counterbalanced by the calf muscles M_t. The resultant R_1 of both forces necessarily passes through the axis of flexion of the ankle for reasons of equilibrium. The weight P tends to tilt the pelvis forward. It is counterbalanced by the hamstring muscles M_i (Fig. 21). The resultant R_2 of both forces P and M_i passes through the centre of the femoral head and acts behind the knee. It tends to flex the knee, as does the force M_g produced by the gastrocnemius. The resultant R_3 of forces R_2 and M_g tends also to flex the knee (Fig. 22). A force in front of the knee is required for balance. It is provided by the patella tendon P_a. The resultant R_4 of forces P_a and R_3 necessarily intersects the axis of flexion of the femoro-tibial joint (Fig. 23). Since the patella keeps its orientation and its distance in relation to the femur and tibia, force P_a balances the force M_v developed by the quadriceps muscle (Fig. 24). Their moments are equal and of opposite direction:

$$M_v \cdot q = P_a \cdot k.$$

The lever arm q of force M_v is the distance between that force and the centre of curvature of the force transmitting surfaces between the patella and the femur (Fig. 25). The lever arm k is the distance between P_a and that centre of curvature. The resultant R_5 of both forces P_a and M_v squeezes the patella against the femur (Fig. 24). Forces R_4 and R_5 create compressive stresses in the femoro-tibial and patello-femoral joints (Fig. 26).

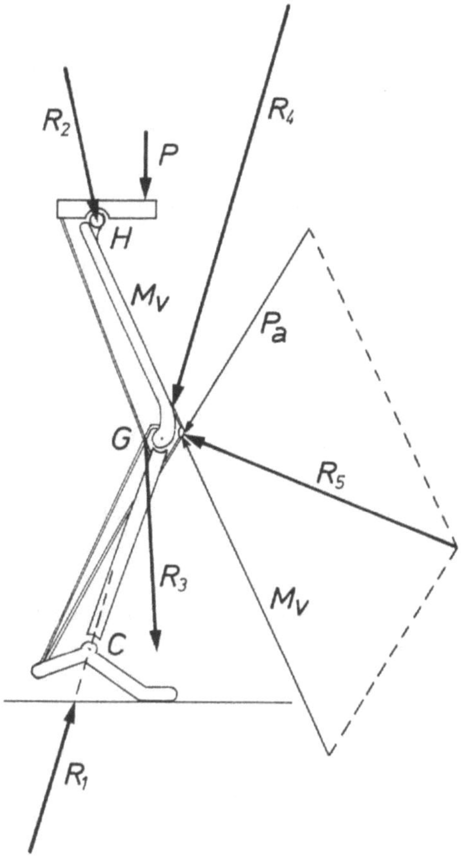

Fig. 24

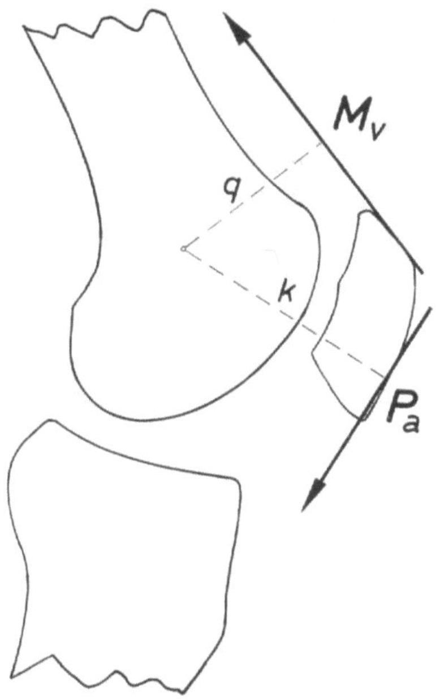

Fig. 25. M_v: quadriceps. P_a: patella tendon. q: lever arm of M_v. k: lever arm of P_a

Fig. 20.* Stance: lateral projection. P: body weight. H: hip. G: knee. C: ankle. M_t: calf muscles. R_1: resultant of P and M_t

Fig. 21.* M_i: hamstrings. R_2: resultant of P and Mi

Fig. 22.* M_g: gastrocnemius. R_3: resultant of R_2 and M_g

Fig. 23.* P_a: patella tendon. R_4: resultant of R_3 and P_a

Fig. 24.* M_v: quadriceps muscle. R_5: resultant of P_a and M_v

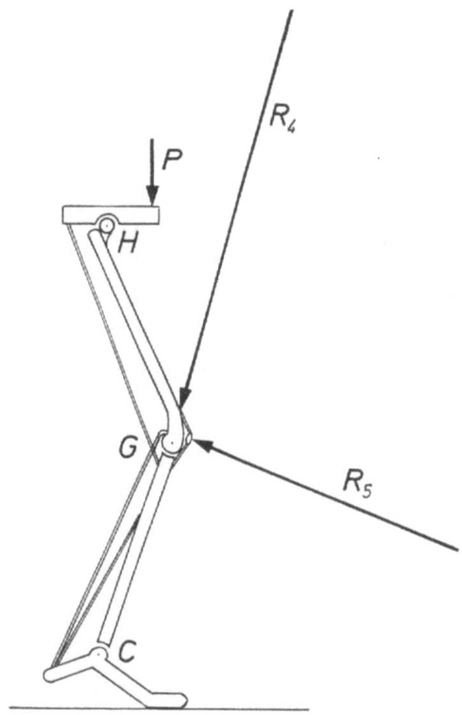

Fig. 26.* Femoro-tibial R_4 and patello-femoral R_5 compressive forces

* Figures 20–24, 26 originate from resolving the forces in a figure of Kummer, 1962.

3. Forces Exerted on the Knee During Gait

During the single support period of gait, the centre of gravity S_6 of the body is practically never above the supporting foot (Fig. 27). It is in front of, behind, or medial to the foot. The dynamic equilibrium is ensured by forces of inertia generated by accelerations and decelerations of the body mass. This mass can be assumed to be concentrated at S_6. It is a dynamic equilibrium since there is movement. The foot transmits to the ground the resultant R of body weight and forces of inertia. V represents the algebraical sum of the body weight and the vertical forces of inertia. H is the horizontal component of the forces of inertia. The line of action of resultant R passes through the centre of gravity S_6 and the support on the ground. The ground reaction force R' is equal but directly opposed to R and can also be resolved into a horizontal force H' and a vertical force V'.

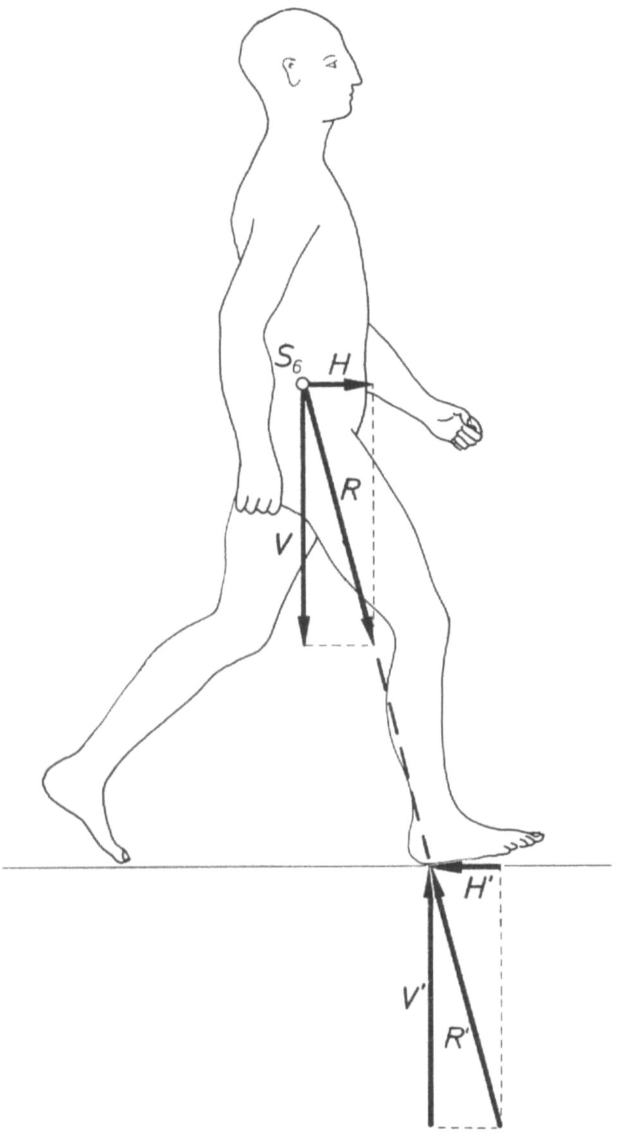

Fig. 27. Walking individual. S_6: centre of gravity of the body. H: horizontal component of the forces of inertia. V: vertical forces (weight and vertical component of inertia). R: resultant of H and V, forces exerted by the body mass. R': resultant of H' and V', reaction of the ground

During the single support period of gait the knee carries the same part of the body as when standing on one foot: head, trunk, upper limbs, loaded thigh and opposite lower limb. The mass of this part can be imagined to be concentrated at S_7. This centre of gravity continually moves in the three planes of space. Accelerations and decelerations of S_7 produce forces of inertia. The mass S_7 thus exerts on the knee not only its own weight but also forces of inertia. The resultant of all these forces will be designated by P.

The line of action of P usually does not intersect the knee. Force P is eccentrically exerted on the joint. It tends to tilt the femur on the tibia and is balanced by muscular and ligamentous forces which maintain the equilibrium. The knee then supports the vectorial sum of all these forces. In order to know them and the load carried by the joint, it is necessary to determine the moment of the force P developed by the partial mass of the body in movement and the moment of the muscular and ligamentous forces which balance it.

The moment of force P is the product of this force P and the distance a between the line of action of P and the point G, the centre of the axis of flexion of the knee (Fig. 19). We therefore have to calculate force P and distance a.

Force P represents the vectorial sum of the weight of the partial mass of the body and of the forces of inertia due to the accelerations and decelerations of S_7. The partial body weight has been calculated above. Its value is 54.560 kg for the subject of Braune and Fischer. The forces of inertia depend upon the movement of S_7. This must be analysed.

a) Displacement of the Centre of Gravity S_7

In their study, Braune and Fischer established the successive co-ordinates of the centres of gravity of the different parts of the body during gait, in three planes at right angles. From these data it is possible to calculate the co-ordinates of the centre of gravity S_7 for each phase of gait. x_{S_7} indicates the position of S_7 in the horizontal direction of gait, y_{S_7} in a horizontal direction at right angles to the gait, and z_{S_7} the vertical distance of S_7 from the ground.

The following formula is applied:

$$P_7 x_{S_7} = P_1 x_1 + P_2 x_2 + P_3 x_3,$$

$$x_{S_7} = \frac{P_1 x_1 + P_2 x_2 + P_3 x_3}{P_7}$$

P_7 = weight of head + trunk + upper limbs + swinging lower limb + loaded thigh.
P_1 = weight of head + trunk + both upper limbs.
P_2 = weight of the swinging lower limb.
P_3 = weight of the loaded thigh.
x_{S_7} = co-ordinate x of the centre of gravity S_7.
x_1 = co-ordinate x of the centre of gravity of head + trunk + both upper limbs.
x_2 = co-ordinate x of the centre of gravity of the swinging lower limb.
x_3 = co-ordinate x of the centre of gravity of the loaded thigh.

Co-ordinates y_{S_7} and z_{S_7} are obtained in the same way:

$$y_{S_7} = \frac{P_1 y_1 + P_2 y_2 + P_3 y_3}{P_7},$$

$$z_{S_7} = \frac{P_1 z_1 + P_2 z_2 + P_3 z_3}{P_7}.$$

The co-ordinates of the centre of gravity S_7 have been calculated in this manner for each phase of gait. They are given in Table 1.

The data of Braune and Fischer make it possible to locate the successive positions of the subject during gait. The following drawings also indicate the position of S_7. Figure 28 represents an A.-P. view of the subject at phase 16. A perpendicular dropped from S_7 passes medial to the loaded knee. At phase 12, S_7 is located far behind the knee (Fig. 29) and at phase 22 far in front (Fig. 30).

Table 1. Co-ordinates of centre of gravity S_7 (trunk + head + both upper extremities + swinging lower extremity + thigh of the loaded extremity)

Phases of gait	x	y	z
1	38.53	−0.86	95.12
2	44.89	−0.89	96.32
3	51.25	−0.84	97.50
4	57.40	−0.77	98.28
5	63.54	−0.71	98.47
6	69.58	−0.65	98.07
7	75.54	−0.56	97.12
8	81.92	−0.43	95.85
9	88.24	−0.25	94.93
10	94.57	+0.08	94.25
11	–	–	–
12	104.16	−0.17	94.36
13	111.20	+0.21	94.76
14	117.59	+0.47	95.62
15	124.10	+0.59	96.97
16	129.81	+0.59	97.97
17	136.10	+0.54	98.58
18	141.92	+0.42	98.53
19	148.13	+0.34	97.95
20	154.19	+0.20	96.99
21	160.21	+0.04	95.95
22	166.44	−0.27	95.18
23	172.84	−0.71	94.61
24	–	–	–
25	182.01	−0.56	94.56
26	189.21	−0.82	94.96
27	195.85	−1.03	95.63
28	202.49	−1.11	96.68
29	208.40	−1.06	97.60
30	214.34	−0.94	98.28
31	220.02	−0.83	98.32

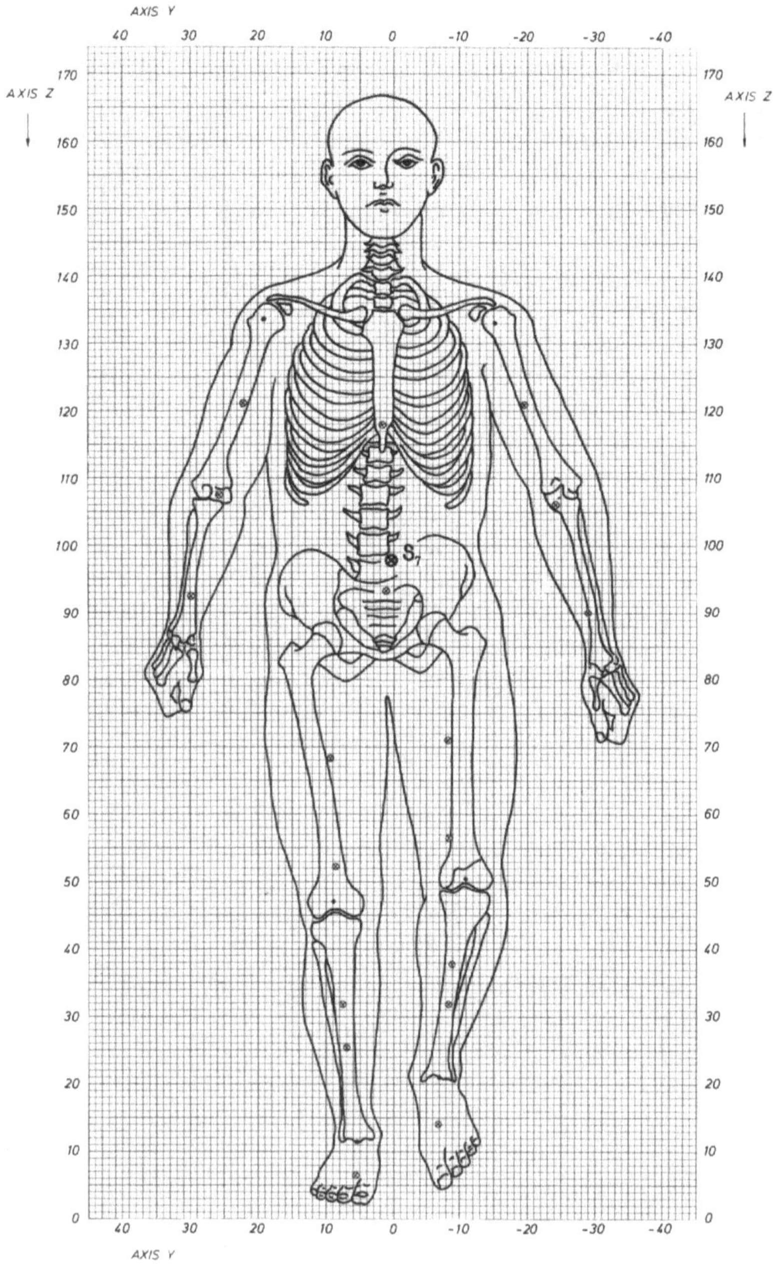

Fig. 28. Subject of Braune and Fischer at phase 16 of gait. S_7: partial centre of gravity. The centre of gravity of the whole body is indicated by a circle surrounding a cross on the sacrum

Fig. 29. Subject of Braune and Fischer at phase 12 of gait

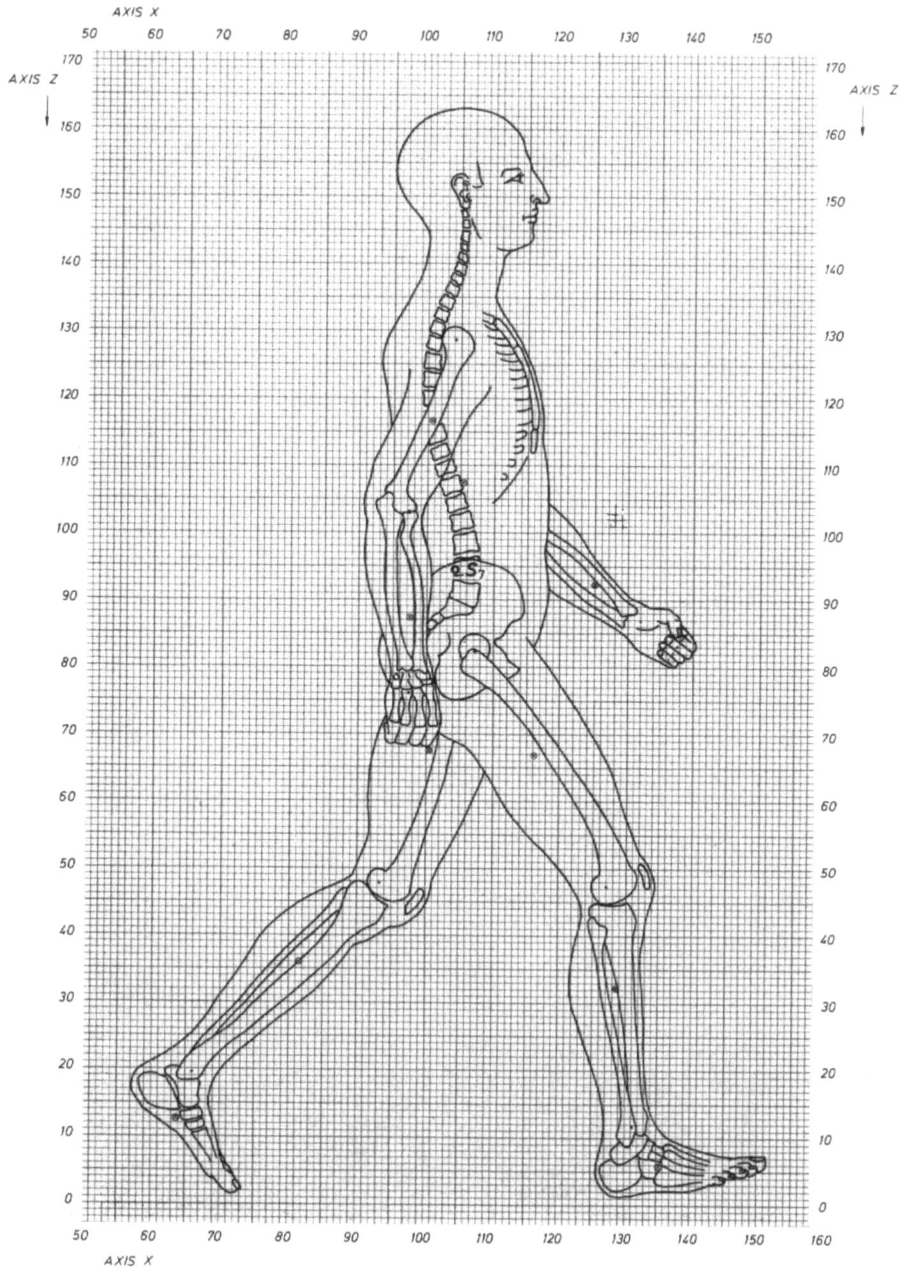

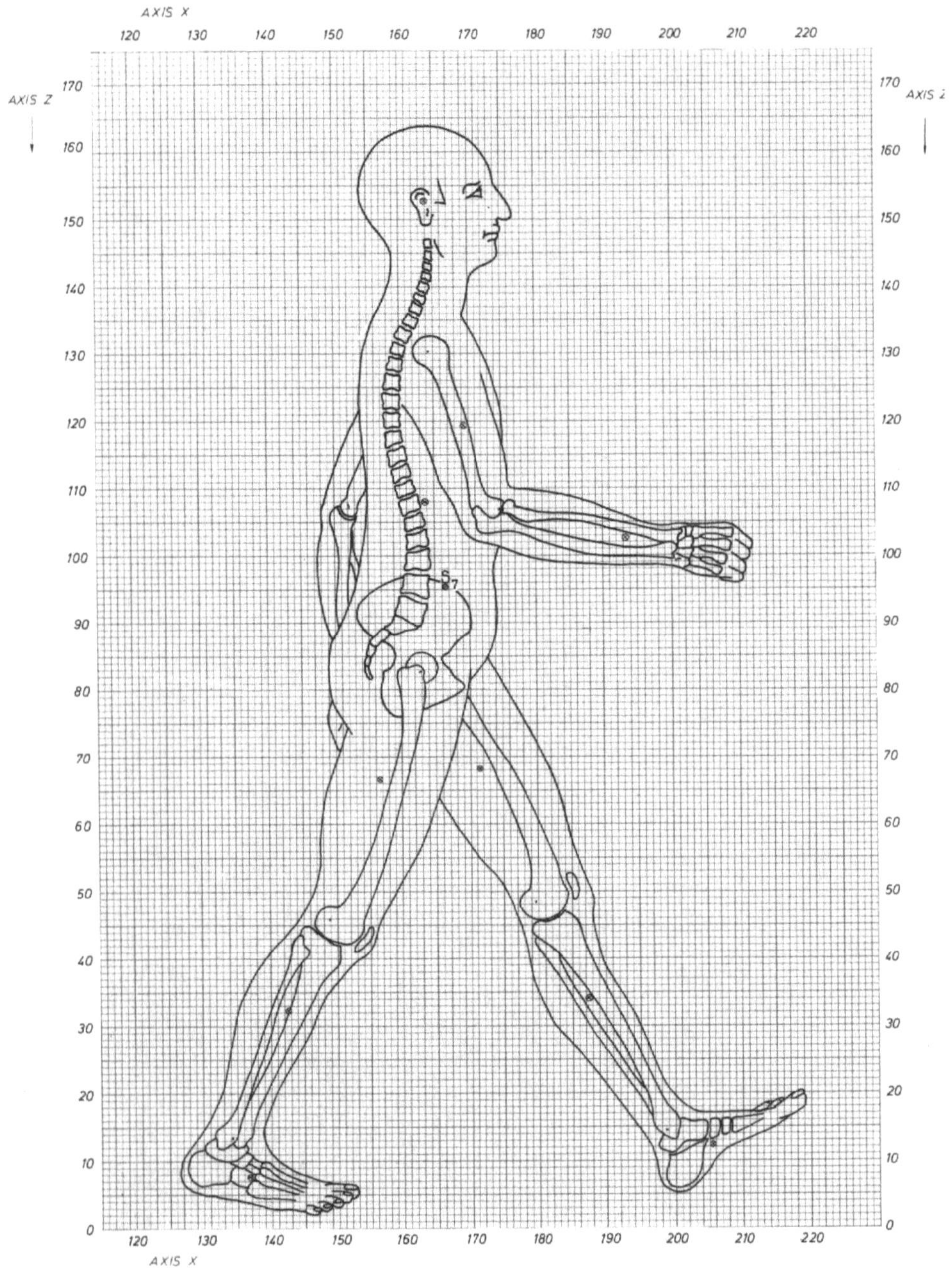

Fig. 30. Subject of Braune and Fischer at phase 22 of gait

b) Forces of Inertia Due to the Accelerations of S_7

The successive co-ordinates of the centre of gravity S_7 allow an analysis of the displacements of S_7 and the determination of the forces of inertia which result from them.

The displacement of S_7 in space during the right single support period of gait (phases 12–23) corresponds to the tridimensional curve, Figure 31. S_7 moves on this curve in a vertical direction in relation to the ground, in a transverse direction in relation to a median sagittal plane and in a longitudinal direction in relation to a coronal plane moving forward with the average speed of the gait. The curve is a solid line when the partial centre of gravity S_7 lies to the right of a vertical median plane, parallel to the direction of the gait ($y+$). It is a dotted line when S_7 lies to the left of this plane ($y-$).

For calculating the forces of inertia produced by the acclerations and decelerations of S_7, the co-ordinates z of S_7 are first entered into a graph. Its ordinate represents in centi-

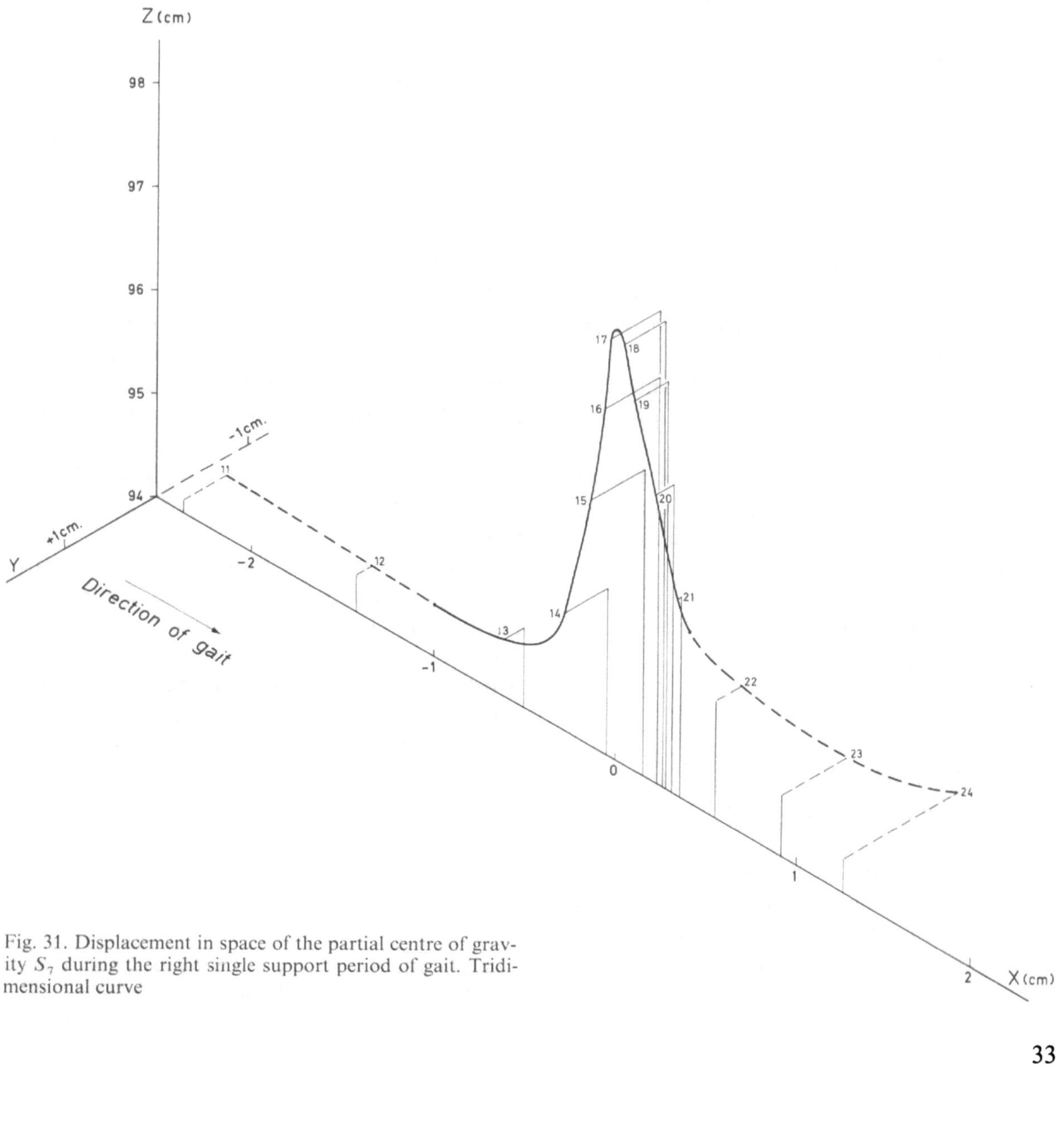

Fig. 31. Displacement in space of the partial centre of gravity S_7 during the right single support period of gait. Tridimensional curve

metres the vertical distance of S_7 from the ground and the abscissa the time in seconds and the phases of gait (Fig. 32). The curve joining the successive co-ordinates z of S_7 illustrates the vertical displacement of S_7 during the step. It may need to be smoothed: points which are aberrant can eventually be slightly displaced to give the curve a regular shape. These slight displacements correspond to the correction of minimal mistakes made in recording during the experiment. They are justified by the fact that we do not normally walk in a spastic way. The centre of gravity S_7 moves without bumps. In this case the curve did not require to be smoothed.

From the curve of the displacement, the velocity can be obtained by graphical differentiation. The curve of the velocity is also drawn. A new graphical differentiation from the latter reveals the curve of the acceleration.

The acceleration for each phase of gait could theoretically be calculated from the tridimensional rectangular co-ordinates of the centre of gravity S_7. However, this calculation would give incoherent results because of the importance of the variations in velocity and acceleration between the successive phases. To obtain sensible results, much shorter intervals must be considered. This is made possible by the graphical differentiation mentioned above.

A positive acceleration has a downward direction, a negative acceleration an upward direction.

The relationship Force = Mass × Acceleration enables one to calculate the vertical component of the acceleration force exerted by S_7 at each phase of gait.

$$\frac{\text{Partial Mass} \times \text{Acceleration}}{\text{Earth Gravity}}$$

$$= \frac{54.560}{981} \cdot \frac{\text{kg}}{\text{cm} \cdot \text{sec}^{-2}} \cdot \text{acc}$$

$$= 0.055 \text{ kg cm}^{-1} \cdot \text{sec}^{+2} \cdot \text{acc}$$

$$= \text{vertical acceleration force.}$$

Since Force of Inertia = − Acceleration Force (d'Alembert) the negative values of the vertical acceleration force represent the vertical component of the forces of inertia. They are given in Table 2.

Table 2. Forces of inertia due to accelerations and decelerations of mass S_7

Phases	D_x horizontal component in direction of movement	D_y horizontal component perpendicular to movement	D_z vertical component
12	+ 1.155	0	6.548
13	+18.102	1.926	17.717
14	+ 9.629	5.392	18.872
15	+ 4.622	4.237	−13.480
16	+ 2.311	1.926	−15.062
17	+ 0.385	1.926	−25.420
18	0	0	−20.413
19	− 0.770	0.770	−14.636
20	− 1.155	1.541	− 3.081
21	− 5.392	5.007	10.399
22	− 6.548	5.007	7.703
23	+ 1.541	3.081	11.169

The vertical component D_z of the forces of inertia is added algebraically to the partial weight P_7 of the body.

In the direction of gait, the successive positions of the centre of gravity S_7 are determined in relation to a coronal plane which moves along with a velocity corresponding to the average speed of gait. The successive distances of S_7 from the mobile plane are entered as the ordinate on the graph. The abscissa represents the time as well as the phases of gait (Fig. 33). The curve joining the points is smoothed if necessary.[4] It represents the horizontal displacement of S_7 in the direction of gait. By graphical differentiation of the curve of displacement, we obtain the curve of velocity. Graphical differentiation of the latter gives the curve of acceleration.

The positive accelerations have a forward direction, the negative a backward direction. Positive accelerations correspond to negative forces of inertia D_x, that is forces with a backward direction. Negative accelerations correspond to positive forces of inertia with a forward direction (Table 2).

[4] In fact only this curve x had to be smoothed. Curves in Figures 32 and 34 were obtained without any smoothing.

Fig. 32. Analysis of the displacement of S_7 in a vertical direction. In ordinate: displacement (cm), velocity (cm/sec), acceleration (cm/sec²). In abscissa: time and phases of gait. R: right heel strike. S_L: swing of left leg. L: left heel strike. S_R: swing of right leg

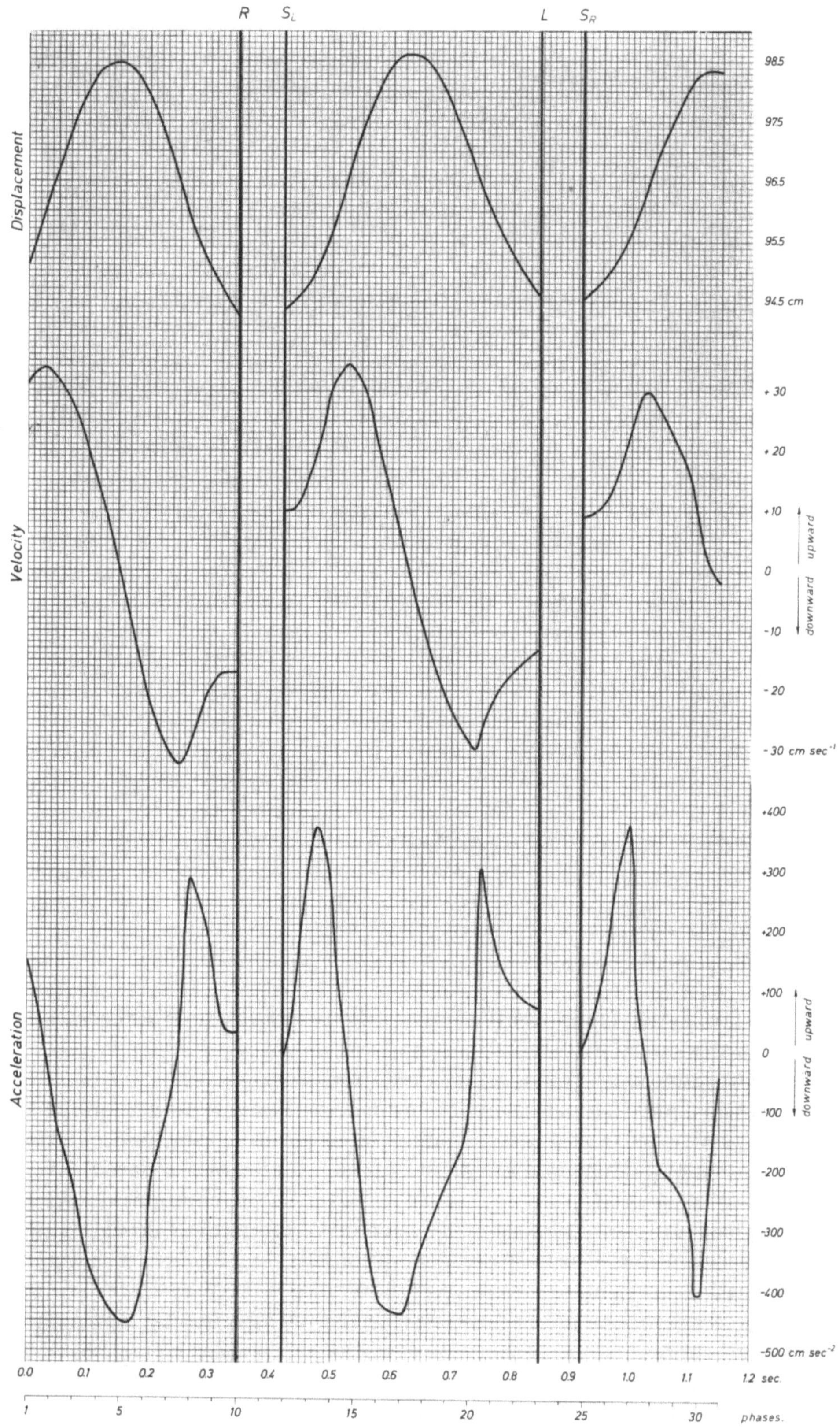

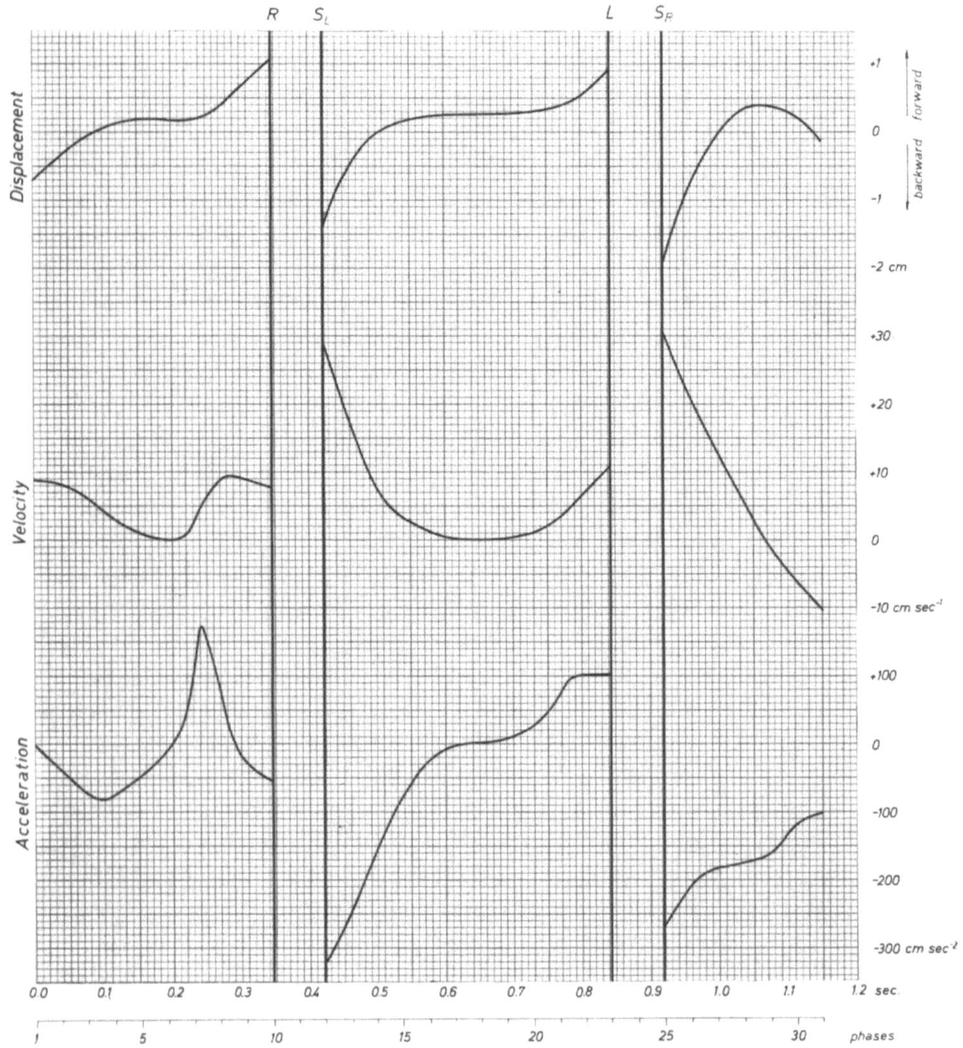

Fig. 33. Analysis of the displacement of S_7 in the direction of gait. In ordinate: displacement (cm), velocity (cm/sec), acceleration (cm/sec^2). In abscissa: time and phases of gait. Abbreviations as for Figure 32

The horizontal component D_y – at right angles to the direction of gait – of the forces of inertia produced by the mass S_7 is obtained in the same manner. The successive co-ordinates y of the partial centre of gravity S_7 are entered as the ordinate, the time and the phases of gait as the abscissa (Fig. 34).

Graphical differentiation gives the velocities and the accelerations. The positive accelerations are directed to the right, the negative to the left.

The y accelerations are negative during the right single support period of gait. The horizontal component D_y of the forces of inertia at right angles to the direction of gait is directed to the right (Table 2).

c) Force P Exerted on the Knee by the Partial Mass S_7 of the Body

Force P exerted on the knee by the partial mass of the body (concentrated at S_7) is the resultant force of the partial weight P_7 of the body, and of the components D_x, D_y, and D_z of the forces of inertia (Fig. 45). It can be easily calculated:

$$P=\sqrt{(P_7+D_z)^2+D_x^2+D_y^2}.$$

Its point of application is S_7 and its direction in space is given by the angles which the line of action of the force P forms with the three axes.

α_P = angle formed by P and axis Ox.
β_P = angle formed by P and axis Oy.
γ_P = angle formed by P and axis Oz.

Fig. 34. Analysis of the displacement of S_7 in a horizontal direction at right angles to the direction of gait. Abbreviations as for Figure 32

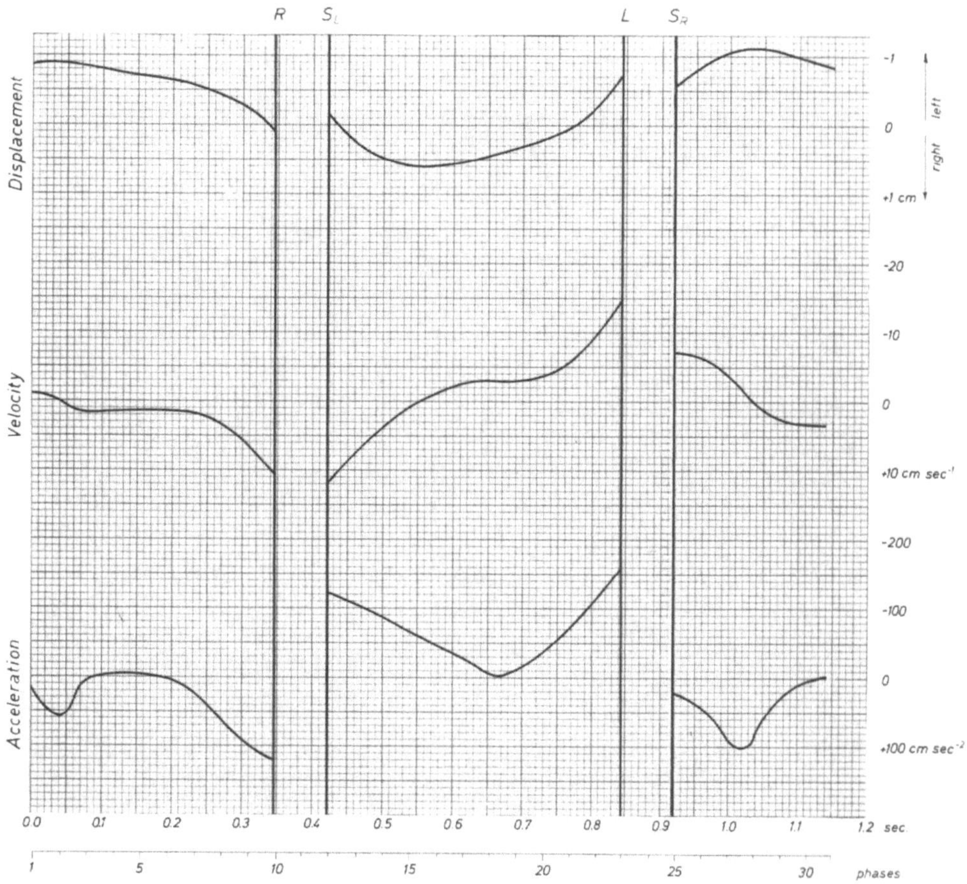

$$\cos\alpha_P = \frac{D_x}{P},$$

$$\cos\beta_P = \frac{D_y}{P},$$

$$\cos\gamma_P = \frac{P_7 + D_z}{P}$$

with the relation:

$$\cos^2\alpha_P + \cos^2\beta_P + \cos^2\gamma_P = 1.$$

$\cos\alpha_P$, $\cos\beta_P$ and $\cos\gamma_P$ are the directional cosines of force P.

The successive values of force P are given in Table 5.

d) Position in Space of Point G Which Lies Centrally on the Axis of Flexion of the Knee

To study the stressing of the joint at each phase of gait the knee will be considered to be in equilibrium. Therefore the resultant of the forces acting on it must pass through the geometric centre of curvature G of the weight-bearing surfaces and be normal to a plane tangential to these surfaces at the point of impact of the resultant force R, since the coefficient of friction is negligible in an animal joint (Fig. 35a).

We designate as axis of flexion the geometric centre of curvature of the weight-bearing surfaces and as "Evolute" the path of the successive axes in the sagittal plane (Fick, 1910) The "instant centre of rotation" of a moving body

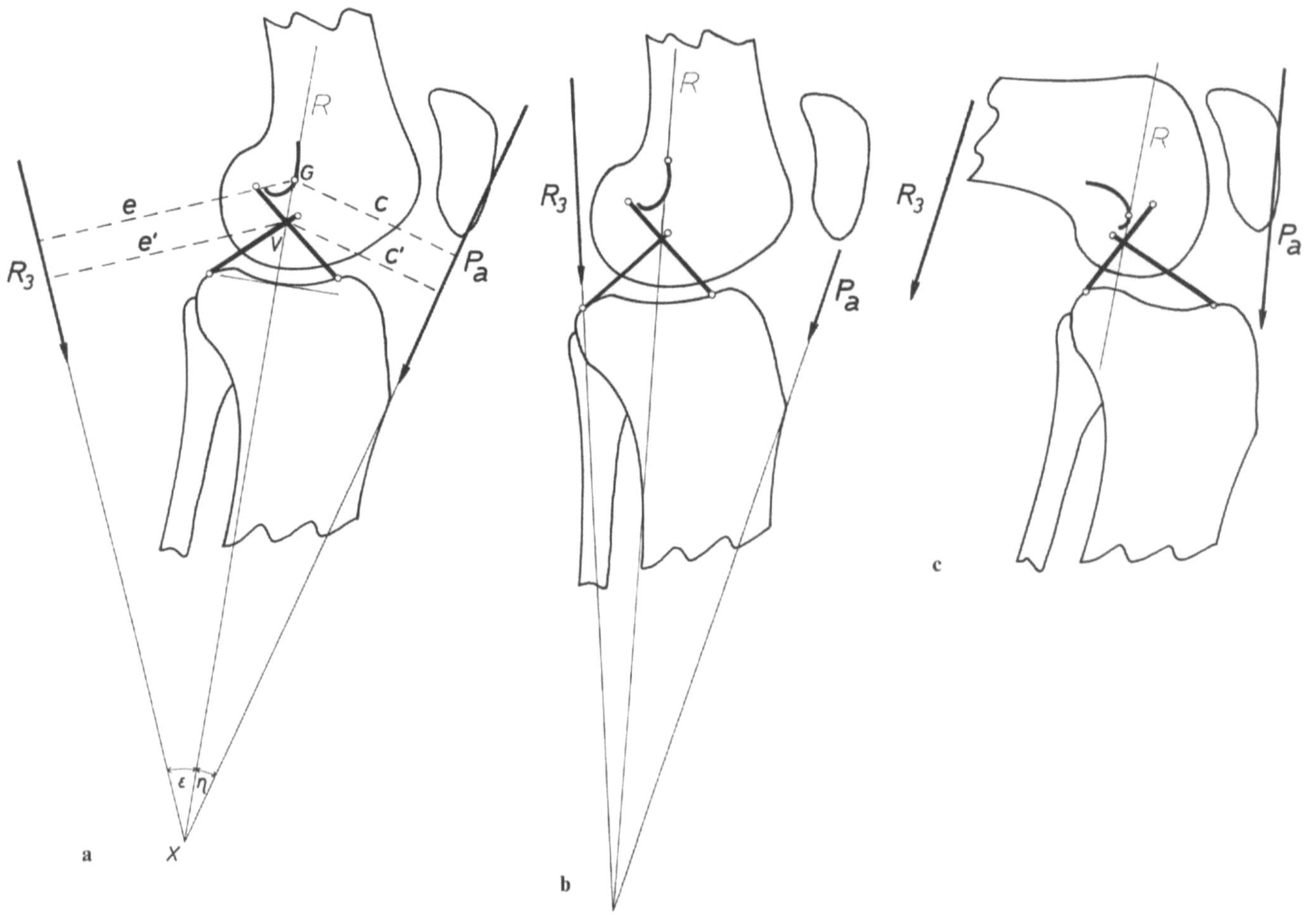

Fig. 35. The geometrical axis G of the femoro-tibial joint lies on the "Evolute". The instant centre of movement V coincides with the intersection of the cruciate ligaments. R_3: force tending to flex the knee. e: lever arm of R_3 from the geometrical centre G as origin. e': lever arm of R_3 from the instant centre V as origin. P_a: force exerted by the patella tendon. c: lever arm of P_a from the geometrical centre G as origin. c': lever arm of P_a from the instant centre V as origin. X: intersection of the lines of action of forces R_3 and P_a. R: line of action of the resultant of forces R_3 and P_a. ε: angle formed by the lines of action of R_3 and R. η: angle formed by the lines of action of P_a and R

(Frankel and Burstein, 1970; Husson, 1973a, b; Menschik, 1974a) is different. It represents a point of zero velocity (lying inside or outside the body) of this body moving in relation to another body. The pathway of the successive instant centres of rotation of the tibia moving in relation to the femur and that of the femur moving in relation to the tibia are curves called polodes. The movement that exists between the two polodes during flexion and extension of the knee is purely rolling. The two curves give a picture of the movement between the femur and the tibia projected on the sagittal plane. The instant centres of rotation are, therefore, useful for studying the *kinematics* of the knee, as did Husson and Menschik. According to them, the instant centre lies at the intersection V of the cruciate ligaments, in the sagittal projection. In order to analyse the *forces* acting on the joint we shall use the geometric centre G as described above. Actually the lever arms of these forces, originating from the geometric centre and those drawn from the instant centre are proportional. As described by Menschik, the perpendicular drawn from the instant centre V (intersection of the cruciate ligaments) to the line joining the distal insertions of the cruciate ligaments and to the tangent to the tibial plateaux also passes through the geometric centre of curvature of the weight-bearing surfaces. The following equations pertain:

$$e = \overline{GX} \cdot \sin\varepsilon$$

$$e' = \overline{VX} \cdot \sin\varepsilon$$

$$\frac{e}{e'} = \frac{\overline{GX}}{\overline{VX}}$$

$c = \overline{GX} \cdot \sin\eta$

$c' = \overline{VX} \cdot \sin\eta$

$\frac{c}{c'} = \frac{\overline{GX}}{\overline{VX}}$

Consequently, the lever arms c and c' of force P_a and e and e' of force R_3 are in the proportion:

$\frac{c}{c'} = \frac{e}{e'}$

Therefore, taking one or the other of these centres as origin of the lever arms does not change anything in the equilibrium of the forces.

$R_3 e = P_a c$ or $R_3 e' = P_a c'$

This is true for any position of the knee since, from full extension to full flexion (Fig. 35b and c), the line of action of the resultant force R must pass through the axis of flexion or geometric centre of the knee and through the instant centre.

Next, the position of the knee in space will have to be determined for each phase of the single support period of gait. Braune and Fischer designated a point in the knee which retains a constant location in the lower extremity of the femur. They established its co-ordinates for each phase of gait. This point, however, cannot be used as such for the calculation of the forces because it does not coincide with the geometrical axis of flexion of the knee. The location of this axis of flexion in the femur changes with movement because the shape of the femoral condyles does not have the same radius of curvature throughout. The radius of curvature of the weight-bearing surfaces of the femoral condyles is longer in extension of the knee and is different for both condyles. During movement, the axis moves in a sagittal plane located between the femoral condyles. The path of this axis forms the "Evolute" (Fick, 1910). Therefore, it is necessary to know for each phase of gait where the axis of flexion lies on the curve. This point must then be plotted within the system of co-ordinates.

The "Evolute" of Fick was first drawn on a sagittal projection of the knee of the subject of Braune and Fischer, standing erect, and its grid of co-ordinates (Fig. 36).

The angle $\beta_{2,4}$ formed by the femur and the tibia is known for each phase of gait . This and the other data of Fischer enable us to locate the axis G on Fick's curve for each degree of flexion. The position of G can then be determined in the system of co-ordinates for each phase of gait.

The centre of the knee in the sense of Braune and Fischer (co-ordinates: $x=0$; $z=48$) is shown in the drawing. It will be designated by O and considered as the origin of the co-ordinates for the next calculation. We know the co-ordinates of O and the position of G on the "Evolute" and must find out the co-ordinates of point G. A line connecting points O and G

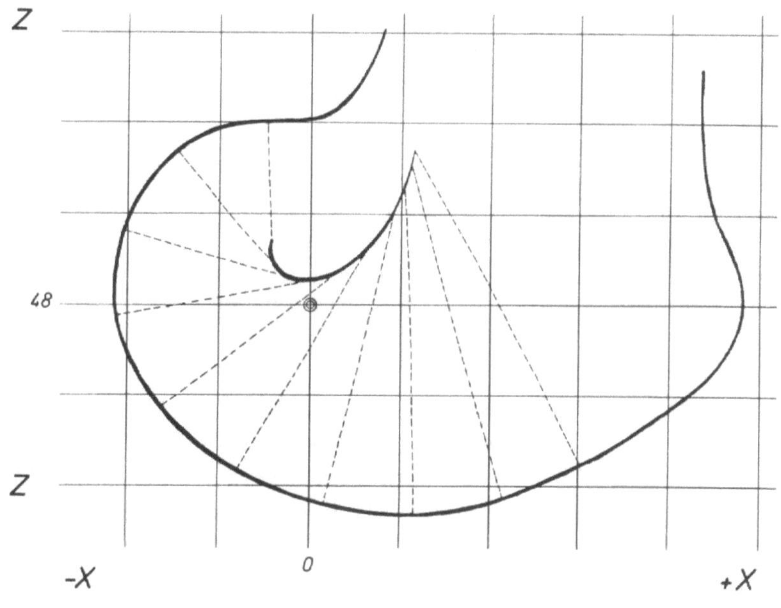

Fig. 36. "Evolute" of Fick and centre of the knee according to Braune and Fischer (co-ordinates: 0,48)

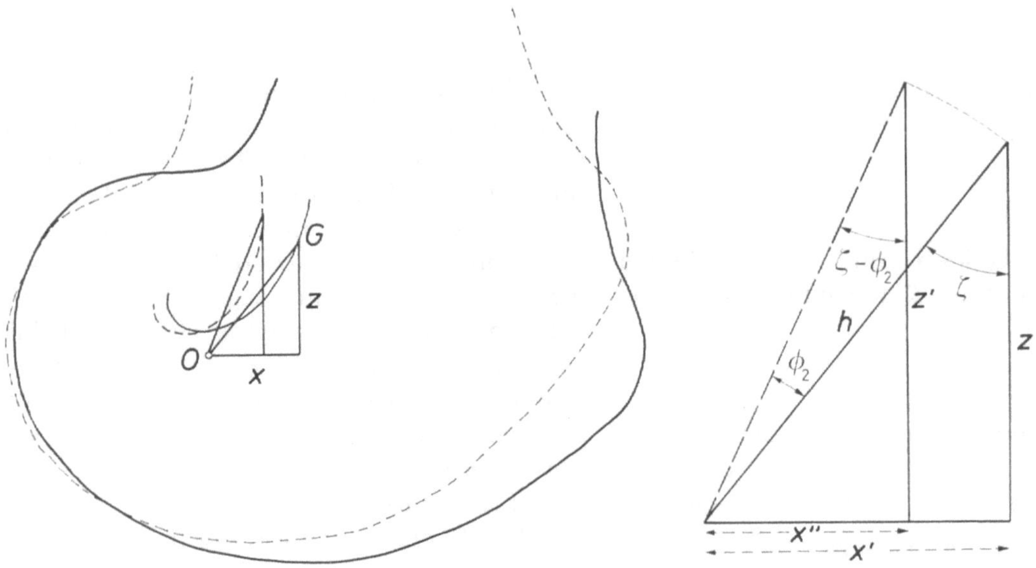

Fig. 37. Displacement of the "Evolute" of Fick in relation to the knee-centre of Braune and Fischer, considering the femur in two different positions in space

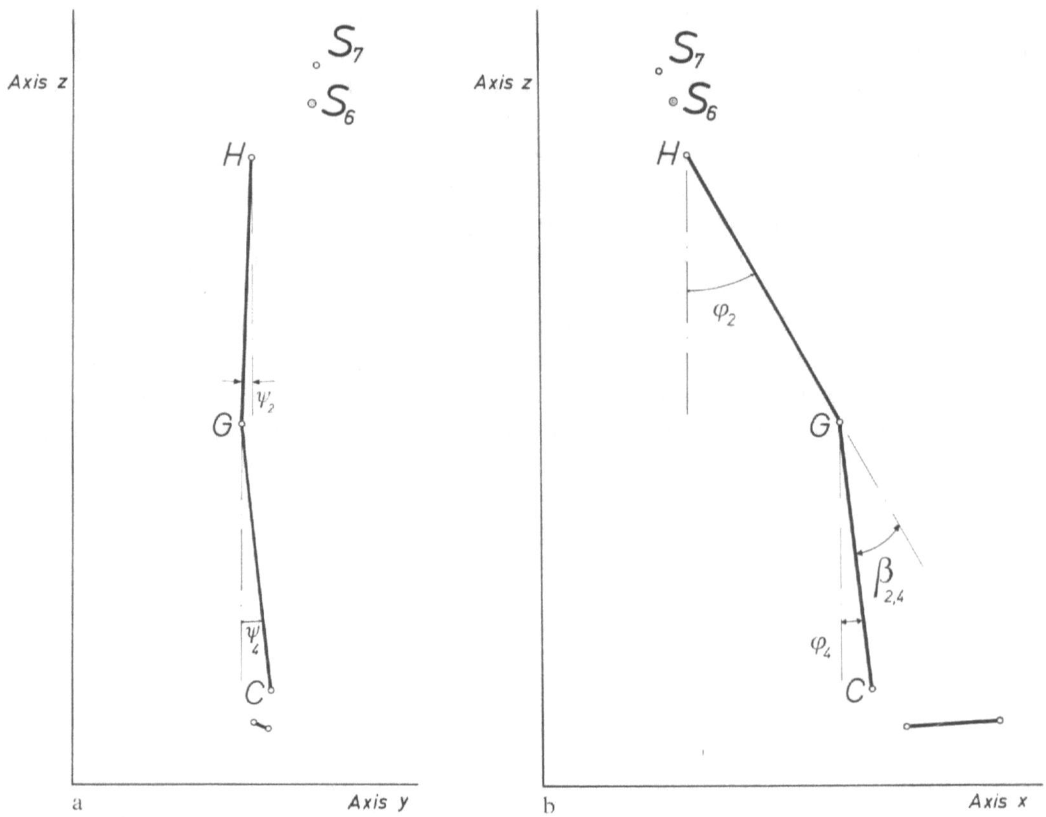

Fig. 38a and b. Angles formed by the vertical and the projections of the femur and tibia on the coronal (ψ_2 and ψ_4) and on the sagittal plane (φ_2 and φ_4); $\beta_{2,4}$ angle of flexion of the knee; H hip; G knee; C ankle

represents the hypotenuse of a right-angled triangle (Fig. 37). The upright z of this triangle is drawn vertically from G, parallel to the axis Z of the system of co-ordinates. The hypotenuse and the upright z of the triangle form an angle ζ. The base x of the triangle is drawn through O, at right angles to z. The distances z and x can be measured on the knee of the subject standing erect. The distance between the projections of O and G on a transverse horizontal line in the coronal plane is designated by y.

The triangle is located in the femur and

changes with the movements of the latter. The projections of the lines z, x and y in the tridimensional system of co-ordinates can be calculated provided that the rotation of the femur about its longitudinal axis and its projection on the two vertical planes are known. The projections of the femur on the sagittal plane and on the coronal plane are given by the angles φ_2 and ψ_2 of Fischer (Fig. 38). It should be possible to calculate the rotation θ of the femur about its longitudinal axis from the projections φ_2, ψ_2 of the femur and φ_4, ψ_4 of the tibia on the two vertical planes.

When we think of the femur as a cylinder, we can draw imaginary longitudinal parallel lines called generates on this cylinder (Fig. 39). The generate lying in front of the subject when the limb is vertical, at rest, is represented by a dotted line. When the femur rotates about its long axis, the generates remain vertical but each of them is replaced by the next one and the dotted one is moved to the right or to the left. The extent of this rotatory movement during gait must be determined. One must realize that the longitudinal axis of the femur remains unchanged by its rotatory movement.

If the longitudinal axis of the femur is in line with the longitudinal axis of the tibia, the angle $\beta_{2,4}$ which they form is zero. The common direction of femur and tibia is not changed by the rotation of the femur.

If the knee is flexed, $\beta_{2,4} \neq 0$ (Fig. 38). The direction of the femur and that of the tibia define a plane, the orientation of which depends on the rotation of the femur. One cannot determine the rotation of the femur when $\beta_{2,4} = 0$. When $\beta_{2,4} \neq 0$ but is fairly small, a very small change of an angle of direction produces an enormous modification of the angle of rotation θ of the femur about its longitudinal axis. It becomes meaningless. Furthermore, if the axis of flexion of the knee is not always perfectly normal to the plane determined by the femur and the tibia, the problem cannot be solved.

Although, theoretically, the rotation of the femur about its long axis can be calculated from the angles φ and ψ, we prefer to use the direct measurements of θ as obtained by Johnston and Smidt (1969).

These authors found that the rotation of the femur about its longitudinal axis varied with every phase of gait. Their measurements produced a curve on which heel strike (phase 10), flat foot (phase 13), heel off (phase 20) and toe off (phase 25) are marked. These points make it possible to determine which part of the curve corresponds to which phase of gait. The rotation θ of the femur about its longitudinal axis can then be readily determined. These values are given in Table 3.

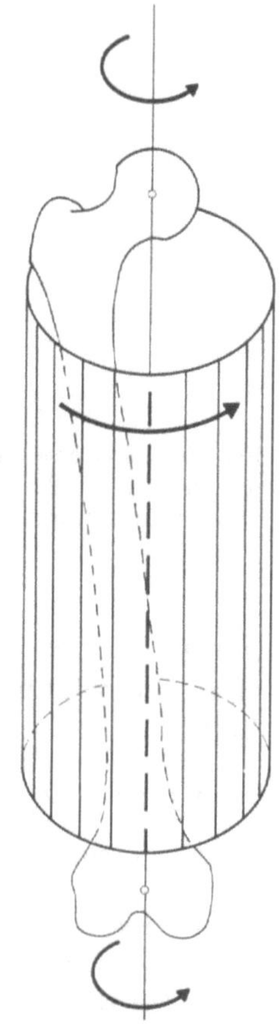

Fig. 39. One can think of the femur as a cylinder. Generates drawn on the cylinder will move from left to right when the femur rotates about its longitudinal axis as indicated by the arrows

Table 3. Rotation θ of the femur about its axis during the right single support period of gait (Johnston and Smidt). Figures are of internal rotation in relation to rest position

Phases	Angle θ	Phases	Angle θ
12	+2°20′	18	+4°25′
13	+3°	19	+3°45′
14	+4°	20	+3°20′
15	+4°30′	21	+3°15′
16	+4°40′	22	+3°10′
17	+4°30′	23	+1°45′

It is now possible to calculate the projections of the distances x, y and z on the three planes, for each phase of gait. O is considered as a fixed point about which the femur moves. In a plane

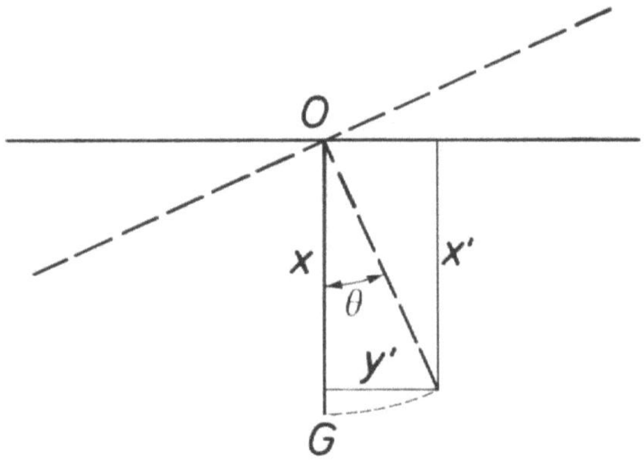

Fig. 40. Transverse cross section. Displacement of point G (on the axis of flexion of the knee) in relation to point O when the femur rotates about its longitudinal axis

normal to the longitudinal axis of the femur (Fig. 40), rotating on itself with an angle θ, one finds:

$$x' = x \cdot \cos\theta,$$

$$y' = x \cdot \sin\theta.$$

In the sagittal plane (Fig. 37), when the femur forms an angle φ_2 with the vertical (Fig. 38b):

$$h = \sqrt{x'^2 + z^2},$$

$$\sin\zeta = \frac{x'}{h}$$

pertains

$$x'' = h\sin(\zeta - \varphi_2),$$

$$z' = h\cos(\zeta - \varphi_2).$$

In the coronal plane:

$$y'' = z' \cdot \sin\psi_2 + y',$$

$$z'' = z' \cdot \cos\psi_2.$$

The values x'', y'', z'' are successively added to the respective co-ordinates x, y, z of point O, centre of the knee as determined by Braune and Fischer. For each phase of the right single support period, the co-ordinates of point G on the curve of Fick and on the axis of flexion of the knee are obtained in this manner. They are given in Table 4.

Table 4. Co-ordinates of point G, central on the axis of flexion of the load bearing knee

Phases	x_G	y_G	z_G
11	120.52	+ 7.87	47.31
12	127.30	+ 9.59	48.02
13	132.27	+ 9.97	48.20
14	134.04	+10.27	48.03
15	135.00	+ 9.70	48.00
16	135.83	+ 8.93	48.00
17	136.93	+ 8.32	47.87
18	138.22	+ 7.87	47.77
19	140.18	+ 7.67	47.52
20	142.94	+ 7.25	47.05
21	145.95	+ 7.18	46.82
22	150.22	+ 7.37	46.70
23	155.84	+ 7.79	46.26
24	163.01	+ 8.55	45.68

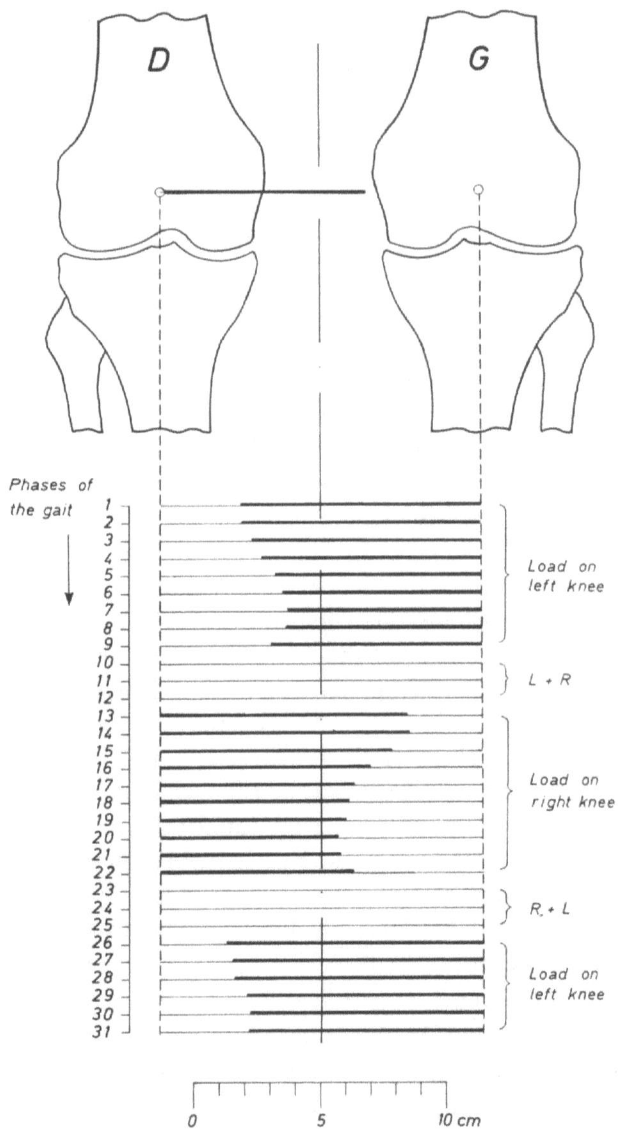

Fig. 41. Distance from the loaded knee to the vertical drawn from the partial centre of gravity S_7, projected on the coronal plane, for each phase of gait

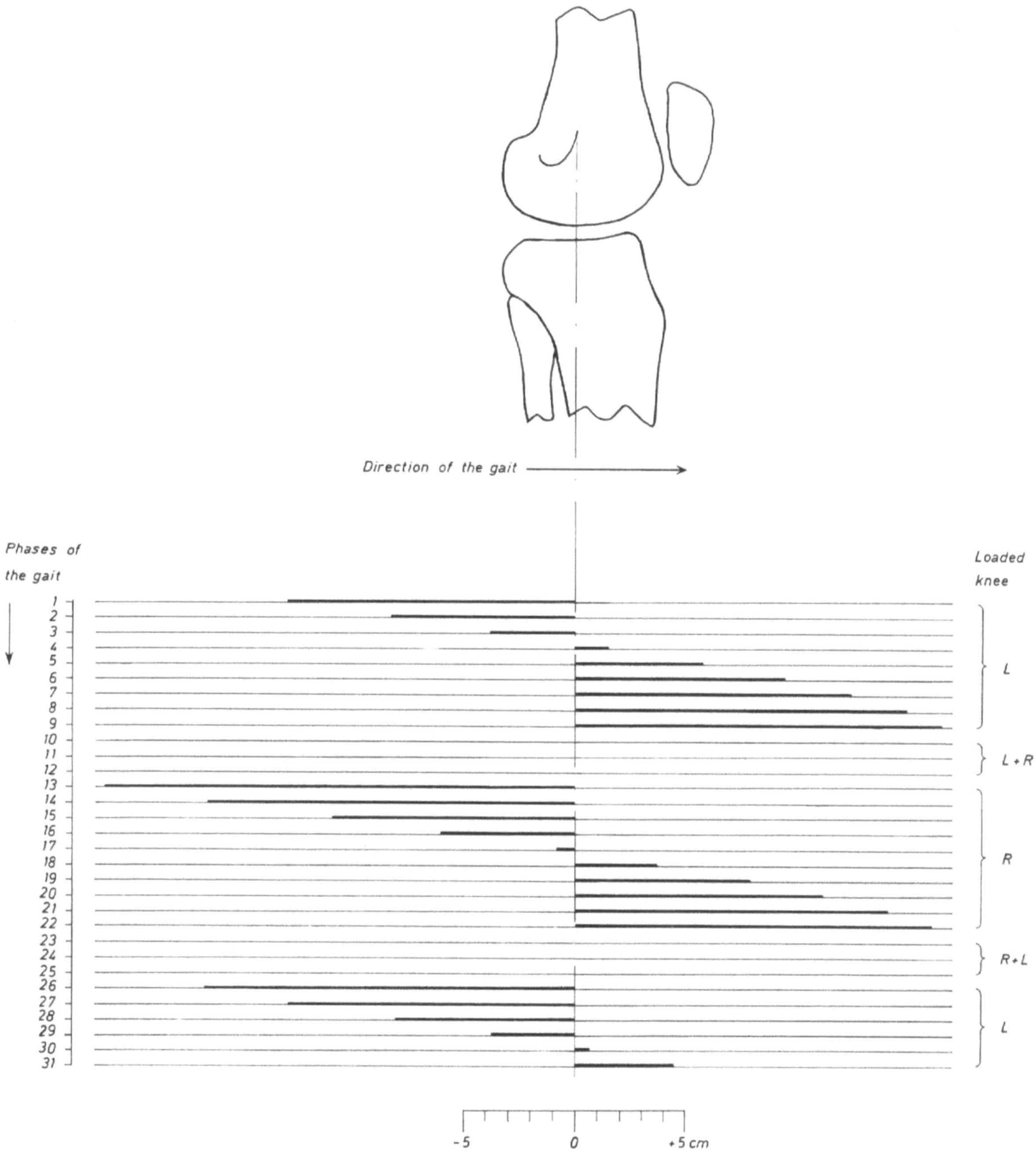

Fig. 42. Distance from the loaded knee to the vertical drawn from the partial centre of gravity S_7, projected on the sagittal plane for each phase of gait

e) Position of the Knee in Relation to the Partial Centre of Gravity S_7

The co-ordinates of the partial centre of gravity S_7 (Table 1) as well as the co-ordinates of point G on the axis of flexion (Table 4) have been calculated. The schematic drawing Figure 41 indicates the successive distances between point G and the vertical drawn from the partial centre of gravity S_7 both projected on a coronal plane. The schematic drawing Figure 42 shows the distances between the vertical drawn from S_7 and

G projected on a vertical plane parallel to the direction of gait. Figure 43 indicates the distances between S_7 and G projected on a horizontal plane. The successive distances between S_7 and G (i.e. the three preceding drawings together) are represented in space by the tridimensional curve Figure 44.

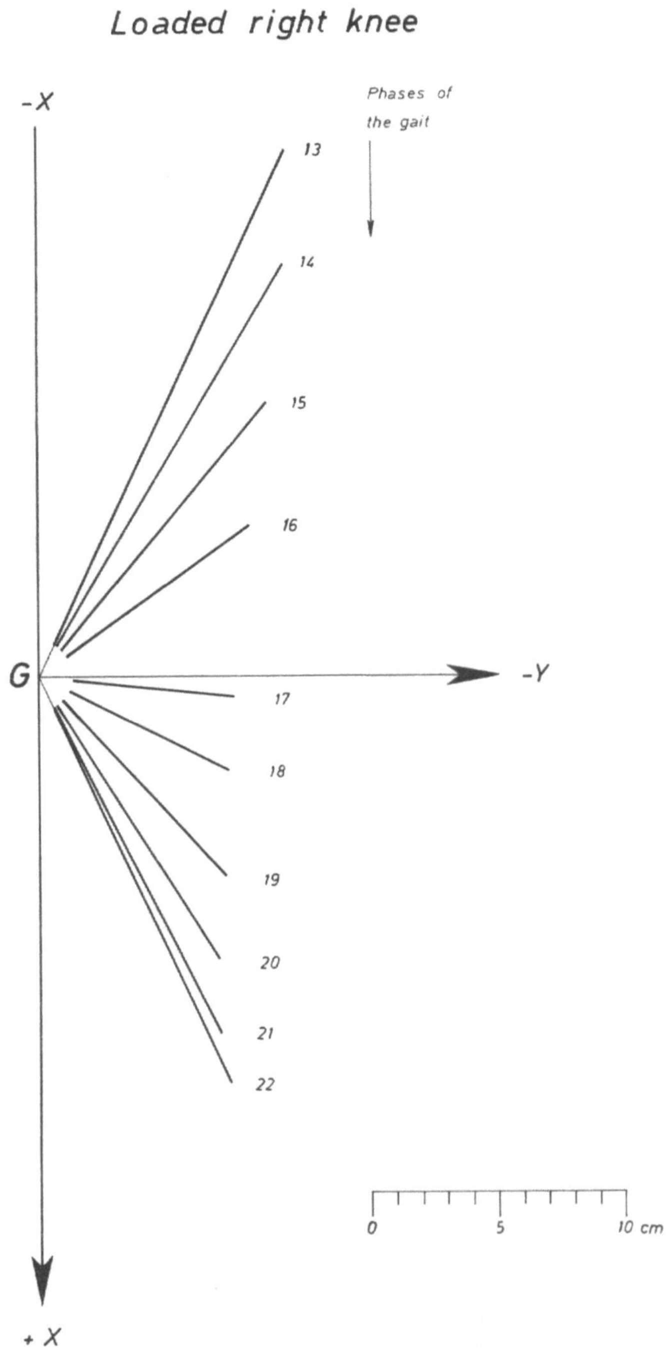

Fig. 43. Distance from the loaded knee to the vertical drawn from the partial centre of gravity S_7, projected on the horizontal plane, for each phase of gait

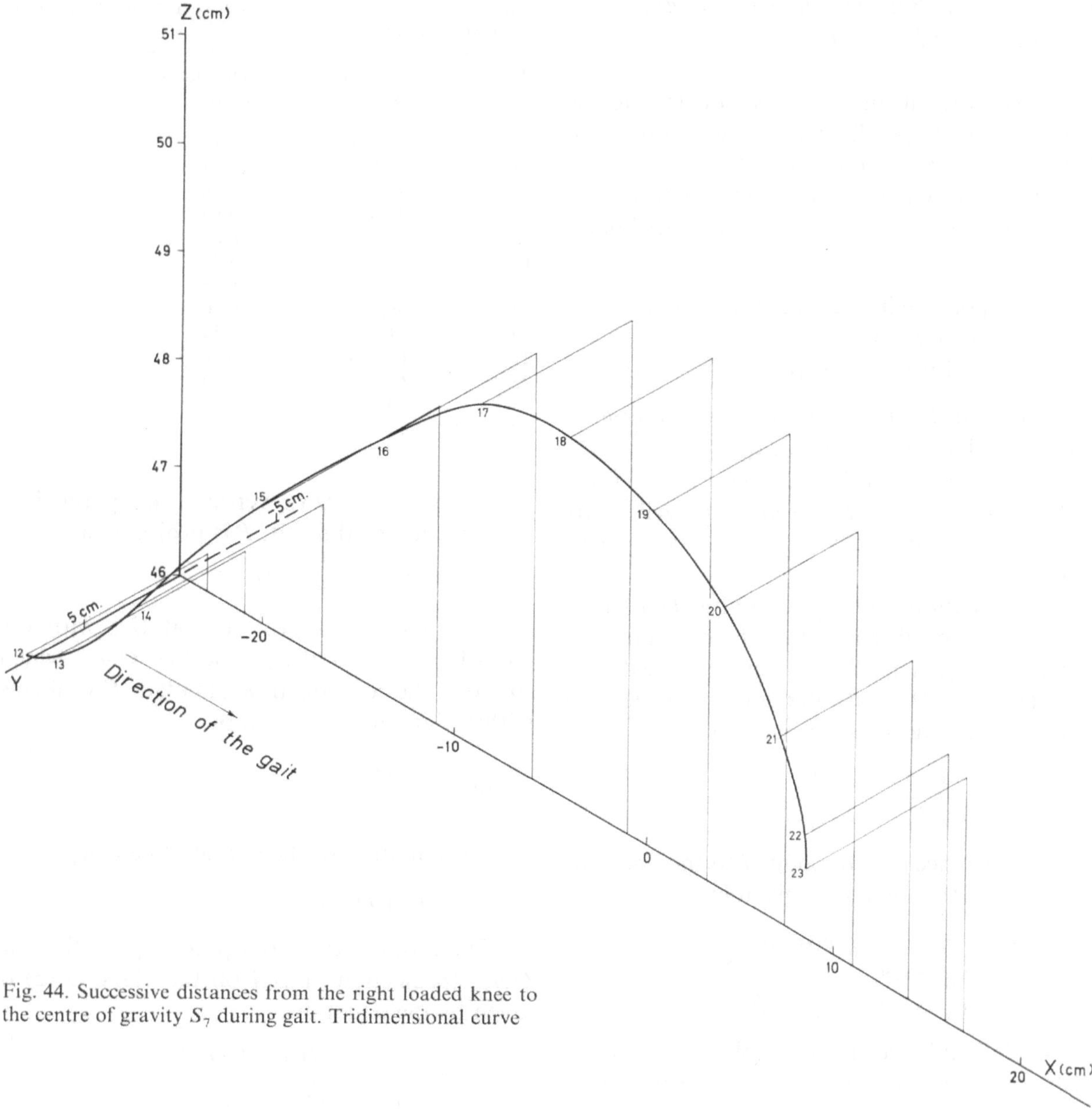

Fig. 44. Successive distances from the right loaded knee to the centre of gravity S_7 during gait. Tridimensional curve

f) Distance a Between the Line of Action of Force P and Point G

We have now all the elements to calculate the length of the perpendicular a drawn from point G to the line of action of force P.

For this calculation let us consider O located in S_7 as origin of the system of co-ordinates (Fig. 45).

Ox is positive in the direction of gait.
Oy is positive to the right.
Oz is positive downward.

The co-ordinates of point G in relation to the triangulate $Oxyz$ are x'_G, y'_G, z'_G.[5]

The components P_7+D_z, D_x, D_y of force P have been determined above, as well as its directional cosines $\cos\alpha_P$, $\cos\beta_P$, $\cos\gamma_P$ (see pages 36–37).

The equations of the straight line OA which bears the force P, goes through the origin S_7, forms the angles $\alpha_P, \beta_P, \gamma_P$ respectively with the axes Ox, Oy, Oz and intersects the horizontal plane containing G, can be written:

$$\frac{D_x}{\cos\alpha_P}=\frac{D_y}{\cos\beta_P}=\frac{P_7+D_z}{\cos\gamma_P}.$$

The distance a from point G to the line OA, direction of the force P, is given by:

$$\begin{aligned}a^2=&(y'_G\cdot\cos\gamma_P-z'_G\cdot\cos\beta_P)^2\\&+(x'_G\cdot\cos\gamma_P-z'_G\cdot\cos\alpha_P)^2\\&+(y'_G\cdot\cos\alpha_P-x'_G\cdot\cos\beta_P)^2.\end{aligned}$$

Table 5 indicates for each phase of gait the distance a from point G to the line of action of P.

The force P and its distance a to the knee G can be projected on the three directional planes (Fig. 45).

[5] In relation to the general system of co-ordinates of Fischer we have substituted:

$$x_G-x_{S_7}=x'_G,$$
$$y_G-y_{S_7}=y'_G,$$
$$z_{S_7}-z_G=z'_G$$

where

x_{S_7} = co-ordinate x of S_7.
x_G = co-ordinate x of G.
y_{S_7} = co-ordinate y of S_7.
y_G = co-ordinate y of G.
z_{S_7} = co-ordinate z of S_7.
z_G = co-ordinate z of G.

Table 5. Force P exerted by mass S_7 in movement and its distance a to point G

Phases	Force P (kg)	Distance a (cm)
12	61.118	24.16
13	74.534	12.74
14	74.257	11.86
15	41.555	6.89
16	39.612	6.63
17	29.206	4.43
18	34.147	8.25
19	39.939	9.30
20	51.515	11.54
21	65.375	10.66
22	62.806	11.64
23	65.820	17.03

In the plane Oxz vertical and parallel to the direction of the gait, P is projected at P_{xz}

$$P_{xz}=\sqrt{(P_7+D_z)^2+D_x^2}.$$

The point G is projected at G_{xz} with the co-ordinates x'_G and z'_G. The line OA_{xz} which supports the component P_{xz} in plane Oxz forms with Oz an angle γ'_P given by:

$$\operatorname{tg}\gamma'_P=\frac{D_x}{P_7+D_z}.$$

The equation of the straight line OA_{xz} is:

$$D_x=(P_7+D_z)\cdot\operatorname{tg}\gamma'_P.$$

The distance a_{xz} from point G_{xz} to the line OA_{xz} (projection of OA on the plane Oxz) is given by the equation:

$$a_{xz}=\frac{|x'_G\cdot(P_7+D_z)-z'_G\cdot D_x|}{\sqrt{(P_7+D_z)^2+D_x^2}}.$$

The projection P_{yz} of force P on the plane Oyz vertical and at right angles to the direction of gait is calculated in the same manner:

$$P_{yz}=\sqrt{(P_7+D_z)^2+D_y^2}.$$

It forms with axis Oy an angle β'_P given by

$$\operatorname{tg}\beta'_P=\frac{P_7+D_z}{D_y}.$$

G is projected at G_{yz} with the co-ordinates y'_G and z'_G. The distance a_{yz} from G_{yz} to the line OA_{yz} is obtained by the equation:

$$a_{yz}=\frac{|y'_G\cdot(P_7+D_z)-z'_G\cdot D_y|}{\sqrt{D_y^2+(P_7+D_z)^2}}.$$

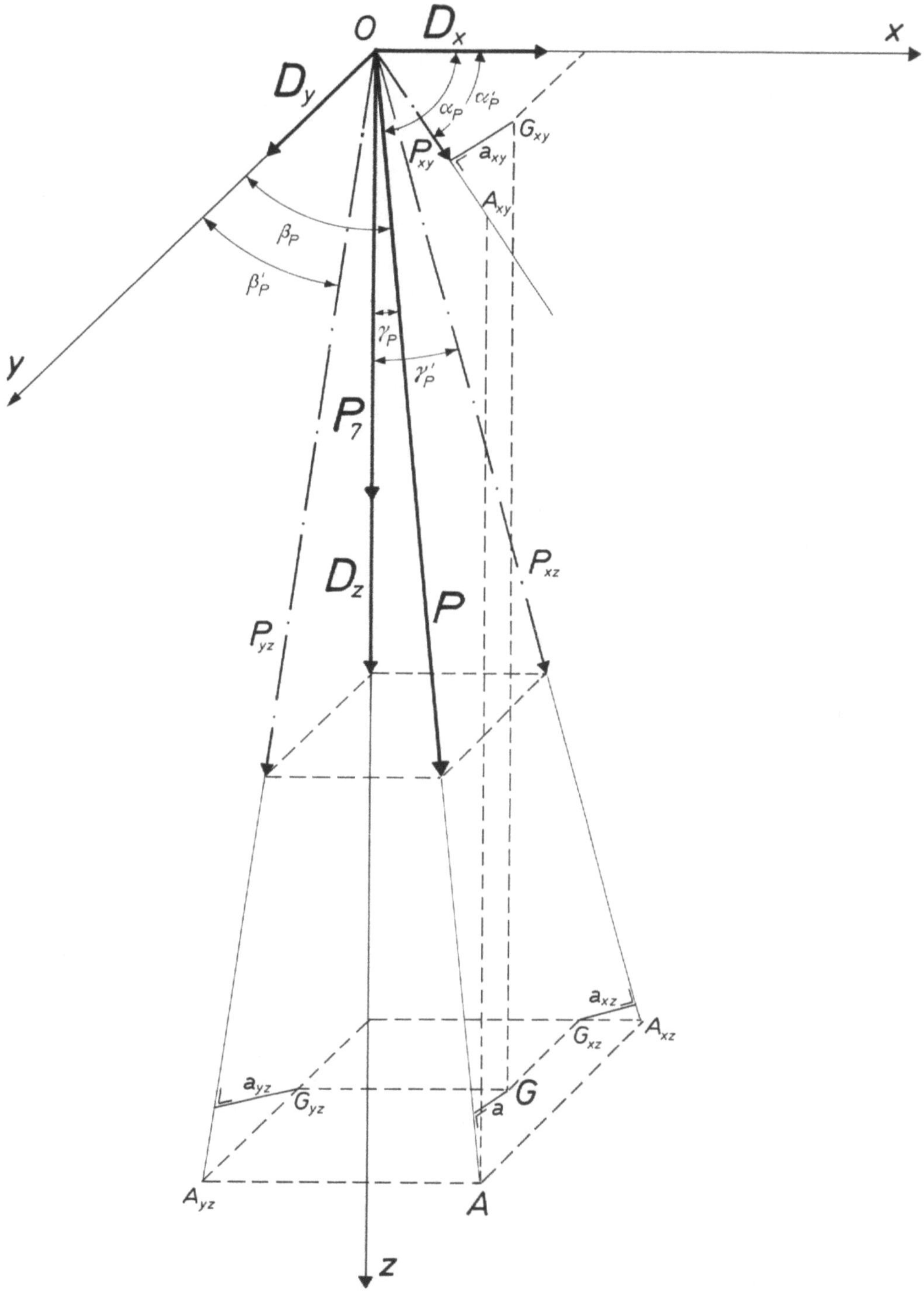

Fig. 45. Line of action of force P related to the knee G. P_7: body weight minus loaded lower leg and foot. D_z: vertical component of the forces of inertia. D_x: horizontal component in the direction of gait, of the forces of inertia. D_y: horizontal component in a direction at right angles to that of gait, of the forces of inertia. P: force eccentrically exerted on the knee by the partial body mass. G: knee

Table 6. Projections of force P and distance a in the system of co-ordinates

Phases	In the vertical plane Oxz		In the vertical plane Oyz		In the horizontal plane Oxy	
	P_{xz}(kg)	a_{xz}(cm)	P_{yz}(kg)	a_{yz}(cm)	P_{xy}(kg)	a_{xy}(cm)
12	61.119	22.26	61.108	9.42	1.155	9.42
13	74.509	9.21	72.303	8.50	18.204	7.45
14	74.061	10.12	73.630	6.28	11.036	0.50
15	41.339	5.63	41.298	4.04	6.270	0.84
16	39.565	3.05	39.544	5.90	3.008	2.58
17	29.142	0.28	29.204	4.42	1.964	0.59
18	34.147	3.58	34.147	7.43	0	0
19	39.932	6.91	39.932	6.23	−1.089	0.48
20	51.492	10.06	51.502	5.53	−1.926	4.73
21	65.183	10.15	65.152	3.34	−7.358	4.47
22	62.206	11.06	62.464	3.73	−8.243	3.78
23	65.748	15.86	65.802	6.23	3.447	11.40

The projection P_{xy} of force P on the horizontal plane Oxy can also be calculated:

$$P_{xy}=\sqrt{D_x^2+D_y^2}.$$

OA_{xy} is the projection of OA on Oxy and supports the component P_{xy}. It forms with axis Ox an angle α'_P given by:

$$\operatorname{tg}\alpha'_P=\frac{D_y}{D_x}.$$

G is projected at G_{xy} with the co-ordinates x'_G and y'_G. The distance from G_{xy} to the straight line OA_{xz} is:

$$a_{xy}=\frac{|x'_G\cdot D_y-y'_G\cdot D_x|}{\sqrt{D_x^2+D_y^2}}.$$

The magnitudes P_{xz}, P_{yz} and P_{xy} of the projections of force P on the three planes are given in Table 6 as well as the distances between the projections of point G and the projections of the line of action of force P.

g) Muscular and Ligamentous Forces Balancing Force P

The components of force P are (P_7+D_z), D_x and D_y. They are known for each phase of the right single support period of gait. They tend to tilt the femur on the tibia, first medially and backward, then medially, finally medially and forward. They are balanced by muscles and ligaments. What the tibia carries on its upper end must consequently be a system of forces in space, equivalent to P.

These forces are constituted by a force of contact with the femur, as well as muscular and ligamentous forces. They must be defined in magnitude and position and present six unknown parameters which are to be determined by the six conditions of equivalence of P in space.

α) Formularization

We begin the analysis by considering the tibia as a straight line GC with G as the superior point (Fig. 46). The directional cosines of line GC can be calculated, since the angles formed by the projection of the tibia and the vertical on the plane xz parallel to the gait, φ_4, and on the plane yz transverse to the gait, ψ_4, are known for each phase of the step (Fig. 38). It is easy to go from these known angles φ_4 and ψ_4 to the directional cosines of the tibia GC with a length t. The directional cosines of the tibia GC are $\cos\alpha_t$, $\cos\beta_t$, $\cos\gamma_t$, or the cosines of the angles (GC,x), (GC,y), and (GC,z) respectively. We have:

$$GA=t\cos\alpha_t,\quad GA=GF\operatorname{tg}\varphi_4=t\cos\gamma_4\cdot\operatorname{tg}\varphi_4,$$
$$GE=t\cos\beta_t,\quad GE=GF\operatorname{tg}\psi_4=t\cos\gamma_t\cdot\operatorname{tg}\psi_4,$$
$$GF=t\cos\gamma_t,\quad FC=GF\operatorname{tg}\gamma_t=t\cdot\sin\gamma_t$$

and

$$\overline{GA}^2+\overline{GE}^2=\overline{FC}^2$$

or

$$t^2\cdot\cos^2\gamma_t\cdot(\operatorname{tg}^2\varphi_4+\operatorname{tg}^2\psi_4)$$
$$=t^2\cdot\sin^2\gamma_t=t^2\cdot(1-\cos^2\gamma_t).$$

Fig. 46. Directional angles. t: tibia. G: knee

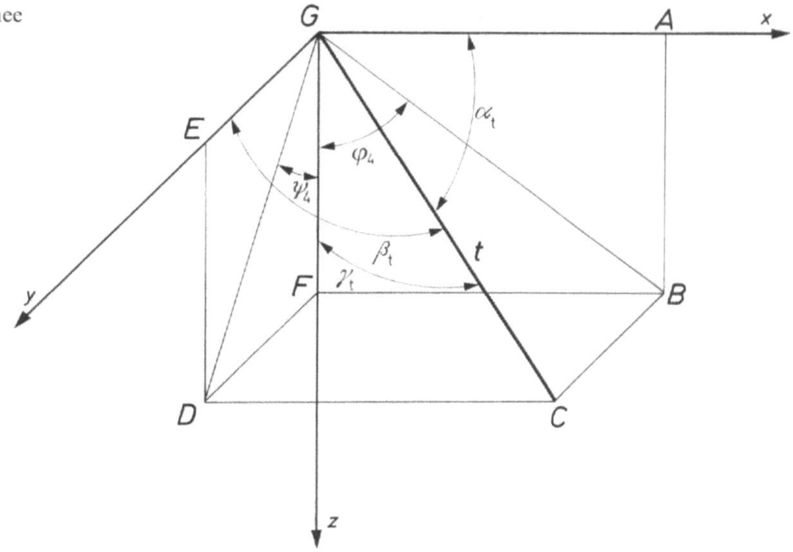

From which we get:

$$\cos^2\gamma_t\cdot(1+\mathrm{tg}^2\varphi_4+\mathrm{tg}^2\psi_4)=1,$$

$$\cos\gamma_t=\frac{1}{\sqrt{1+\mathrm{tg}^2\varphi_4+\mathrm{tg}^2\psi_4}},$$

$$\cos\beta_t=\frac{GE}{t}=\cos\gamma_t\cdot\mathrm{tg}\,\psi_4,$$

$$\cos\beta_t=\frac{\mathrm{tg}\,\psi_4}{\sqrt{1+\mathrm{tg}^2\varphi_4+\mathrm{tg}^2\psi_4}},$$

$$\cos\alpha_t=\frac{GA}{t}=\cos\gamma_t\cdot\mathrm{tg}\,\varphi_4,$$

$$\cos\alpha_t=\frac{\mathrm{tg}\,\varphi_4}{\sqrt{1+\mathrm{tg}^2\varphi_4+\mathrm{tg}^2\psi_4}}.$$

The directional cosines of the tibia are thus known for each phase of gait. Calculation shows that $\cos\alpha_t$ is negative except for phase 12. Cosine β_t is always negative. Cosine γ_t is always positive.

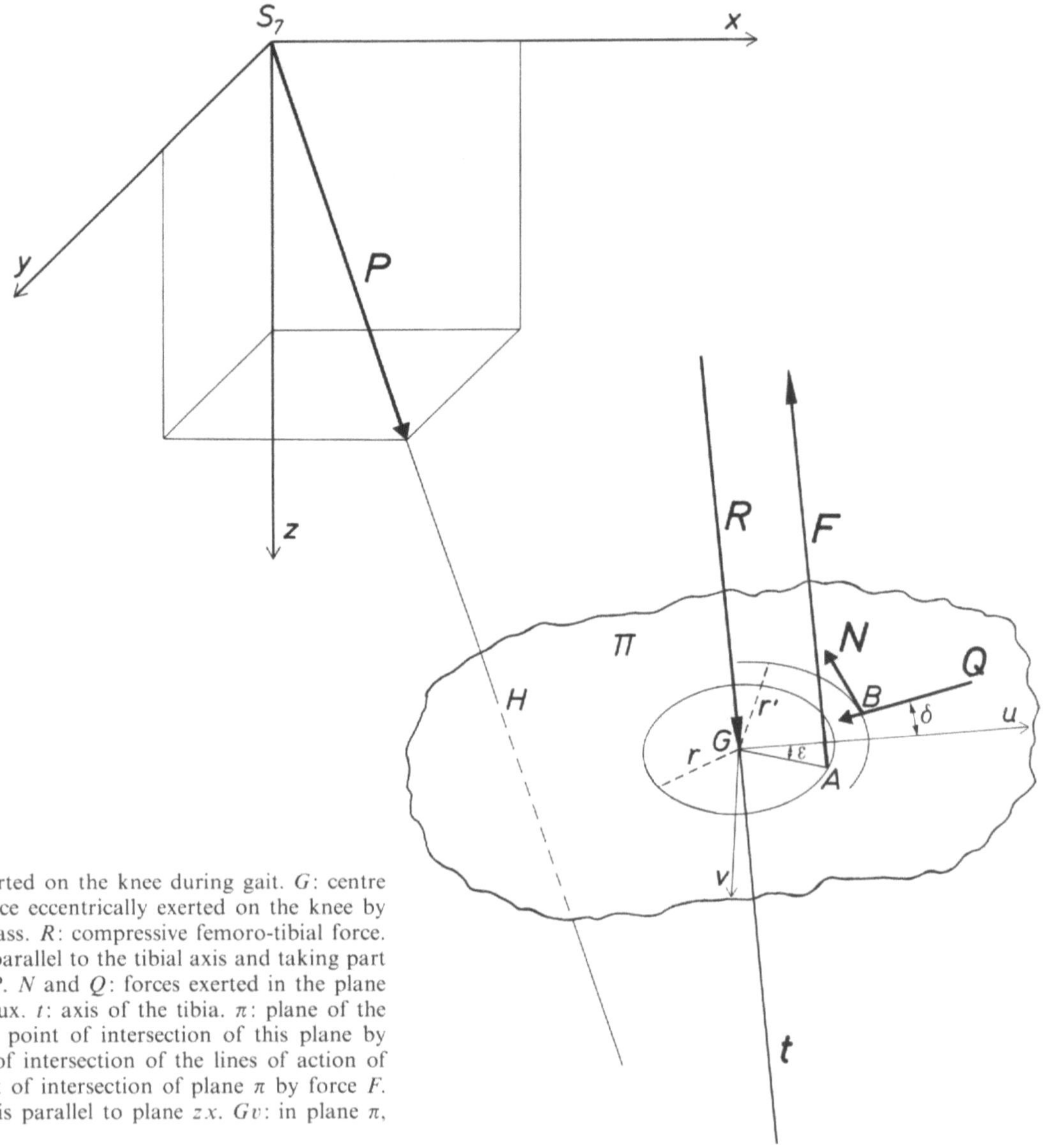

Fig. 47. Forces exerted on the knee during gait. *G*: centre of the knee. *P*: force eccentrically exerted on the knee by the partial body mass. *R*: compressive femoro-tibial force. *F*: muscular force parallel to the tibial axis and taking part in the balance of *P*. *N* and *Q*: forces exerted in the plane of the tibial plateaux. *t*: axis of the tibia. π: plane of the tibial plateaux. *H*: point of intersection of this plane by force *P*. *B*: point of intersection of the lines of action of *N* and *Q*. *A*: point of intersection of plane π by force *F*. *Gu*: in plane π, axis parallel to plane *zx*. *Gv*: in plane π, axis normal to *Gu*

Let us now imagine a disc tightly welded to the tibia in *G*. The plane π of this disc is at right angles to *t* (Fig. 47).[6] The forces acting on the rigid disc-tibia construction are:

1. Force *R* acting at point *G* in the direction of *t*. This force can be roughly considered as the force compressing the femur on the tibia.

2. Force *F*, parallel to *R*, produced by the tension of muscles inserted in the tibia. This force *F* acts at right angles to plane π at point *A* on a circle with centre *G* and radius *r*. Point *A* is determined on the circle by the angle ε formed by *GA* and an axis of reference *Gu*. *Gu* is chosen in the plane π and parallel to plane *xz*.

3. Two forces acting in the plane of the disc at point *B* on a circle with centre *G* and radius *r'*:

force *N* tangential to a circle with radius *r'*,

force *Q* radial with the direction *BG*.

Point *B* is determined by the angle δ formed by the axes *Gu* and *GB*.

Force *F* can be considered as the component parallel to *t* of the resultant of muscular tensions. Measurements on anatomical cross sections, between point *G* and the muscles and tendons which surround the knee, let us give *r* a

[6] The obliquity of the tibial plateaux to the long axis of the tibia varies with the individuals and is not important (less than 10°). Considering it would only very slightly change the results of calculation.

mean value of 5 cm. For each phase the muscular force is thus located by the angle ε.

Muscular and ligamentous forces have also a component in the plane of the disc. All the forces in this plane are brought to N and Q with the characteristic that their point of intersection B is on a circle of radius r'.

The following calculation shows that the basic interesting values R, F, ε are independent from r' which only intervenes in the values of N, Q and δ.

The value of r' has a minimum for each phase. Without much risk of error, r' can be considered as a little greater than this minimum in order to evaluate, N, Q, and δ.

In our formularization the data are:

1. components D_x, D_y and P_7+D_z of P in S_7,

2. the directional cosines of the tibia $\cos\alpha_t$, $\cos\beta_t$, $\cos\gamma_t$,

3. radii r and r'.

The unknowns are R, F, ε and N, Q, δ. They are the six unknowns which make the problem possible to solve.

The axis of reference Gu of the disc is chosen parallel to plane xz. Axis Gv is at right angles to Gu in the plateau as indicated in Figure 47.

The directional cosines of axis Gu, at right angles to the tibia and parallel to plane xz are thus:

$$a_u=\frac{\cos\gamma_t}{\sin\beta_t},\qquad b_u=0,\qquad c_u=\frac{\cos\alpha_t}{\sin\beta_t}$$

and the directional cosines of axis Gv are:

$$a_v=-\frac{\cos\alpha_t}{\mathrm{tg}\beta_t},\quad b_v=\sin\beta_t,\quad c_v=-\frac{\cos\gamma_t}{\mathrm{tg}\beta_t}.$$

We have taken into account the partial body weight, the forces of inertia and the muscular and ligamentous forces. The torques of inertia have been neglected. Consequently, terms corresponding to the differentiation of the kinetic moments of the articulated parts are not included. The calculation of the torques of inertia would require the knowledge of the moments of inertia of the different parts of the body. These moments of inertia are difficult to determine with accuracy because of the inhomogeneity of living structures. But the movement of the different parts of the body in relation to each other is slow during gait. Moreover, the movement of one part compensates somewhat that of another. Consequently, the effect of the torques of inertia is very small, much smaller than that of the forces of inertia, and it can be neglected without changing the final result significantly.

β) Calculation

In identifying P and the group of forces R, F, N, Q which have been defined, the axes u, v, t are adopted. If M_x, M_y, M_z are the moments of P in relation to the axes passing through G and parallel respectively to x, y, z (in the sense of a corkscrew), we have for G the co-ordinates x'_G, y'_G, z'_G of which are known:

$$\left.\begin{aligned}M_x&=D_y\cdot z'_G;(P_7+D_z)\cdot y'_G\\ M_y&=(P_7+D_z)\cdot x'_G-D_x\cdot z'_G\\ M_z&=D_x=D_x\cdot y'_G-D_y\cdot x'_G\end{aligned}\right\}\ \textit{known quantities.}$$

The equations of projection on the axes u, v, t can be written:

$$\begin{aligned}-N\cdot\sin\delta-Q\cdot\cos\delta&=a_u\cdot D_x+b_u\cdot D_y\\&\quad+c_u\cdot(P_7+D_z)=A,\\ -N\cdot\cos\delta+Q\cdot\sin\delta&=a_v\cdot D_x+b_v\cdot D_y\\&\quad+c_v\cdot(P_7+D_z)=B,\\ R-F&=D_x\cdot\cos\alpha_t+D_y\cdot\cos\beta_t\\&\quad+(P_7+D_z)\cdot\cos\gamma_t=C.\end{aligned}$$

The equations of the moments in relation to (u, v, t) are:

$$\begin{aligned}F_r\cdot\sin\varepsilon&=a_u\cdot M_x+b_u\cdot M_y+c_u\cdot M_z=D,\\ F_r\cdot\cos\varepsilon&=a_v\cdot M_x+b_v\cdot M_y+c_v\cdot M_z=E,\\ -Nr'&=M_x\cdot\cos\alpha_t+M_y\cdot\cos\beta_t\\&\quad+M_z\cdot\cos\gamma_t=K.\end{aligned}$$

A, B, C, D, E, K are known. We can thus obtain:

$$N=-\frac{K}{r'},\qquad \mathrm{tg}\varepsilon=\frac{D}{E},$$

$$F^2=\frac{D^2+E^2}{r^2},\qquad R=C+F,$$

$$Q^2=A^2+B^2-N^2,\qquad \sin\delta=\frac{BQ-AN}{Q^2+N^2}.$$

Since Q^2 must be positive we must have:

$$A^2+B^2>N^2=\frac{K^2}{r'^2}$$

thus:

$$r'^2>\frac{K^2}{A^2+B^2},$$

$$r'>\frac{K}{\sqrt{A^2+B^2}}.$$

This is the minimum value of r'. It varies from 3 to 8 cm. We can adopt $r'=10$ cm for all cases in order to calculate N, Q, and δ. This can be done since the value of r' does not intervene for R, F, ε, and r' can obviously be such that the point B is outside the limb. The results of these calculations are given in the three first columns of Table 7.

Thus we know the force R, the compression force transmitted from the femur to the tibia through point G. But it can intersect the axis of flexion of the tibia on the femur anywhere between the centres of curvature O_1 of the medial condyle and O_2 of the lateral condyle, without disrupting the equilibrium. Therefore, we have calculated the extreme values of R and F. They are obtained for R passing through the centre of curvature of the medial condyle O_1 (R_1 and F_1) and the centre of curvature of the lateral condyle O_2 (R_2 and F_2) (Table 7).

When the compression force R_1 passes through the centre of curvature O_1 of the medial condyle, the distance r between muscular force F_1 and point O_1 is $r=7.6$ cm. It is longer than the distance between F and G. On the contrary, if force R_2 passes through the centre of curvature O_2 of the lateral condyle, the distance r between the muscular force F_2 and O_2 is $r=2.4$ cm. It is shorter than the distance between F and G.

γ) Critical Analysis of the Chosen Solution

We direct the muscular force F with ε and find the value for each phase of gait. This corresponds to reality. The force P intersects the plane π at the point H with the co-ordinates

$$u=-\frac{a_v\cdot M_x+b_v\cdot M_y+c_v\cdot M_z}{D_x\cdot\cos\alpha_t+D_y\cdot\cos\beta_t+(P_7+D_z)\cdot\cos\gamma_t},$$

$$v=\frac{a_u\cdot M_x+c_u\cdot M_z}{D_x\cdot\cos\alpha_t+D_y\cdot\cos\beta_t+(P_7+D_z)\cdot\cos\gamma_t},$$

$$t=0.$$

This point H is oriented in relation to axis u by ε' and $\operatorname{tg}\varepsilon'=-\frac{v}{u}$. We find again $\operatorname{tg}\varepsilon=\frac{D}{E}$ seen before. Tilting of the plateau due to P logically provokes a muscular action which is diametrically opposed to P since actually: $\varepsilon=\varepsilon'+180°$.

Figure 48 shows disc π seen from above. Axis u is in the direction of the gait. The crosses indicate the successive points of application of force F on the circle with radius r. The circles

Table 7. Normal muscular force and compressive femoro-tibial force exerted on the loaded knee

Phase	R(kg)	F(kg)	ε	R_1(kg)	F_1(kg)	R_2(kg)	F_2(kg)	R_f(kg)
	$r=5$ cm			$r=7.6$ cm		$r=2.4$ cm		
12	353	292.54	− 23° 4′	252.92	192.46	669.92	609.47	346.18
13	257.3	185.94	− 41°17′30″	193.69	122.33	458.73	387.37	256.62
14	247.03	175.55	− 31°43′	186.97	115.49	437.21	365.73	245.94
15	96.95	57.13	− 35°42′	77.41	37.59	158.85	119.03	97.56
16	91.17	52.45	− 62° 1′30″	73.23	34.50	147.99	109.26	91.73
17	54.3	25.69	− 85°44′30″	45.52	16.90	82.14	53.52	54.88
18	89.24	55.56	−115°20′30″	70.24	36.55	149.44	115.75	90.03
19	112.13	72.97	−138° 2′30″	87.17	48.01	191.18	152.02	112.61
20	165.76	115.93	−151°54′30″	126.1	72.27	291.35	241.52	166.02
21	199.14	136.4	−163°38′30″	152.48	89.74	346.92	284.18	199.49
22	201.3	142.3	−163°43′	152.6	93.6	355.46	296.47	203.54
23	297.66	242.55	−164°51′30″	214.7	159.57	560.43	505.32	307.71

Fig. 48. Plane of the tibial plateaux. Orientation of forces P and F. G: central point of the knee. ε: angle formed by the postero-anterior straight line Gu and the line joining G to the point of application of the muscular force F in the plane of the tibial plateaux. ε': angle formed by line Gu and the radius passing through the point of application of force P in the plane of the tibial plateaux

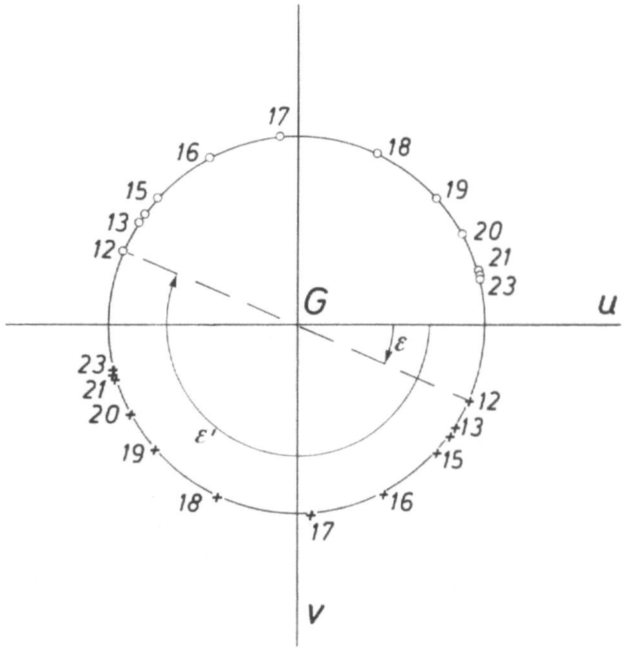

indicate the azimuth of point H where P intersects π. Electromyographic studies (Basmajian, 1967; Blaimont et al., 1971) confirm the successive positions of muscular force F diametrically opposed to P in relation to G.

Let us meet an objection which could be made to this conception. We have placed the disc π at right angles to the tibia and accepted that R has the same direction as t to which F is parallel. Why not weld a plateau π' normal to the femur and put R and F parallel to the femur? Calculation of this case is simple. The directional cosines of the femur, $\cos\alpha_f$, $\cos\beta_f$, $\cos\gamma_f$, must be calculated with φ_2 and ψ_2 (Fig. 38) instead of φ_4 and ψ_4. The results of the calculation are suprisingly similar to those obtained previously. The values of R obtained in this way are given in Table 7 under the designation R_f.

In fact, R (or R_f) is a compressive force. Its direction is along t (tibia) and f (femur) when $\beta_{2,4}$ (angle femur-tibia) $=0$. It is a little oblique to t or f when $\beta \neq 0$ but it practically consists of action and reaction and this confirms the results already obtained.

But the equilibrium requires the action of forces Q and N in the plane of the plateau π. With reference to the preceding results, the value of resultant R_π of these forces is

$$R_\pi = \sqrt{Q^2 + N^2} = \sqrt{A^2 + B^2}.$$

Table 8 gives the values of R_π for each phase of the right single support period of gait if disc π is considered as fixed to the tibia. Force R_π is the component of muscular and ligamentous forces tangential to the tibial plateaux. It is es-

Table 8. Tangential force exerted on the loaded knee

Phases	R_π(kg)	r' (cm)
12	9	−22.99
13	21.50	2.45
14	20.12	− 3.64
15	11.89	− 1.86
16	8.32	1.85
17	5.83	2.56
18	5.61	8.23
19	7.83	8.79
20	13.07	8.86
21	18.38	7.84
22	21.55	7.83
23	35.98	9.70

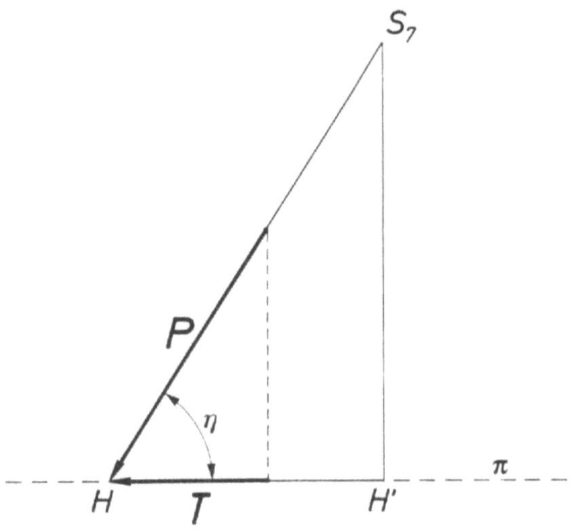

Fig. 49. Plane normal to the tibial plateaux and determined by the line of action of P. H: point of intersection of the line of action of force P with the plane π. H': projection of the partial centre of gravity S_7 on plane π

sentially produced by the cruciate ligaments and by the muscles most oblique to the long axis of the tibia.

Simple reasoning shows that force R_π passes through point H where P is resolved into a component $P_\perp$ normal to plane π and a component P_π located in the plane π. The component $P_\perp$ normal to plane π must balance the resultant of R and F. The component P_π situated in the plane π balances the resultant of N and Q. Because it does so it must pass through the point H.

It can also be seen that the projection T of P on the plane π gives the true value of R_π (Fig. 49). One must find out the point where the perpendicular drawn from S_7 intersects the plane. Let it be H', the co-ordinates of which are:

$$u_{H'} = -(a_u \cdot x'_G + b_u \cdot y'_G + c_u \cdot z'_G),$$
$$v_{H'} = -(a_v \cdot x'_G + b_v \cdot y'_G + c_v \cdot z'_G),$$
$$t_{H'} = 0.$$

As shown in Figure 49,

$$T = P \cdot \cos\eta \quad \text{and} \quad \overline{HH'} = \overline{S_7H} \cdot \cos\eta.$$

Calculation gives:

$$\overline{HH'} = \sqrt{A^2+B^2} \cdot \frac{x'_G \cdot \cos\alpha_t + y'_G \cdot \cos\beta_t + z'_G \cdot \cos\gamma_t}{D_x \cdot \cos\alpha_t + D_y \cdot \cos\beta_t + (P_7+D_z) \cdot \cos\gamma_t},$$

$$\overline{S_7H} = \sqrt{D_x^2+D_y^2+(P_7+D_z)^2} \cdot \frac{x'_G \cdot \cos\alpha_t + y'_G \cdot \cos\beta_t + z'_G \cdot \cos\gamma_t}{D_x \cdot \cos\alpha_t + D_y \cdot \cos\beta_t + (P_7+D_z) \cdot \cos\gamma_t},$$

$$T = \sqrt{A^2+B^2} \text{ since } \sqrt{D_x^2+D_y^2+(P_7+D_z)^2} = P.$$

One again obtains the value of R_π.

One can finally find out the value of distance d_1 from point G to line $\overline{HH'}$, line of action of R_π. The angular coefficient of line $\overline{HH'}$ in the plane (u, v) is first calculated:

$$\operatorname{tg}\theta = \frac{v_H - v_{H'}}{u_H - u_{H'}}.$$

This can be put in the form:

$$\operatorname{tg}\theta = \frac{a_v \cdot D_x + b_v \cdot D_y + c_v \cdot (P_7+D_z)}{a_u \cdot D_x + b_u \cdot D_y + c_u \cdot (P_7+D_z)} = \frac{B}{A}$$

(Fig. 50).

The equation of line HH' will then be written, such that it passes through one point H or the other H' since its angular coefficient is known, for instance:

$$v - v_H = \frac{B}{A} \cdot (u - u_H)$$

or

$$Av - Bu - Av_H + Bu_H = 0.$$

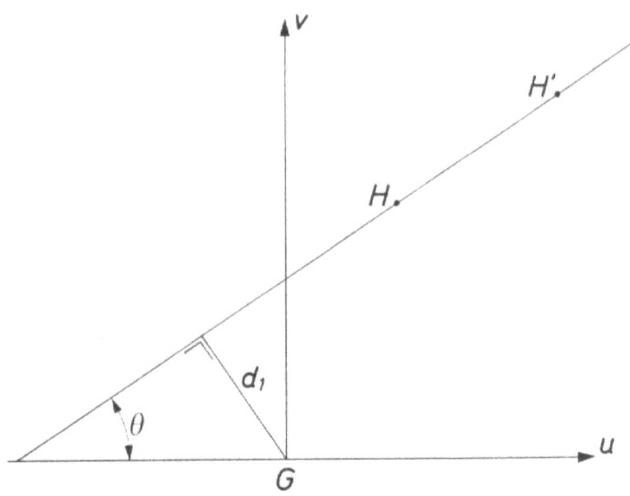

Fig. 50. Plane of the tibial plateaux. θ: angle formed by the line joining H to H' and the postero-anterior line Gu in plane π of the tibial plateaux. Gv: normal to Gu in plane π. d_1: distance from point G to line HH'

The distance d_1 from G to this line is:

$$d_1 = \frac{Av_H - Bu_H}{\sqrt{A^2 - B^2}}$$

and this value d_1 takes the successive forms:

$$d_1 = \frac{AD + BE}{c \cdot \sqrt{A^2 + B^2}} = -\frac{K}{\sqrt{A^2 + B^2}}$$

which is actually the minimum value attributable to r' as explained above. This value of d_1 fixes the value of r' and eliminates Q. The values of r' for each phase of the right single support period of gait are given in Table 8 for the plateau π fixed to the tibia.

h) Curves Illustrating the Forces Transmitted Across the Femoro-Tibial Joint

Force R normal to the joint surfaces of the tibia and force R_π tangential to the tibial surfaces during the right single support period of gait are illustrated by the curves Figure 51.

The upper full line (G) represents force R, the load transmitted from the femur to the tibia, during the right single support period of gait when this force passes through point G, the centre of the axis of flexion of the knee.

It is between dotted curves. These give the extreme values which R can attain by displacement over the axis of flexion of the knee between the centres of curvature O_1 of the medial condyle and O_2 of the lateral condyle.

The results allowing one to draw the curve have been obtained by assuming that the full load is exerted on the knee at phases 12 and 23. Since the left foot takes off the ground during the 0.038 sec of duration of phase 12 and strikes the ground during phase 23, the actual maximum values of R must be between 353 kg (phase 12) and 257.3 kg (phase 13) at the beginning of the right single support period of gait and between 201.3 kg (phase 22) and 297.66 kg (phase 23) at the end of the right single support period of gait.

For the studied subject, of 58.700 kg, the force R transmitted from the femur to the tibia quickly attains a maximum of approximately six times the body weight. One must remember that in standing with symmetrical loading on both feet each knee supports only 43% of the body weight as illustrated by the transverse dotted line (Fig. 51).

During the single support period of gait the force R changes considerably, due to accelerations of the centre of gravity S_7 and to the variations of distance between the knee and the line of action of force P exerted by the mass S_7. From six times the body weight, force R falls to less than the body weight and goes up again to four times. During gait the knee supports a considerable load comparable with hammering on the joint.

The lower solid curve illustrates force R_π exerted in the plane of the joint surfaces of the tibial plateaux by the component of the ligamentous and muscular forces tangential to these joint surfaces.

Fig. 51. R: magnitude of the femoro-tibial compressive force during gait. G: when its line of action passes through the centre of the axis of flexion of the knee. O_1: when its line of action passes through the centre of curvature O_1 of the medial condyle. O_2: when its line of action passes through the centre of curvature O_2 of the lateral condyle. R_π: magnitude of the musculo-ligamentous force exerted tangentially to the tibial plateaux ▷

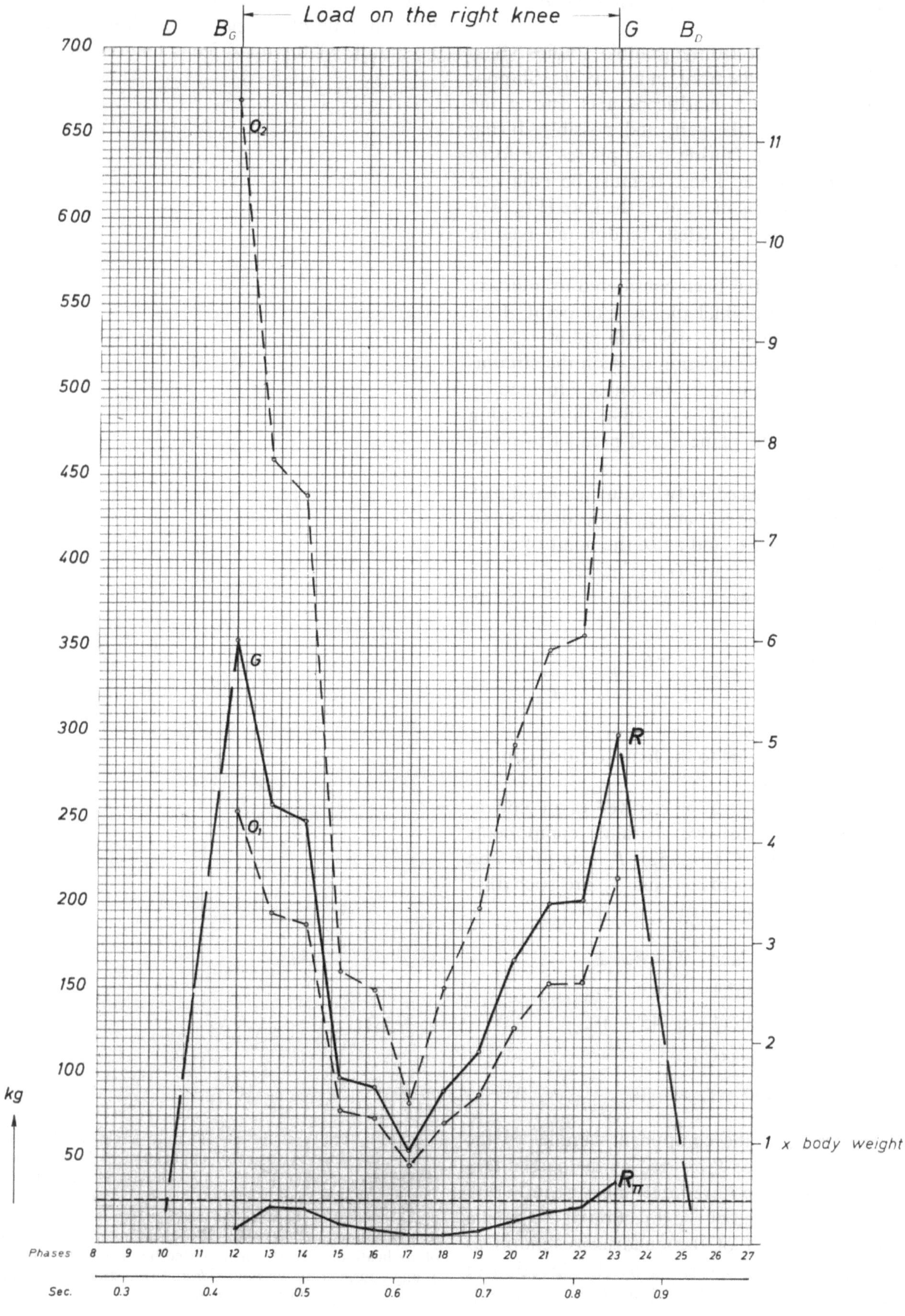

Load on the right knee
D
B_G
G
B_D
700
650
600
550
500
450
400
350
300
250
200
150
100
50
kg
11
10
9
8
7
6
5
4
3
2
1 x body weight
O_2
G
O_1
R
R_π
Phases
8 9 10 11 12 13 14 15 16 17 18 19 20 21 22 23 24 25 26 27
Sec.
0.3 0.4 0.5 0.6 0.7 0.8 0.9

i) Patello-Femoral Compressive Force

According to electromyographic studies (Basmajian, 1967; Blaimont et al., 1971; Shinno, 1961, 1968b), the quadriceps acts during the phases 12–14 of the gait, and possibly during phase 15 as well. During these phases the force P acts behind the knee. It must be balanced by a force F acting in front of the knee (Fig. 47). In what way is the patella tendon involved in force F?

The force F has been considered to intersect the plane of the tibial plateaux at point A on a circumference of the circle with radius r (5 cm) and centre G. The position of A on the circumference is determined by the angle ε formed by the radius AG and a straight line passing through G in the plane π and parallel to plane xz.

F can be replaced by two equivalent and parallel forces, S and F', acting respectively at G, centre of the circle, and at D, the point of intersection of lines CB and GA (Fig. 52).

S acts downward at G and $F'=F+S$ upward at D.

$$F'=\frac{rF}{e}, \quad S=F'-F=\frac{rF}{e}-F=F\frac{r-e}{e}$$

(Figs. 52 and 53).

Force S is added to the femoro-tibial contact force R.

Force F' at D can be resolved in an equivalent manner into two parallel components acting at B and C (Fig. 52). F_a acts at B and F_l at C. Components F_l and F_a are such that (Figs. 52 and 53):

$$d \cdot F_l = a \cdot F_a \quad \text{and} \quad F_l + F_a = F',$$
$$a = r \cdot \sin 45°(1-\operatorname{tg}(45°-\varepsilon)),$$
$$d = r \cdot \sin 45°(1+\operatorname{tg}(45°-\varepsilon)),$$
$$e = \frac{r \cdot \cos 45°}{\cos(45°-\varepsilon)}.$$

Force F_a is the component parallel to the tibial axis of force P_a exerted by the patella tendon. To calculate P_a we firstly measure in X-rays the angle β formed by the direction of the patella tendon and that of the quadriceps tendon as well as the angle α formed by the direction of the patella tendon and the axis of the tibial shaft (Fig. 54). We have assumed that P_a is exerted in the middle of the patella tendon and M_v in the middle of the quadriceps tendon. We have

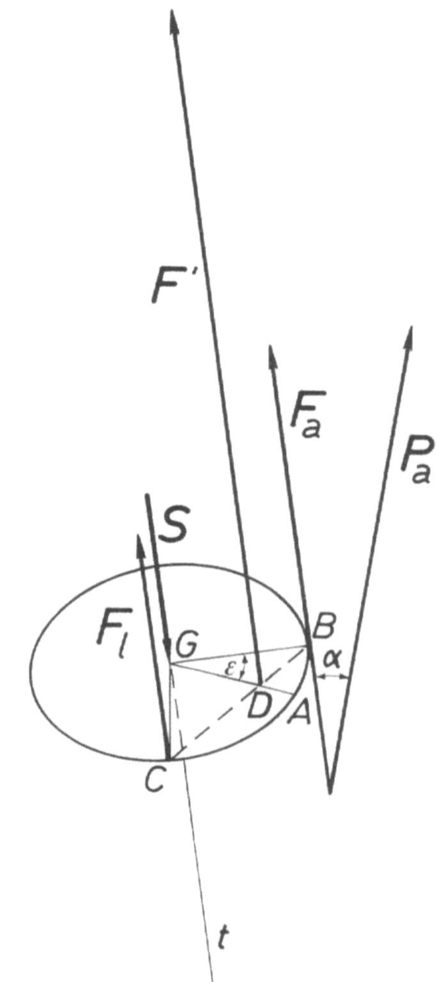

Fig. 52. G: centre of the disc at right angles to the long axis of the tibia. A: point of application of force F (not represented, see Fig. 47). F': muscular force. F_a: anterior component of F', parallel to the tibia. F_l: lateral component of F', parallel to the tibia. S: force exerted in G. D: point of application of F'. B: point of application of F_a. C: point of application of F_l. P_a: force exerted by the patella tendon. ε: angle formed by the radii GB and GA. α: angle formed by the force P_a and its component F_a

Fig. 53. Same signs as for Figure 52. a=distance BD. b= distance between the straight line f and point D. d=distance CD. e=distance GD. f=median of triangle CGB

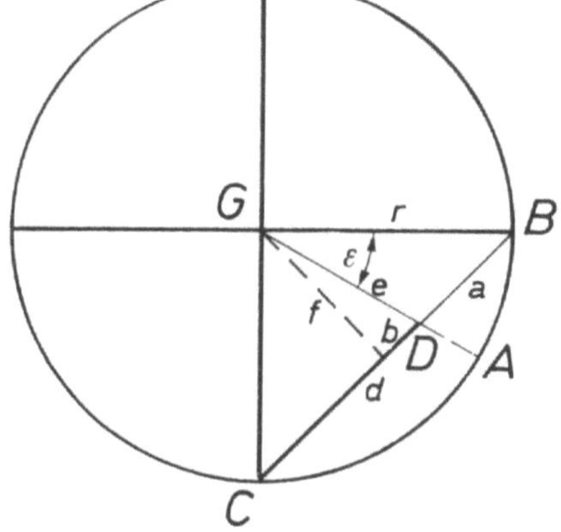

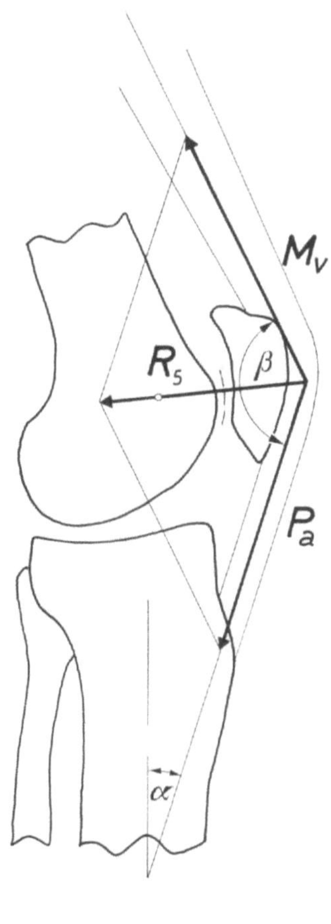

Fig. 54. Force R_5 pressing the patella on the femur is the resultant of the pull of the quadriceps M_v and the patella tendon P_a. β: angle formed by the lines of action of P_a and M_v. α: angle formed by the line of action of P_a and the tibial axis

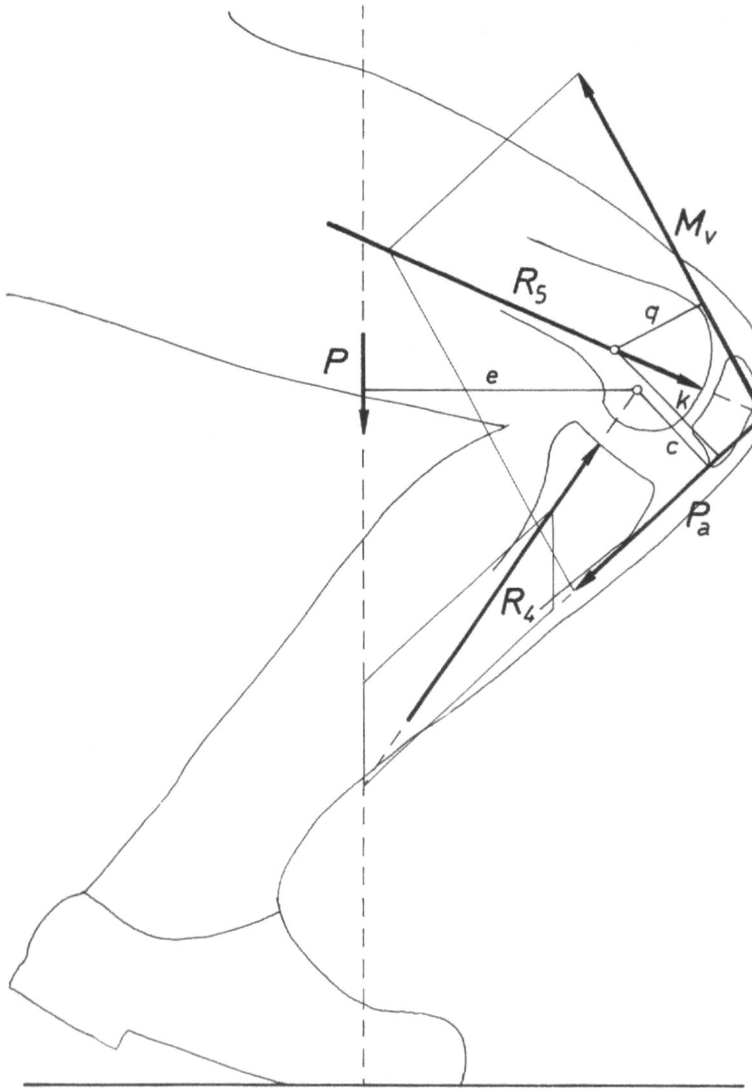

Fig. 55. Forces exerted on the knee of a subject when squatting. P: partial body weight. P_a: force exerted by the patella tendon. R_4: reaction force to the resultant of forces P and P_a or femoro-tibial compressive force. M_v: force exerted by the quadriceps tendon. R_5: reaction force to the resultant of forces P_a and M_v or patello-femoral compressive force. e: lever arm of force P. c: lever arm with which force P_a acts on the femoro-tibial joint. k: lever arm with which force P_a acts on the patello-femoral joint. q: lever arm with which force M_v acts on the patello-femoral joint

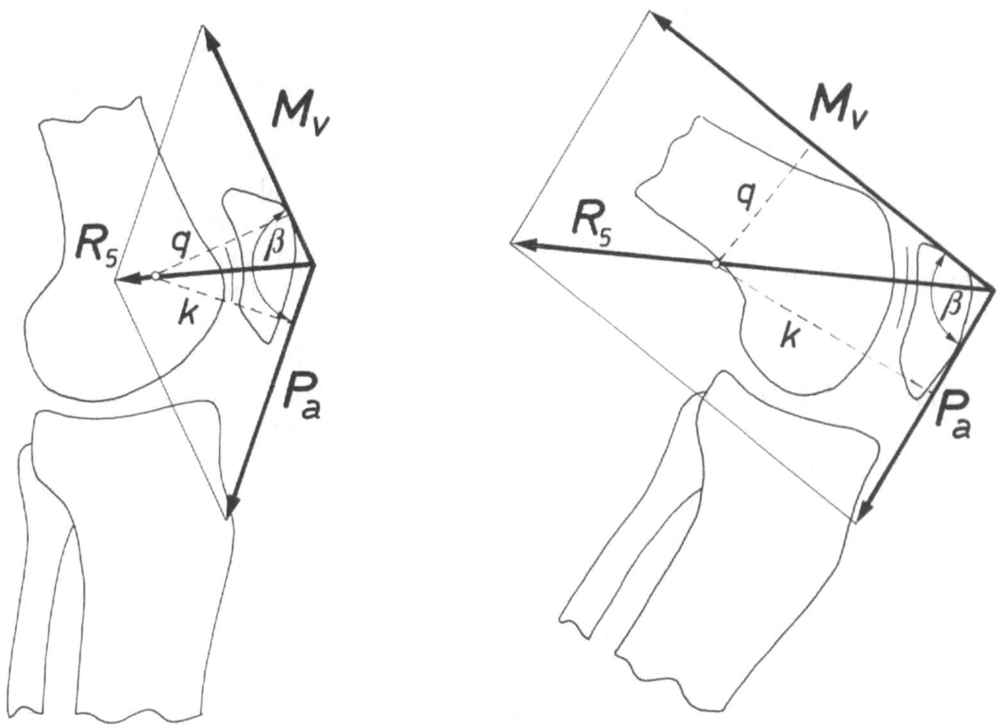

Fig. 56. The force R_5 considerably increases when the knee is flexed. This results from the closing of angle β and the shortening of the lever arm q of force M_v in relation to that k of P_a

then measured the lever arms of both forces in relation to the centre of curvature of the contact area of the patello-femoral joint. (k for P_a and q for M_v) (Fig. 55). The lever arms k of force P_a and q of force M_v are not necessarily equal and can differ considerably. Forces M_v and P_a vary accordingly since

$$M_v \cdot q = P_a \cdot k.$$

Force P_a acts on the patella with the lever arm k and counterbalances force M_v. Force P_a extends the tibia on the femur and counterbalances the flexing force R_3 with the lever arm c. These two lever arms of force P_a must not be confused, as is often the case in the medical literature.

Force P_a can now be calculated as well as force M_v exerted by the quadriceps tendon

$$P_a = \frac{F_a}{\cos\alpha}, \quad M_v = \frac{P_a \cdot k}{q}.$$

Resultant R_5 of P_a and M_v compresses the patella against the femur (Fig. 56). It can also be calculated that

$$R_5 = \sqrt{P_a^2 + M_v^2 + 2 P_a \cdot M_v \cdot \cos\beta}.$$

Its values for phases 12 through 15 are given in Table 10 (page 70).

The patello-femoral compressive force R_5 attains three times the body weight in the beginning of the single support period of gait and rapidly decreases.

The resultant compressive force R_5 can also be determined graphically. Its direction is given by the line joining the intersection point of forces M_v and P_a and the centre of curvature of the patello-femoral weight-bearing surface since R_5 must be normal to this surface. The force R_5 considerably increases when the knee is flexed. This is due primarily to the closing of the angle β formed by the lines of action of forces M_v and P_a and also to the shortening of the lever arm q of force M_v (Fig. 56).

B. Weight-Bearing Surfaces of the Joint

We know the forces acting on the knee and particularly the load R transmitted from the femur to the tibia and the patello-femoral compressive force R_5. We must now define the joint surfaces which transmit these forces.

1. Femoro-Tibial Joint

The spherical form of the hip joint has enabled Kummer (1968, 1969) to calculate the weight-bearing surfaces of the hip in its different positions. As far as the knee is concerned, the geometry of the joint and the presence of menisci make such a determination impossible. For this reason we had directly to measure the loaded joint surfaces in X-rays of anatomical specimens (Maquet et al., 1975).

a) Technique

Ten knees from amputated limbs or from fresh corpses were dissected and maintained in glycerine in order to keep their soft parts pliable. The ligaments and capsule were carefully spared. The tibia and fibula were sawn at right angles to the diaphysial axis, immediately distal to the insertions of the collateral ligaments and of the patella tendon. The femur was cut transversely immediately proximal to the condyles. The capsule was opened only at the level of the suprapatellar pouch. Through this opening the subpatellar fat was excised. The patella was retained. Coloured water was injected into the cavity. The knee was then moved manually, showing leaks through the lower part of the capsule and through the sheath of the popliteus tendon. Each rent was carefully sutured, making the capsule watertight.

The prepared specimen was tightly fixed by a screw to a rectangular sheet of plexiglass (Fig. 57). There were two slits in the plexiglass: one medial to the knee, the other lateral. The sheet rested on two supports between which an X-ray cassette could be inserted. The specimen was then subjected to a systematic radiological exploration.

A tracing from a first lateral X-ray made it possible to determine the curve ("Evolute") formed by the successive axes of flexion of the knee (Fig. 58 and see page 37). A Steinmann pin was then inserted through the femur across the superior point of the "Evolute" of each condyle. A second Steinmann pin, located underneath the plastic support, was fixed to the first with two compression clamps through the slits made for that purpose in the plexiglass on each side of the knee. The connecting bars of the clamps were only partly threaded. Their lower part was equipped with strain-gauges fixed to the bar with araldite glue. The strain-

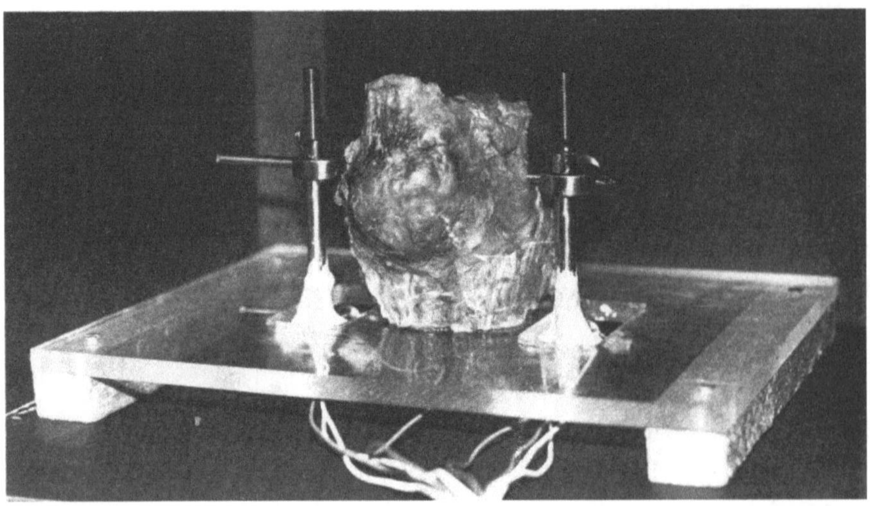

Fig. 57. The specimen is fixed by a screw on a plexiglass support. Compression clamps equipped with strain gauges are placed through slits on either side of the knee

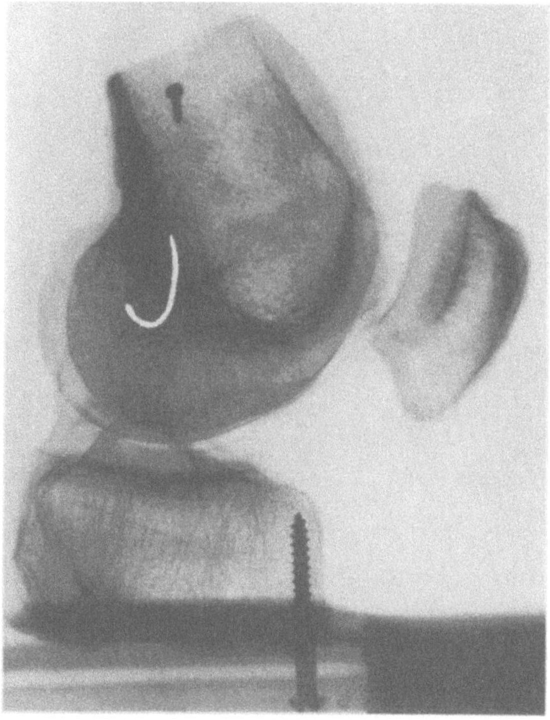

Fig. 58. Lateral X-ray of the specimen with the curve formed by the successive axes of flexion

gauges were plugged into a Wheatstone bridge. The measurement of the tensile stresses exerted on each clamp immediately indicated the load supported by the joint. Through the suprapatellar pouch a suspension of barium sulphate was injected and spread throughout by moving the joint to diffuse the opaque substance.

The clamps were then symmetrically tightened until a compression of 200–250 kg was attained, which represented the load. Greater compression would have deformed the device. An X-ray, taken with the beam directed from above downward, showed radiolucent areas where the opaque substance had been expelled by the compression of the articular surfaces, i.e. between the surfaces transmitting the load. These radiolucent areas were surrounded by a dense area consisting of the expelled barium. The limit between dark and light areas was well defined, except in some instances in the anterior part of the flexed knee. From the X-ray, the margins between translucent and dense areas were traced on transparent paper. They outlined the surfaces through which the load was being transmitted. They were drawn as shown by an example (Fig. 59). The surfaces were then measured with a planimeter. Three determinations were made and the average recorded. A lateral X-ray showed the degree of flexion of the joint.

After releasing the compression, the femoral pin was extracted and reinserted lower on the curve at a point corresponding to another position of flexion of the knee. The clamps were again symmetrically tightened until a compression of 200–250 kg was attained. The joint assumed a new position corresponding to the chosen axis. It is obvious that the equilibrium was rather unstable. For that reason in some instances a Steinmann pin was inserted from behind forward in the femoral fragment. It was supported by a piece of wood of the proper height depending on the degree of flexion. The surface areas were again determined for this position (Fig. 59b).

The use of an additional pin and wooden block altered the load transmitted across the knee joint. The compressive force exerted through the compression clamps was 250 kg. The wooden block stood 70 mm behind the axis of flexion of the knee. If the transverse Steinmann pin, instead of intersecting the "Evolute" of Fick, was situated 5 mm behind, the force exerted on the wooden block by the additional pin was 18 kg $\left(=\frac{250 \times 5}{70}\right)$. The compressive force across the joint was then 232 kg instead of 250 kg. Consequently, its order of magnitude was not changed by the additional pin and wooden block.

The force exerted through the compression clamps and the force exerted on the wooden block were both directed from above downward. Consequently, the line of action of the compressive force acting on the knee was also vertical. Since there was equilibrium, this line of action intersected the axis of flexion of the knee and was normal to the plane of the tibial plateaux.

This explains why the measurements of the weight transmitting areas in the five instances with an additional Steinmann pin were not different from those on the knees in equilibrium without any additional pin.

The experiment was repeated several times with the femoral pin inserted lower and lower on the curve of the axes of flexion. Every time

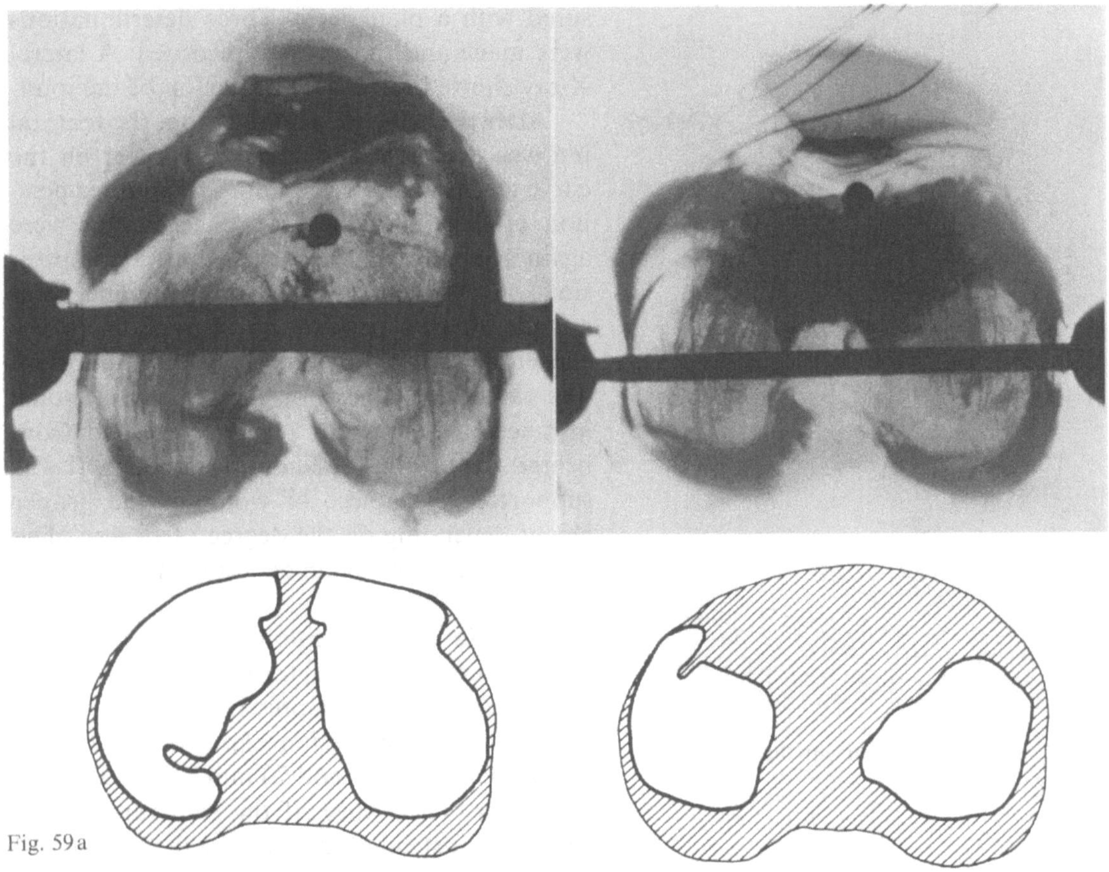

Fig. 59a

Fig. 59a–d. X-rays showing a cross section of the knee filled with barium sulphate suspension and tracings of the weight-bearing areas. (a) Under compression, hyperextended knee (5°). (b) Under compression, knee flexed at 45°. (c) Under compression, knee flexed at 75°. (d) Without compression

an X-ray, beam directed from above downward, showed the weight-bearing surfaces. Their contours were carefully traced (Fig. 59b and c). Each time a lateral X-ray showed the exact degree of knee flexion.

The quantity of barium sulphate was not large enough to fill the suprapatellar pouch and we made sure that some suspension was not collected in bursae connected with the joint and overlying compressed areas.

The Steinmann pins were then extracted. After putting the joint through a full range of movement, a final X-ray was taken to show the appearance of the articular surfaces without compression. These were found to be covered by the barium sulphate suspension (Fig. 59d).

At the conclusion of the tests the anatomical appearance of the knees was examined. In all of the knees the menisci were intact and no macroscopic sign of osteoarthritis could be seen. The anatomical integrity of the specimens established that the experiments were carried out on normal knees.

As a complementary experiment, in four of the knees the menisci were excised through anterior and posterior incisions. After careful closure of the incisions, the experiment was repeated. The outline and location of the areas of contact were mapped as the knees were gradually flexed.

The surface area of all the joints examined was reduced to a common ideal size. The surface area of subject of Braune and Fischer was adopted. The reduction of the surface areas of all the knees to a common nominal size not only eliminates anatomical variations among the samples used but also makes differences in radiological magnification irrelevant.

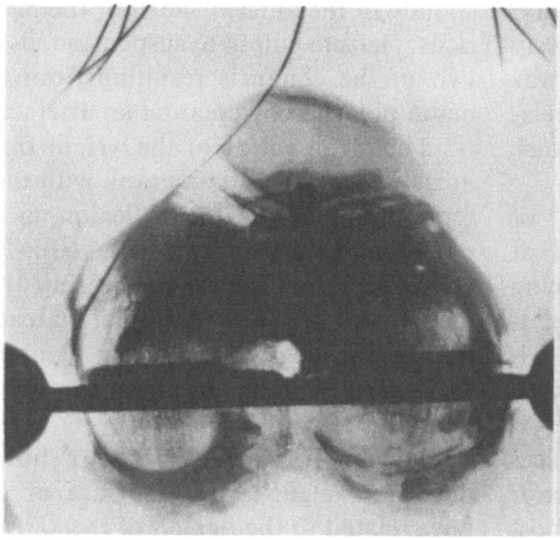

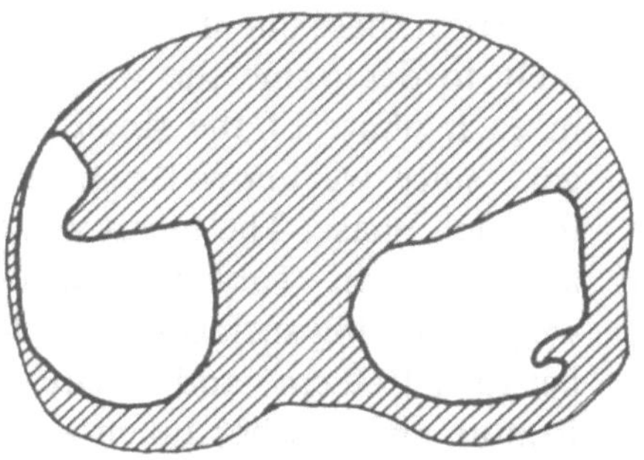

Fig. 59c

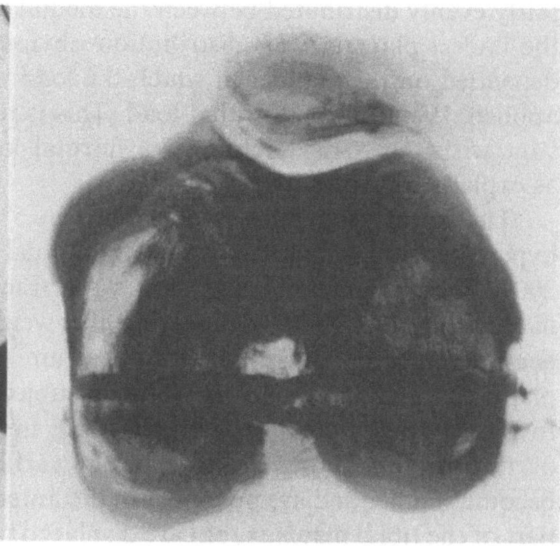

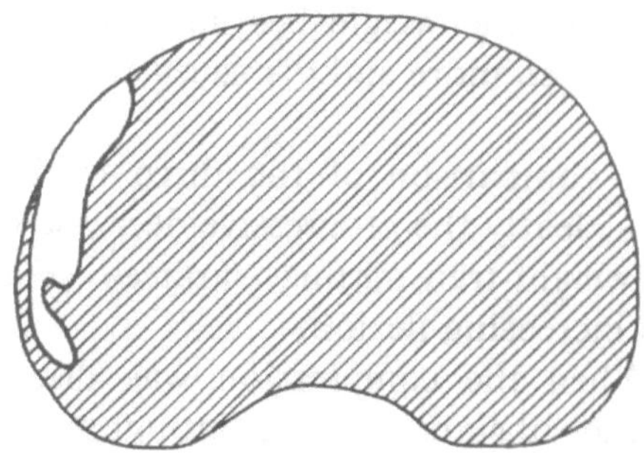

Fig. 59d

b) Results

The joint being fixed, a compression of 200–250 kg was exerted through the Steinmann pin inserted into the femoral condyles. Under this compression, the knee with intact ligaments and capsule spontaneously assumed the position of flexion corresponding to the position of the Steinmann pin on the "Evolute" of Fick.[7] This is in keeping with the law of mechanics which states that in a state of equilibrium the resultant force intersects the geometrical axis of flexion.

The X-rays show the projection of the weight transmitting surface on a plane normal to the line of action of the load. This plane, in our experiment, corresponds to the plane of the tibial plateaux. In order to calculate the compressive stresses in the knee joint, the projection of the weight-bearing surfaces is required and not the actual curved surfaces. Because of the small depth of the tibial plateaux, the curved surfaces do not significantly differ from their projection. Thus, the experiment produced values which did not require remanipulation except that they had to be adjusted to the common denominator of surface area.

When the load was applied with the joint in full extension, the measured surfaces varied between 18.22 cm^2 and 21.95 cm^2 with a mean value of 20.13 cm^2. During flexion the weight-bearing surfaces were found to move backward over the tibial plateaux and become progressively smaller, diminishing to a mean value of 11.61 cm^2 between 90° and 110°. They were

[7] In the five instances in which we succeeded in putting the Steinmann pin exactly through the curve of Fick of both condyles

fairly evenly distributed between the medial and the lateral plateau. This distribution obviously depended on the manner in which the load was applied. We applied a centred load. This is relevant with the X-ray picture of a normal knee as explained further (page 71).

The X-ray (Fig. 59a) is of a knee in 5° of hyperextension, supporting a centred load of 200–250 kg. Figure 59a also shows the tracing made from this X-ray delineating the weight-bearing surfaces, surrounded by barium sulphate suspension. When the joint is subjected to the same compression with the knee in 45° of flexion (Fig. 59b), the translucent surfaces become smaller and are no longer in the anterior part of the tibial plateaux, but are displaced posteriorly. With the knee in 75° of flexion (Fig. 59c) the weight-bearing surfaces are reduced even more and displaced further posteriorly.

A comparison of the three tracings illustrates clearly the progressive decrease in size of the weight-transmitting surfaces (Fig. 59a–c) with increasing flexion.

When the compression was discontinued, opaque substance flowed throughout the space between the femur and the tibia but for one small zone (Fig. 59d), part of which corresponds to the anterior horn of the medial meniscus. Barium sulphate suspension also invaded two of the channels resulting from the Steinmann pin, thereby creating an artefact.

The mean values of the weight transmitting areas were plotted on a graph with the abscissa representing the degree of flexion and the ordinate the surface in cm^2 (Fig. 60). The upper curves represent the weight-bearing areas of the medial and of the lateral tibial plateau depending on the degree of flexion. It appears that these surface areas are about the same in both plateaux.

The upper curve of Figure 61 shows in ordinate the weight-bearing surface of the whole knee related to the flexion of the joint in abscissa.

From these curves we are now able immediately to determine the weight-bearing surfaces in each phase of the single support period of gait, since we know the position of the joint for each phase. The mean values thus obtained are given in Table 9. During walking, the weight-bearing surfaces of the loaded knee vary between 19.9 cm^2 and 17 cm^2 in the subject of Braune and Fischer taken as an example. The load is distributed almost evenly over both plateaux.

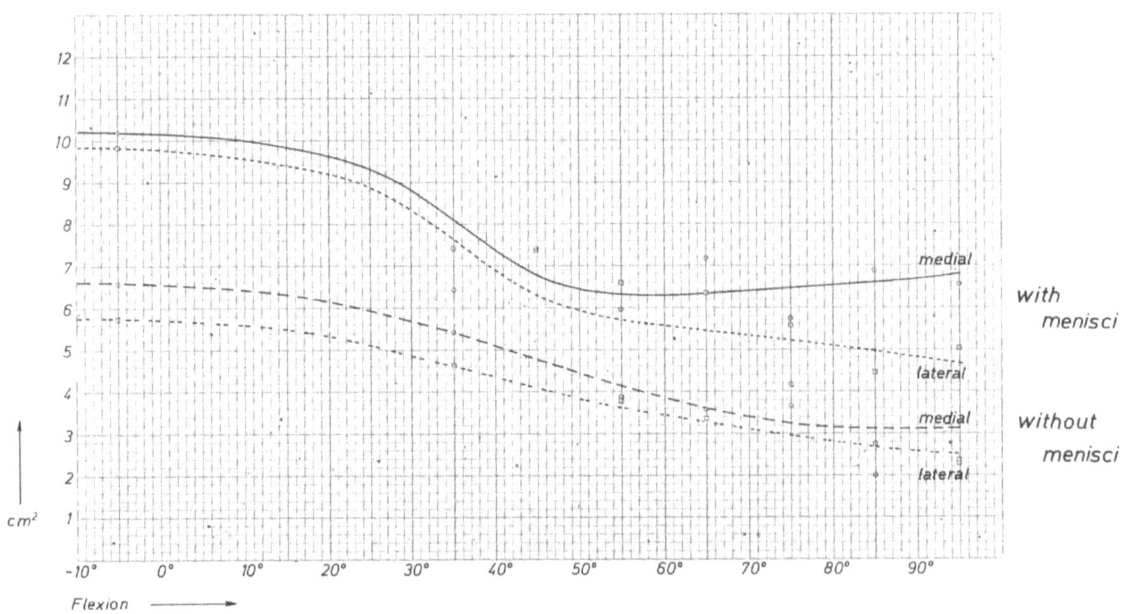

Fig. 60. In ordinate the weight-bearing surfaces. In abscissa the degree of flexion. Upper curves: intact knee, — medial plateau, ... lateral plateau. Lower curves: after removal of the menisci, ---- medial plateau, -- -- -- lateral plateau

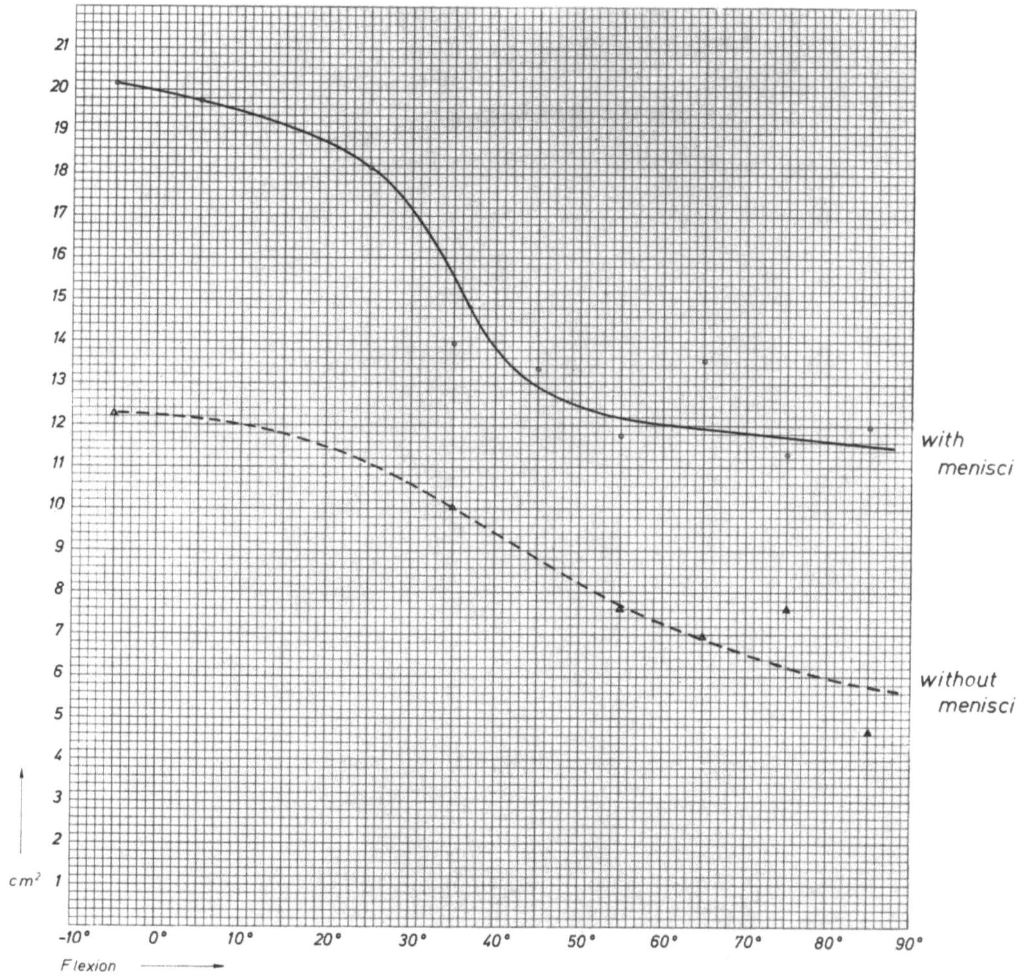

Fig. 61. In ordinate, weight-bearing surfaces of the femoro-tibial joint supporting a load of about 225 kg. In abscissa, the degree of flexion of the joint. Upper curve: intact knee. Lower curve: after removal of the menisci

Table 9. Weight-bearing femoro-tibial surfaces and mean compressive joint stresses

Phases	Angle $\beta_{2,4}$ of Fischer	Femoro-tibial weight-bearing area (cm^2)			Mean compressive articular stresses (kg/cm^2)
		Medial plateau	Lateral plateau	Total	
12	23°35′	9.34	8.92	18.25	19.342
13	28°35′	8.74	8.26	17	15.135
14	24°45′	9.26	8.84	18.1	13.716
15	18°31′	9.65	9.25	18.9	5.130
16	12°29′	9.94	9.46	19.4	4.699
17	7°42′	10.03	9.62	19.65	2.763
18	3°48′	10.10	9.70	19.8	4.507
19	1°26′	10.14	9.76	19.9	5.635
20	0°38′	10.14	9.76	19.9	8.330
21	0°43′	10.14	9.76	19.9	10.007
22	3°29′	10.10	9.70	19.8	10.153
23	9°38′	10	9.55	19.55	15.226

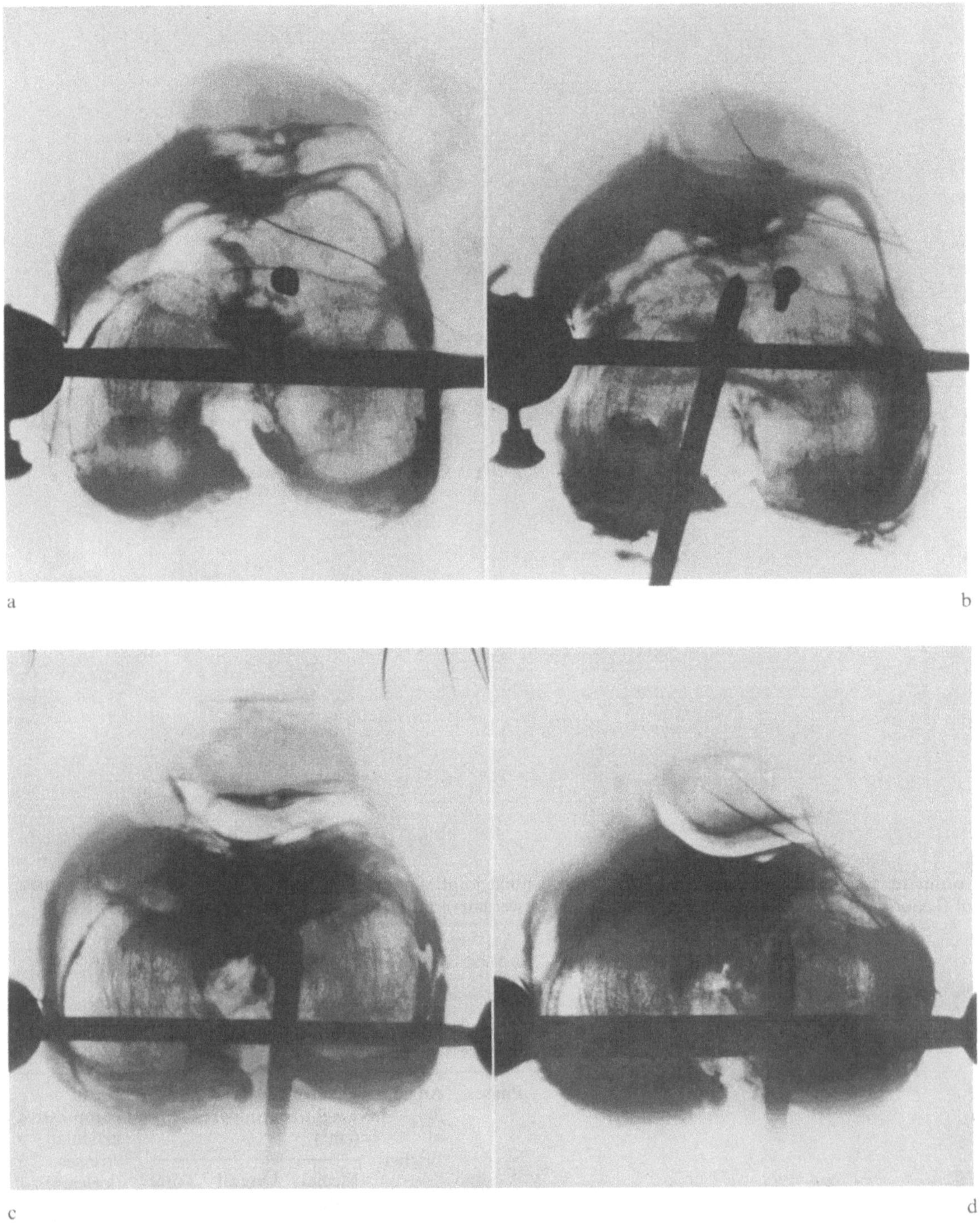

Fig. 62a–d. X-rays showing a cross section of the knee filled with barium sulphate suspension. (a) With menisci, in extension. (b) Without menisci, in extension. (c) With menisci, in flexion. (d) Without menisci, in flexion

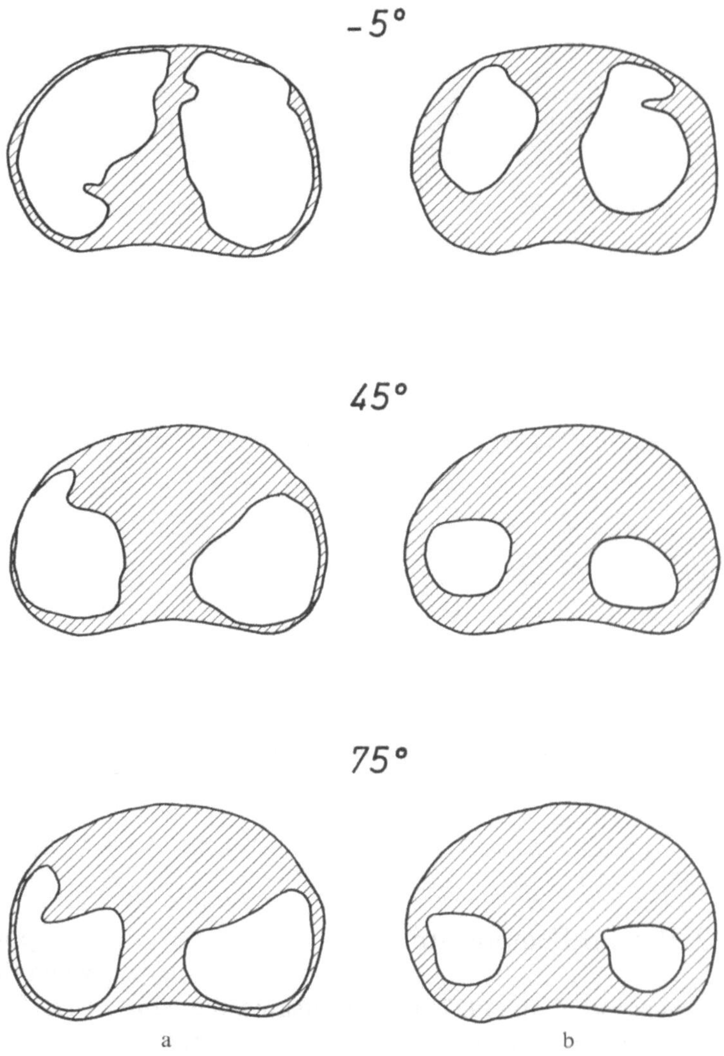

Fig. 63a and b. Tracings of the weight-bearing areas at 5° hyperextension, 45° flexion and 75° flexion. Left: knee with menisci. Right: knee without menisci

The areas determined before and after meniscectomy clearly demonstrate that the menisci do indeed contribute to the transmission of the load (Fig. 62a–d). With the joint in the same position, the weight-bearing surface of a knee with the menisci is much greater than that after meniscectomy. The barium sulphate suspension then fills the space previously occupied by the menisci. In Figure 63a the weight-bearing surface areas of an intact knee are traced when the joint is loaded at 5° hyperextension, 45° flexion, and 75° flexion. On the right (Fig. 63b), the weight-bearing areas of the knee, loaded in the same conditions, are shown after removal of the menisci.

Comparison of the upper and lower curves of Figures 60 and 61 emphasizes the reduction of the weight-bearing surfaces after the menisci have been removed.

The menisci represent an important part of the weight-bearing surface of the knee in any position of knee flexion. Meniscectomy reduces the weight-bearing area to the direct contact between the femur and the tibia. In view of the fact that the radii of curvature of the tibial plateaux and of the femoral condyles are different, the extent of contact is only dependent on the elasticity of the articular cartilage, which is slightly depressed under the effect of compression.

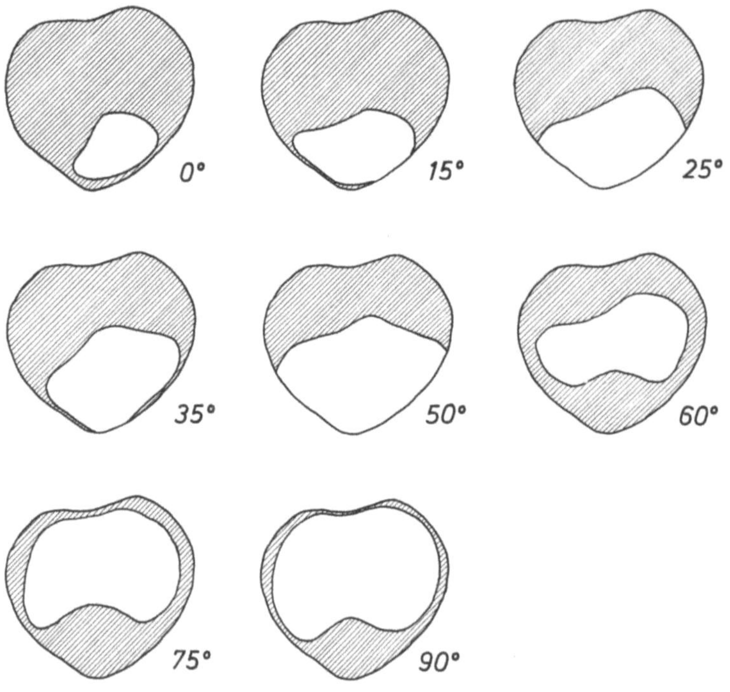

Fig. 64. Patello-femoral contact surfaces as measured by Townsend et al. (Redrawn after Townsend et al. 1977)

2. Patello-Femoral Joint

Townsend et al. (1977) have measured the contact surfaces between the patella and the femur in different positions of the knee from extension to 90° flexion, with increments of 10° or 15°, after applying a tension of 50 kg on the quadriceps tendon (Fig. 64). Actually, a much larger force of the quadriceps seems necessary to counterbalance in vivo the partial mass of the body acting eccentrically on the knee. In these conditions, the patello-femoral contact surfaces are probably larger than in the experiment. The surface areas published by Townsend et al. thus must be considered as minimum. These surfaces have been measured and multiplied by an appropriate factor to correspond to the patello-femoral joint of the subject of Braune and Fischer (Table 10). The mechanics of the patello-femoral joint suggest that the force-transmitting surfaces coincide with the surfaces actually measured.

The positions chosen by Townsend et al. to measure the contact surfaces did not exactly correspond to the positions of the knee in the first phases of the single support period of gait, when the quadriceps is acting. We have assumed that the force-transmitting surfaces vary proportionally to the position of the flexion between two successive measurements. The approximation entailed in this assumption should hardly affect the results since the successive measurements were separated only by 10° or 15°.

Table 10

Phases	Angle $\beta_{2,4}$ of Fischer	Force R_5 (kg)	Patello-femoral contact area (cm^2)	Average compressive articular stresses (kg/cm^2)
12	23°33′	218.992	5.562	39.373
13	28°35′	126.847	6.065	20.915
14	24°45′	126.646	5.900	21.126
15	18°31′	33.468	4.876	6.864

C. Contact Articular Stresses

1. Femoro-Tibial Joint

In part II A of this chapter, we have determined the load R exerted on the knee during the right single support period of gait. In part II B, we have measured the projection of the femoro-tibial weight-bearing surfaces when properly loaded, in different positions from extension to flexion. Since Fischer (1900) gives the angle of flexion between the femur and the tibia (angle $\beta_{2,4}$) for the different phases of gait, we have all the elements necessary to calculate the average contact stresses.

They are obtained by dividing the supported load by the surface areas transmitting the load, projected on a plane normal to the axis of the tibia. Their values are given in Table 9 for each phase of the right single support period of gait.

Obviously the actual stresses can correspond to the calculated average stresses only if they are uniformly distributed over the whole weight-bearing surfaces. As shown by Pauwels for the hip, the subchondral dense bone is a materialization of the diagram of the articular stresses. At knee level this dense area consists of two flat cups, each of them underlining a tibial plateau (Fig. 65). These flat cups have a nearly even thickness throughout. From their shape one can deduce that the compressive stresses are evenly distributed over the weight-bearing surfaces of a normal knee and, consequently, that the resultant force R is exerted through the centre of gravity of the weight-bearing surfaces. The joint stresses thus correspond to the calculated average stresses.

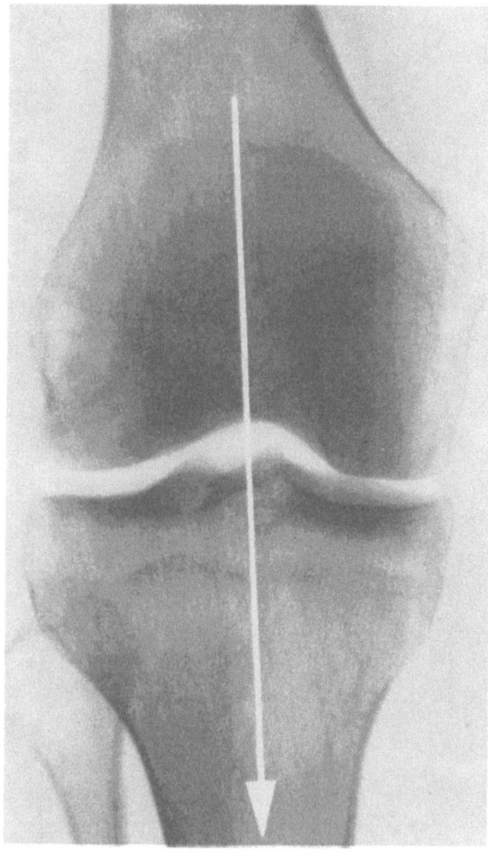

Fig. 65. The outline of the dense subchondral bone underlining the tibial plateaux corresponds to the stress diagram

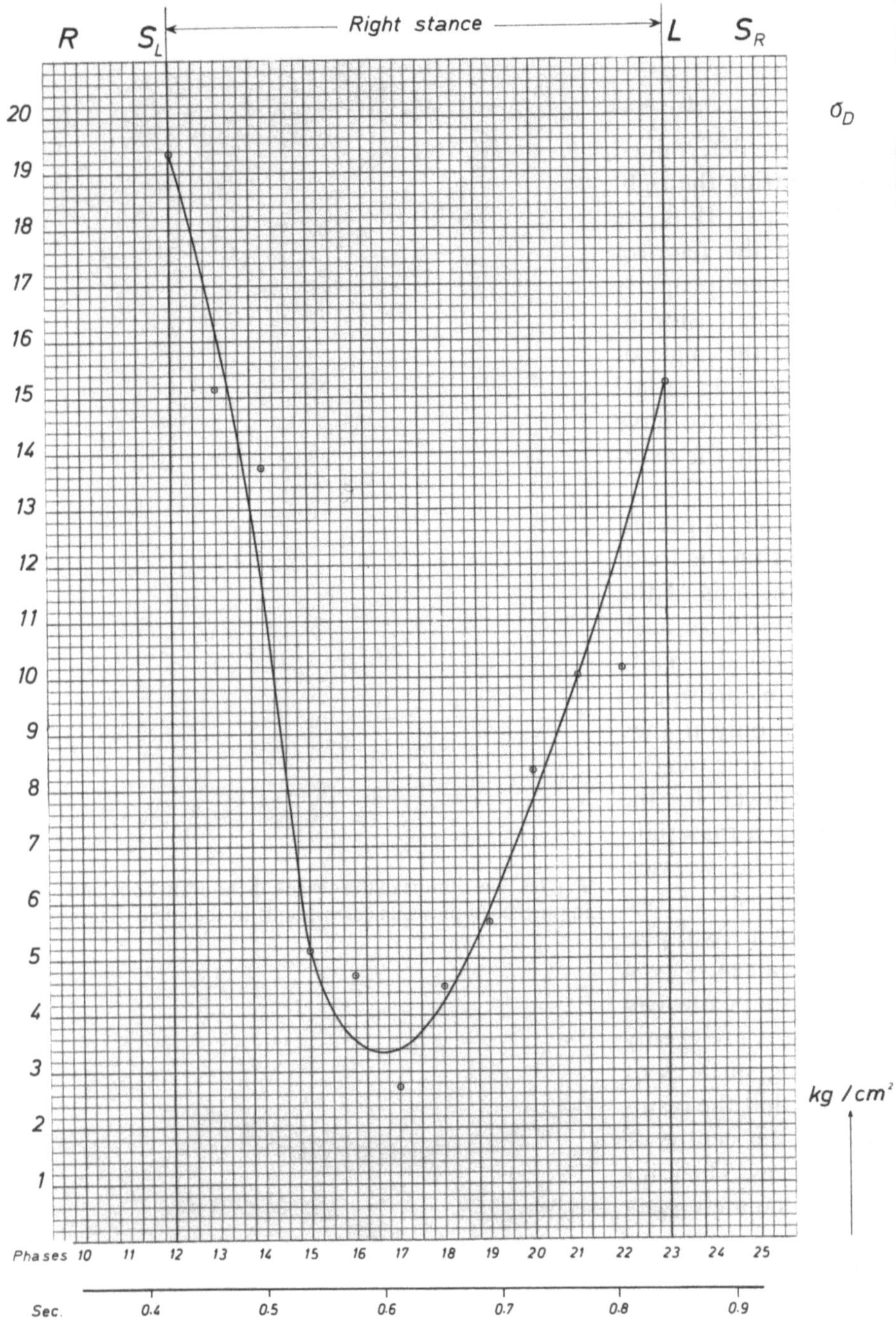

Fig. 66. Compressive articular stresses during the right single support period of gait. In ordinate: compressive stresses in kg/cm^2. In abscissa: time and phases of gait. *R*: right heel strike. S_L: swing of left leg. *L*: left heel strike. S_R: swing of right leg

The joint pressure in the knee varies between 19.3 kg/cm^2 and 3 kg/cm^2 during the successive phases of the single support period of gait. Attaining 19.3 kg/cm^2 as soon as the opposite foot takes off from the ground (phase 12), it quickly goes down to about 3 kg/cm^2 when the partial centre of gravity S_7 is in the same coronal plane as the knee (phase 17). It then increases to 15.1 kg/cm^2 at the end of the single support period (phase 23). Its evolution is illustrated by the curve in Figure 66.

From the work of Kummer (1968, 1969) and of Amtmann and Kummer (1968), it can be deduced that a maximum pressure of 16–20 kg/cm^2 is exerted in the hip joint during gait. Concordance of these and our results, obtained in completely different ways, sustains the validity of our results.

2. Patello-Femoral Joint

The subchondral sclerosis in a normal patella presents the same thickness throughout. This points to an even distribution of the articular compressive stresses over the weight-bearing surfaces. If periodical increases occur, there is no sign of persistent peaks of stresses. We have calculated the patello-femoral compressive force in part II A of this chapter and its magnitudes in the first phases of the single support period of gait are given in Table 10. The force-transmitting surfaces have been measured. We can thus calculate the average compressive stresses in the joint (Maquet, 1981). These would reach almost 40 kg/cm^2 at phase 12, but it must be remembered that the forces arrived at for phase 12 are probably somewhat excessive because the opposite foot is taken off the ground during the 0.038 sec of this phase (see page 56). The actual stresses are probably less than those calculated.

For the following phases, the stresses attain 20 kg/cm^2. This matches the magnitude of the femoro-tibial contact stresses. This coincidence makes sense if cartilage grows and is maintained in between a narrow range of well-determined compressive stresses, as suggested by Pauwels (1965b) and Kummer (1963).

III. Conclusion

We now know all the parameters which allow us to define the mechanical stress exerted in a normal knee under physiological conditions in the standing position and when walking. The force exerted on the knee in the standing position with symmetrical support on both feet is half the body weight minus both lower legs and feet. When standing on one leg, the knee supports the partial body weight (body weight minus the loaded lower leg and foot) and the muscular forces necessary to balance it.

During gait it is subjected to static and dynamic forces developed by the mass of the same part of the body and to muscular and ligamentous forces necessary to keep equilibrium and allow walking.

The surface of the joint weight-bearing areas has been measured. In this way we could determine the physiological articular pressure. Its order of magnitude is 20 kg/cm^2.

Knowing the mechanical stress in a normal knee and the factors which have made possible its definition, we can now systematically investigate the pathogenesis of osteoarthritis of the knee.

Chapter V. The Pathomechanics of Osteoarthritis of the Knee

I. Theoretical Analysis of the Causes of Knee Osteoarthritis

In order to study the possible causes of a displacement of the line of action of the load *R* exerted on the knee and to consider its consequences, one must first undertake a geometrical analysis of the forces acting on the joint. We shall study a projection of the forces on a coronal plane, then on a sagittal plane, and finally on a horizontal plane.

In a normal knee (Fig. 67a), the line of action of force *P*, resulting from the mass of the body minus the loaded leg and foot, is medial to the knee. It is balanced by a lateral force *L*. Construction of a parallelogram of forces determines the resultant force *R* which normally passes through the centre of gravity of the weight-bearing surfaces of the knee. The line of action of force *P* is indicated by the straight line prolonging vector *P*. The line of action of the lateral muscular tension band *L* is known. Vector *L* is drawn and prolonged by a straight line which intersects the line of action of *P*. From the point of intersection a third line is drawn which passes through the centre of gravity of the weight-bearing surfaces of the knee and intersects the axis of flexion of the joint in *G* (Fig. 19). The magnitude of force *P* can be calculated for each phase of the single support period of gait. From this it is possible to determine, either by drawing or by calculation, the magnitude of force *L* and of resultant force *R*.

From this normal schematic drawing of the forces projected on the coronal plane, one can analyse the possible causes of a displacement of *R*.

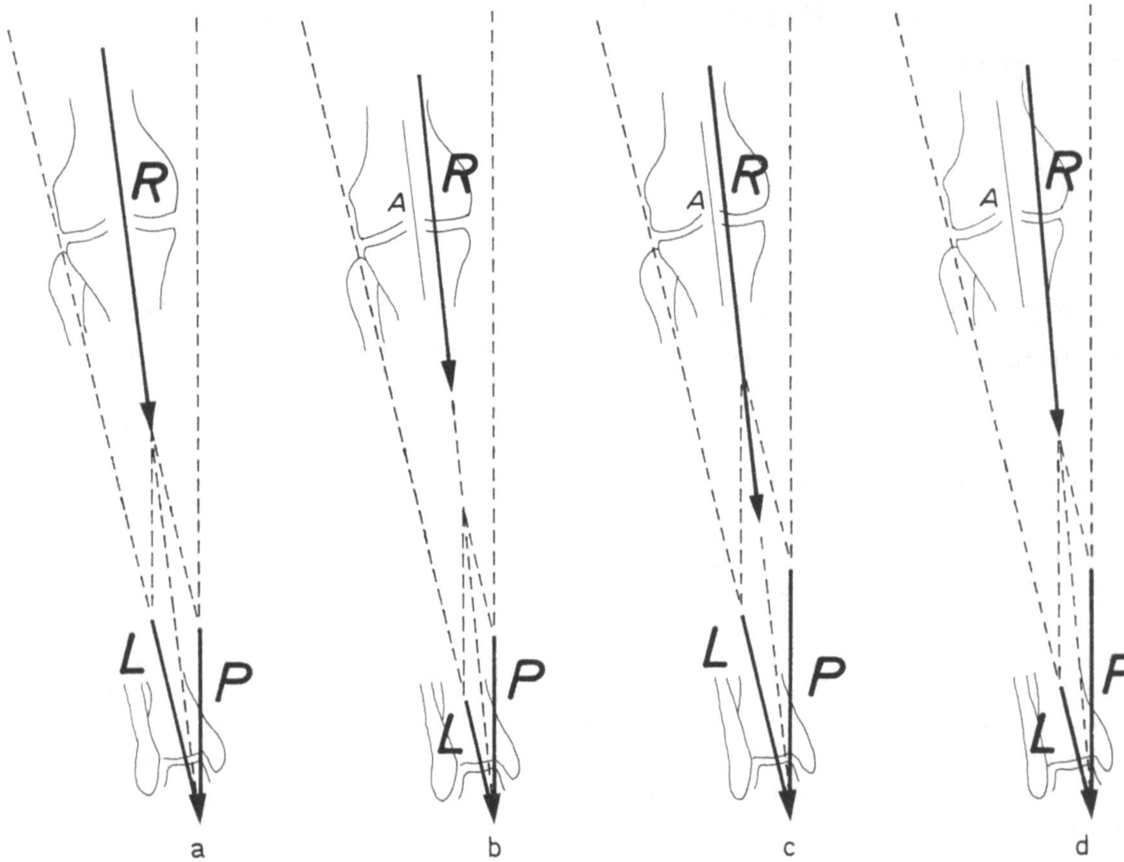

Fig. 67. (a) Normal knee. *P*: force exerted by the mass of the body eccentrically supported. *L*: lateral muscular tension band. *R*: resultant of forces *P* and *L* or load exerted on the knee. *A*: so called mechanical axis of the leg. (b) Decrease of magnitude of force *L*. (c) Increase of body weight. (d) Decrease of power of the lateral muscular tension band *L* combined with an increased body weight

A. Medial Displacement of Force *R*

A diminution of force *L*, a release of the lateral muscles, displaces the line of action of resultant *R* medially (Fig. 67b). At the same time the line of action of *R* comes slightly near the vertical.[8]

An increase of force *P*, if not compensated by a corresponding augmentation of muscular force *L*, produces the same result (Fig. 67c). If, as may happen after the menopause, the muscular tension band *L* is loosened and the body weight is increased, the medial shifting of *R* would be even greater (Fig. 67d).

Varus deformity of the knee modifies the direction of the muscular tension band *L* and of the femur. It increases the distance between the line of action of *P* and the knee (Fig. 68b). Change in the direction of force *L* and lengthening of the distance between the line of action of *P* and the knee move the point of intersection of forces *P* and *L* further from the joint. If these forces are not modified, the displacement of their point of intersection shifts *R* medially, as shown by the parallelogram of forces. The medial displacement of *R* is still more pronounced if the muscular tension band *L* is additionally loosened (Fig. 68c).

[8] Blaimont et al. (1971) experimentally verified this pathogenesis of osteoarthritis of the knee with varus deformity. Applying their test, they measured the strength of the lateral muscular tension band, with the patient lying on the opposite side. They found that the lateral muscular tension band was much weaker in cases of osteoarthritis of the knee with varus deformity than in a normal knee. For three patients with osteoarthritis of the knee with varus deformity, the potential of muscular effort had an average moment of 95 kg·cm as compared to an average value of 391.6 kg·cm in 10 normal individuals

Fig. 68. (a) Normal knee. (b) Knee with a varus deformity. (c) Knee with a varus deformity and decreased power of the lateral muscular tension band L

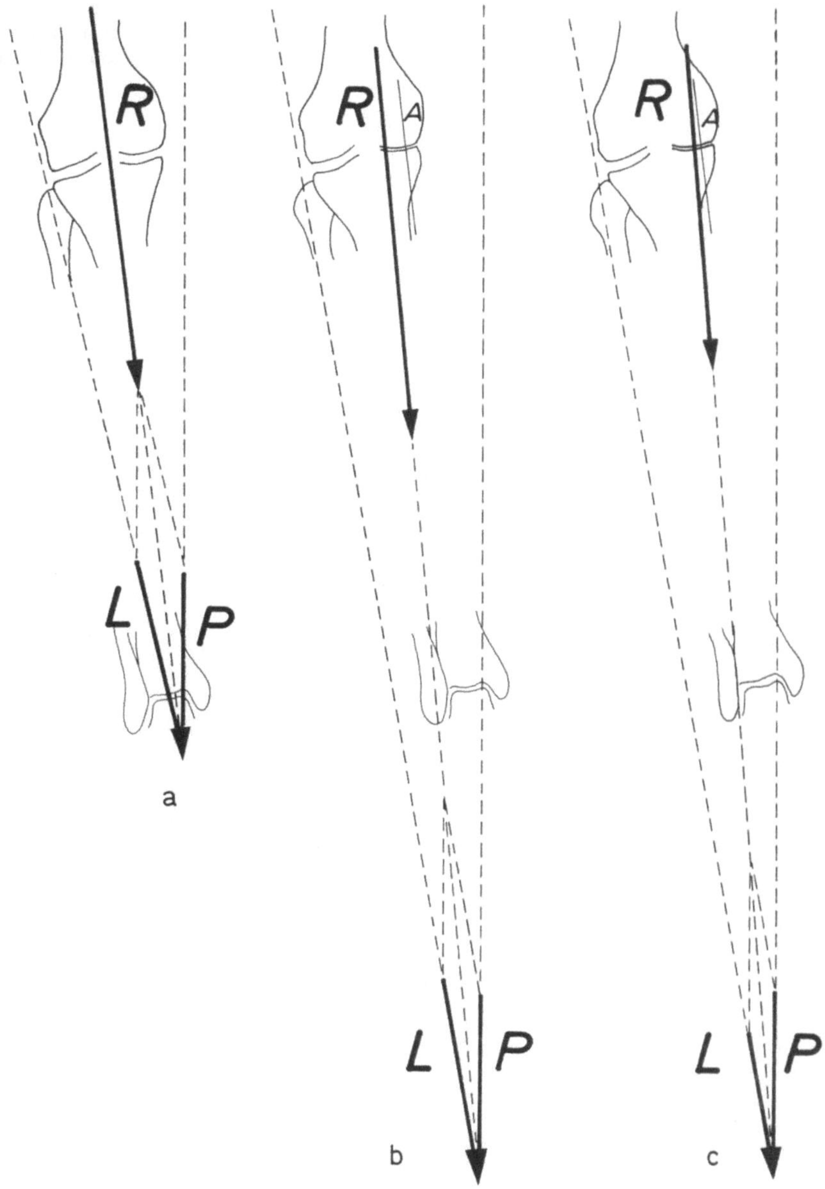

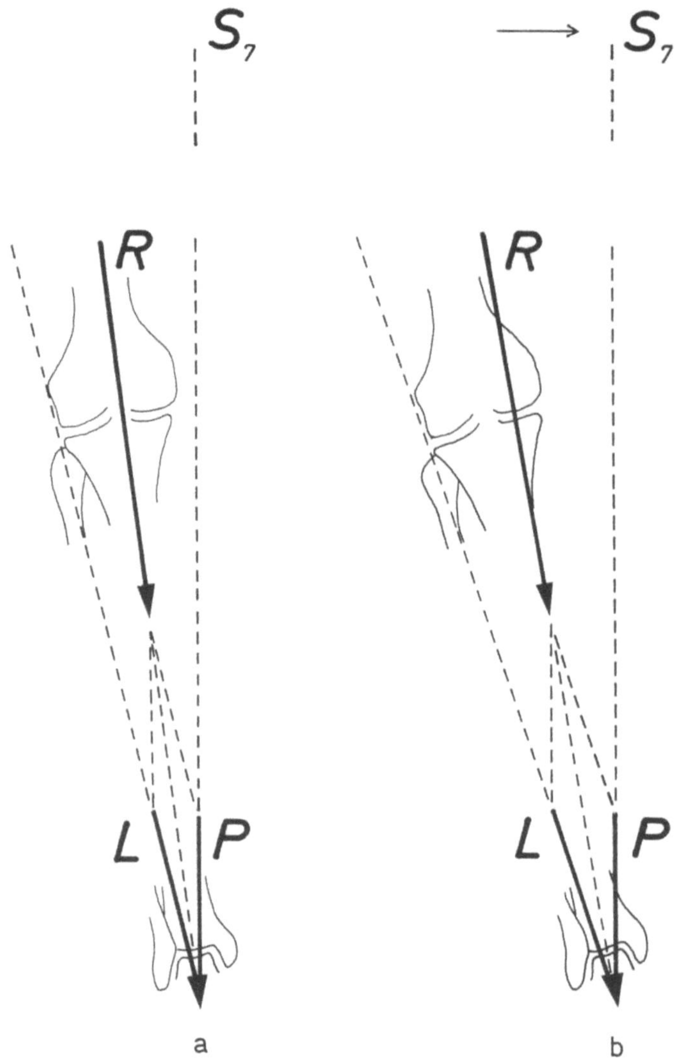

Fig. 69. (a) Normal knee. (b) Line of action of P brought away from the knee by a displacement of centre of gravity S_7

Medial displacement of R can also result from a shifting of the centre of gravity of the body, which increases the distance between force P and the knee (Fig. 69b). This displacement can be the consequence of an important discrepancy in length of the lower limbs tilting the pelvis and causing scoliosis. In such conditions, if the magnitude of force L does not change, the resultant R is shifted medially.

When the resultant force R is displaced medially, it provokes significantly increased compressive stresses in the medial part of the knee

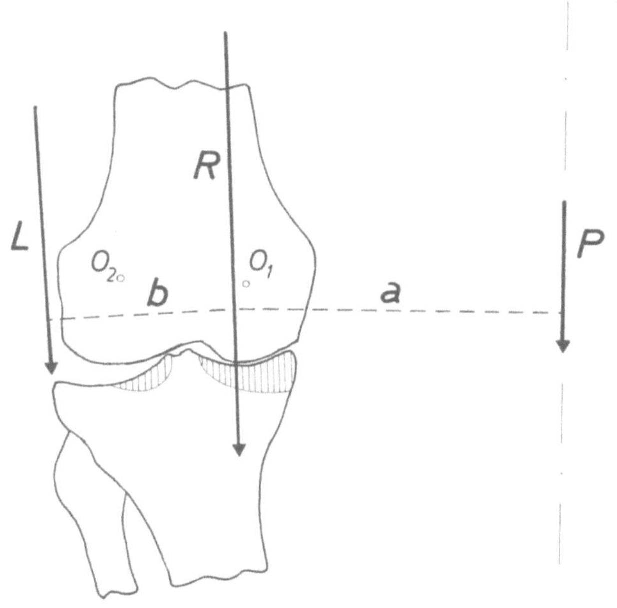

Fig. 70. Increased stresses in the medial part of the knee resulting from a medial displacement of resultant R. P: force exerted by the mass of the body supported by the knee. L: lateral muscular tension band. R: resultant of P and L. O_1: centre of curvature of the medial condyle. O_2: centre of curvature of the lateral condyle. a: lever arm of P. b: lever arm of L

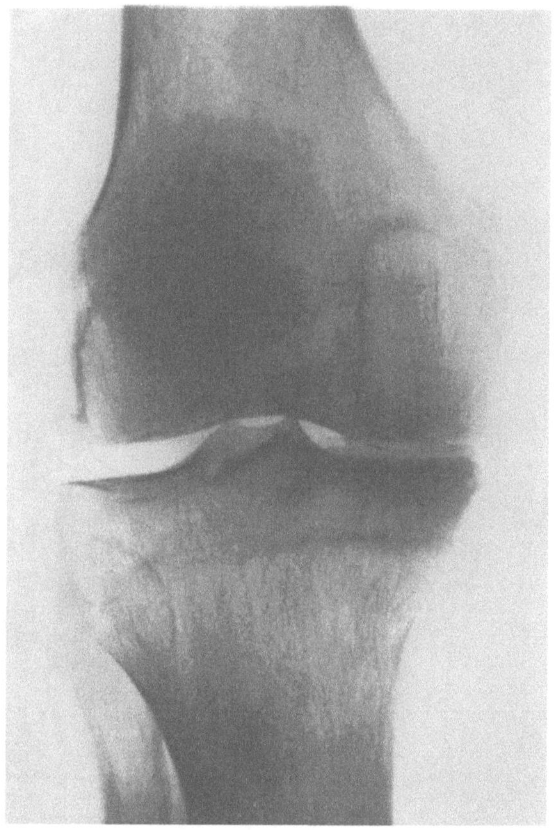

Fig. 71. Dense triangle beneath the medial plateau, corresponding to the stress diagram

(Fig. 70). According to Pauwels' law of functional adaptation (1965b, 1973a)[9], an increase of compressive stresses must cause apposition of bony tissue. This occurs. The thickness of the subchondral dense cup underlining the medial plateau increases (Fig. 71). The dense bone underlining the tibial plateau progressively takes on a triangular shape and the density of the femoral condyle is accentuated. Finally, increased compression destroys the articular cartilage. Thinning and disappearance of articular cartilage narrows the medial joint space and causes or aggravates the varus deformity of the knee. The varus deformity furthers the medial displacement of load R. Thus a vicious circle has been created: progressive worsening is the rule.

[9] Although Wolff (1892) is generally credited with establishing the relationship between stress and bone formation and resorption, he, in fact, only observed a change in the trabecular pattern of deformed bones and assumed that this change was related to an alteration in function. It was Pauwels (1965b, 1973a) who first clearly enunciated the direct relationship between the total amount of bone and the magnitude of the stress to which it is subjected

border of the joint. The medial displacement of force *R* evokes a triangular distribution of the articular stresses (Fig. 71).

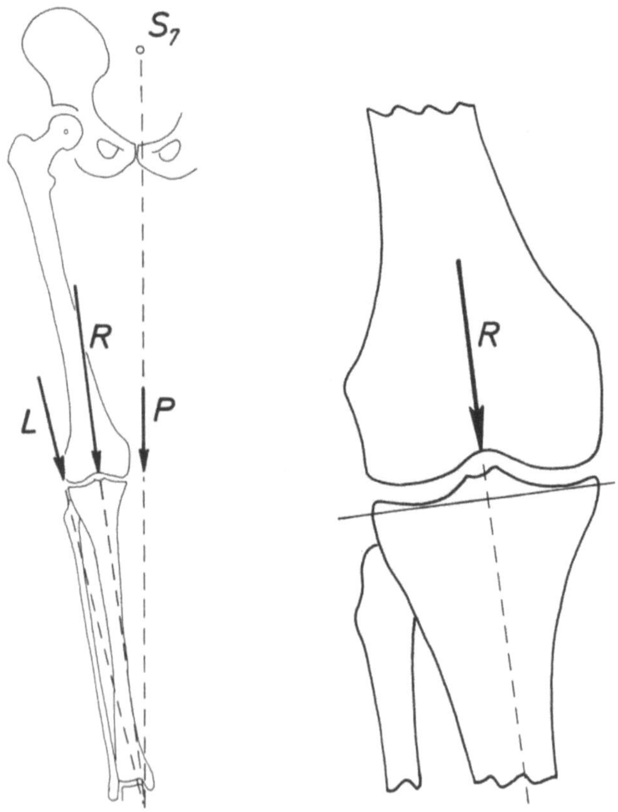

Fig. 72. Normal knee. Force *R* acts at right angles to the plane tangential to the tibial plateaux

In a normal knee the resultant force *R* acts at right angles to the plane tangential to the tibial plateaux (Fig. 72). If the leg is deformed into varum, the femur and hence the lateral muscles are less inclined to the line of action of force *P* than in a normal knee. In a further stage the direction of this inclination may even be reversed. The tibia is more inclined to the line of action of force *P* (Fig. 73). The resultant force *R*, which in a normal knee is directed from above downward and medially, tends to become more vertical or may even become directed downward and laterally. Overstressing due to the medial displacement of the resultant force *R* finally erodes the medial border of the plateau which is often somewhat enlarged by osteophytes. Because of this change in shape, the plateau remains normal to the line of action of the resultant force *R* (Fig. 73). Equilibrium is maintained and the knee is stable as long as *R* does not act further medially than the medial

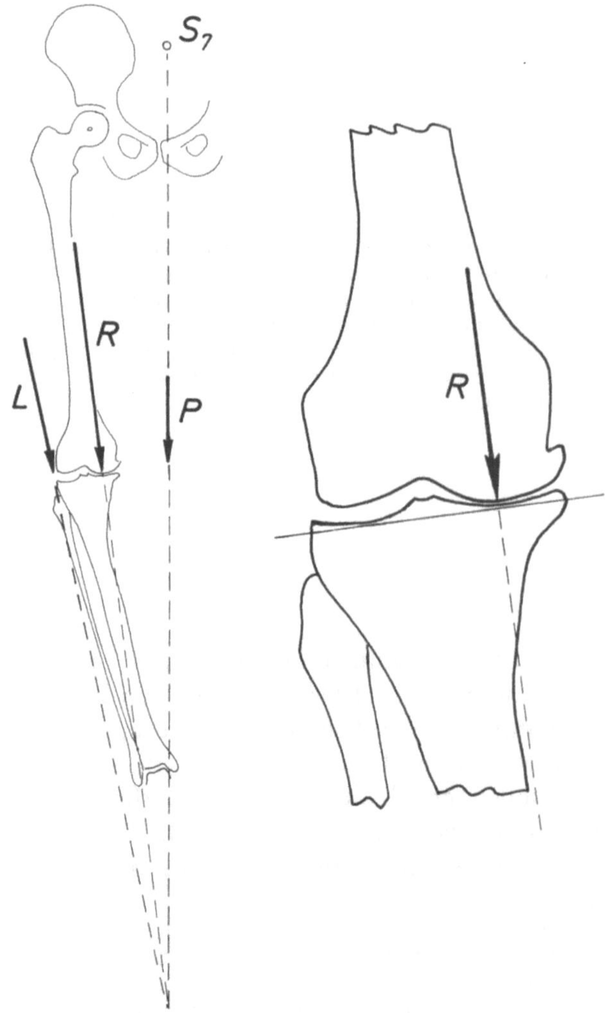

Fig. 73. Osteoarthritis with varus deformity. The resultant force *R* has come closer to the vertical but keeps acting at right angles to the plane tangential to the tibial plateaux

B. Lateral Displacement of Force R

In an otherwise normal knee, augmentation of the muscular force L laterally displaces the resultant R (Fig. 74b). An increase of the muscular force L may result from the muscles which span the knee and the hip. This action of the biarticular muscles would be necessary in order to compensate the weakness of the monarticular abductor muscles of the thigh, gluteus medius and minimus, and maintain equilibrium at hip level. The muscles which develop the force L can also take part in a contracture of all the abductors of the thigh. Their abnormally increased strength shifts R laterally.

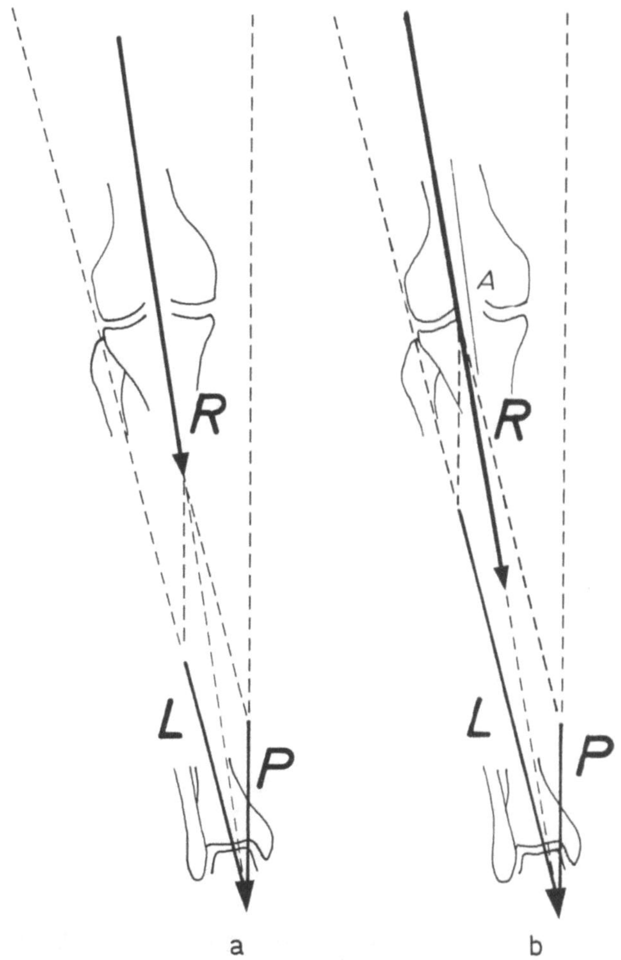

Fig. 74. (a) Normal knee. (b) Increased power of lateral muscular tension band L

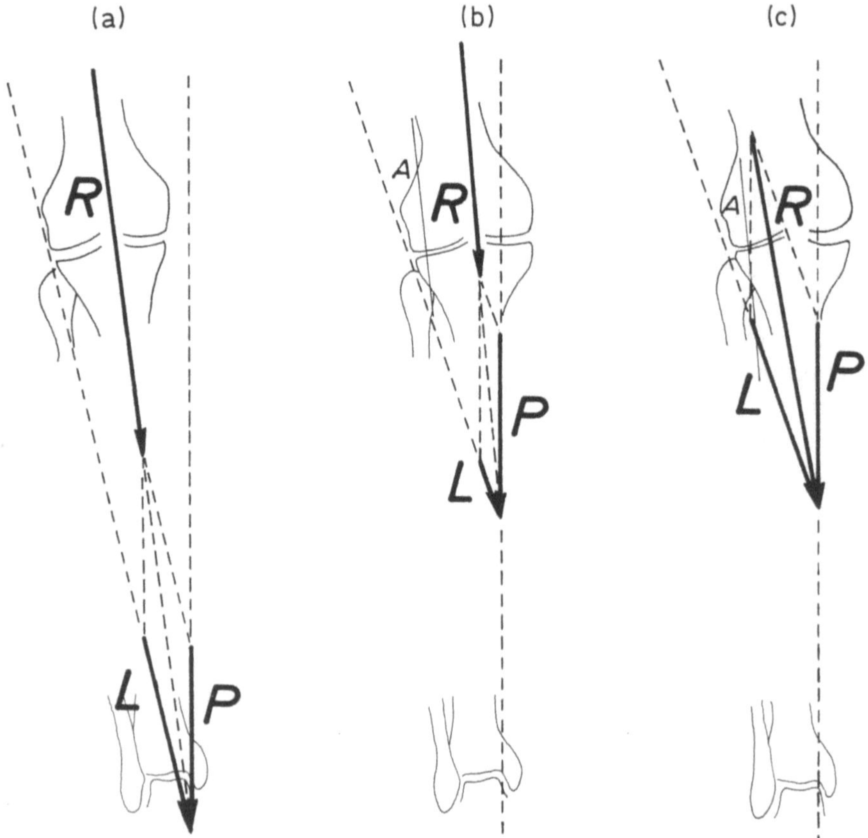

Fig. 75. (a) Normal knee. (b) Valgus knee. To keep resultant force R centred, the power of the lateral muscular tension band L must be decreased. (c) Valgus knee with P and L of normal magnitude

At first glance, a valgus knee (Fig. 75b) should be subjected to smaller stresses than a normal knee (Fig. 75a) since the valgus knee (Figs. 75b and 76) is nearer the line of action of force P than the normal knee. If the resultant R remains at the centre of gravity of the weight-bearing surfaces, the lever arm a of force P is shorter and the moment $P \cdot a$ smaller. Since the lever arm b of muscular force L is unchanged, a force L smaller than normal is enough to counterbalance P. Consequently, the resultant R, vectorial sum of forces P and L, is smaller in a valgus knee than in a normal knee if its line of action passes through the centre of gravity of the weight-bearing surfaces. The stresses exerted in the joint are thus smaller. This is the reason why constitutionally valgus knees do not usually develop osteoarthritis.

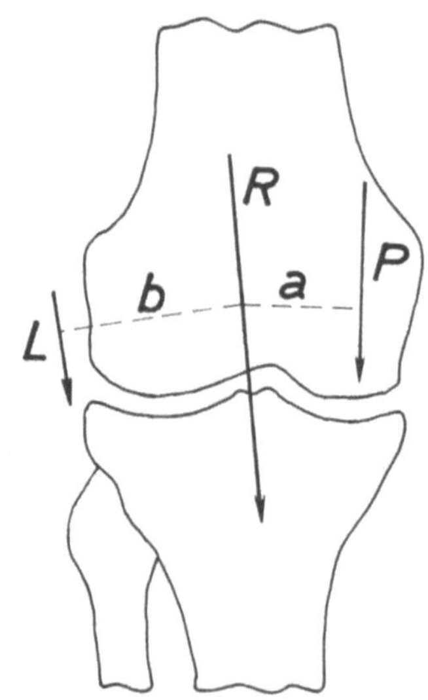

Fig. 76. Valgus knee. If resultant R remains centred, the power of the lateral muscular tension band L must be decreased and R is smaller than normal

Fig. 77. (a) Normal knee. (b) Line of action of P brought closer to the knee by a displacement of the centre of gravity S_7

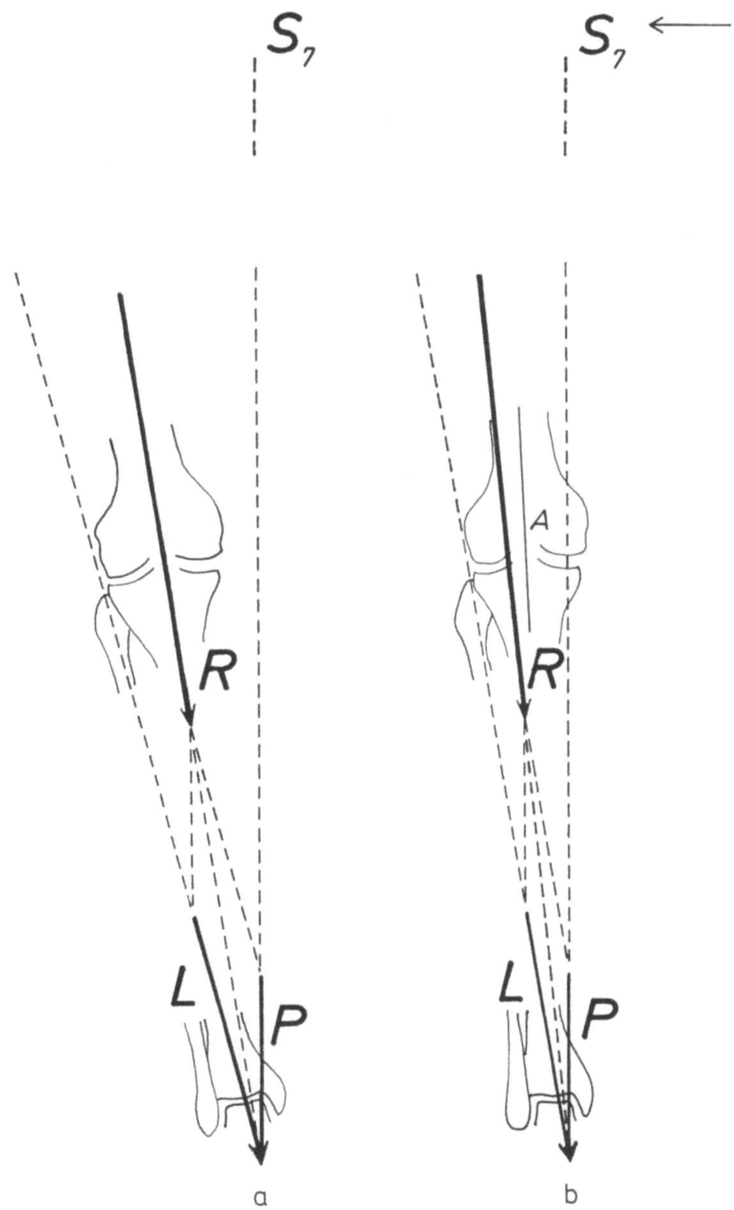

However, osteoarthritis in the lateral compartment of the knee is observed despite a valgus deformity. To explain this one must remember that a part of the muscles developing the force L are biarticular. They span the knee and also the hip (Figs. 17 and 18). If, in order to keep equilibrium at hip level, any diminution of L is made impossible, force L keeps its normal magnitude despite the valgus deformity and consequently the resultant R is displaced laterally (Fig. 75c).

A displacement of the centre of gravity of the body toward the side of the loaded knee shortens the distance between the latter and the line of action of P. If this reduction of the lever arm of P is not compensated by a corresponding loosening of the muscular tension band L, this tension band, because its magnitude remains normal, necessarily displaces the resultant R laterally (Fig. 77b).

Laterally shifted, the resultant *R* evokes asymmetrically distributed and abnormally high joint compressive stresses in the corresponding part of the knee (Fig. 78). This concentration of very high stresses provokes a thickening of the dense subchondral cup beneath the lateral plateau and increased bone density in the corresponding femoral condyle (Fig. 79). The density underlining the lateral tibial plateau spreads toward the intercondylar eminence. Here again, Pauwels' law is verified. The articular cartilage is destroyed under the effect of the localized overcompression. Its thinning and later its disappearance narrows the lateral joint space, which produces or potentiates the valgus deformity (Fig. 79). Such deformity aggravates the lateral displacement of *R*. A vicious circle is created and joint degeneration results.

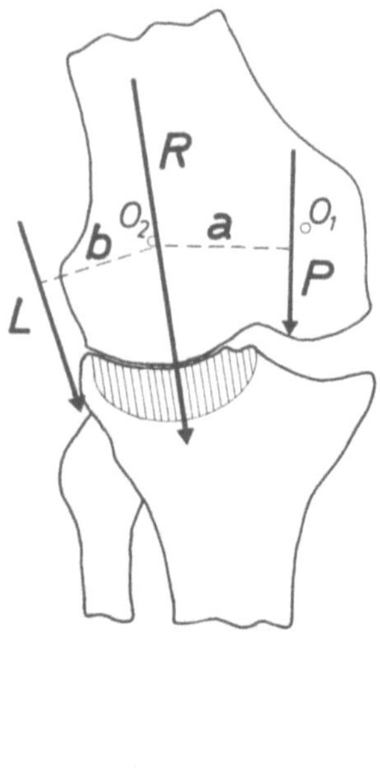

Fig. 78. Increased stresses in the lateral part of the knee resulting from a lateral displacement of resultant *R*. Abbreviations as for Figure 70

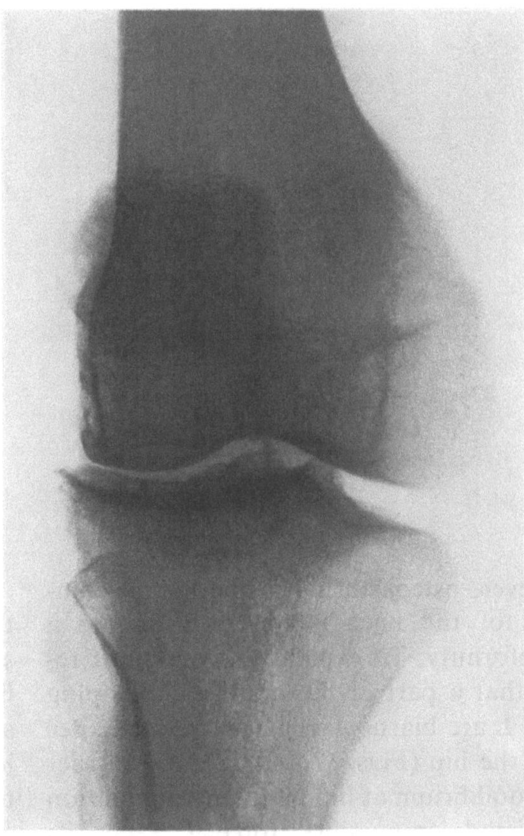

Fig. 79. Big cup-shaped sclerosis beneath the lateral plateau corresponding to the stress diagram

In a valgus deformity, the femur and hence the muscles *L* are more inclined to the line of action of force *P* than in a normal knee. Therefore the resultant *R*, directed from above downward and medially, is also more oblique than in a normal knee. It is normal to the medial aspect of the lateral part of the joint, close to the intercondylar eminence, no longer at right angles to the transverse plane tangential to both plateaux (Fig. 80). These mechanical conditions concentrate and increase the compressive stresses on the lateral plateau toward the tibial spines rather than in the lateral aspect of the joint. Erosion permits the femoral condyle to sink into the deepened tibial plateau (Fig. 82).

This difference of evolution between medial and lateral osteoarthritis can be observed in the X-rays and is decisive for the treatment (Maquet, 1980a). A dense triangle with its base medial underlies the medial plateau in osteoarthritis with a varus deformity (Fig. 71). In osteoarthritis with a valgus deformity, the pathological subchondral sclerosis is more centrally localized beneath the lateral plateau, sometimes even extending into the region of the intercondylar eminence (Fig. 79). Lateral osteoarthritis with valgus deformity can be compared with osteoarthritis of the hip with protrusio acetabuli, whereas medial osteoarthritis with varus deformity develops more like osteoarthritis of the hip with progressive subluxation of the femoral head.

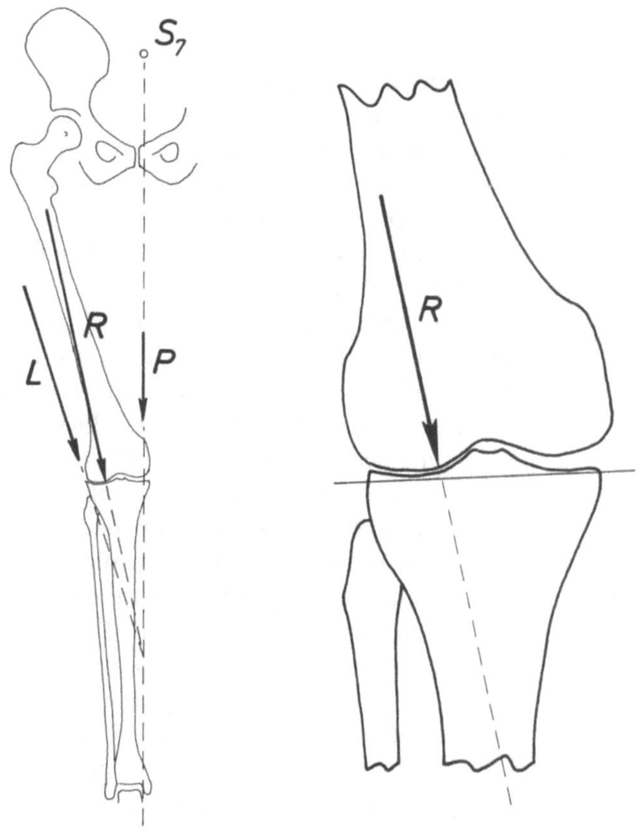

Fig. 80. Osteoarthritis with valgus deformity. The line of action of the resultant force *R* is oblique to the plane tangential to the tibial plateaux. Force *R* acts at right angles to the articular surfaces close to the intercondylar eminence

C. Unstable Knees

As long as the line of action of resultant R lies between the centres of curvature O_1 and O_2 of the femoro-tibial articular surfaces, the knee is stable (see page 21). As soon as R becomes lateral to O_2 the femur tilts on the tibia. There is instability. This is regularly observed, as the following instances illustrate.

A patient with a congenital dislocation of the right hip developed osteoarthritis in the medial part of her left knee and in the lateral part of her right knee (Fig. 81). From 33 years of age her gait became painful. At 48 years of age this patient became unable to walk. Her right knee was unstable. What happened? The resultant R has been progressively displaced to the centre of curvature O_2 of the lateral condyle while the valgus deformity increased (9 cm between the malleoli despite the varus deformity of the left knee). During the 15th year of evolution the line of action of resultant R became lateral to the centre of curvature O_2. The femur is held only by the medial ligaments, which are stretched and give way, as does any ligament when stretched long enough. The femur tilts on the tibia. The knee has become unstable. In this case, when standing on one foot, the centre of gravity of the supported part of the body is above the knee and the line of action of force P still passes between the centres of curvature O_1 and O_2.

In another patient (Fig. 82) the right knee, with a valgus deformity, has become unstable through the same process.

Instability happens much more often in a knee with a valgus deformity than in one with a varus deformity. This difference results from the geometry of the forces in varus and in valgus deformity. In a valgus knee, the line of action of the resultant force R is oblique to the plane tangential to the tibial plateaux, concentrating the compressive stresses towards the intercondylar eminence. The articular surfaces remain cup-shaped. The femoral condyle may even penetrate the tibial plateau. As soon as the line of action of force R passes outside the centre of curvature of the lateral articular surfaces the knee is held only by its medial ligaments. In a varus knee, the line of action of the resultant force R remains normal to the plane tangential to the tibial plateaux, the medial border of which becomes eroded in severe cases of osteoarthritis (Fig. 83). Therefore, equilibrium is maintained as long as the line of action of the resultant R is not displaced medially beyond the edge of the medial plateau. The lateral muscles must be very weak or the deformity very severe to reach this degree.

Nevertheless, instability can be observed in a knee with a varus deformity. The drawing Figure 83 represents a knee of a 75-year-old patient who has developed osteoarthritis with a progressive varus deformity (same knee, Fig. 188a). The knee is unstable and the patient cannot walk one step without leaning heavily on two crutches. How did the instability appear? With progressive deformity of the knee, the resultant R has been displaced medially to beyond the edge of the medial plateau. The femur, held only by the lateral ligaments, tilts on the tibia. The joint subluxates as a consequence of the instability.

Common to the several examples, instability can be explained only by a displacement of resultant R laterally beyond the centre of curvature O_2 of the lateral femoro-tibial articular surfaces or medially beyond the edge of the medial plateau.

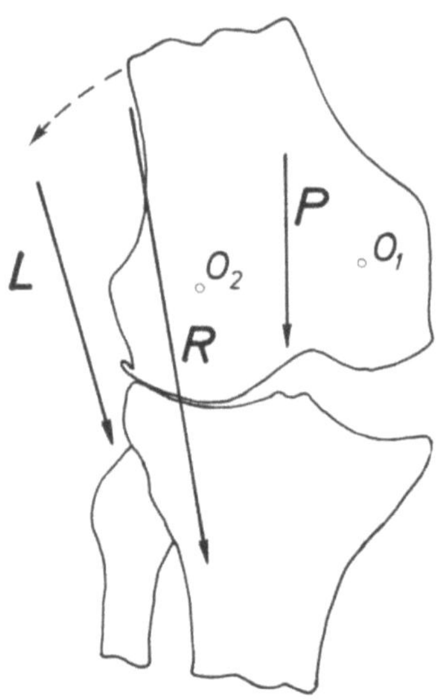

Fig. 81. Unstable valgus knee. The line of action of resultant R is outside the centre of curvature O_2 of the lateral condyle. R is the resultant of force P and muscular force L. The definitive resultant force acting on the knee includes, besides P and L, the force exerted by the medial collateral ligament. Abbreviations as for Figure 70

Fig. 82. Unstable valgus knee

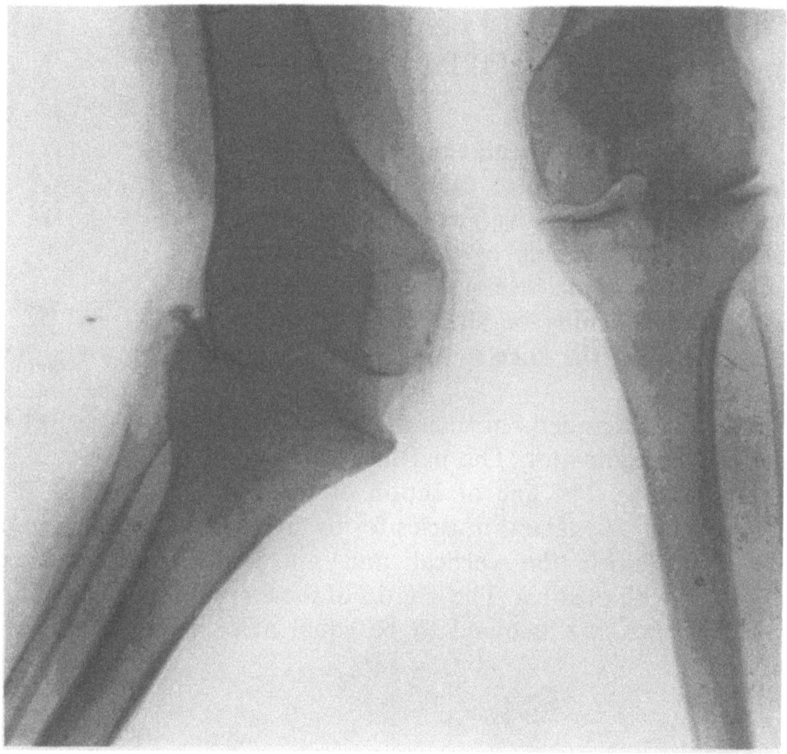

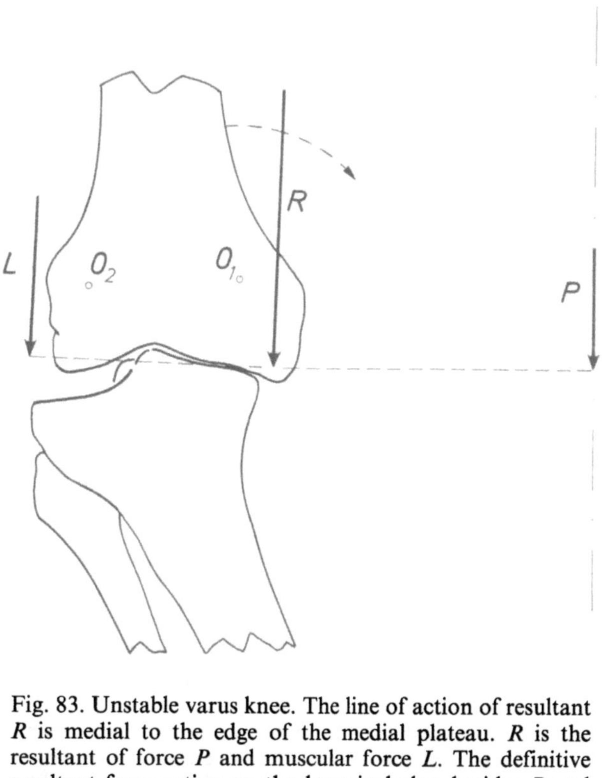

Fig. 83. Unstable varus knee. The line of action of resultant *R* is medial to the edge of the medial plateau. *R* is the resultant of force *P* and muscular force *L*. The definitive resultant force acting on the knee includes, besides *P* and *L*, the force exerted by the lateral collateral ligament. Abbreviations as for Figure 70

D. Evolution of the Maximum Stress in Relation to Several Parameters

The evolution of the maximum stress can be calculated in relation to each of the following parameters: varus or valgus deformity, strengthening or weakening of the lateral muscular tension band L, increase or decrease of the body weight, lengthening or shortening of the distance between the knee and the line of action of force P.

We first consider a normal individual standing on the right foot. The partial body weight P is 54.56 kg. The line of action of the force L exerted by the lateral muscles forms an angle $\psi = 10°47'$ with the vertical and an angle $\beta = 4°37'$ with the tibia. The lengths of the femur and the tibia are assumed to be equal at $l =$ 41 cm.

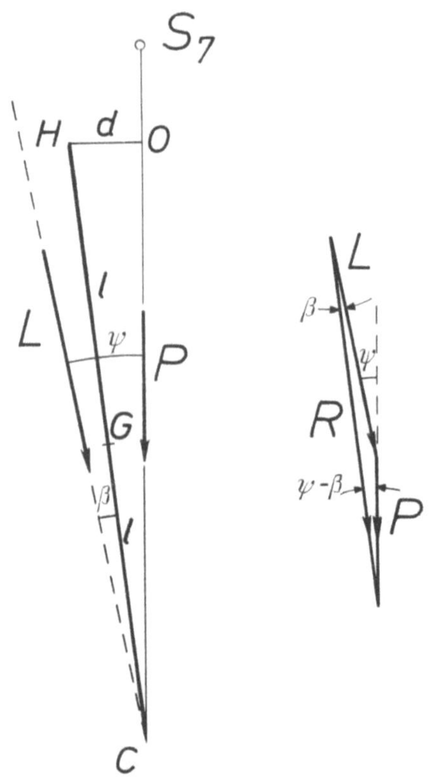

Fig. 84. P: partial body weight. L: lateral muscles. R: resultant femoro-tibial compressive force. H: hip. G: knee. C: ankle. OC: vertical line through the partial body weight S_7. d: distance of the hip from the line of action of force P. ψ: angle formed by the lines of action of P and L. β: angle formed by the line of action of L and the longitudinal axis of the tibia. l: length of the femur and of the tibia

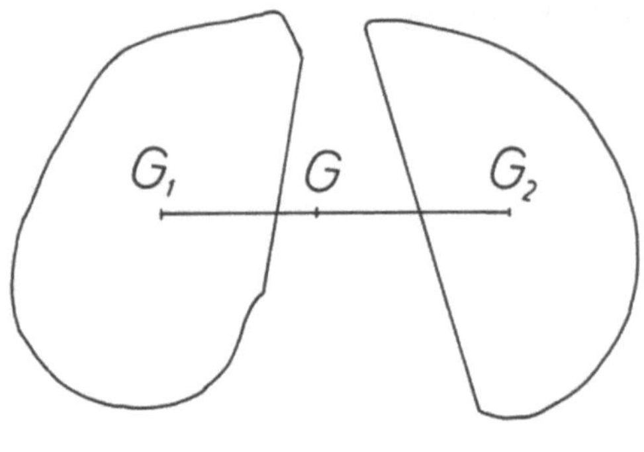

Fig. 85. Weight-bearing surfaces of the femoro-tibial joint with the centre of gravity G. G_1: centre of gravity of the medial plateau. G_2: centre of gravity of the lateral plateau

The femoral axis, i.e. the line joining the centre of the femoral head and the intercondylar space, is in line with the tibial axis. The distance between the centre of the hip joint H and the vertical line of action of P is d (Fig. 84).

The forces P and L balance each other at knee level. Their resultant R passes through the intersection C of their lines of action and through G, the centre of the knee.

The triangle of forces Fig. 84 gives:

$$L = P\frac{\sin(\psi - \beta)}{\sin\beta} = 72.816 \text{ kg},$$

$$R = P\frac{\sin\psi}{\sin\beta} = 126.824 \text{ kg},$$

$$d = 2l\sin(\psi - \beta) = 8.8 \text{ cm}.$$

The weight-bearing surface of the knee is formed by two distinct parts, the sizes of which are 11.034 cm² and 9.096 cm². The partial centres of gravity G_1 and G_2 of both areas and the centre of gravity G of the knee are in line (Fig. 85).

It appears that in a normal knee the stresses are evenly distributed over the whole weight-bearing surface. The stress is then

$$\sigma = \frac{R}{S} = \frac{126 \cdot 824}{20 \cdot 13} = 6.300 \text{ kg/cm}^2.$$

This uniform distribution requires the resultant R to pass through the centre of gravity of the weight-bearing surfaces.

Four parameters can be modified: the angle formed by the axis of the femur and that of the tibia in the coronal plane, the magnitude of the muscular force L, the magnitude of the force P (body weight), the distance between the knee and the line of action of the force P. Any variation of one of these parameters, if not compensated, will change the magnitude and the line of action of the resultant R and, consequently, the distribution of the joint stresses.

The effect of changing each of the four parameters will be studied, the three others being considered as constant.

1. Varus or Valgus Deformity

a) *Magnitude and Line of Action of R*

The varus deformity is represented in Figure 86.

The vertical OC supporting P, the distance d and the magnitude of L are constant.

α represents the angle of deformity between the femur and the tibia. α is positive in a varus deformity, negative in a valgus deformity. The tibia forms an angle x and the femur an angle y with the vertical. $x+y=\alpha$.

The new resultant R forms an angle ε with the vertical. It is exerted on the knee through the point G' located on the axis G_1G_2 at a distance u from G.

The line of action of L forms a constant angle β with the tibia.

We can now write:

$$d=2l\sin(\psi-\beta)=l(\sin x-\sin y)$$
$$=l(\sin x-\sin(\alpha-x))$$

from which:

$$\sin\left(x-\frac{\alpha}{2}\right)=\frac{\sin(\psi-\beta)}{\cos\frac{\alpha}{2}}.$$

In these equations, ψ and β have the same values as in the normal knee.

Forces P and L and angle α being known, the triangle of forces acting on the knee can be drawn (Fig. 86).

The following relations are deduced:

$$\operatorname{tg}\varepsilon=\frac{L\sin(x+\beta)}{P+L\cos(x+\beta)},$$
$$R=\sqrt{P^2+L^2+2PL\cos(x+\beta)},$$
$$u=l\cdot\operatorname{tg}(x-\varepsilon).$$

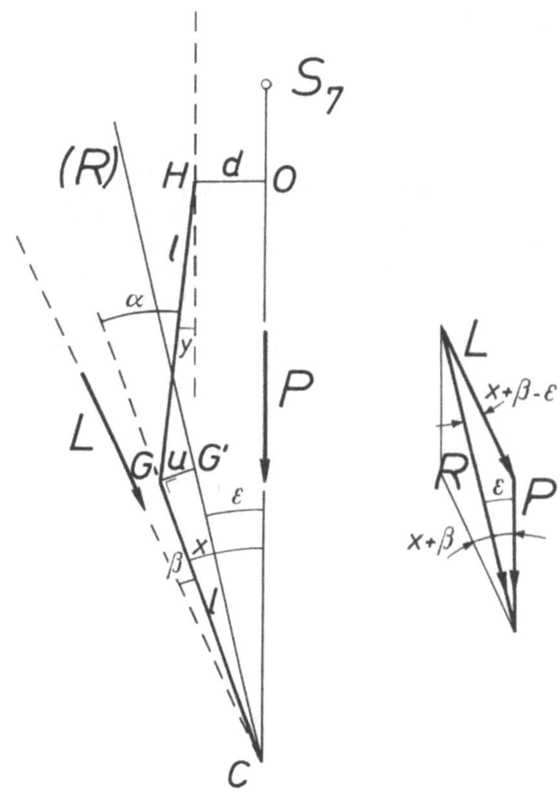

Fig. 86. Varus deformity. The line of action of R intersects G' at a distance u from G

Table 11 shows the variations of these magnitudes if α changes from 0° (normal knee) to 35° (varus knee) with an increment of 5°. If L remains unchanged, the magnitude of resultant R is very little modified by the deformity.

Table 11. $\beta=4°37'$; $\psi=10°47'$; $l=41$ cm; $P=54.56$ kg; $L=72.816$ kg

α	x	$x+\beta$	ε	u (cm)	R (kg)
0°	6°10′	10°47′	6°10′	0	126.824
5°	8°28′21″	13°05′21″	7°29′12″	0.71	126.563
10°	11°11′25″	15°48′25″	9°02′36″	1.54	126.191
15°	13°43′12″	18°20′12″	10°29′36″	2.31	125.782
20°	16°15′44″	20°52′44″	11°57′07″	3.09	125.311
25°	18°49′01″	23°26′01″	13°25′10″	3.87	124.777
30°	21°36′06″	26°00′06″	14°53′46″	4.66	124.179
35°	23°58′02″	28°35′02″	16°22′59″	5.46	123.516

The problem of a valgus deformity is treated in a similar way. It is illustrated in Figure 87. The results are given in Table 12. Again the valgus deformity changes the magnitude of R very little.

Table 12. $\beta=4°37'$; $\psi=10°47'$; $l=41$ cm; $P=54.56$ kg; $L=72.816$ kg

α	x	x+β	ε	u (cm)	R (kg)
0°	6°10′	10°47′	6°10′	0	126.824
5°	3°28′21″	8°05′21″	4°37′31″	0.82	127.065
10°	1°11′25″	5°48′25″	3°19′12″	1.52	127.216
15°	− 1°16′48″	3°20′12″	15°04′27″	2.28	127.323
20°	− 3°44′16″	0°52′44″	0°30′09″	3.04	127.372
25°	− 6°10′59″	−1°33′59″	−0°53′43″	3.79	127.364
30°	− 8°36′54″	−3°59′54″	−2°17′09″	4.55	127.300
35°	−11°01′58″	−6°24′58″	−3°40′06″	5.30	127.181

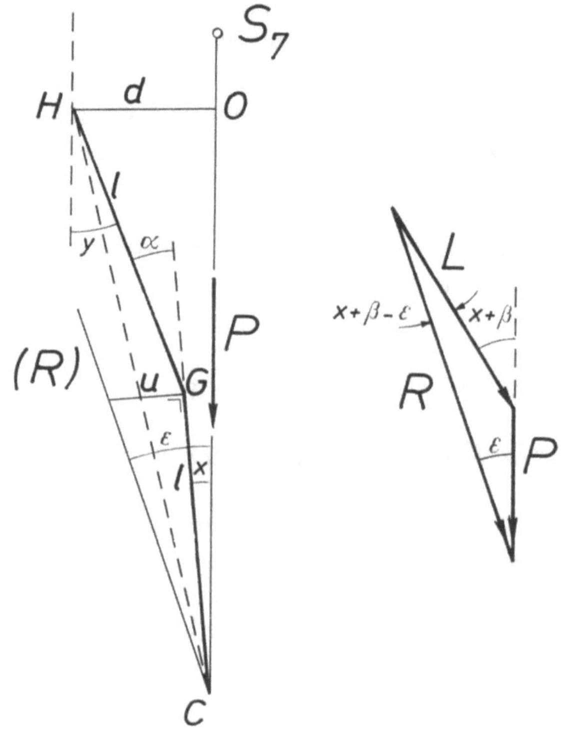

Fig. 87. Valgus deformity. *d* is the same as in the normal and in the varus knee. It has been drawn longer for clarity

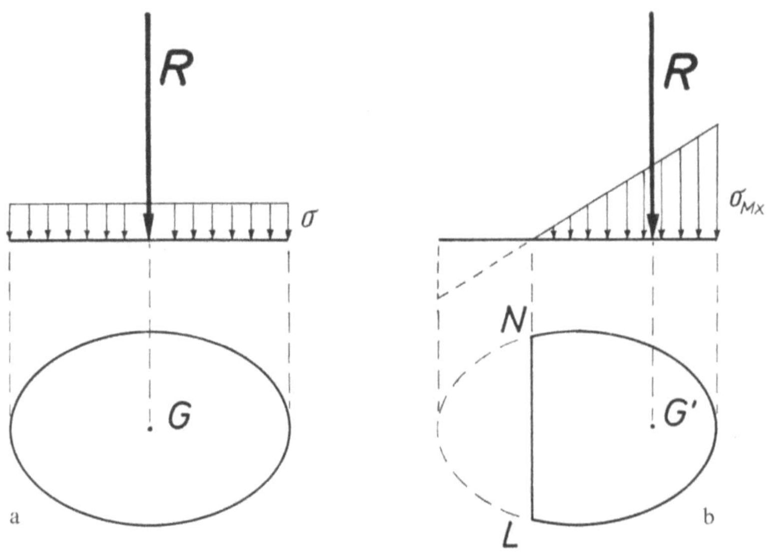

Fig. 88. (a) The compressive force *R* acts at the centre *G* of the weight-bearing surface. σ: stress. (b) The compressive force *R* acts eccentrically through G'. *NL*: line of zero compression. σ_{Mx}: maximum stress

b) Articular Compressive Stresses

The stresses have been analysed by a method used to evaluate the stress in masonry (Pirard and Sibille, 1954). Like cement, joints cannot withstand tension.

When any cross section of a mason's work is subjected to compressive forces the resultant R of which passes through the centre of gravity G of the surface, the stresses are uniform compressive stresses $\sigma = R/S$ (Fig. 88a).

If the resultant R is displaced from G to G' (Fig. 88b), the stresses change simultaneously and are distributed according to a triangular diagram. With sufficient displacement there is no compression in some areas. Whatever the shape of the cross section, the limit LN is always a straight line. It is called the *line of zero compression* or neutral line and its position can be precisely determined.

The area beyond the line of zero compression can no longer be considered as weight-bearing even if it remains in contact.

This corresponds to a deformed knee in which the resultant R is moved over the axis $G_1 G_2$ sufficiently to bring the line of zero compression into the joint.

We assume the problem to be solved. In a cross section S, a force R, normal to the surface, acts at a point G', distinct from the centre of gravity. The line of zero compression is LN (Fig. 89). The surface is referred to two rectangular axes xy. The first is at right angles to LN and passes through G'. The second corresponds to LN. We can divide the surface in elements dS parallel to LN with a centre of gravity g (x,y), x varying from nil to x_1. For each of these surface elements the stress is $\sigma = \sigma_1 \dfrac{x}{x_1} = x \operatorname{tg}\alpha$.

The total stress in a surface element is then:

$$\sigma \cdot dS = \sigma_1 \frac{x}{x_1} dS = \operatorname{tg}\alpha \cdot x \cdot dS.$$

The resultant of all the stresses must balance R.

1. The projection on the perpendicular to the surface gives

$$\int_0^{x_1} \sigma \, dS = \operatorname{tg}\alpha \int_0^{x_1} x \, dS = R.$$

This means that the static moment of the whole surface in relation to LN is $R/\operatorname{tg}\alpha$.

2. The static moment of the summation of the stresses in relation to axis x must be zero since R intersects this axis.

$$\int_0^{x_1} (\sigma \cdot dS) \, y = \operatorname{tg}\alpha \cdot \int_0^{x_1} xy \, dS = \operatorname{tg}\alpha \cdot J_{xy} = 0.$$

Consequently, the product of inertia of the surface in relation to the axes x and y is zero and these axes are called principal axes of inertia.

3. The static moment of the stresses in relation to y must be equal to the moment of R in relation to this axis.

$$\int_0^{x_1} (\sigma \cdot dS) \, x = \operatorname{tg}\alpha \cdot \int_0^{x_1} x^2 \, dS = \operatorname{tg}\alpha \cdot I_y = R \cdot d.$$

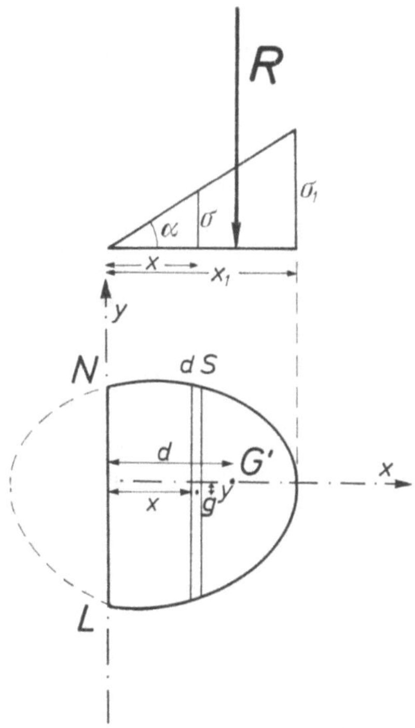

Fig. 89. Compressive force acting eccentrically. Calculation of the line of zero compression

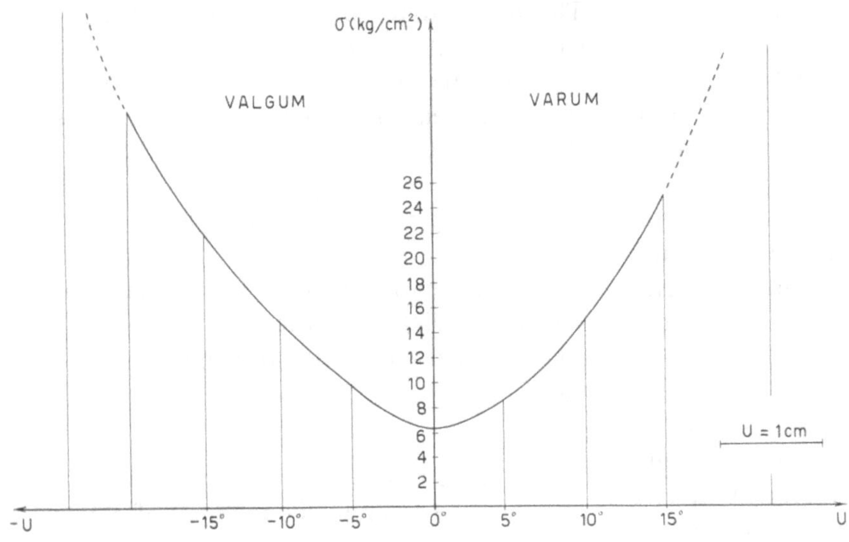

Fig. 90. Evolution of the maximum stress σ_{Mx} depending on the deformity

The moment of inertia of the surface in relation to *LN* is equal to the static moment of *R* in relation to this axis divided by tgα.

These three equations express the equilibrium and must permit the position of *LN* to be known and σ_1 to be calculated. Their resolution is not immediate since neither the orientation of *LN* nor its distance *d* from point *G'* is known. Successive approximations are used to discover the position of *LN* and the value of σ_1 which simultaneously satisfy the three conditions of equilibrium.

The problem can be solved by a computer or, with good precision, by a graphic method (Pirard and Sibille, 1954). We chose the graphic method, starting with the known surface of a normal knee.

In order to improve the accuracy, the actual dimensions have been doubled. But the measured surfaces were surrounded by irregular contours. These were smoothed without altering the general shape, without changing the total sizes, and without moving the partial centres of gravity. This smoothing makes the graphic work possible.

Since the displacement *u* of force *R* is known in relation to the angle of deformity, it is possible to locate the line *LN*. The static moment of the remaining surface in relation to this axis is then calculated. Finally the magnitude of the maximum stress is deduced. The maximum stress is always located on the part of the contours the furthest away from *LN*.

For small displacements of the line of action of *R*, the maximum stress is regularly increased but within acceptable limits. When *u* becomes longer than 1 cm, the line of zero compression comes to lie within the contact surfaces. Then some part of the latter is no longer weight-bearing. The maximum stress then increases quickly. Soon the values which could be calculated are no longer relevant. It is understandable that the tissues subjected to such stresses, after an attempt at functional adaptation (dense triangle, osteophytes), deteriorate. Cartilage disappears, bone is resorbed.

When the subject is standing on one foot, the stress in a normal knee is about 6.3 kg/cm^2. During gait it attains 19 kg/cm^2. A 15° varus deformity would increase the maximum stress to 24.74 kg/cm^2 when standing on one foot and consequently to about 75 kg/cm^2 when walking.

The curve in Figure 90 illustrates the evolution of the maximum stress. Beyond 15° of deformity this would soon tend to the infinite. Obviously, at this stage walking without additional support or without limping, even standing on one foot, would be impossible. The calculations for deformities beyond 20° are thus irrelevant.

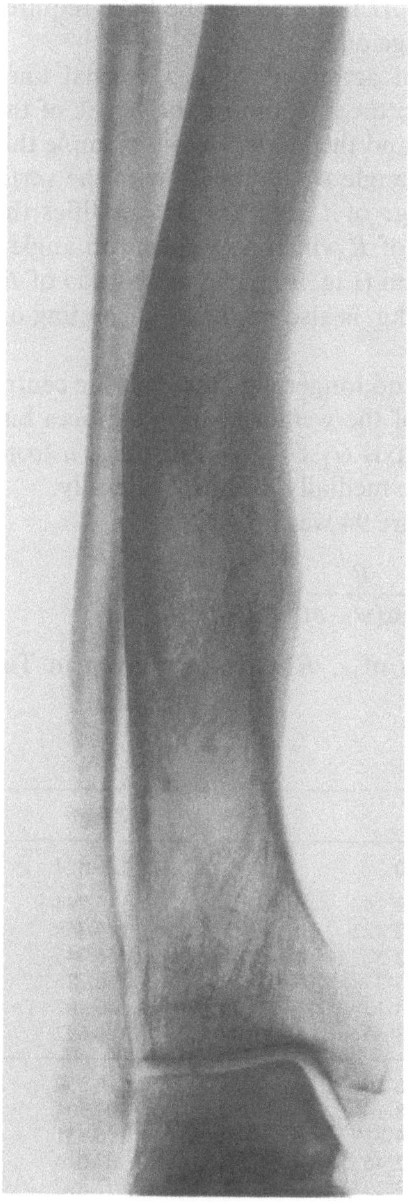

Fig. 91. The patient sustained a fracture of the lower leg thirty years previously

Fig. 92. The fracture healed with a varus deformity which brought the knee away from the line of action of *P*

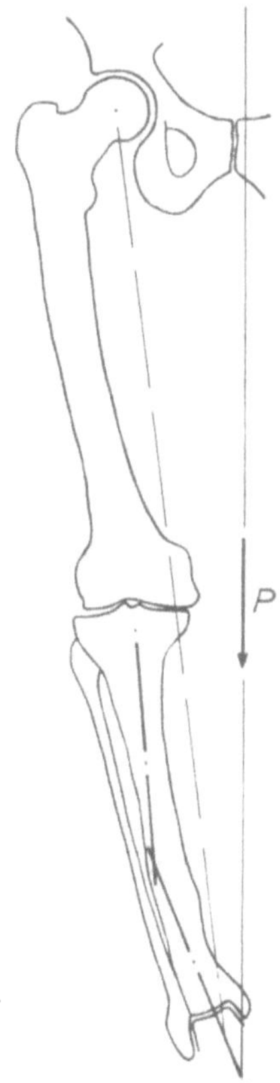

The effect of a deformity alone and eventually the result of its correction are demonstrated by the follow-up of fractures of the femur and of the tibia healed with a deformity in the coronal plane. A patient (Fig. 91) suffered a compound fracture of a lower leg during the war, at the age of 36. He was treated by immobilization in a plaster splint. Healing occurred with a varum of the leg (Fig. 92). Later the patient developed osteoarthritis in the medial part of the knee subjected to exceedingly high stresses (Fig. 93).

An example of osteoarthritis, the consequence of a fracture of the femur, is described on page 248.

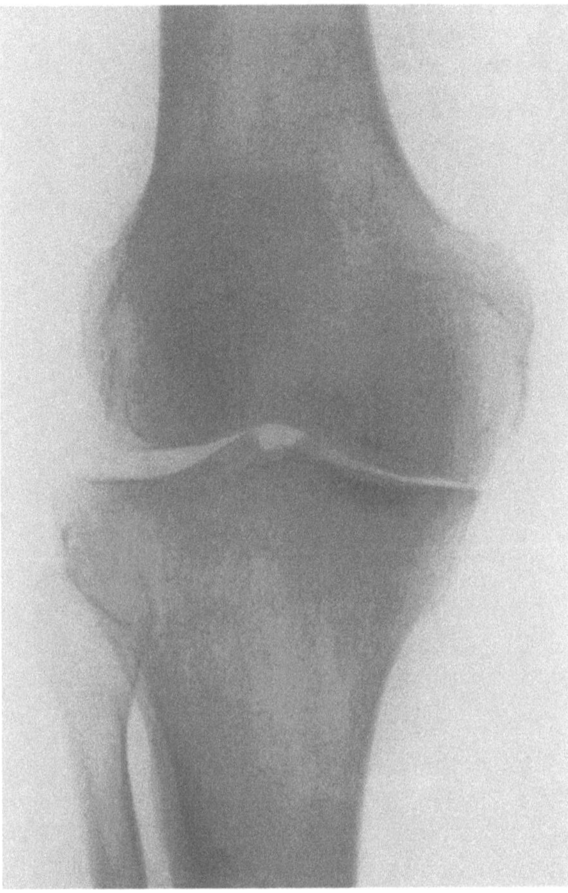

Fig. 93. Consequently the patient developed medial osteoarthritis of the knee

2. Strengthening or Weakening of the Muscular Force L

Strengthening or weakening of the lateral muscular tension band L shifts the point of application of the resultant R. Consequently, the distribution and the magnitude of the articular stresses are modified.

We shall analyse here the influence of a change of L alone. The other parameters – femoro-tibial angle, body weight, position of S_7 – are assumed to be constant. The anatomical characteristics are the same as in the previous paragraph.The resultant R and the joint stress are here also considered when the subject is standing on one foot.

Force L is modified by a quantitiy ΔL (positive or negative) but keeps its angle $\psi = 10°47'$ with the vertical. The balance between the forces P, R and $(L+\Delta L)$ exerted on the knee requires a related change of R.

The line of action of R in a normal knee corresponds to the line joining the centre of the femoral head and the ankle. In our example this line forms an angle $\psi - \beta = 6°10'$ with the vertical. The change of L into $L+\Delta L$ modifies the line of action of R which now forms an angle ε with the vertical (Fig. 94). The magnitude of R, initially 126.8 kg, is also modified depending on ΔL.

Finally, R no longer acts through the centre of gravity G of the weight-bearing surfaces but glides on the axis $G_1\ GG_2$ at a distance u from G. u is positive medially, negative laterally.

From Figure 94 we can write:

$$\frac{L+\Delta L}{\sin\varepsilon} = \frac{P}{\sin(\psi-\varepsilon)} = \frac{R}{\sin\psi}.$$

The values of ε, R and u are given in Table 13.

Table 13

ΔL(kg)	ε	u(cm)	R(kg)
0	6°10′	0	126.824
+10	6°30′14″	−0.24	136.794
+20	6°47′42″	−0.45	146.768
+30	7°02′57″	−0.63	156.745
+40	7°16′23″	−0.79	166.725
+50	7°28′13″	−0.93	176.707
+60	7°38′56″	−1.06	186.692
−10	5°46′19″	0.28	116.859
−20	5°18′14″	0.62	106.901
−30	4°44′22″	1.02	96.951
−40	4°02′46″	1.52	87.014
−50	3°10′27″	2.14	77.092
−60	2°02′42″	2.95	67.193

From Table 13 it appears that the inclination of the line of action of R is not much modified. But the magnitude of R considerably changes with ΔL, as does the distance u between the point of application of R and the centre of the knee.

The weight-bearing areas and the line of zero compression in relation to u have been determined above (pages 91–92). Graphical interpolation gives, with good accuracy, the characteristics of the weight-bearing surface for some values of ΔL. In each case the component of R normal to the contact surfaces has been considered. Since the angles of deviation are small, the normal component differs very little from the total value of R. The maximum stress has been calculated in these conditions. The results are illustrated by the curve in Figure 95.

A strengthening of the lateral muscular tension band ($+\Delta L$) increases the resultant R and moves its line of action u cm laterally. Consequently, the maximum stress σ_{Mx} is increased.

A weakening of the lateral muscles ($-\Delta L$) decreases the magnitude of the resultant R and moves its line of action u cm medially. This displacement u is much greater for the negative values of ΔL than for the corresponding positive values. Consequently the line of zero compression is further displaced and the weight-bearing surface more rapidly reduced by a weakening ($-\Delta L$) of the lateral muscles than by their strengthening ($+\Delta L$).

A weakening of the muscular tension band ($-\Delta L$), although it decreases the resultant R, paradoxically increases the maximum stress because of the considerable medial displacement of the line of action of R. There is no total compensation of the effects.

Finally, a variation of L, positive ($+\Delta L$) or negative ($-\Delta L$), always increases the maximum stress, an increase of L sooner, a decrease later. This may explain why a knee affected by lateral osteoarthritis usually deteriorates much more quickly than a knee with medial osteoarthritis.

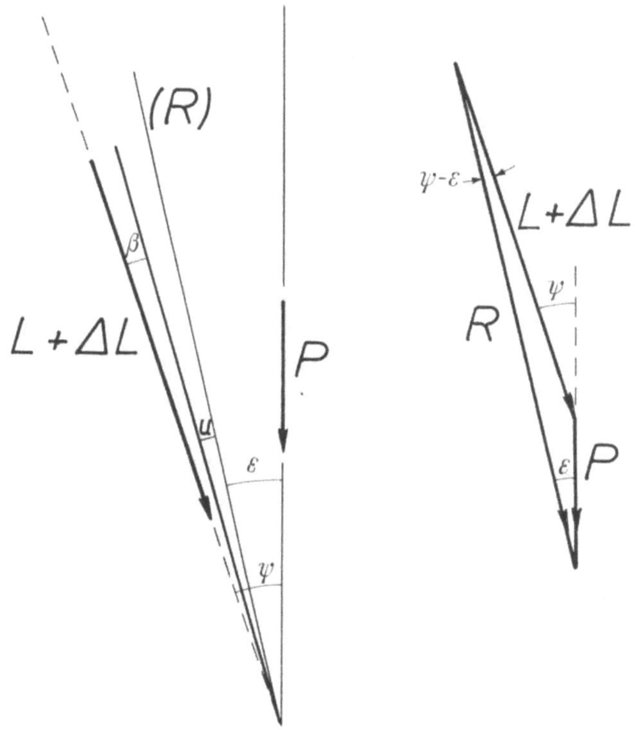

Fig. 94. Strengthening or weakening of the lateral muscles L

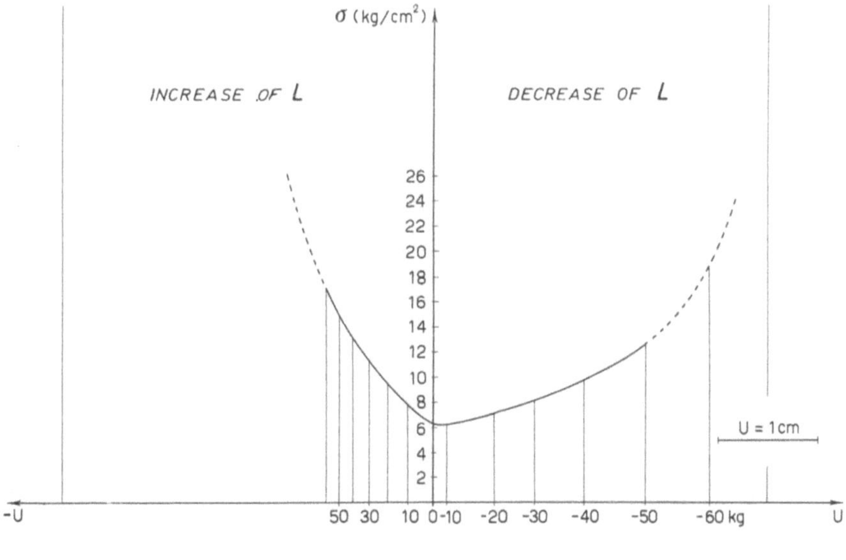

Fig. 95. Evolution of the maximum stress σ_{Mx} depending on the change of magnitude of force L

3. Cumulative Effect of a Change of the Force *L* and a Deformity of the Leg

Modifying the force *L* changes the point of application, the magnitude and the line of action of the resultant *R*. Weakening of the muscles *L* displaces the force *R* medially. Moving the point of application of the femoro-tibial compressive force *R* medially correspondingly increases the maximum stress. The concentration of pressure progressively destroys the articular cartilage. The joint space is locally narrowed. Narrowing of the medial joint space provokes a varus deformity. The varus deformity medially displaces the line of action of *R* and additionally increases the maximum joint stress. Medial displacement of *R* and varus deformity aggravate each other.

Similarly, a strengthening of the muscles *L* displaces the force *R* laterally. It produces localized destruction of the cartilage with narrowing of the lateral joint space. Narrowing of the lateral joint space causes a valgus deformity which in turn displaces the line of action of *R* laterally. Lateral displacement of *R* and valgus deformity aggravate each other.

The follow-up of a patient illustrates the simultaneous effect of a valgus deformity and a strengthening of the lateral muscular force *L*. The 65-year-old patient has been subjected to a cup arthroplasty for osteoarthritis of the hip (Fig. 96). The result is good at hip level. But in order properly to insert the cup, the femoral neck has been resected. Consequently:

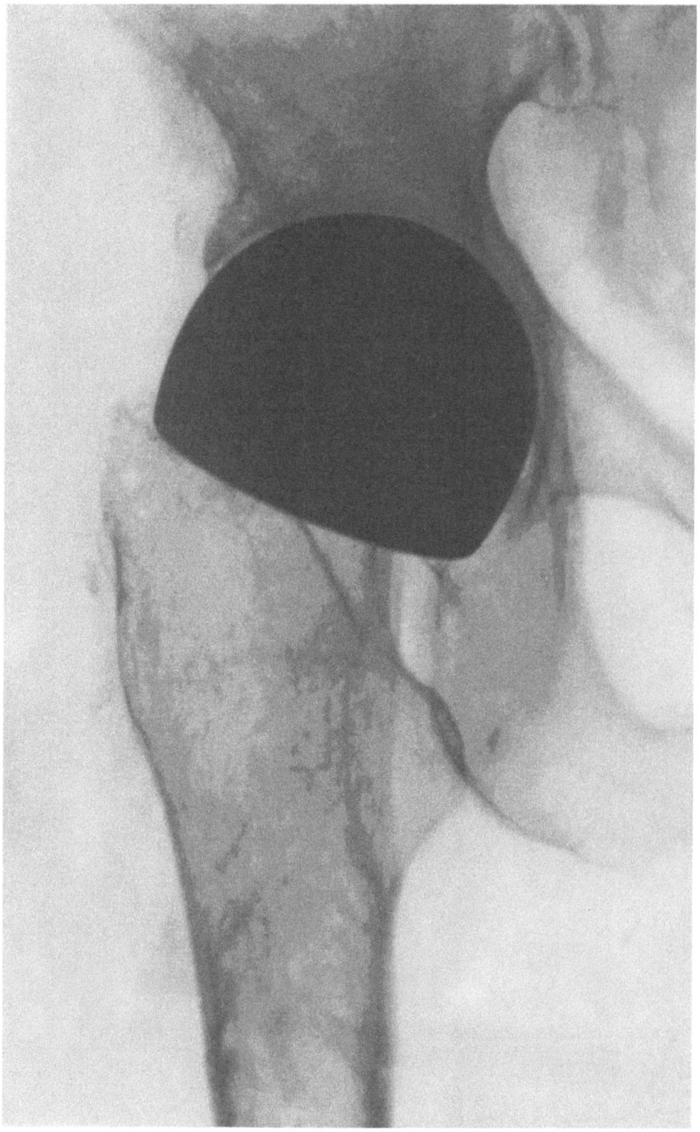

Fig. 96. Cup arthroplasty with resection of the femoral neck

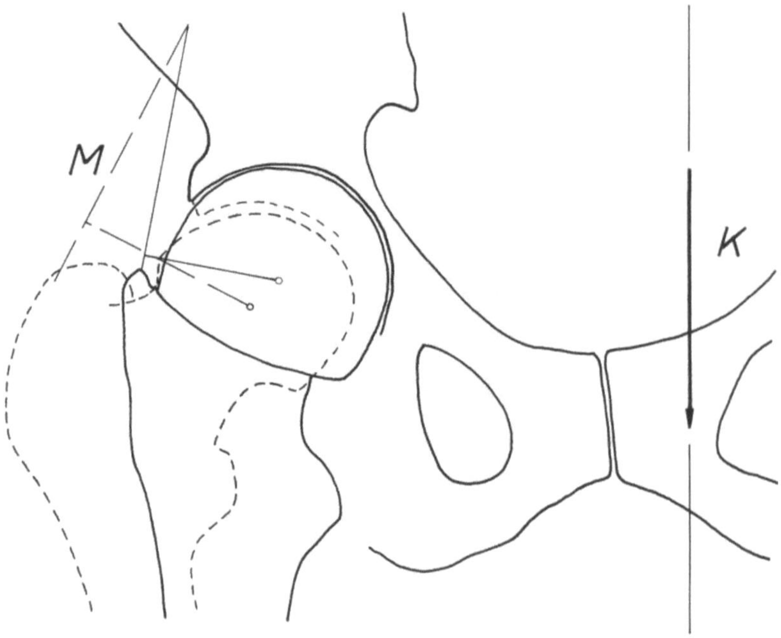

Fig. 97. The resection of the neck shortens the lever arm of the abductor muscles *M*. *K*: body weight acting on the hip

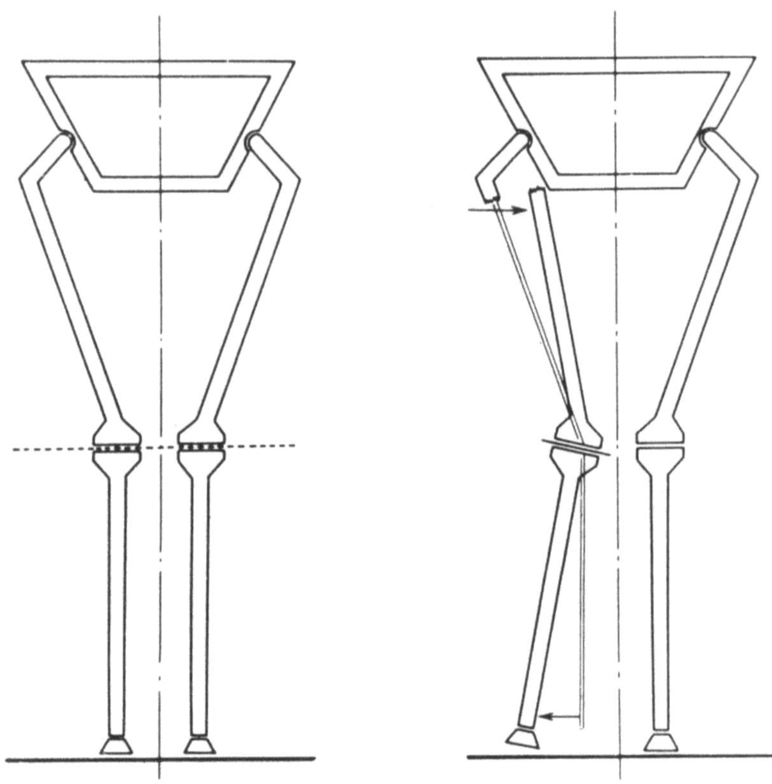

Fig. 98. Medial displacement of the upper extremity of the femoral diaphysis causes a valgus deformity. (From Pauwels, 1963)

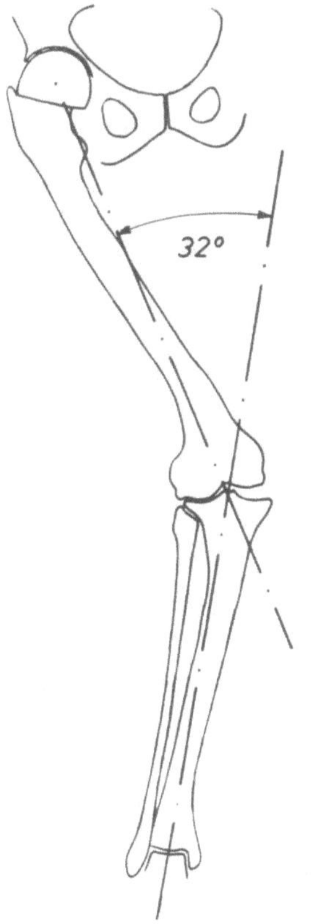

Fig. 99. Valgus deformity

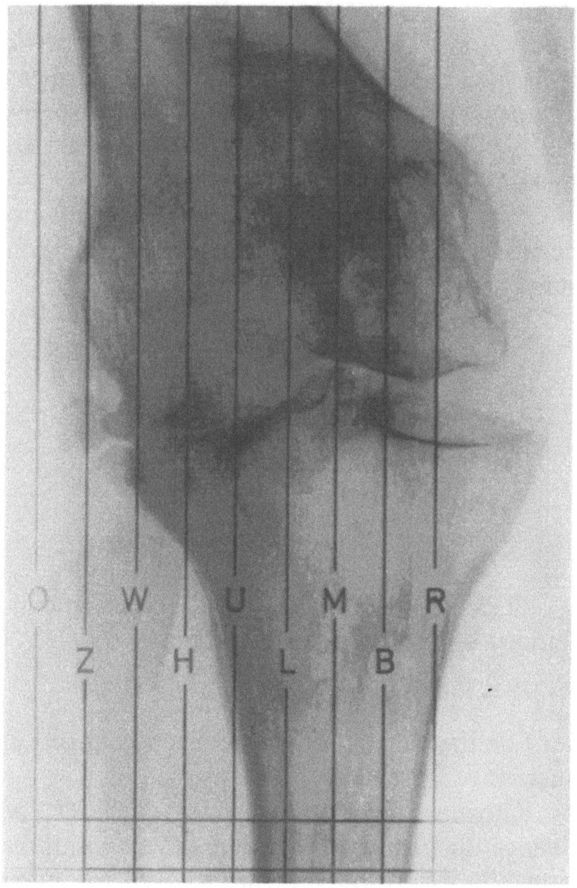

Fig. 100. Lateral osteoarthritis of the knee due to strengthening of the lateral muscles and to valgus deformity

1. The upper end of the femoral shaft is brought closer to the pelvis (Fig. 97). Therefore, the limb is deformed in valgum as demonstrated in Figure 98.

2. The gluteus minimus and probably the gluteus medius have been divided. If the gluteus medius has been spared or sutured, its lever arm is considerably shorter than normal (Fig. 97).

3. The lever arm of the "pelvic deltoid" (gluteus maximus and tensor fasciae latae) in relation to the centre of the femoral head is shortened because these muscles are no longer pushed laterally by the greater trochanter.

Because the gluteus medius and the gluteus minimus have no remaining or at least a much smaller action and because the lever arm of the "pelvic deltoid" is shortened, the biarticular muscles necessary for balance at hip level must exert a much greater force.

Both the increase of the force L exerted by the lateral muscles and the valgus deformity of the limb laterally shift the resultant of the forces acting on the knee. Consequently, the patient has developed a lateral osteoarthritis of the knee, obvious 2 years after the cup arthroplasty (Figs. 99 and 100).

4. Modification of Force P

Changing the body weight without proportional compensation by the muscles L will also displace R, modify its magnitude and increase the maximum stress.

P is changed by a quantity ΔP (positive or negative). The other parameters remain constant (Fig. 101).

We have:

$$\operatorname{tg}\varepsilon = \frac{L \sin\psi}{L \cos\psi + (P + \Delta P)},$$

$$u = l \operatorname{tg}(\psi - \beta - \varepsilon);$$

$$R = \frac{L \sin\psi}{\sin\varepsilon}.$$

The component of R normal to the articular femoro-tibial surfaces is given by:

$$R' = R \cos(\psi - \beta - \varepsilon).$$

For the calculation P has been changed by successive increments of 5 kg.

Knowing the distance u for each ΔP, we deduce the position of the line of zero compression, the weight-bearing surface and the maximum stress σ_{Mx}. Table 14 gives the figures for some values of ΔP.

Table 14

P(kg)	ε	u(cm)	R(kg)
+ 5	5°55′59″	0.17	131.796
+10	5°43′	0.32	136.770
+20	5°19′39″	0.60	146.724
+30	5°59′17″	0.84	156.684
+50	4°25′26″	1.25	176.616
+80	3°46′55″	1.71	206.540
− 5	6°25′09″	−0.18	121.854
−10	6°41′36″	−0.38	116.887
−20	7°19′03″	−0.82	106.961
−30	8°04′10″	−1.36	97.051
−50	10°09′03″	−2.86	77.300

The direction of R and its point of application are little changed by an increase of P, the body weight. For a positive ΔP the increase of the maximum stress is mainly due to an increase of R.

On the contrary, for a negative ΔP, R becomes smaller but its point of application is shifted laterally. The weight-bearing surface is quickly reduced, resulting in increased maximum stress. The curve in Figure 102 illustrates the evolution of σ_{Mx} in relation to ΔP.

Fig. 101. Increasing the force P

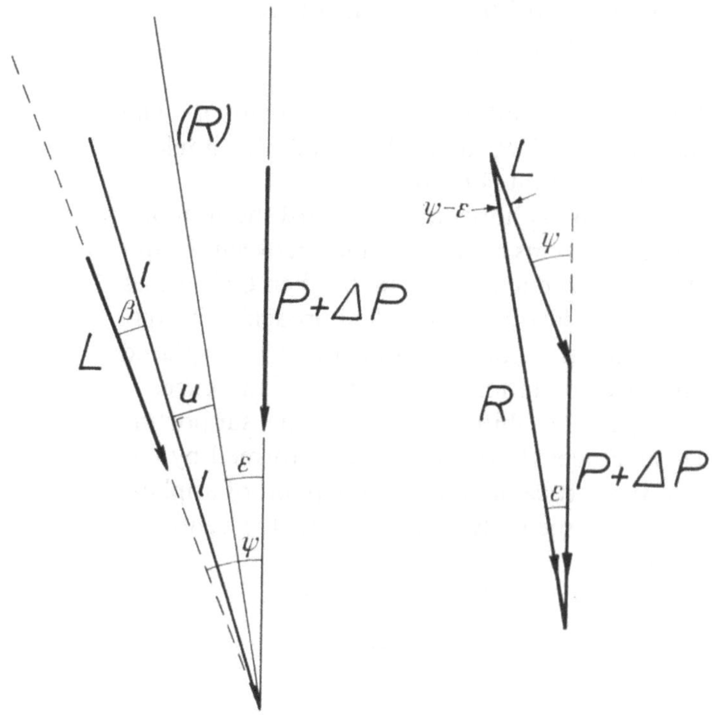

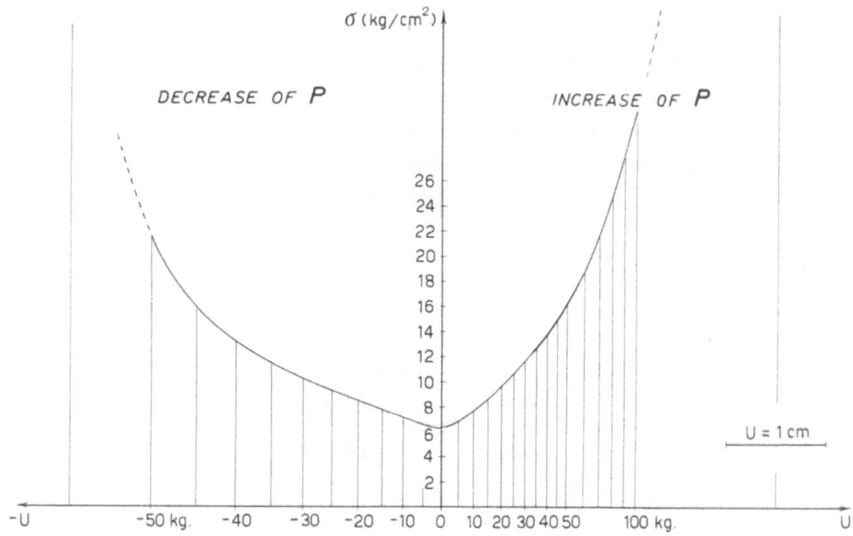

Fig. 102. Evolution of the maximum stress σ_{Mx} depending on the change of magnitude of force P

5. Horizontal Displacement of S_7 in the Coronal Plane

Forces P, L, and R converge at point C. The magnitude of P, that of L and the angle β remain constant (Fig. 103).

Shifting S_7 horizontally toward the supporting leg shortens d_1, the distance between S_7 and the prolonged axis of the limb. Then CH rotates about C. The direction of R approaches the vertical and the line of action of R is displaced laterally in relation to the centre of the knee.

Inversely, shifting S_7 from the supporting leg lengthens d_1, opens the angle formed by CH and the vertical and moves the point of application of R medially in relation to the centre of the knee.

From Figure 103 we can write:

$$d_1 + \Delta d_1 = \overline{S_7 C} \cdot \mathrm{tg}(\psi - \beta)$$

$$\mathrm{tg}\,\varepsilon = \frac{L \cdot \sin\psi}{P + L \cdot \cos\psi}$$

$$u = l \cdot \mathrm{tg}(\psi - \beta - \varepsilon)$$

$$R = \sqrt{L^2 + P^2 + 2L \cdot P \cdot \cos\varepsilon}$$

Fig. 103. Horizontal displacement of the partial centre of gravity S_7 in the coronal plane

Table 15 gives some figures. The values of σ_{Mx} in relation to the displacement of S_7 are demonstrated in the curve in Figure 104.

The curve illustrating the maximum stress σ_{Mx} is at a minimum when the line of action of force P is close to the centre G of the knee.

Table 15

Δd(cm)	ε	u(cm)	R(kg)
+ 1	6°43′32″	0.19	126.110
+ 2	7°19′00″	0.37	126.047
+ 3	7°53′22″	0.55	125.981
+ 4	8°27′21″	0.72	125.912
+ 5	9°01′49″	0.90	125.840
+10	11°50′57″	1.74	126.090
− 1	5°35′23″	−0.16	126.227
− 2	5°00′43″	−0.34	126.282
− 3	4°25′57″	−0.52	126.333
− 4	3°51′09″	−0.69	126.381
− 5	3°16′19″	−0.87	126.426
−10	0°21′37″	−1.78	86.709
−11	−0°13′22″	−1.96	76.572
−13	−1°23′19″	−2.32	127.326

Fig. 104. Evolution of the maximum stress σ_{Mx} depending on the displacement of S_7

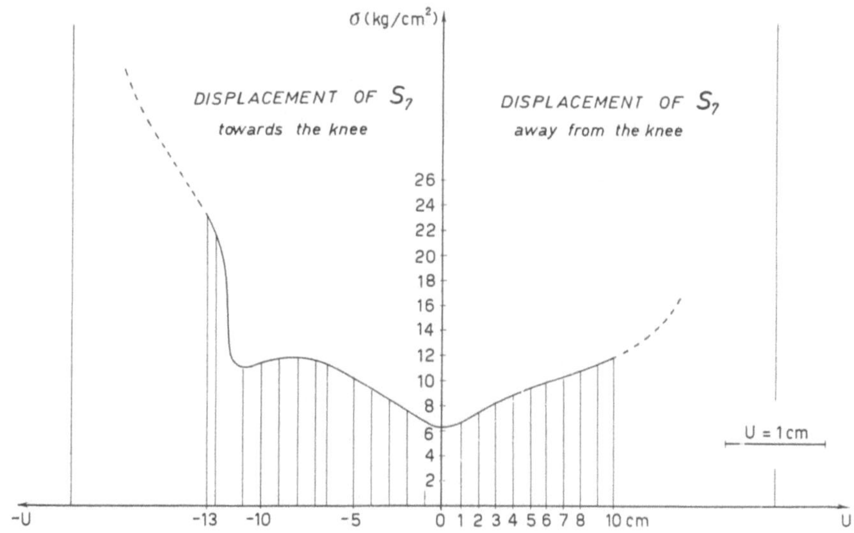

6. Conclusion

We have analysed the effect on the maximum stress σ_{Mx}

1. of a varus or valgus deformity,
2. of a change of the magnitude of the muscular force L,
3. of a change of the force P due to the partial body mass,
4. of a horizontal displacement of the partial centre of gravity S_7 in the coronal plane.

Any change of a single parameter from normal increases the maximum stress σ_{Mx}. This is true even for a weakening of the lateral muscles and for a decrease of body weight. Such a paradox results from the displacement of the point of application of the femoro-tibial compressive force R. This displacement shifts the line of zero compression and produces an uneven distribution of the joint stresses.

In a normal individual, the change of one parameter is immediately counterbalanced by a modification of one or several of the others. For instance, if somebody has lost weight, his muscles will work less, just enough to ensure equilibrium. Only when such a compensation no longer takes place does osteoarthritis develop. Then a modification of one of the parameters which ensure the balance of the knee causes successive and cumulative alterations of the joint. If nothing is done, the situation cannot but deteriorate.

The mathematical analysis confirms the theory proposed to explain the pathogenesis of osteoarthritis of the knee.

E. Posterior Displacement of Force R

In the sagittal plane, flexion of the knee usually brings the joint away from the line of action of force P. P acting more eccentrically must be counterbalanced by increased muscular forces. The analysis of the forces projected on a sagittal plane shows that flexing the knee provokes an increase of the forces R_4 compressing the femur on the tibia and R_5 compressing the patella against the femur. Such an augmentation is necessary to maintain equilibrium (Fig. 105). On the other hand the force R_4 is exerted on smaller weight-bearing surfaces. When the normal knee is completely extended (Fig. 106a) the femoral condyles rest on the tibial plateaux through contact surfaces with a long radius of curvature r. The lower parts of the condyles are in contact with the whole surface formed by the menisci and the tibial plateaux. At the beginning of flexion an anterior gap opens between the lower end of the femur and the upper extremity of the tibia (Fig. 106b). When the knee is flexed the posterior part of the femoral condyles rests on the back of the tibial plateaux (see pages 64 and 65). The radius of curvature r' of the femoral condyles is then shorter than the radius of curvature of the tibial plateaux. This results in a reduction of the weight-bearing articular surfaces despite the presence of the menisci (page 66). Decrease of the weight-bearing surfaces and augmentation of the load combine to increase the compressive articular stresses.

But flexion of a normal knee is accompanied by gliding of the femoral condyles on the tibial plateaux. For this reason the increase in pressure when going from extension to flexion does not remain localized. It is intermittent for each part of the contact surfaces because of the constantly changing contact area. The overall even distribution of the articular stresses is made apparent in X-rays by the subchondral sclerosis of even thickness which underlies the tibial plateaux of a normal knee (Fig. 107a).

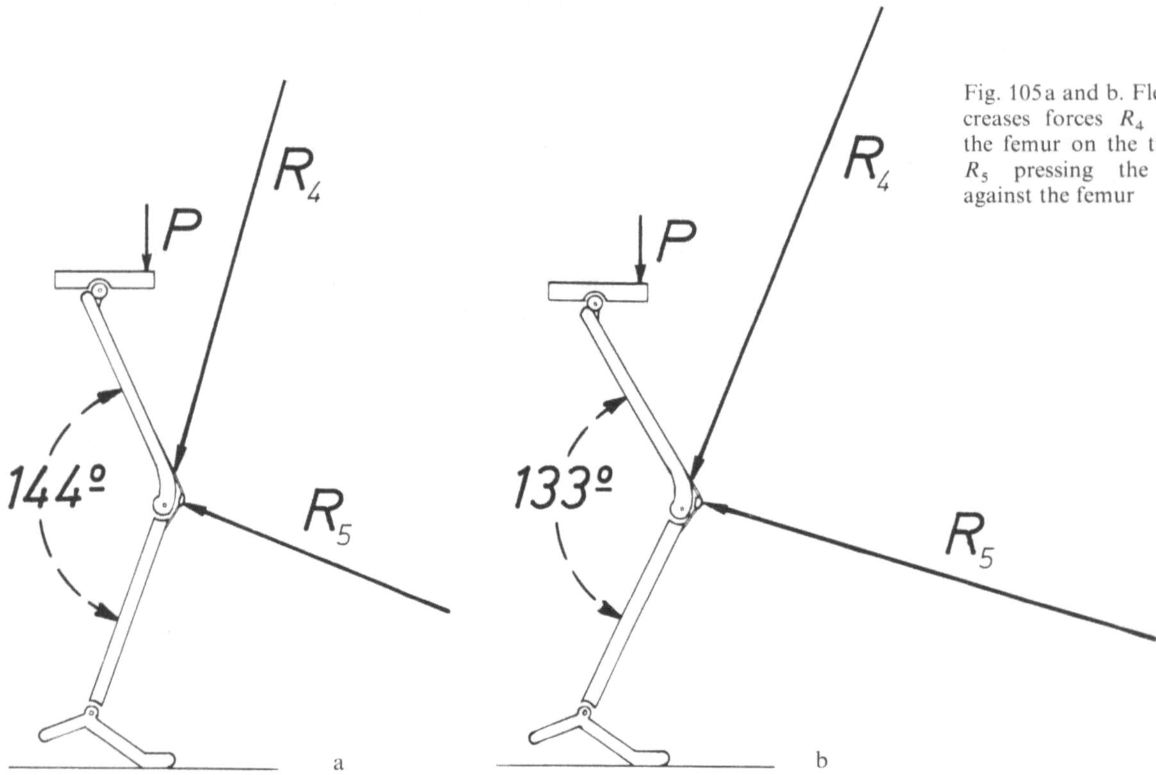

Fig. 105a and b. Flexion increases forces R_4 pressing the femur on the tibia and R_5 pressing the patella against the femur

Fig. 106a and b. Flexion decreases the weight-bearing surface of the knee. r and r': radii of curvature of the femoral condyles

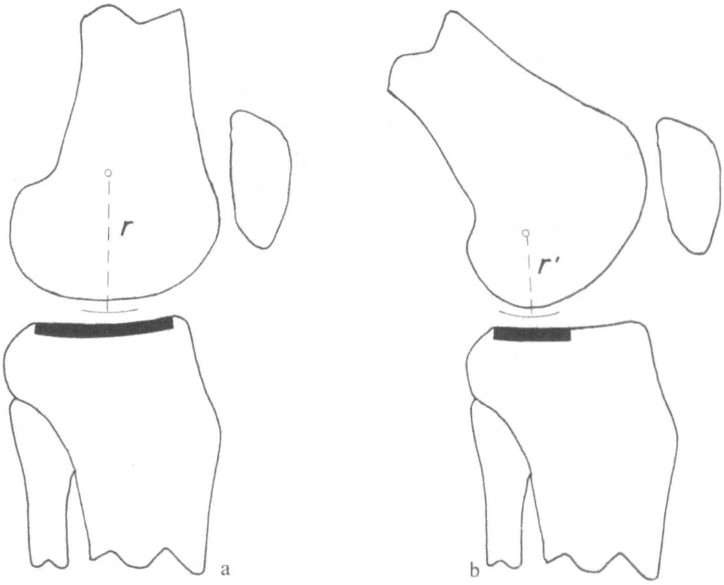

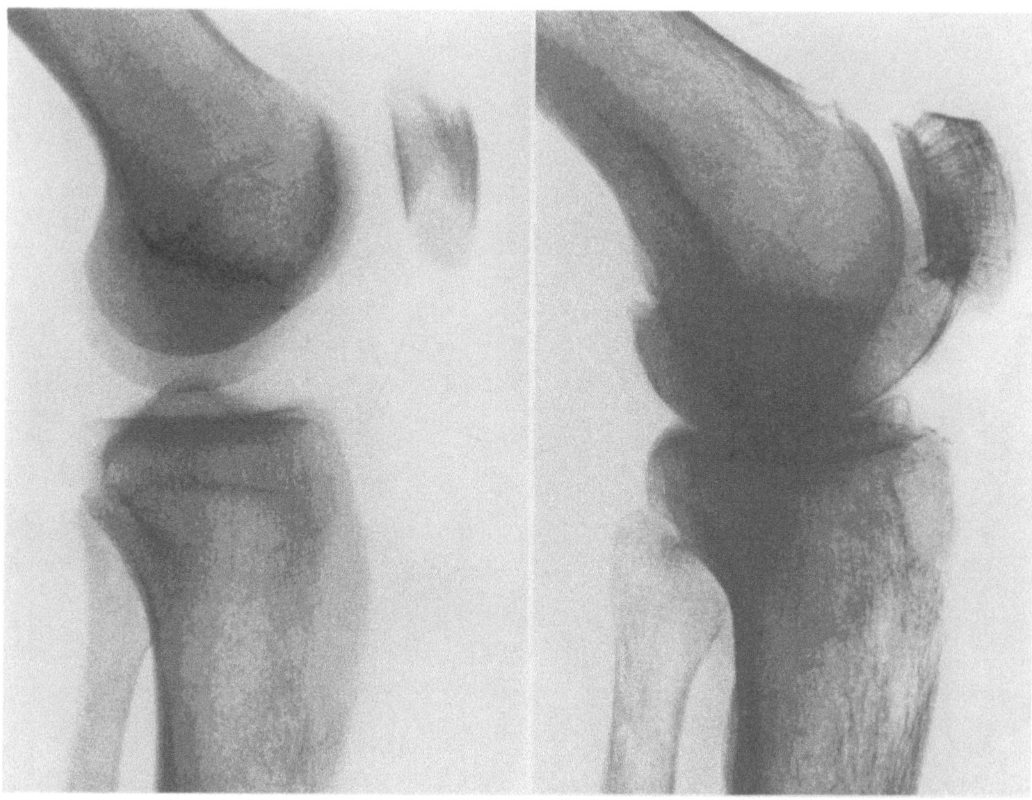

a b

Fig. 107. (a) Normal knee. A dense strip of even thickness underlines the tibial plateaux. (b) Osteoarthritic knee. A dense triangle results from the locally increased stresses

If the knee is unable to extend completely (as occurs in osteoarthritis) or if it is kept in slight flexion (as for some women wearing high-heeled shoes), not only is the femoro-tibial force R_4 increased but it is also transmitted through the posterior part of the condyles, through a cylindrical surface with a smaller radius of curvature r' (Fig. 106b). In this area the compressive articular stresses are thus strongly and permanently increased. They are concentrated on the posterior part of both tibial plateaux or, if the load is displaced medially or laterally, on one of the plateaux. They are thus abnormally high and create in the X-ray an appearance of a dense triangle with a posterior base and an anterior apex (Fig. 107b). The outline of this triangle corresponds to the stress diagram in the theoretical example of a monocylindrical joint where the force R eccentrically intersects the contact surfaces (Fig. 13b and c). Pauwel's law is verified in this projection as well.

F. Anterior Displacement of Force *R*

Post-traumatic or postoperative anterior tilt of the tibial plateaux also places the knee in the conditions of the models in Figure 13b and c. The forces R_3 and P_a acting on the knee are little changed if at all. Their resultant force R has to pass through the centre of curvature of the weight-bearing surfaces. Because of the anterior tilt of the plateaux force R will intersect the latter in their anterior aspect. This results in an uneven distribution of the stresses. Their diagram is triangular and underlies the anterior aspect of the knee (Fig. 108). Again the stress diagram corresponds to the sclerotic triangle with an anterior base, beneath the tibial plateaux (Fig. 115).

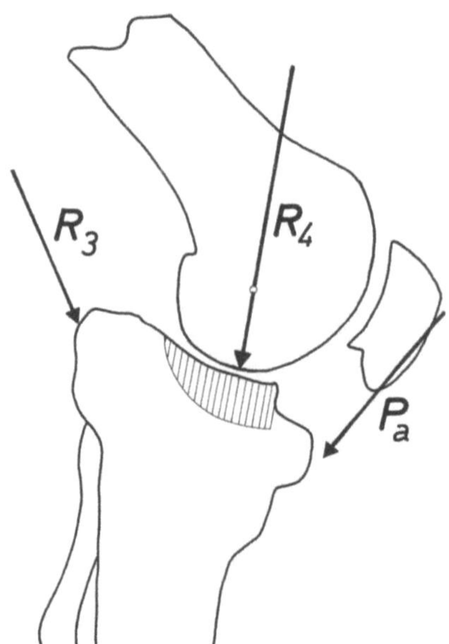

Fig. 108. Osteoarthritis with postoperative recurvatum of the knee. R_3: flexion force. P_a: force exerted by the patella tendon. R_4: resultant compressive force of R_3 and P_a. Concentration of the stresses in the anterior aspect of the joint

G. Increase of the Patello-Femoral Compressive Force

The force R_5 compressing the patella against the femur is also increased considerably by flexion of the knee. In flexion the knee is usually brought away from the line of action of force P. The quadriceps muscle must develop a stronger force M_v to counterbalance the moment of P (Figs. 105 and 109). On the other hand, the angle between the lines of action of force M_v and of the patella tendon P_a closes. The consequence is an augmentation of their resultant R_5 (Fig. 56 and 105). In a normal knee, the augmentation of R_5 is intermittent and occurs only in flexion. But when the knee can no longer be completely extended – as in osteoarthritis and in some women with high-heeled shoes – instead of decreasing each time the knee is extended, the resultant R_5 remains permanently elevated. The permanent increase of force R_5 compressing the patella against the femur causes higher intra-articular compressive stresses and subsequent osteoarthritis. The augmentation of the patello-femoral stresses is represented by the dense bone which appears in the patella in the X-ray (Fig. 117b and c).

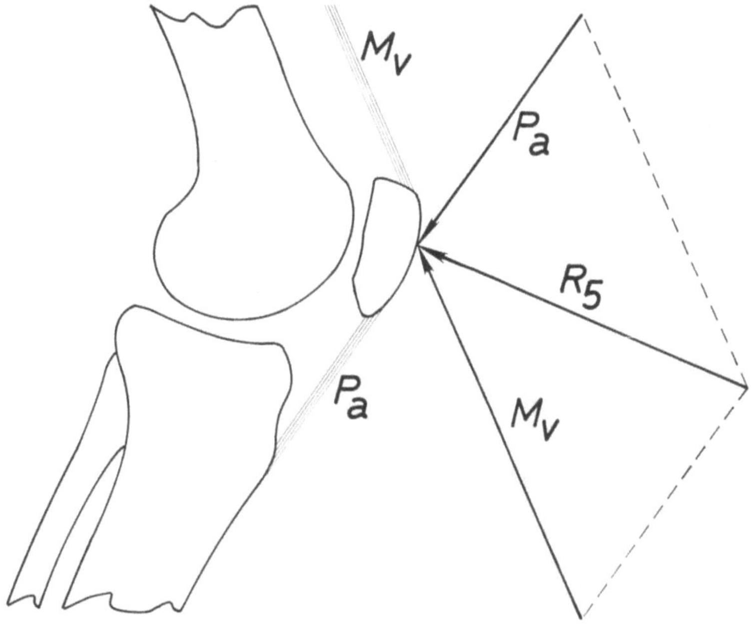

Fig. 109. The force R_5 presses the patella against the femur. It is the resultant of forces M_v exerted by the quadriceps tendon and P_a exerted by the patella tendon

H. Lateral Displacement of the Patello-Femoral Compressive Force

A horizontal cross section of a normal knee shows an even thickness of subchondral sclerosis underlying the articular surface of the patella (Fig. 110a). We conclude that to represent an even distribution of joint pressure. The patello-femoral compressive force R_5 can be resolved into a component R_L normal to the joint surface of the lateral condyle and one R_M normal to that of the medial condyle. Since the pressure is evenly distributed and the coefficient of friction between two articular cartilage layers is very low, each of these components, R_L and R_M, is then proportional to the surface area across which it is transmitted (Fig. 110b). The parallelogram of the forces shows the line of action of R_5 projected on a transverse plane through the knee.

In its projection on a coronal plane, force R_5 appears much smaller than in its projection on a sagittal and on a transverse plane. On the coronal plane the quadriceps tendon M_v pulls upward and laterally and the patella tendon pulls downward and laterally. The magnitude and direction of their resultant R_5 depend on their magnitude and on the angle formed by the lines of action of M_v and P_a (Fig. 111a).

A valgus deformity of the knee closes the angle formed laterally by the forces M_v and P_a. This entails an increase of the resultant R_5 in the coronal projection (Fig. 111b) if forces M_v and P_a do not change in magnitude. An increase in the strength of the vastus lateralis or a decrease in that of the vastus medialis (Fig. 111c) also results in an increase of force R_5 in the coronal projection. Lateral displacement of the tibial tuberosity closes the angle formed by the lines of action of forces M_v and P_a and increases their resultant force R_5 in the coronal projection as well (Fig. 111d).

An increase in magnitude of resultant force R_5 in the coronal projection tends to subluxate the patella laterally.

A progressive lateral subluxation of the patella is observed in many cases of osteoarthritis. It is necessarily accompanied by a displacement of the resultant force R_5, which squeezes the patella against the femur. Force R_5 is then transmitted by only a part of the lateral facet of the patella and by the corresponding area of the lateral condyle (Fig. 112b), instead of being transmitted by both patellar facets and both sides of the intercondylar groove as in a normal knee (Fig. 110).

Diminution of the weight-bearing surfaces causes locally increased stresses with bone condensation (Fig. 112a), destruction of the articular cartilage, narrowing of the joint space, reshaping of the bone and osteophytes.

By analysing the causes of osteoarthritis of the knee theoretically, we have deduced the mechanical disturbances which they provoke in the joint. These changes induce chondral lesions and bony reactions which obey the general laws defined by Pauwels. They are visible in the X-ray and can be interpreted from the radiographic pictures.

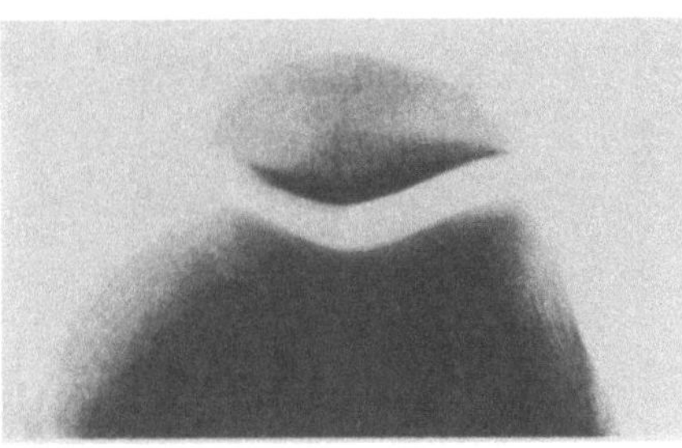

a

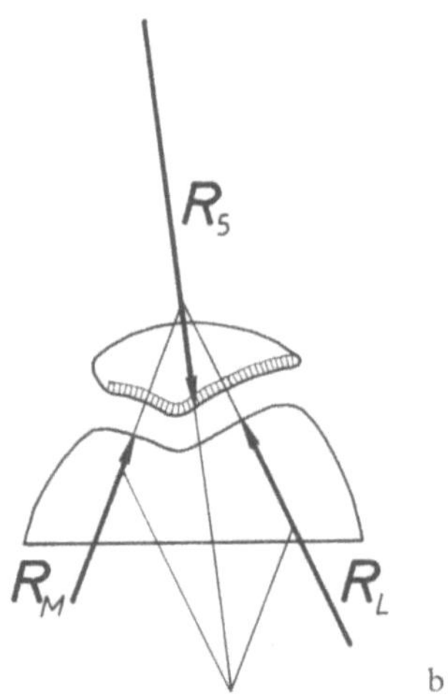

b

Fig. 110. (a) In a normal patello-femoral joint, the subchondral sclerosis evenly underlines both articular facets of the knee-cap. (b) R_5: force pressing the patella against the femur. R_L: reaction force to the lateral component of R_5. R_M: reaction force to the medial component of R_5

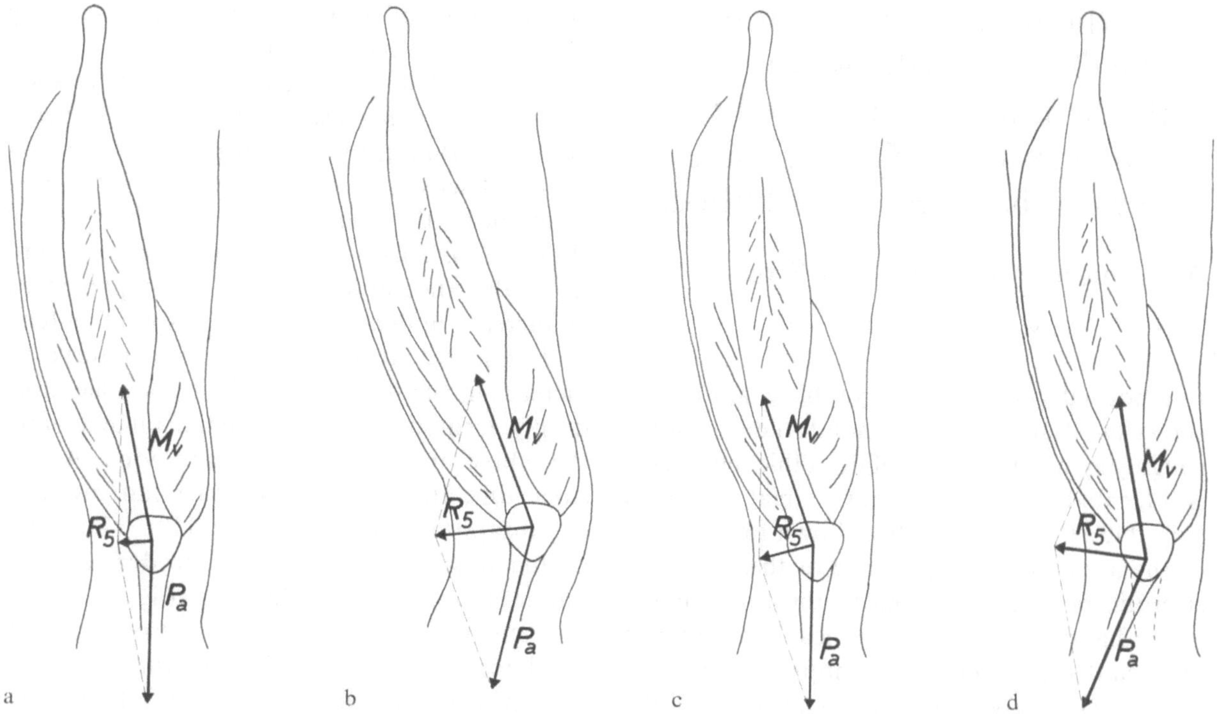

Fig. 111a–d. Coronal projection. M_v: force exerted by the quadriceps tendon. P_a: force exerted by the patella tendon. R_5: resultant of M_v and P_a. (a) Normal knee. (b) Valgus knee. (c) Weakening of the vastus medialis. (d) Lateral displacement of the tibial tuberosity

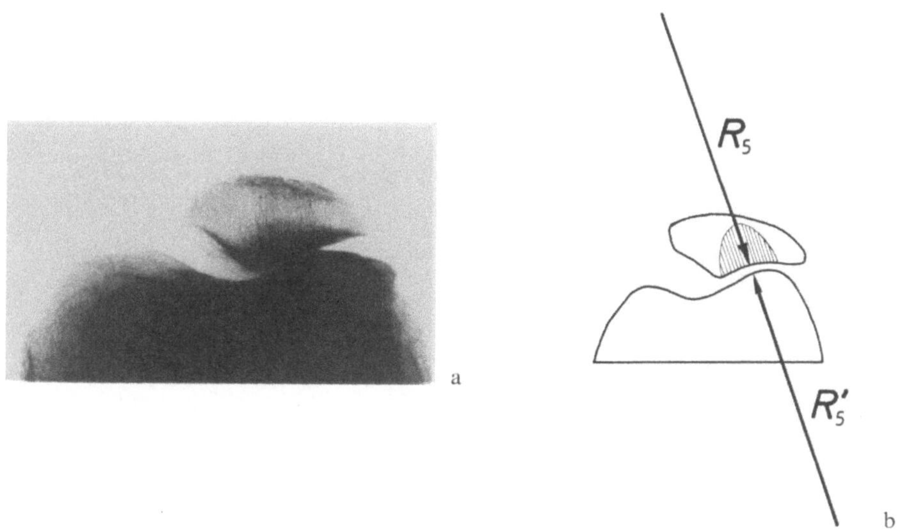

Fig. 112. (a) In a subluxated patella, the sclerosis is much thicker beneath the lateral facet. (b) R_5: force pressing the patella against the femur. R_5: force pressing the femur against the patella

II. Radiographic Examination of the Osteoarthritic Knee with Demonstration of the Effect of Changes in the Compressive Force on the Stress Distribution

A. Demonstration of Joint Stresses

1. A.-P. View

In a normal knee the radiographic examination shows two flat cups of subchondral dense bone, roughly symmetrical and of even thickness. One is located beneath each tibial plateau (Fig. 113a). As Pauwels has shown for the roof of the acetabulum, the outline of these cups of sclerotic bone corresponds to the form of the stress diagram. These symmetrical cups indicate an even distribution of the compressive stresses over the whole weight-bearing surfaces of the normal joint and make possible the conclusion that the femoro-tibial compressive force R acts at the centre of gravity of the weight-bearing surfaces of the knee (Fig. 65, page 71).

A medial displacement of load R is soon followed by an increase in thickness of the dense bone underlining the medial plateau. The subchondral sclerosis becomes progressively triangular near the medial edge of the plateau (Fig. 113b). This is the first sign of osteoarthritis of the knee and it can be accompanied by an increase of density in the corresponding femoral condyle. At a later stage, we observe an increase in the thickness of the trabeculae more deeply located under the medial plateau until the whole area becomes a dense zone. At the same time the medial joint space narrows and can disappear completely (Fig. 113c). The edge of the tibial plateau becomes eroded and the femur is subluxated on the tibia (Fig. 113d). Simultaneously, the subchondral sclerosis beneath the lateral plateau decreases in thickness and the cancellous trabeculae seem to fade. Undergoing smaller stresses, bone is resorbed in the lateral aspect of the bone (Pauwels, 1973).

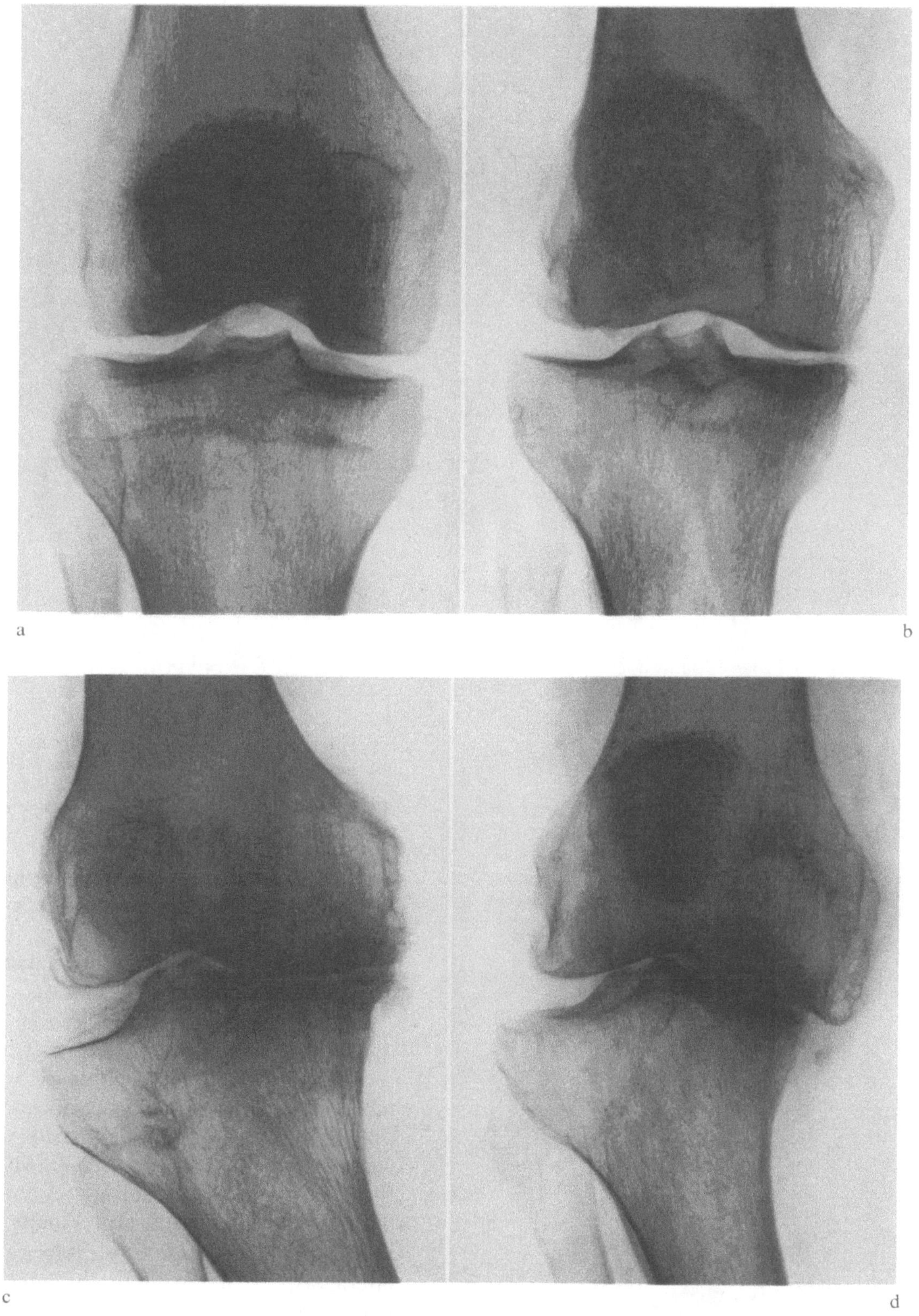

Fig. 113. (a) Normal knee. (b) Dense triangle underlining the medial tibial plateau. (c) Larger triangle and narrowing of the joint space. (d) Sclerosis beneath the eroded medial tibial plateau and subluxation of the joint

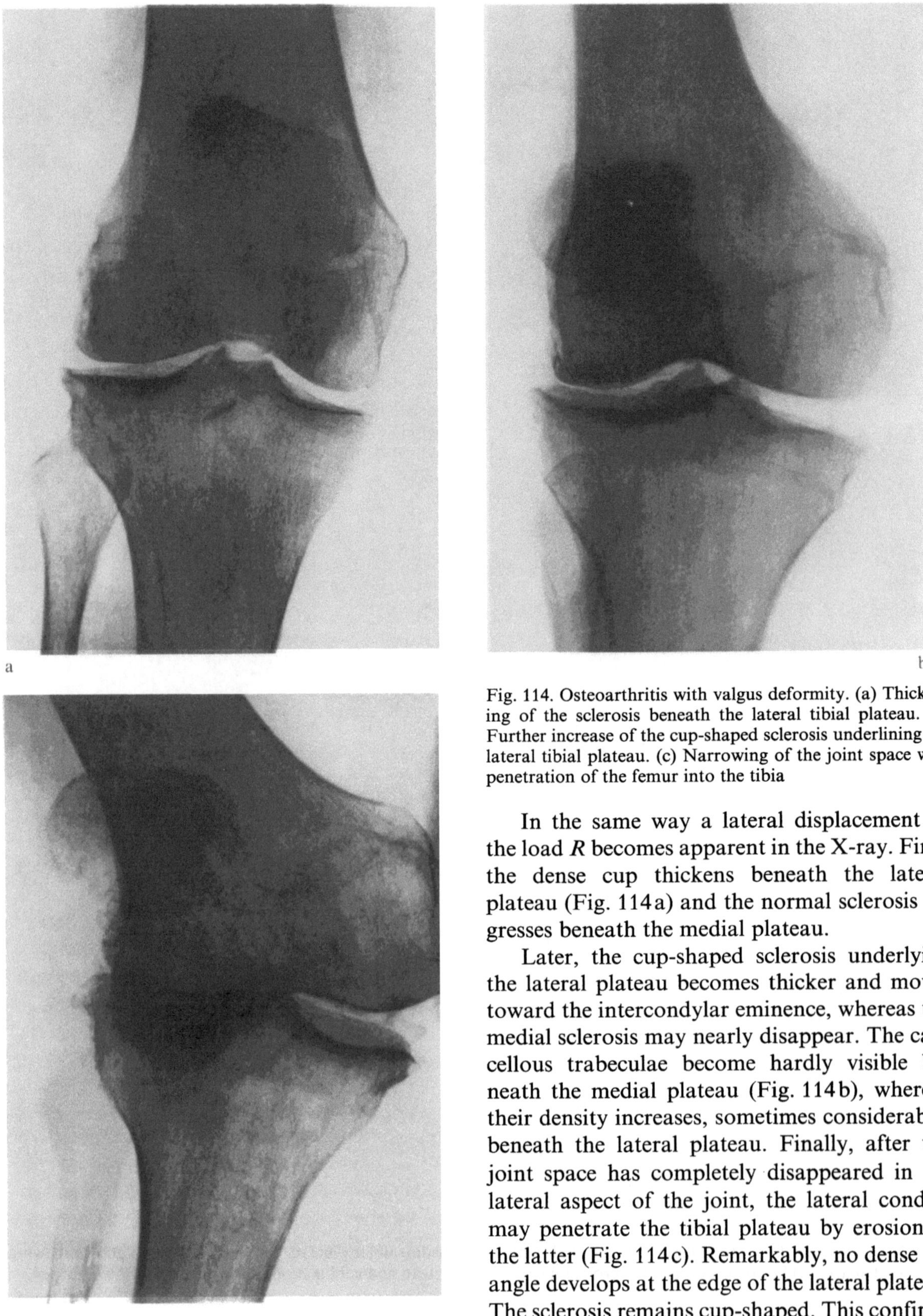

a

b

c

Fig. 114. Osteoarthritis with valgus deformity. (a) Thickening of the sclerosis beneath the lateral tibial plateau. (b) Further increase of the cup-shaped sclerosis underlining the lateral tibial plateau. (c) Narrowing of the joint space with penetration of the femur into the tibia

In the same way a lateral displacement of the load *R* becomes apparent in the X-ray. First, the dense cup thickens beneath the lateral plateau (Fig. 114a) and the normal sclerosis regresses beneath the medial plateau.

Later, the cup-shaped sclerosis underlying the lateral plateau becomes thicker and moves toward the intercondylar eminence, whereas the medial sclerosis may nearly disappear. The cancellous trabeculae become hardly visible beneath the medial plateau (Fig. 114b), whereas their density increases, sometimes considerably, beneath the lateral plateau. Finally, after the joint space has completely disappeared in the lateral aspect of the joint, the lateral condyle may penetrate the tibial plateau by erosion of the latter (Fig. 114c). Remarkably, no dense triangle develops at the edge of the lateral plateau. The sclerosis remains cup-shaped. This confirms our geometrical analysis of the forces acting on the valgus knee (Fig. 80).

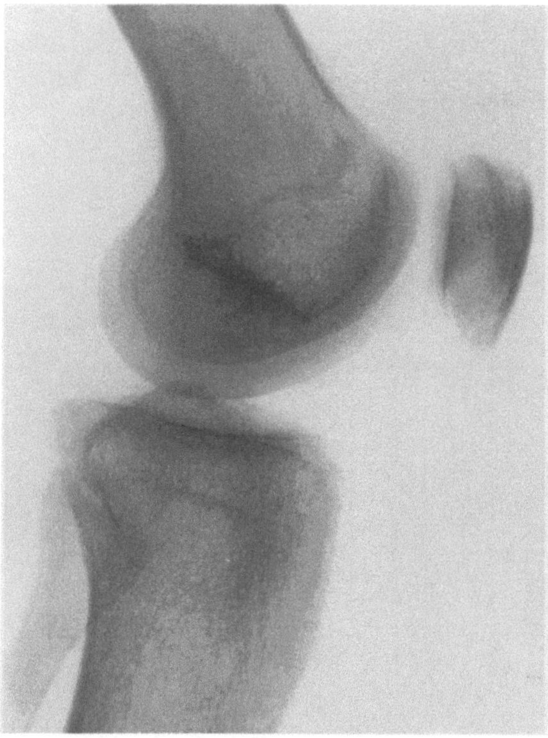

a

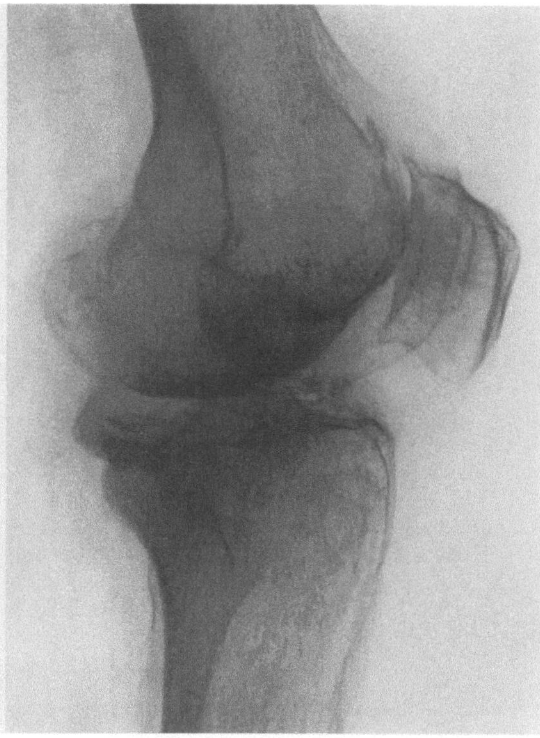

b

Fig. 115. (a) Subchondral sclerosis of even thickness throughout underlining the tibial plateaux of a normal knee. (b) Osteoarthritis with flexum of the knee. A dense triangle underlines the posterior part of the joint. (c) Osteoarthritis with recurvatum of the knee. A dense triangle underlines the anterior part of the plateaux

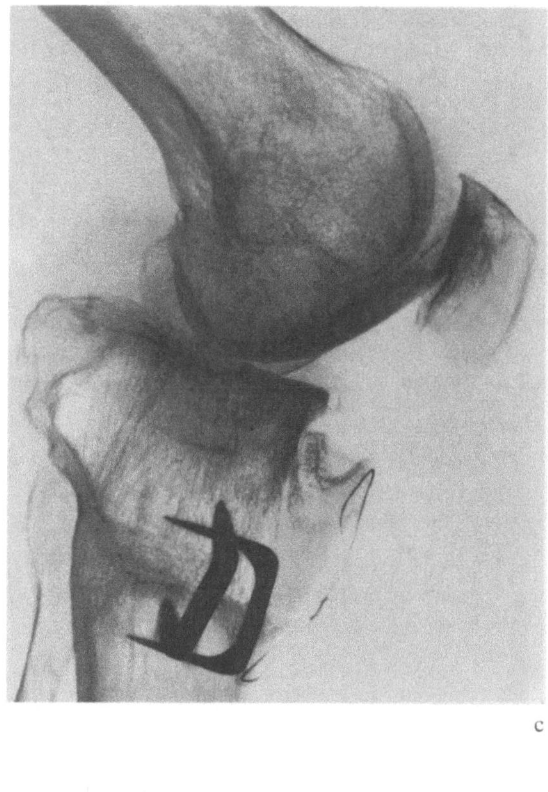

c

2. Lateral View

In the lateral view, the subchondral scleroses of even thickness throughout underlining the normal tibial plateaux are superimposed (Fig. 115a). In osteoarthritis a dense triangle appears beneath the posterior part of the joint. Again this triangle represents the diagram of the stresses considerably and unevenly increased in the posterior part of the knee (Fig. 115b). In a case of postoperative osteoarthritis with recurvatum of the joint, a dense triangle was observed beneath the anterior part of the plateaux (Fig. 115c). It demonstrates the increase of the joint pressure in this part of the knee (see pages 106 and 246–247).

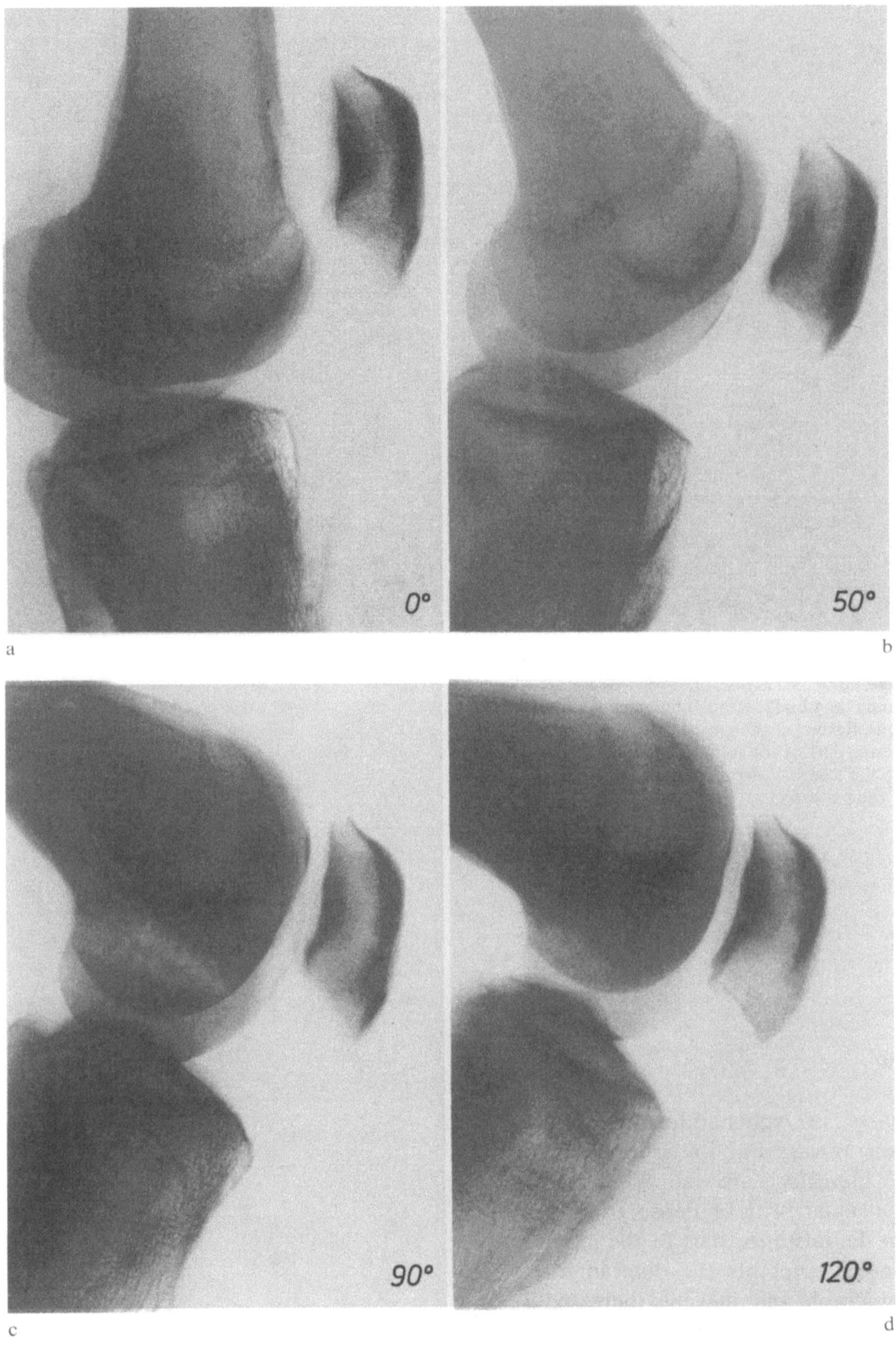

Fig. 116a–d. Variations of the patello-femoral force-transmitting surfaces during flexion of the knee

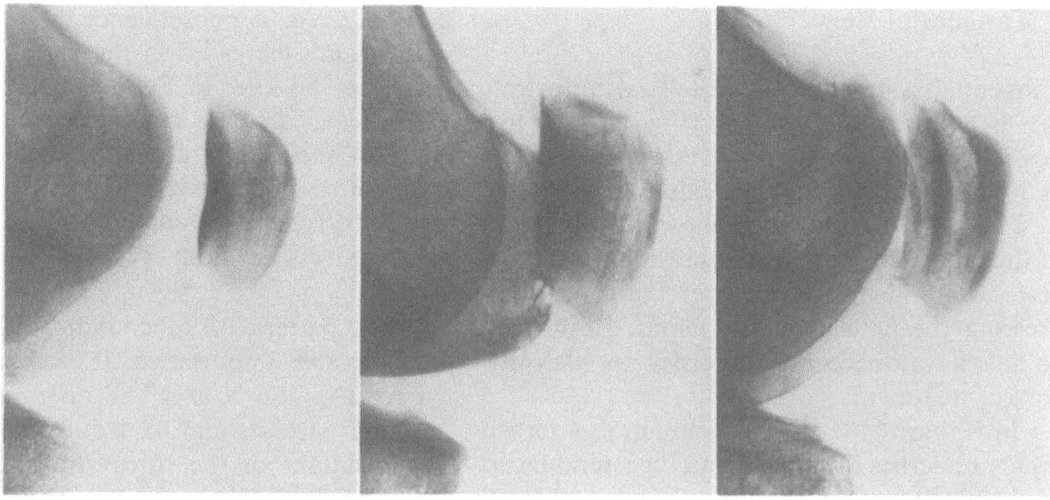

Fig. 117. (a) Patella of a normal knee. (b) Thick subchondral sclerosis in patello-femoral osteoarthritis. (c) Thickening of the anterior cortex of the knee-cap

The lateral view also shows how the patello-femoral contact surfaces change during flexion. Practically nonexistent in full extension (Fig. 116a), they are located at the middle facet of the patella at a flexion of 50° (Fig. 116b), at its upper facet near 90° (Fig. 116c), and simultaneously at both upper and middle facets at a more pronounced position of flexion (Fig. 116d).

The subchondral part of the patella physiologically presents a thin dense ribbon corresponding to the stress diagram of the patello-femoral joint (Fig. 117a). This dense bone can increase considerably in thickness and take the form of a deep cup in cases of patello-femoral osteoarthritis (Fig. 117b). We have already seen why. The anterior cortex of the patella may also become thickened (Fig. 117c). These changes result from the increase of tensile and compressive stresses in the bone. The stresses are mainly compressive in the posterior aspect of the patella, due to the compressive force R_5 pressing the knee-cap against the femur. They are mostly tensile in the anterior aspect of the bone, resulting from the opposite pulls of the quadriceps tendon and of the patella tendon. The patella is actually stressed in bending (Tillmann and Brade, 1980).

3. Tangential View

Tangential X-rays of the patella demonstrate the patello-femoral groove (Fig. 118). When possible the X-rays should be taken successively at 30°, 60° and 90° of knee flexion, as suggested by Ficat (1970). They may show a subluxation of the patella or a trabecular pattern which suggests a lateral displacement of the force compressing the patella on the femur. Both must be taken into account in order to plan the correct treatment.

In Figure 118a the subchondral sclerosis evenly underlies the medial and the lateral facets of the patella in a normal patello-femoral joint. In Figure 118b this sclerosis is thicker beneath the lateral facet and demonstrates that the patello-femoral compressive force is displaced laterally. As a consequence, the compressive stresses are increased in the lateral part of the joint. In the knee in Figure 118c the compressive stresses are even greater between the lateral facet of the patella and the lateral condyle of the femur. The cartilage is destroyed. In Figure 118d functional adaptation has ended. Excessive stress has destroyed the bone.

The pattern of subchondral sclerosis is of considerable importance. It makes it possible to read in the X-rays the distribution of the articular stresses and to deduce the position of the resultant of the forces transmitted across the joint. It allows one to time surgical intervention correctly and to appreciate its results postoperatively.

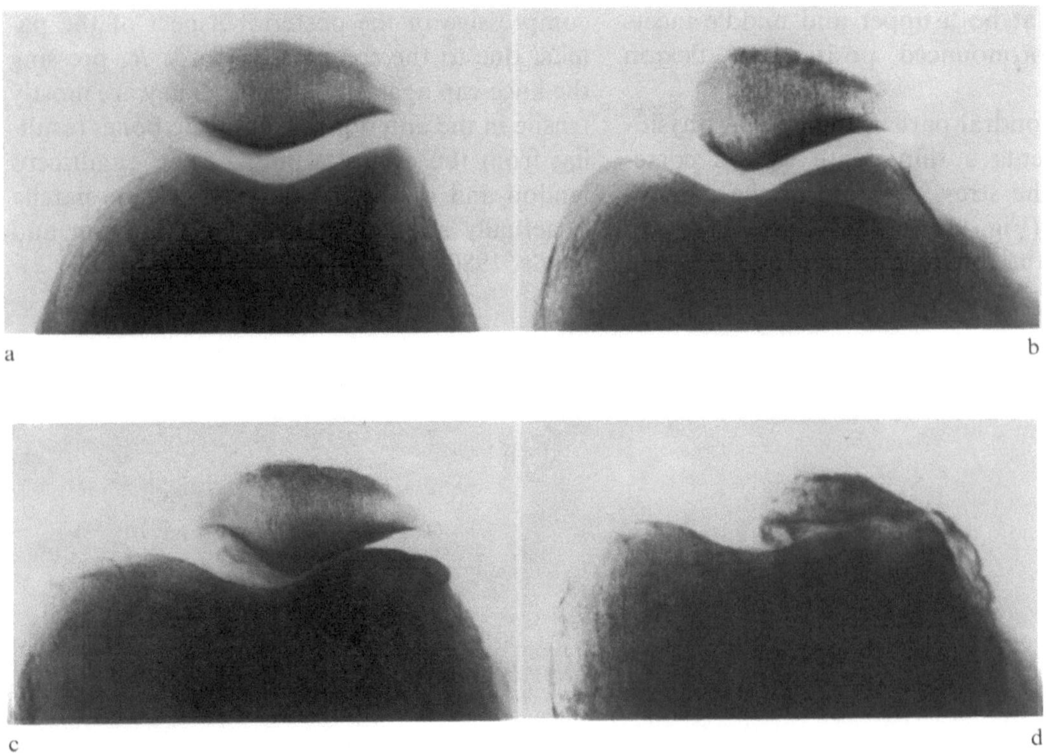

Fig. 118. (a) Normal patello-femoral joint. (b) Thicker sclerosis beneath the lateral facet. (c) Narrowing of the lateral joint space and thicker sclerosis beneath the lateral facet. (d) Destruction of the bone

B. Utility of X-Rays in the Standing Position

Only a radiological examination of the loaded knee gives a good idea of the conditions during gait, when the joint is stressed functionally. The appearance of the same knee is often very different in a lying position, with all the muscles at rest, than when standing with full weight on the limb. For instance, in the lying patient represented in Figure 119a, the X-ray shows a dense triangle beneath the medial tibial plateau but with an adequate joint space. The varus deformity does not look important. In the standing position, however (Fig. 119b), the medial joint space completely disappears; the varus deformity is much more pronounced.

The *A.-P.* X-ray of the loaded knee, which we have recommended since 1963, gives more information than the usual recumbent picture. But in order to make a satisfactorly theoretical analysis and to determine the angle formed by the femur and the tibia in the coronal plane, another radiograph, which shows the whole limb, is necessary (de Marchin et al., 1963a; Duparc and Massare, 1967; Maquet et al., 1967a, b; Ramadier, 1967). The leg is X-rayed with the patient standing and putting his whole weight on it. The opposite foot may touch the floor for equilibrium but without exerting any force. The A.-P. view is taken on a long film, 90 cm × 30 cm or even 120 cm × 30 cm, from a distance of at least 3 m. The image shows the whole leg and part of the pelvis. The centres of the hip, the knee and the ankle normally lie

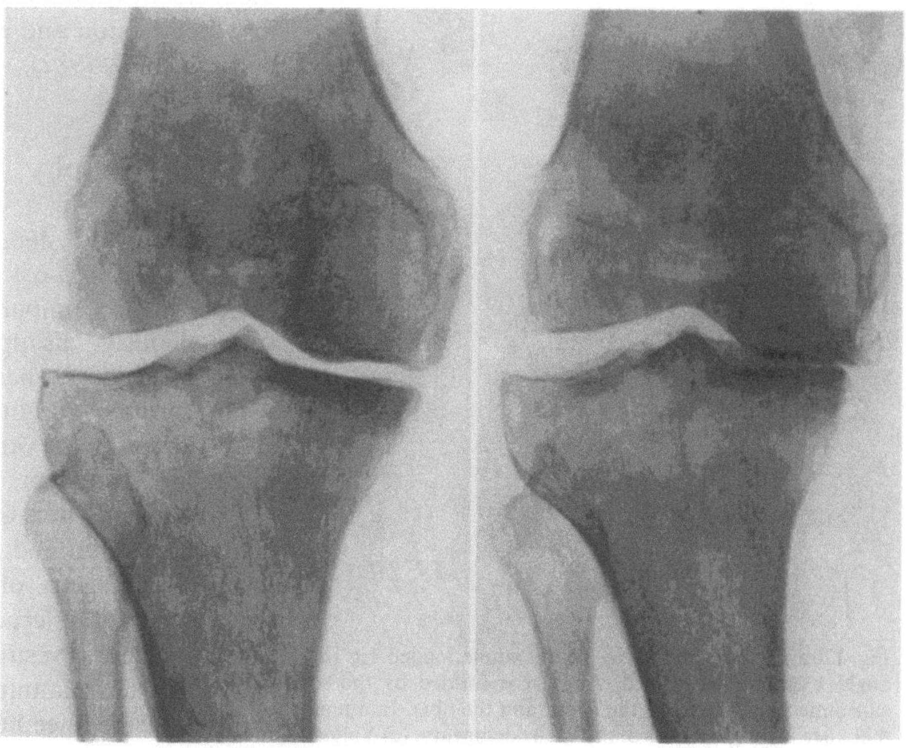

Fig. 119a and b. X-ray of the knee. (a) Patient lying. (b) Patient standing

on a straight line (Fig. 120a). When the tibial spines appear lateral to the line joining the centres of the hip and of the ankle, the leg is deformed in varum. When they appear medial to this line, the leg is deformed in valgum. The deformity must be measured exactly. When planning surgical treatment, a line drawn from the centre of the femoral head to the midpoint of the cross section at the level of the osteotomy (femoral or tibial) defines the axis of the thigh. Another line drawn from the midpoint of the cross section at the level of the osteotomy to the centre of the ankle represents the axis of the lower leg. For practical purposes the angle α which they form measures the varus (Fig. 120b) or valgus (Fig. 120c) deformity. The angle α is thus measured at the level of the osteotomy in order to plan the latter accurately.

The actual deformity corresponds to the angle formed between (a) the line joining the centre of the femoral head and a point between the tibial spines a little closer to the medial than to the lateral and (b) the line joining this point between the tibial spines and the centre of the ankle joint. This measurement is used after surgery.

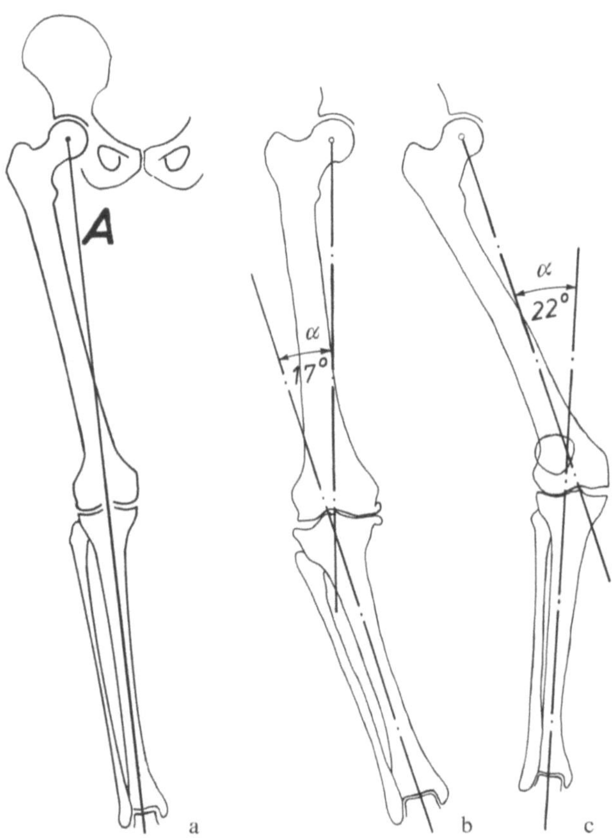

Fig. 120a–c. From the X-ray of the whole loaded leg the angle α can be determined. Angle α is formed by the so called mechanical axes of the femur and the tibia. In a normal knee (a) angle α is zero. (b) Varus deformity. (c) Valgus deformity

The inherent problem with these long X-rays consists in positioning the patient properly to obtain a true A.-P. view of the deformity. Delouvrier et al. (1981) describe a method of defining the lower limb deformity and the plane in which it appears at its maximum. The deformity thus defined combines flexion or recurvatum, valgum or varum, and axial rotation. However, what we need is to measure the deformity in the coronal plane, since we carry out its correction in this plane and use other means for eliminating flexion contracture.

Ramadier et al. (1982) have suggested taking first a lateral view in which the posterior outlines of the two femoral condyles are superimposed exactly and then the A.-P. view precisely at right angles to the lateral view. We recommend this technique.

Flexion contracture and laxity of the knee certainly constitute sources of error which remain unavoidable. We try to minimize them by positioning the knee to be X-rayed in such a way that its plane of function is parallel to the incident beam. The plane of function is that in which flexion and extension take place (Van de Berg et al., 1982).

C. Arthrography

Arthrography to demonstrate the menisci can be useful but in most cases the meniscus of the osteoarthritic compartment of the knee has spontaneously disappeared, crushed by the overcompression it has sustained.

In proper conditions, arthrography can show fissures in the cartilage of the patella (Maldague and Malghem, 1976, 1978). Van de Berg (1976) currently uses tomography after injecting contrast medium and air into the knee. He can show the contours of the articular cartilages, the menisci, the cruciate ligaments and the capsule. These investigations are not requested routinely when planning the treatment of osteoarthritis of the knee but they are useful in some specific cases.

D. Computerized Axial Tomography

Computerized axial tomography should provide reliable information on the rotation deformities of the segments of the lower limbs. Malrotations may well appear as a frequent cause of osteoarthritis of the knee. Computerized axial tomography may then indicate the appropriate treatment, although so far surgical derotations about the knee seem to have rather unpredictable results. More time and experience are needed before essential conclusions on this subject can be drawn. We have not used computerized axial tomography systematically for femoro-tibial osteoarthritis, but it seems most useful in demonstrating the position of the patella in or near full extension of the knee (Despontin and Thomas, 1978) and in showing the structure of the bone in the transverse plane.

In a normal knee (Fig. 121), the anterior articular surface of the femoral condyles appears underlined by a subchondral sclerosis of even thickness throughout. A bundle of cancellous trabeculae lies at right angles to this subchondral sclerosis in each condyle and runs backward and medially in the medial condyle and backward and laterally in the lateral condyle. Both of these bundles cross the bone and end at right angles to the posterior aspect of the condyles. They appear equal in density and they surround a clear, roughly triangular space.

The first stage of osteoarthritis of the patello-femoral joint is often called chondromalacia because the standard X-rays look normal. However, in computerized axial tomography the bundle of cancellous trabeculae may appear much denser in the lateral than in the medial condyle (Fig. 122). This corresponds to an increase of the compressive stresses in the lateral aspect of the joint which is not yet seen in the standard X-rays.

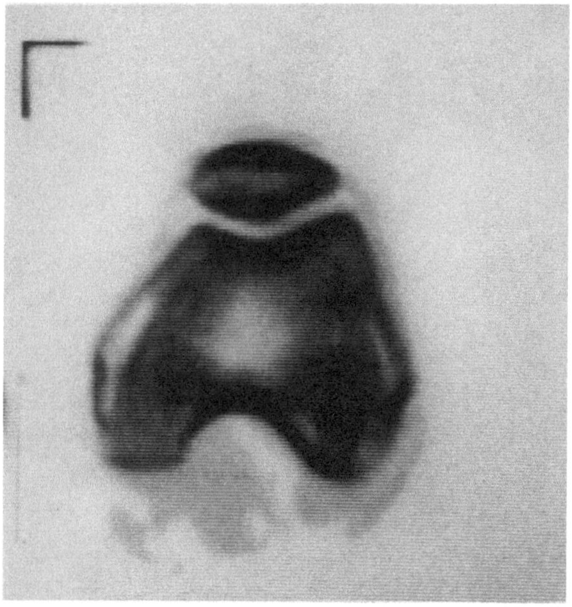

Fig. 121. Computerized axial tomography of a normal knee. Two bundles of trabeculae of even density run from the anterior aspect of the condyles to their posterior aspect. (Collection A. Van de Berg)

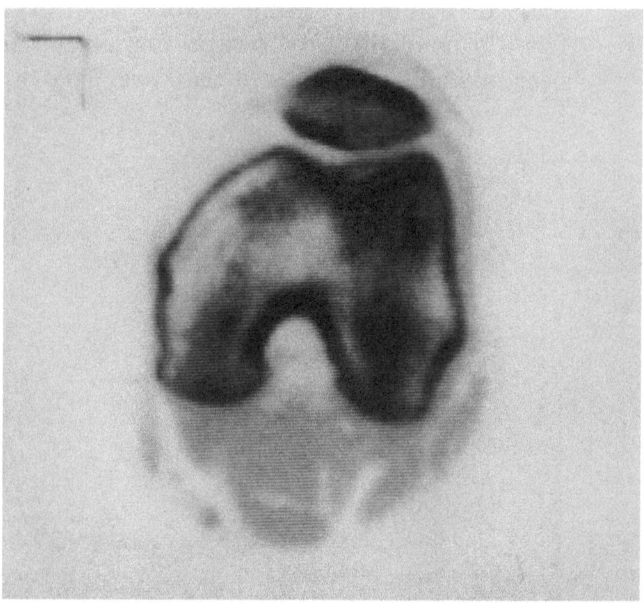

Fig. 122. Computerized axial tomography of a knee with so-called chondromalacia patellae. The lateral bundle of trabeculae is much more pronounced than the medial. This is not apparent in the standard X-rays. (Collection A. Van de Berg)

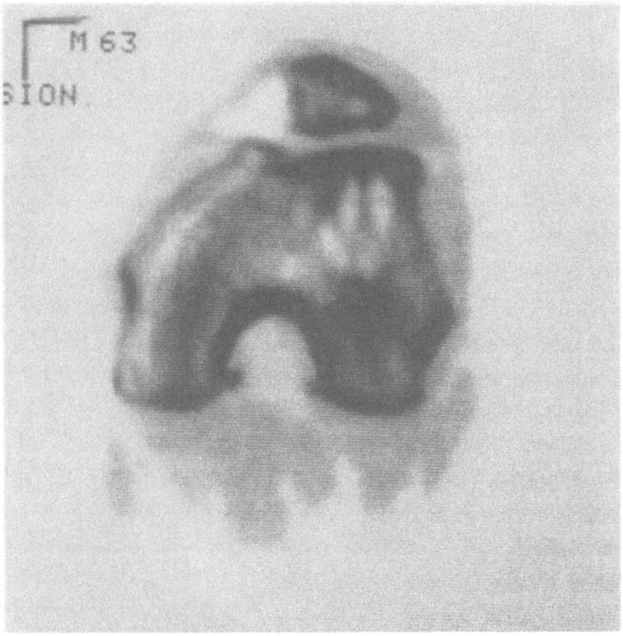

Fig. 123. Computerized axial tomography of a knee with dysplasia of the patella and patello-femoral osteoarthritis. Cysts exist in the patella and in the lateral condyle. They do not appear in the standard X-rays. (Collection A. Van de Berg)

In a more advanced stage of patello-femoral osteoarthritis, large cysts develop in the areas of overpressure in the lateral condyle and in the patella, exactly as they do in the roof of the acetabulum and in the femoral head in osteoarthritis of the hip. They do not usually appear in standard X-rays of the knee. That was the case for the knee the computerized axial tomography of which is presented in Figure 123. Moreover, the medial bundle of trabeculae fades away, corresponding to a decrease of the stresses in this area.

Figures 121, 122 and 123 represent cross sections at the same anatomical level in relation to the femoro-tibial joint space.

In summary, in order correctly to analyse the actual state of an osteoarthritic knee and to plan treatment, the following X-rays are necessary:

(1) *A.-P.* view of the knee, the patient standing on the affected limb; (2) lateral view; (3) X-rays of the patello-femoral groove; (4) X-ray showing the whole lower limb while it sustains the body weight.

Computerized axial tomography can be very useful, mainly in patello-femoral osteoarthritis.

Such a radiological examination makes deduction of the mechanical stress in the knee possible. It makes the chondral and bony lesions apparent and shows evidence of the abnormal mechanical stress due to changes in the equilibrium of the forces acting on the knee, particularly the permanent displacements of contact forces R between the femur and the tibia and R_5 between the patella and the femur.

The changes of mechanical stress in the joint due to displacements of R or R_5 can also be illustrated by photoelastic models.

The reader who does not wish to read about physics and laboratory experiments should turn to page 131.

III. The Use of Photoelastic Models to Illustrate How the Position of Compressive Femoro-Tibial and Patello-Femoral Forces Affects the Distribution of Articular Stresses

A. Femoro-Tibial Joint

Displacement of the line of action of force *R* medially or laterally causes an asymmetrical distribution and a concentration of the compressive stresses in the joint. It is possible to illustrate these changes by analysing photoelastic models of the knee. This is the way we proceed. A plane model is cut from a sheet of plexiglass 1 cm thick to show the isoclinics, a second one with the same dimensions of araldite to show the isochromatics and a third one of lucite to show the isopachics. Each model represents a cross section of the knee in a coronal plane. Its outlines correspond exactly to the outlines of an anatomical specimen. It consists of two parts: the upper one, the femur, is supported by the lower one, the tibia, through rubber sheets which represent the articular cartilage and menisci and distribute the forces. The load exerted on the model represents the projection on a coronal plane of the resultant *R* exerted on the knee. It is applied on each model successively in four different ways:

(1) normal load, centred; (2) normal load off centre; (3) inclined load, centred; (4) inclined load, off centre.

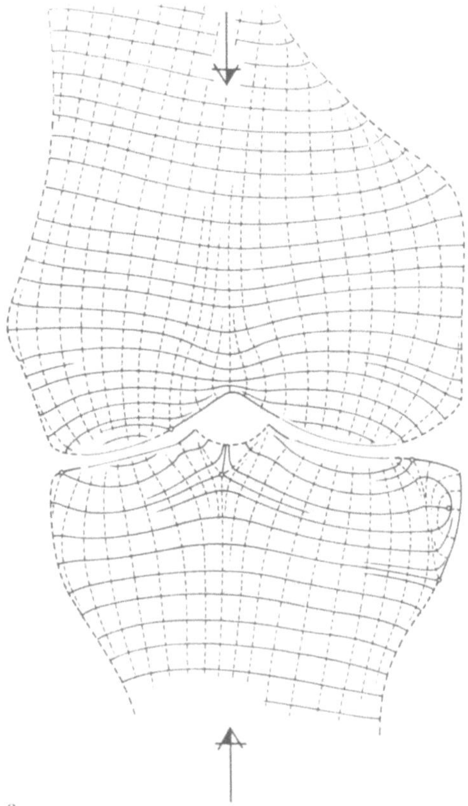
a

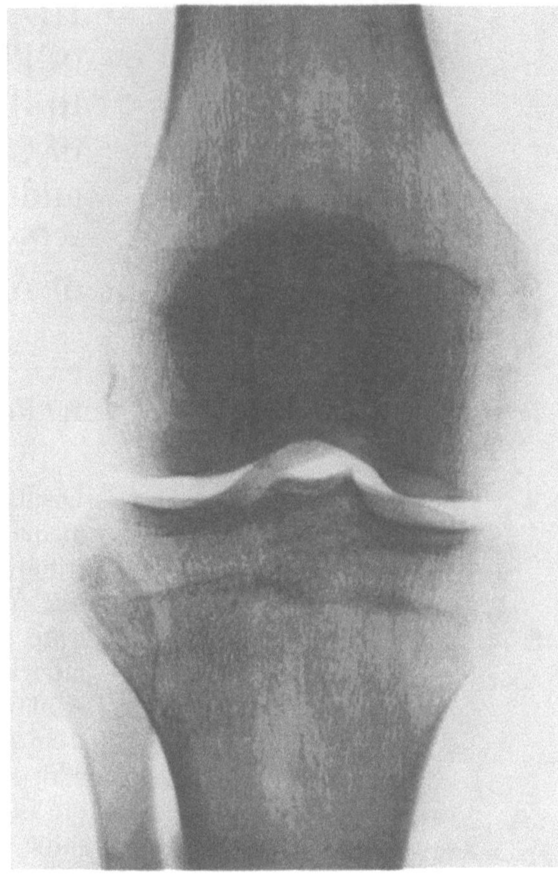
b

Fig. 124. (a) Pattern of isostatics drawn from the isoclinics appearing in a photoelastic model of the knee when subjected to a centred load, normal to the tibial plateaux. (b) X-ray of a normal knee

1. Normal Load, Centred

In this experiment the load is normal to the plane of the tibial plateaux. It is exerted through the centre of gravity of the weight-bearing surfaces.

In a first qualitative study the isoclinics are observed at every 5° of rotation of the model. From these it is possible to build the pattern of isostatics (Fig. 124a). Roughly, the dotted lines indicate the trajectories of the principal compressive stresses, the solid lines the direction of the principal tensile stresses.

All the compressive isostatics leave the weight-bearing surfaces of the joint and are practically at right angles to them. They then have a course roughly parallel to the long axes of the femur and the tibia. The tensile isostatics intersect the compressive ones at right angles and intersect the two pieces of the model transversely.

These trajectories of compression and of tension can be superimposed on the cancellous trabecular structure in the distal femoral and proximal tibial areas (Fig. 124b). A pattern of longitudinal trabeculae exists which become perpendicular to the joint surfaces as they approach them. These trabeculae are stressed in compression. Other trabeculae intersects the first ones transversely. They correspond to the isostatics of tension.

The quantitative study relies on isochromatics (Fig. 126a) which appear in a transverse cross section in the immediate vicinity of the joint surfaces of the femoral condyles and tibial plateaux. Isochromatics attain there the order of 2 and 3. Isochromatics illustrate the relative distribution of the stresses between both tibial plateaux and femoral condyles. We observe that the compressive stresses are distributed on both halves (right and left) of the model symmetri-

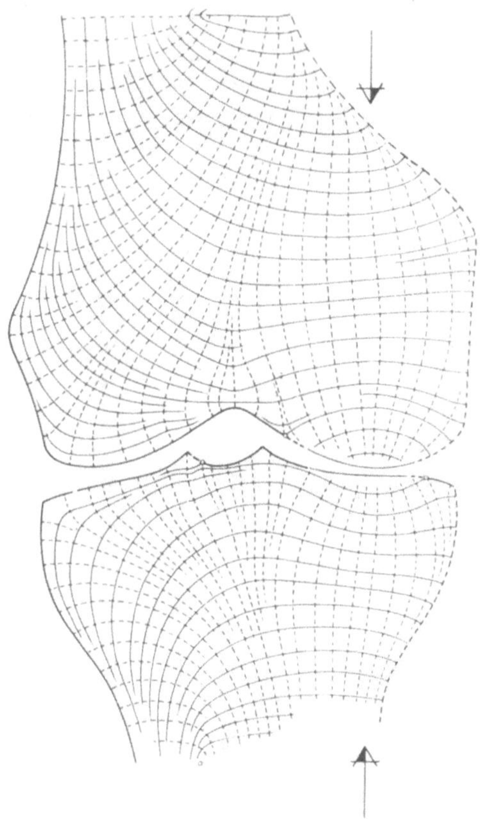

a

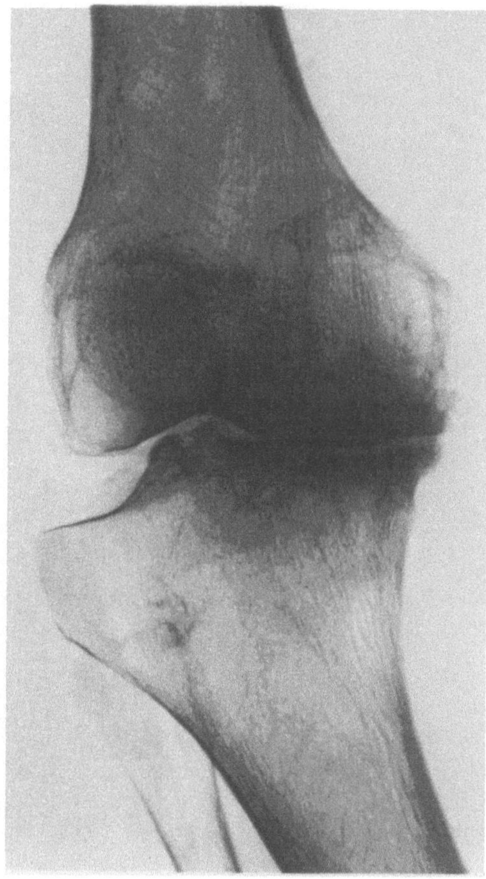

b

Fig. 125. (a) Pattern of isostatics in a model of the knee subjected to an eccentric load normal to the tibial plateaux. (b) X-ray of a knee eccentrically loaded

cally relative to the load (Figs. 126a and 127).[10] The quantitative distribution of the stresses on both tibial plateaux corresponds to what appears in the X-ray of a normal knee. Each tibial plateau is underlined by a cup-shaped sclerosis of even thickness (Fig. 124b).

[10] The magnitude of the stresses is not exactly the same in the femoral as in the tibial part of the model because there is necessarily some joint friction in this type of model. Shearing stresses appear. The resultant of the compressive and of the shearing stresses is the same in the femur and in the tibia

2. Normal Load, Off Centre

In this experiment the load is also normal to the tibial plateaux but it does not act at the centre of gravity of the weight-bearing surfaces. It is displaced to the right (Fig. 125a). Therefore, it acts off centre on the femur and the tibia of the model.

In the pseudo-femur of the model, the isostatics of compression come down from the right diaphysial area and arch toward the weight-bearing surfaces of the joint and also toward the left metaphysial area, which is extra-articular. The isostatics of tension pass downward tangentially to the left border of the femur and arch gently to intersect the isostatics of compression at right angles. Isostatics of compression and isostatics of tension are bow-shaped in the left part of the femur. They form gothic arches with their apices directed downward.

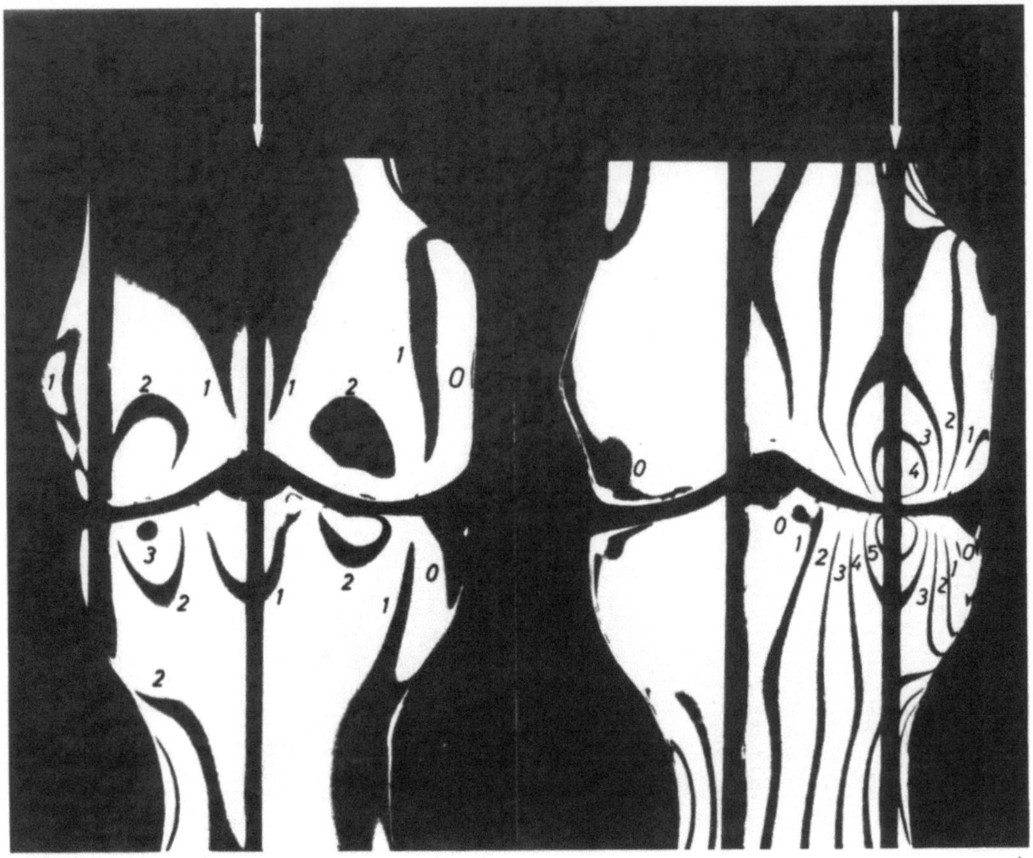

a b

Fig. 126. (a) Isochromatics appearing in a photoelastic model supporting a centred load. (b) Model supporting an eccentric load

A similar disposition is found in the tibia of the model. The isostatics of compression arise not only from the articular surfaces but also from the left extra-articular outline. The isostatics of tension are tangential to the left outline and bend to become tangential to the weight-bearing surfaces. Isostatics of compression and of tension form gothic arches in the left part of the tibia with their apices directed upward. It is the characteristic feature of bending stresses superimposed on shearing stresses. It corresponds to what appears in the X-ray of the lower end of the femur and of the upper end of the tibia when subjected to bending and shearing stress.

It is important to emphasize the similarities between the radiological appearance and the pattern of isostatics deduced from the model. The X-ray taken as an example represents an osteoarthritic knee with a severe varus deformity of both femur and tibia (Fig. 125b. See also Fig. 184). These bones are loaded eccentrically. The compressive force acts on their metaphysis not only in bending but also with a shearing component because of the obliquity of the axis of the metaphysis in relation to the line of action of the load. The medial part of the joint itself is stressed in compression. The bow-shaped arrangement of the bony trabeculae is particularly obvious in the femur but is also present in the tibia. It corresponds to the gothic arches formed by the isostatics in the model when eccentrically loaded and thus subjected to bending and shearing stress (Pauwels, 1954b).

Due to the effect of the load displaced to the right, isochromatics appear in much larger number in the right part of the model. They attain the order of 6 (Fig. 126b). The same load, well centred, produced only isochromatics of the order of 3 (Fig. 126a). On the contrary the only isochromatic in the left part of the model is of the order of 0. The X-ray shows a large

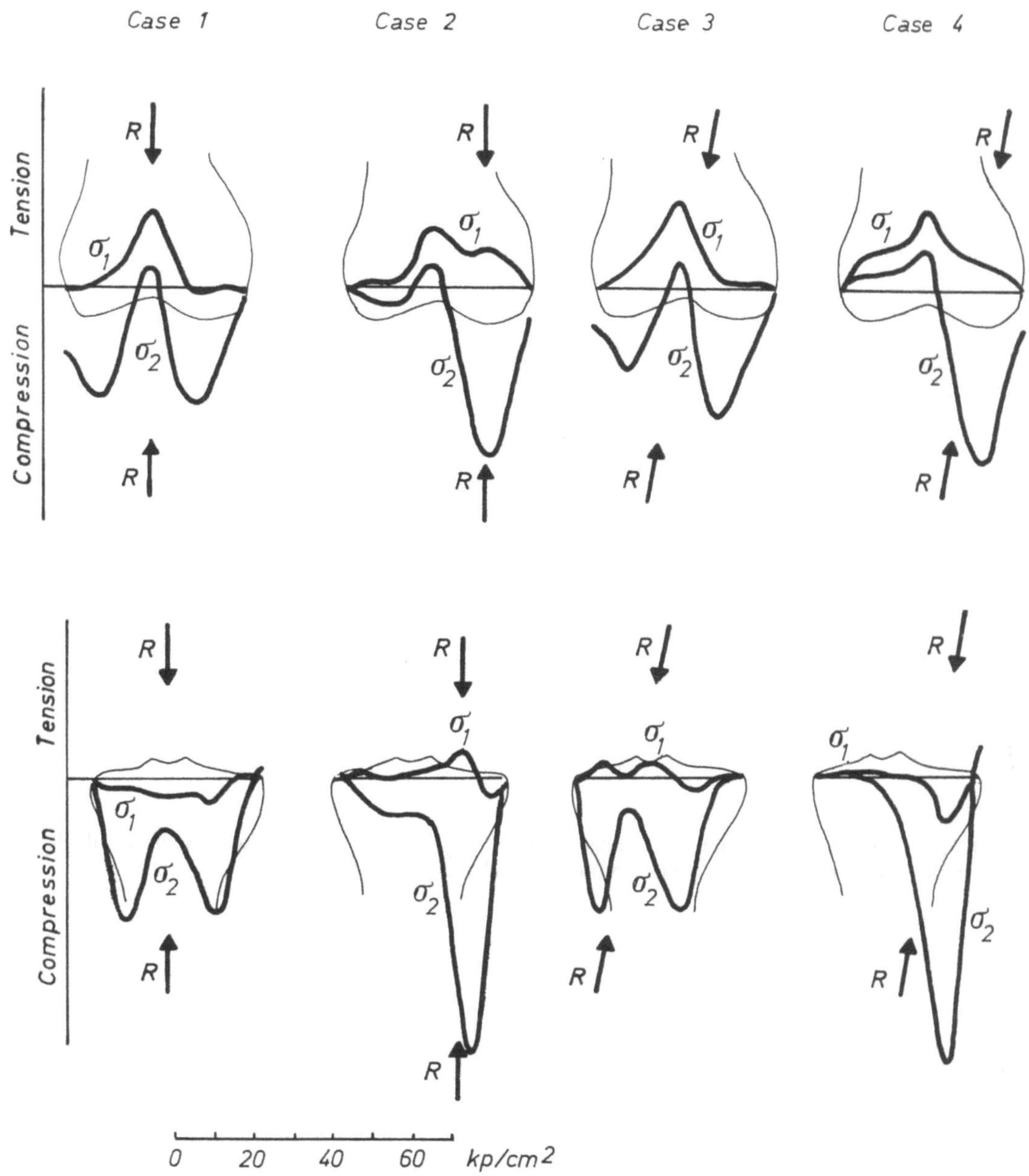

Fig. 127. Quantitative distribution of stresses in the lower end of the femur (above) and in the upper end of the tibia (below)

dense area beneath the medial tibial plateau and in the medial femoral condyle. The bony trabeculae can no longer be seen under the lateral plateau. The eccentric load provokes high and asymmetrically distributed stresses. The maximum stresses arise in the part of the joint toward which the load has been displaced.

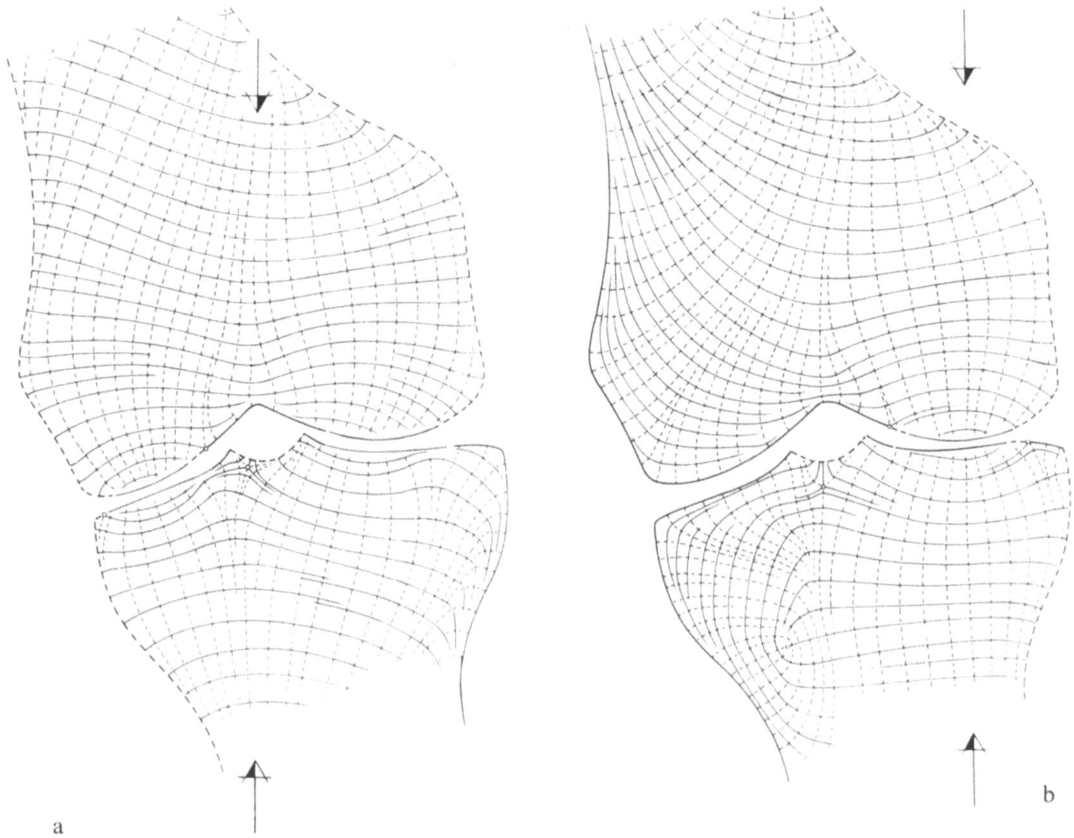

Fig. 128. (a) Pattern of isostatics in a model subjected to a centred load, oblique to the tibial plateaux. (b) Pattern of isostatics in a model subjected to an eccentric load, oblique to the tibial plateaux

3. Inclined Load, Centred

The load is inclined at 10° in relation to the plane tangential to the tibial plateaux. It is exerted at the centre of gravity of the weigh-bearing surfaces. The inclination of the load does not significantly modify the direction of the principal stresses in the vicinity of the joint (Fig. 128a). It also brings about very little change in the quantitative distribution of the articular stresses in both halves of the model as shown by the pattern of isochromatics (Fig. 129a) and confirmed by the calculation based on the isochromatics and isopachics (Fig. 127).

4. Inclined Load, Off Centre

The load inclined to the tibial plateaux is displaced to the right. Compared with a normal and eccentric load, the isostatics showing the direction of the stresses are not changed much in the immediate vicinity of the joint (Fig. 128b). The quantitative distribution of the stresses between both halves of the model is also very similar to the second experiment (Figs. 129b and 127).

In summary, a centred load provokes compressive and tensile trajectories in the model. These trajectories correspond to the cancellous trabeculae seen in the X-rays. An eccentric position of the load fundamentally changes the isostatics. The trajectories of compression and of tension in an eccentrically loaded model correspond to the cancellous trabeculae of an osteoarthritic knee with a varus deformity, eccentrically loaded. Such a correlation between isostatics and bone trabeculae confirms the trajectorial architecture of cancellous bone (Pauwels, 1965b, 1973b).

When off centre, the load causes a greater number of isochromatics in the part of the joint toward which it has been displaced. An eccentrically placed load produces very high and concentrated compressive stresses in the part of the joint toward which the load has been shifted. A slight obliquity of the articular surfaces in relation to the line of action of the load has little effect on the distribution of the compressive stresses in the joint of the model.

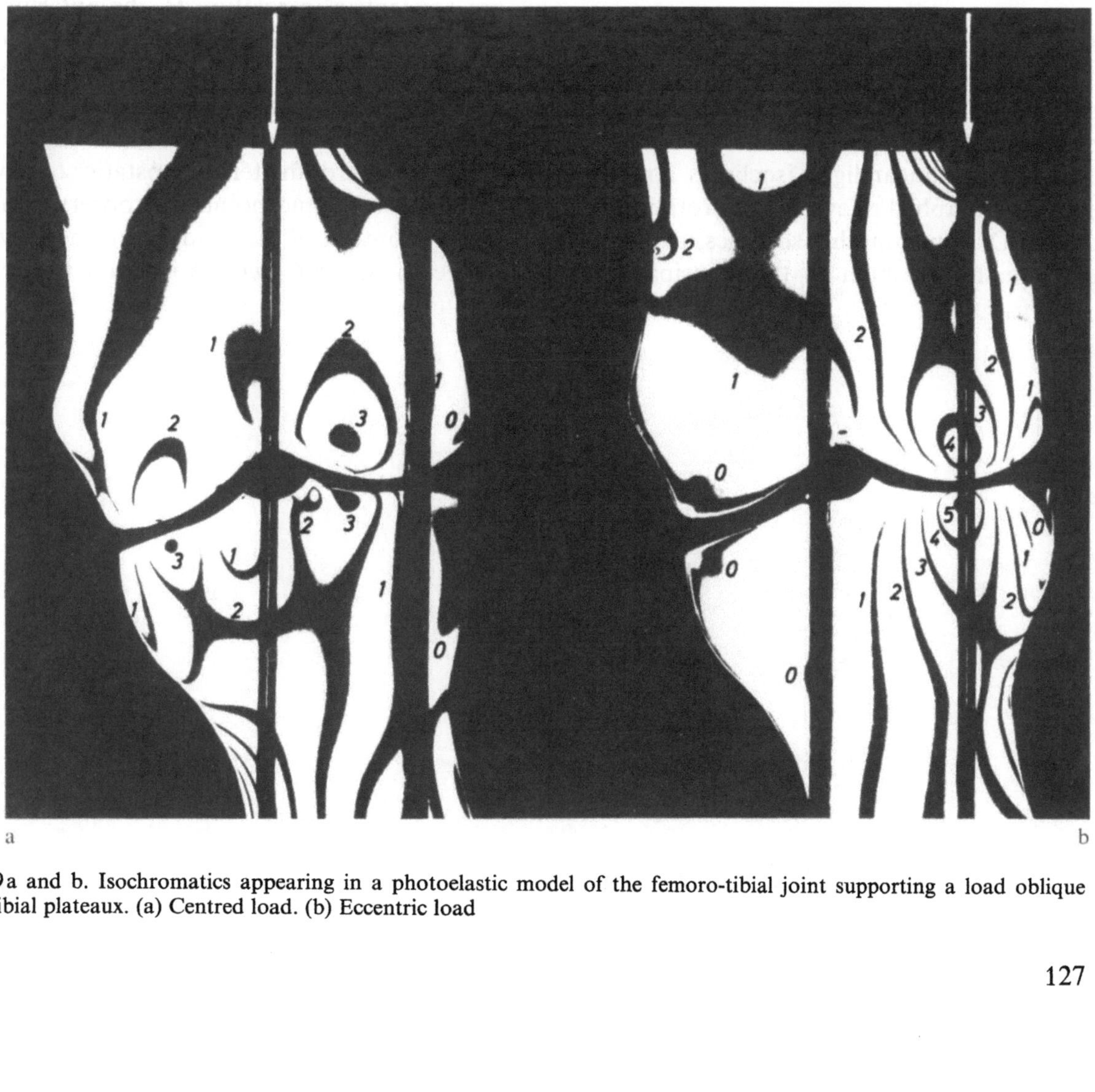

Fig. 129a and b. Isochromatics appearing in a photoelastic model of the femoro-tibial joint supporting a load oblique to the tibial plateaux. (a) Centred load. (b) Eccentric load

B. Patello-Femoral Joint

Plane models have been cut with the same outline as a transverse cross section of the knee at the level of the femoral condyles and kneecap. Each condyle rests with its posterior part on a wooden support covered by a layer of rubber. The load applied to the anterior aspect of the patella is transmitted to the femur through another layer of rubber. It represents the projection of the patello-femoral compressive force R_5 on a transverse plane. The rubber washers represent the cartilage and distribute the pressure over the parts of the model as the cartilage distributes it over the bones.

When loaded, a model of araldite will show isochromatics, a model of plexiglass isoclinics. From the isoclinics the pattern of isostatics will be drawn.

1. Directional Distribution of the Stresses

The model of plexiglass is first loaded. The line of action of the load corresponds to that of the patello-femoral compressive force exerted in a normal knee as determined above (page 108). In polarized linear light isoclinics appear and are photographed at every 5° of rotation of the model. Drawn from the isoclinics, the isostatics indicate the direction of the principal stresses (Fig. 130a). The compressive isostatics cross the patella and the femoral condyles from the front backward. The tensile isostatics transversely cross both bones. Isotropic points appear on the outline of the femoral piece of the model, two positive isotropic points near the patella, two negative in the posterior intercondylar space and two negative at the posterior limit of each condyle.

The pattern of isostatics (Fig. 130a) corresponds to the pattern of trabeculae appearing in the normal patella in X-rays of the patello-femoral groove (Fig. 118a). It can be fairly accurately superimposed on the pattern of cancellous bone visible in X-rays of a transverse cross section of a normal cadaveric knee (Fig. 131a). This correlation suggests a trajectorial architecture of the cancellous bone.

The load is then applied obliquely on the patella laterally subluxated (Fig. 130b). The compressive isostatics then converge toward the lateral condyle. The tensile isostatics intersect the compressive trajectories at right angles and are concentric in relation to the anterior limit of the lateral condyle. In the medial part of the patella there is an appearance of isotropic lines and the stresses fall practically to 0. The compressive isostatics cross the lateral condyle from behind forward, the tensile isostatics transversely. Three isotropic points are observed on the anterior outline of the condyle, one negative isotropic to the left, two positive on the contact

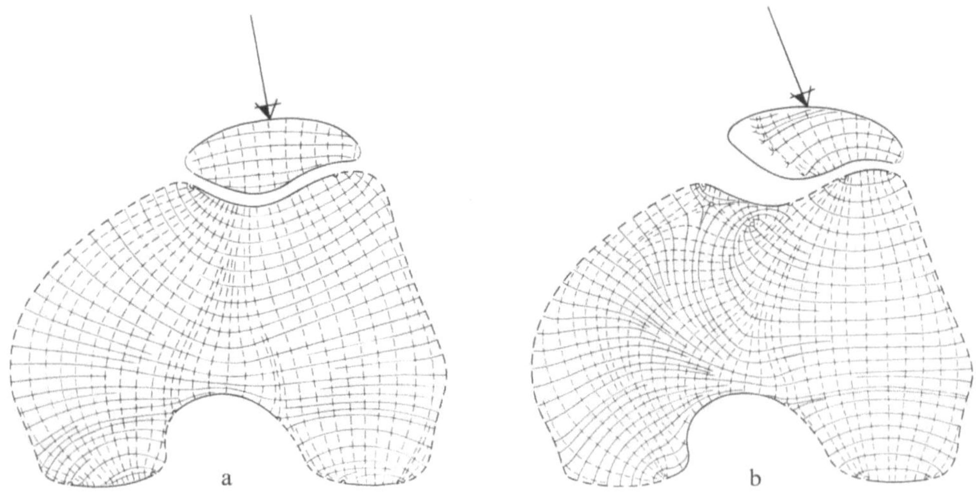

Fig. 130a and b. Isostatics in models of the patello-femoral joint. (a) Normal load. (b) Load on the lateral part of the joint only

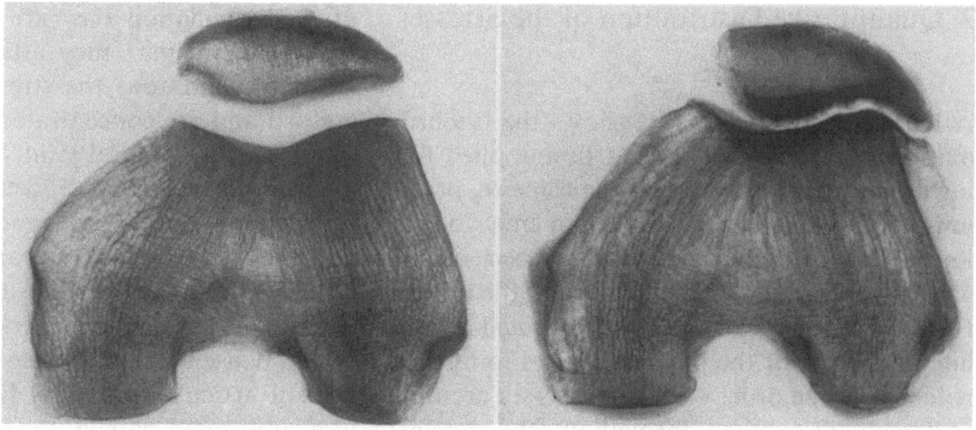

a b

Fig. 131. (a) Cross section of a normal knee. The pattern of trabeculae is very similar to the pattern of isostatics in the model Figure 130a. (b) Cross section of a knee with patello-femoral osteoarthritis

surface. Two more positive isotropic points exist on the posterior outline of the condyle and one negative in the posterior intercondylar groove. In the medial condyle, compressive and tensile isostatics build a much more complicated pattern with three isotropic points in the anterior part of the condyle: one negative point between two positive points. There is another positive isotropic point on the posterior outline of the condyle.

In a subluxated patella (Fig. 118b, c and Fig. 131b) the pattern of cancellous trabeculae corresponds to that of the istostatics of the model (Fig. 130b). The direction of the cancellous trabeculae is no longer antero-posterior as in a normal knee-cap (Fig. 131a) but oblique toward the lateral condyle.

Thus, in a transverse cross section of a normal knee, as well as in subluxation of the knee-cap, the correspondence between the pattern of the isostatics and that of the bony trabeculae confirms the trajectorial structure of cancellous bone, already demonstrated by Pauwels (1973a) in the hip.

2. Quantitative Distribution of the Stresses

A loaded model of araldite shows the isochromatics. A load of 225 kg is at first applied following the same direction as the force R_5 pressing the knee-cap against the femur in a normal knee. The isochromatics can be seen in the whole cross section of the femur and the patella (Fig. 132a). In the vicinity of the medial and lateral aspects of the patello-femoral joint they reach the same order of 4. The stresses are thus uniformly distributed throughout the patello-femoral joint.

The same load is then applied eccentrically on the patella subluxated laterally. The isochromatics are only present in the lateral condyle of the femur (Fig. 132b). In the knee-cap their distribution is not regular. In the medial part of the articulation they are of the order of 0. In the lateral part they attain the order of 9. In these conditions the stresses are greatly increased and are concentrated in the lateral part of the patello-femoral joint.

The correspondence between the distribution of the stresses in the model on the one hand and the radiographic appearance on the other is also striking here. The articular surface of a normal knee-cap is underlined by a thin ribbon of dense bone indicating a uniform distribution of articular pressure. On the other hand, in a subluxated patella, the subchondral sclerosis has the appearance of a thick cup localized opposite the lateral condyle where articular compressive stresses are concentrated. The subchondral dense bone tends to disappear beneath the medial facet of the knee-cap (Fig. 118).

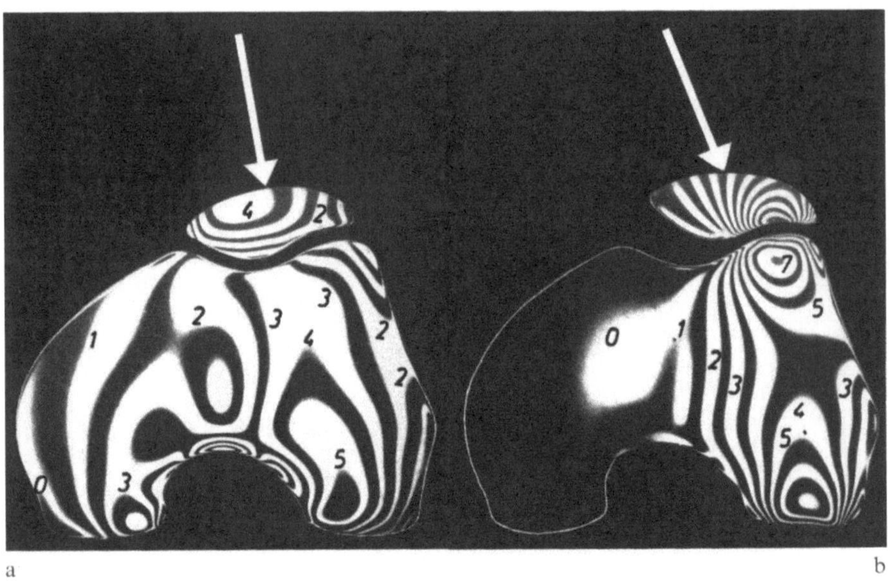

Fig. 132a and b. Isochromatics in a model of the patello-femoral joint. (a) Normal load. (b) Load on the lateral part of the joint only

IV. Osteoarthritis of the Knee of Mechanical Origin

In the first part of this chapter we have analysed how changing the system of forces acting on the knee can modify the femoro-tibial load R and the patello-femoral compressive force R_5. In the second and third parts we have shown that the changes produced by an abnormal distribution of the articular pressure, with augmentation and concentration of stresses in a part of the joint, can be seen in X-rays and in photoelastic models.

When the increase of articular stresses overtakes the capacity of resistance of bone and cartilage tissues, osteoarthritic lesions appear and develop. As a rule, the causes of displacement of the femoro-tibial compressive force R either medially, laterally, anteriorly or posteriorly and the causes of an increase or lateral displacement of the patello-femoral force R_5 are also the causes of osteoarthritis of the knee. More particularly, osteoarthritis with varus deformity is provoked by all the factors which displace the femoro-tibial compressive force R medially and osteoarthritis with valgus deformity by the factors which displace force R laterally. Patello-femoral osteoarthritis is produced by everything that increases the patello-femoral compressive force R_5 or causes it to be exerted eccentrically on the patella.

These different types of osteoarthritis, all of mechanical origin, have a common denominator: an abnormal distribution and a pathological increase of the joint compressive stresses in a part of the knee. The local overcompression provokes hypertrophic and atrophic reactions. The hypertrophic phenomena consist in the formation of osteophytes in the areas which are subjected to less compression and in an apposition of bone tissue in the area supporting the load. This apposition is made visible in X-rays by an increase of the subchondral sclerosis. On the other hand, the abnormally high pressure destroys the articular cartilage and, beyond a certain limit, provokes a resorption of bone and a deformity of the loaded tibial plateau and femoral condyle. These processes are accompanied by pain, a symptom the patient wants to be rid of.

Since osteoarthritis of the knee is finally characterized by joint pressure concentration in some area where it becomes too high, treatment of the disease necessitates redistribution and diminution of the pressure sufficiently to make it tolerable to the articular structures.

But before dealing with the surgical treatment, let us see how the patient by himself empirically tries to decrease the mechanical stress in the diseased knee.

Chapter VI. Instinctive Mechanisms Which Reduce Stress in the Knee

Pain in a lower limb usually provokes a symptomatic change in gait, which becomes asymmetrical, with displacement of the trunk toward the diseased limb during the single support period. The patient limps.

On the other hand, many osteoarthritic patients rely on a supplementary support, using a walking stick usually on the sound side. Some cannot go without its help.

Pauwels (1935) has explained the effects of limping and of the stick on the stress in the hip. We shall extend his results and show how, in a similar way, limping and the use of a stick diminish the load acting on the diseased knee.

I. Effects of Limping

When limping, the upper part of the trunk is displaced toward the diseased limb every time this limb is the only support on the ground. This displacement of the upper part of the trunk is accompanied by another displacement of the pelvis and of the loaded limb in the opposite direction to keep balance. At a first glance the limping individual brings the whole body weight onto the diseased limb. This looks like an increase of the stress in the affected knee. A mechanical analysis shows that, on the contrary, the load exerted on the pathological knee is reduced.

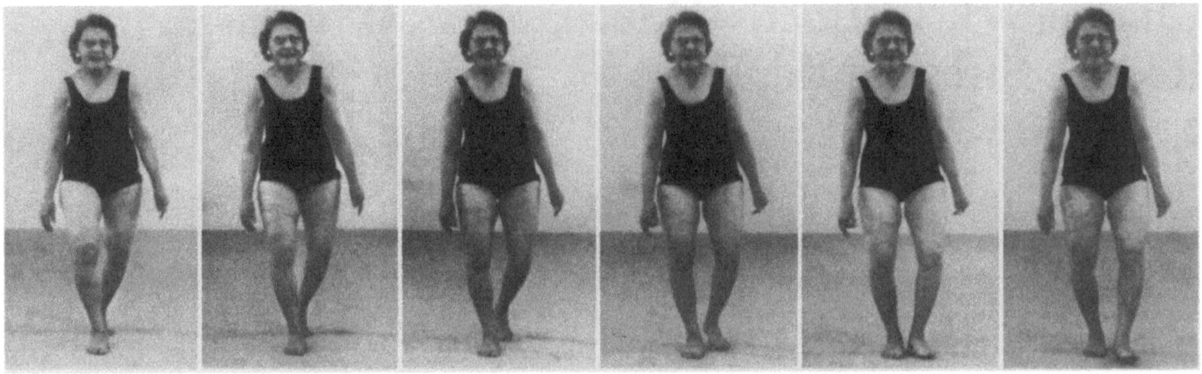

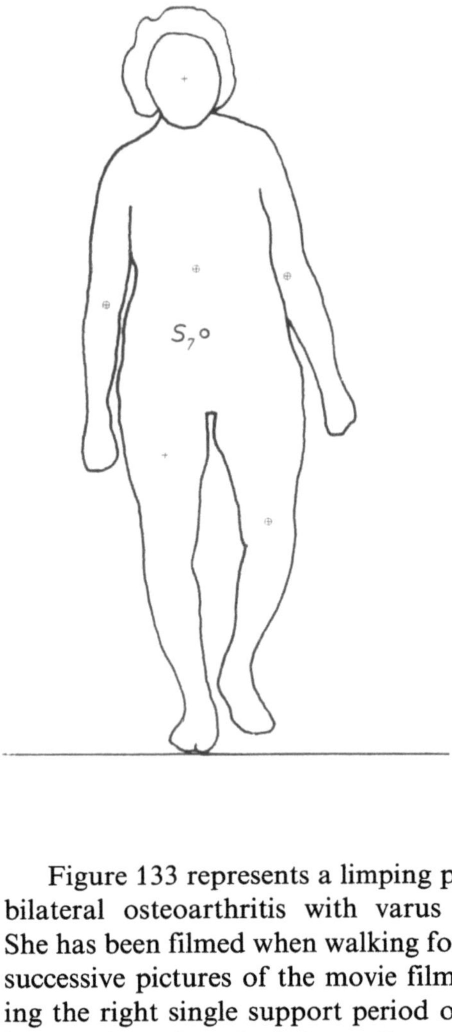

Fig. 133. Limping. +: centre of gravity of a segment of the body. ⊕: centre of gravity of several segments. S_7o: centre of gravity of the body minus the loaded leg and foot

Figure 133 represents a limping patient with bilateral osteoarthritis with varus deformity. She has been filmed when walking forward. The successive pictures of the movie film representing the right single support period of gait have been enlarged and their outlines drawn. By using the data of Braune and Fischer (1889), the centres of gravity have been determined and put into the drawing for the head, the trunk, the upper limbs, the loaded thigh and the opposite lower limb. The projection on a coronal plane (plane Oyz) of the outline of the body and of the centres of gravity has thus been completed and is shown for one phase in Figure 133. This projection can be compared with the corresponding projection of a normal individual during gait (see Fig. 28). In the limping patient, at each phase of the step, the partial centre of gravity S_7 is much closer to the vertical passing through the supporting foot than in the normal individual. The conclusion is that the forces of inertia are smaller when limping. We lack the data to define them with accuracy. Their precise calculation is of little value since the magnitude of the forces of inertia depends mainly on the velocity of the displacement, which changes in the course of the disease. Nevertheless, for the case of the filmed patient, we have calculated these forces of inertia by supposing that, on one hand, the displacement is three times slower than the normal gait described by Braune and Fischer (1895, 1899, 1900, 1901, 1903, 1904) and, on the other hand, the displacement of S_7 is the same as during the physiological gait. In these conditions, for a total weight of 58.7 kg, the force P exerted by the partial body mass attains a maximum of 56.5 kg and a minimum of 51.8 kg. It must be remembered that, for the individual walking normally, the maximum of P attains 74.5 kg and the minimum 29.2 kg. Limping thus decreases the maximum force developed by the body mass.

In the normal individual studied above, the distance between the knee G and the vertical drawn from the centre of gravity S_7 is 8.3 cm at phase 16 of gait (Figs. 28 and 41). Because

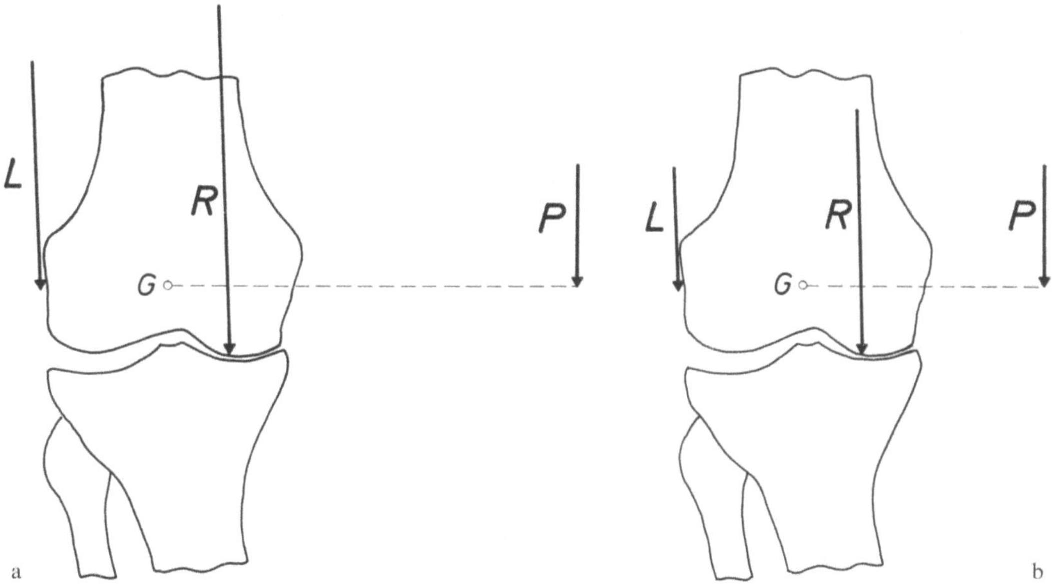

Fig. 134a and b. Changes of the forces, resulting from limping. (a) Without limp. (b) With limp

of the varus deformity it would be 13.3 cm in the osteoarthritic patient if she did not limp (Fig. 134a). Limping brings this distance back to 8.3 cm despite the deformity of the knee (Fig. 134b). The drawing in Figure 134a is based on Figure 28. The knee of the individual has been deformed in varus as is the knee of the filmed patient of Figure 133. For the drawing of Figure 134b the outline of the limping patient Figure 133 has been brought into the system of co-ordinates after her dimensions, had been reduced to the dimensions, of the subject of Braune and Fischer.

When limping, the displacement of the partial centre of gravity S_7 toward the pathological knee can be demonstrated for each phase of the single support period of gait. It reduces the distance between the knee G and the line of action of P. The combined reduction of force P and of the distance between the knee and the line of action of P considerably decreases the moment of P in relation to the knee. Consequently, the muscular force L necessary to balance P is diminished, as is the load R acting on the joint. Moreover, because of the diminution of the variations of P (difference between maximum and minimum) during the single support period of gait, the hammering of force R on the articular surfaces is considerably softened.

As shown by the analysis of this typical case, the displacement of the trunk toward the diseased side due to limping expresses a modification of equilibrium in a loaded joint. For the knee, force P acting at a shorter distance can be balanced by a weaker muscular tension band. This diminishes the femoro-tibial compressive force R and, consequently, the joint pressure.

Limping is observed in two main circumstances. It can be due to a muscular weakness or to an automatic reaction which tends to protect a deficient skeleton and, eventually, to lessen pain.

a) If muscular power is diminished by paresis or paralysis and becomes insufficient to balance the eccentrically placed body weight, the individual must adopt an unnatural gait.

During the single support period of the weakened side, he displaces his trunk toward that side. This tends to bring the centre of gravity above the loaded joint, requiring less muscular force to balance the body weight P. By changing his gait, the subject compensates for the insufficiency of his muscles. Such a limp must be considered a functional adaptation.

b) Limping can also be observed in the presence of normal muscular power. In this case, it is due to the tendency of the body to protect a painful bone or joint. The body tries to reduce the load exerted on the suffering limb. In this way, limping appears with lesions of joints and bones of the lower limb. It progressively disappears as the diseased limb recovers strength and resistance. Its disappearance depends upon the effectiveness of treatment in restoring a normal mechanical equilibrium.

II. Use of a Walking Stick

Many patients with a painful knee use a stick, usually on the opposite side. It reduces the load on the affected limb and can prevent limping or can at least reduce the displacement of the trunk to the diseased side during gait.

The stick transmits a part of the body weight to the ground. Force C exerted on it by the upper limb acts with a lever arm f, the distance between the stick and the knee. In the drawing Figure 135, C tends to tilt the part of the body supported by the knee counterclockwise whereas the weight P of this part acting with a lever arm a tends to tilt it clockwise.

Both static forces C and P are vertical with opposite signs. Their resultant K is also vertical: $K=P-C$. Its line of action is at a distance s from the knee such that:

$$K \cdot s = P \cdot a - C \cdot f,$$

$$(P-C) \cdot s = P \cdot a - C \cdot f,$$

$$s = \frac{P \cdot a - C \cdot f}{P-C}.$$

The force P which normally tends to tilt S_7 clockwise with the moment $P \cdot a$ is now replaced by force K. The moment $K \cdot s$ is smaller because the force K is less and it acts with a lever arm s which is shorter than a.

Since the stick allows the trunk to straighten without increasing the stress on the knee, it makes shifting the trunk to the diseased side superfluous. This is theoretically demonstrated in the same drawing (Fig. 135).

Force K is balanced by the lateral muscular tension band L, the force of which is easily calculated since its lever arm b is known:

$$L = \frac{K \cdot s}{b} = \frac{P \cdot a - C \cdot f}{b}.$$

The knee supports the resultant R of forces K and L:

$$R = \sqrt{K^2 + L^2 + 2K \cdot L \cdot \cos\psi}$$

where ψ is the acute angle formed by the lines of action of forces L and K.

The angle $(\widehat{RK})$ formed by the lines of action of R and of K is also calculated:

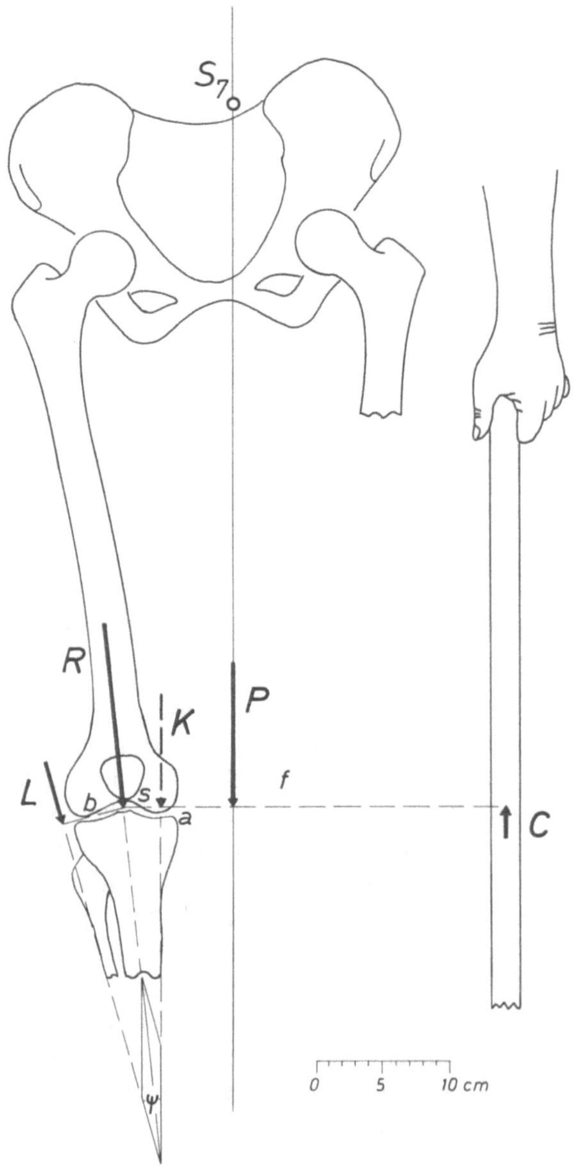

Fig. 135. Using a walking stick. P: force exerted by the partial mass of the body. L: lateral muscular tension band. C: force exerted on the stick. K: resultant of forces P and C. a: lever arm of P. f: lever arm of C. s: lever arm of K. b: lever arm of L. ψ: angle formed by the line of action of L and that of K

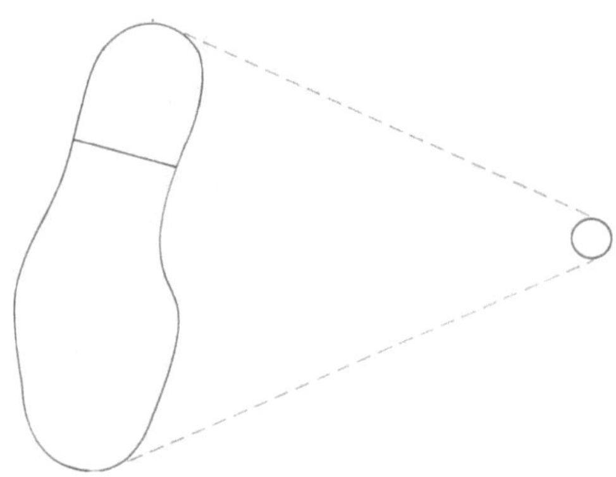

Fig. 136. Widening of the support when using a stick

$$\sin(\widehat{RK}) = \frac{L}{R} \sin \psi.$$

The values of the static forces and of the angle ($\widehat{RK}$) are given in Table 16 for different magnitudes of the force exerted on the stick by the hand which is opposite to the affected side.[11] As is shown, in these conditions the stick markedly reduces the load acting on the knee joint.

Table 16. Consequences of using a walking stick

Force C exerted on the stick (kg)	Lateral muscular force L (kg)	Resultant force R exerted on the knee (kg)	Inclination of the line of action of R to the vertical
0	72.816	126.824	6°38′
2	60.335	112.320	6°12′
4	47.801	97.858	5°38′
6	35.268	83.410	4°53′
8	22.735	68.982	3°48′
10	10.201	54.591	2°09′

Taking the forces of inertia into account would complicate the calculation but would not alter the result significantly because the patient with a stick moves slowly and thus diminishes to a minimum the forces of inertia. These may therefore be neglected.

On the other hand, the stick enlarges the base of support. Normally the latter roughly corresponds to the surface of the foot or shoe in contact with the ground, but with a stick, the area of support takes the shape of a large triangle with the lateral edge of the foot or shoe on the ground as its base and the tip of the stick as its apex (Fig. 136). Considerably enlarging the support, the stick obviously ensures better balance for the subject.

In summary, using a stick diminishes the load exerted on the knee and brings its line of action a little closer to the vertical. Furthermore, it considerably enlarges the support of the subject and ensures a better equilibrium.

[11] For this calculation we started from the data of phase 16 (subject of Braune and Fischer) which we corrected in order to bring the foot under the centre of gravity of the body, thus eliminating the forces of inertia. Then the lever arm a of P is 6.01 cm and that b of L is 4.50 cm

III. Comment and Conclusion

As shown by the preceding analysis, both the use of a stick and limping reduce the load acting on an affected lower limb. Both diminish the moment of the force eccentrically exerted by the mass of the body on the loaded knee. The stick acts by transmitting a part of this force directly to the ground. The muscular forces necessary for balancing the remaining part are therefore considerably reduced and the magnitude of the resultant force acting on the joint is decreased.

Limping shortens the lever arm of the force exerted by the mass of the body by shifting the centre of gravity of the body toward the loaded knee. Consequently, equilibrium can be ensured by a smaller muscular force. The compressive force acting on the knee is therefore reduced.

Chapter VII. Biomechanical Treatment of Osteoarthritis of the Knee

There are basically two types of osteoarthritis: primary, resulting from a lowered resistance of the articular tissues when the load exerted on the knee is physiological, and secondary, resulting from a change of the mechanical stress which overwhelms the resistance of otherwise normal tissues. In the primary form, with insufficient tissues of the knee, the physiological stress provokes an overall osteoarthritis attacking the femoro-tibial and the patello-femoral joints. It is medial as well as lateral. In its early phase, the joint space is narrowed neither in the lateral nor in the medial part of the knee. The dense cups underlying the tibial plateaux remain symmetrical. In secondary osteoarthritis, the mechanical stress becomes abnormally high and overtakes the physiological capability of the tissues to resist stress. A permanent displacement of the load *R* medially, laterally, anteriorly or posteriorly produces a concentration of the articular compressive stresses which, once localized, can be extremely high. The causes and consequences of displacements of *R* have been analysed in Chapter V. Any fracture which results in uneven joint surfaces can also concentrate the articular compressive stresses and cause subsequent osteoarthritis. Meniscectomy has the same effect by decreasing the weight-bearing surface of the joint. It is to avoid the inevitable osteoarthritis following meniscectomy that a meniscus freshly torn through the area close to its capsular insertion should be sutured.

In the end, in osteoarthritis of the knee, from any cause, the articular compressive stresses are too high for the ability of the knee tissues to resist.

I. Rationale of Biomechanical Treatment

As one can conclude from the previous discussion, there are two possible ways to restore an equilibrium between mechanical stress and tissue integrity. One would consist in strengthening the ability of the tissues to resist mechanical stress. Drugs and some surgical operations aim at achieving this. For instance, Palazzi (1961) drills a hole through the lower epiphysis of the femur and through the upper epiphysis of the tibia and puts a fragment of muscle in each.

Benjamin (1969) carries out a double osteotomy, dividing the lower end of the femur and the upper end of the tibia in cases of rheumatoid arthritis and osteoarthritis. This double osteotomy is aimed at "the mere division of various structures, such as blood vessels, nerves and bone". Deliss (1977) divides the patella in a coronal plane with the same goal in mind.

Such operations tend to modify the blood supply of the area and, through that, to increase the mechanical resistance of the tissues. But at this time we cannot increase the mechanical resistance with certainty because we dot not know its determining factors. At best, the treatments with a biological aim could bring back to normal the lowered resistance to stress of the tissues. But once the mechanical stress in the knee has become abnormally high, bringing the mechanical resistance of the tissues back to normal is insufficient.

The second method of restoring a physiological equilibrium would be to diminish the articular compressive stresses as much as possible to make them tolerable even for tissues with a lowered resistance to stress. The articular compressive stresses can be decreased either by reducing the load which provokes them or by enlarging the surface which transmits them (see Chapter IV). The example of the columns, which we borrow from Pauwels (1963), demonstrates this. The cylindrical column (Fig. 137a) supports a well-centred load of 200 kg. Its cross section is indicated below. The load of 200 kg evokes in the material of the column compressive stresses of 255 kg/cm², shown in the diagram. The reduction of the load to 100 kg (Fig. 137b) lowers the compressive stresses to 127.5 kg/cm². If the first load of 200 kg is supported by a column (Fig. 137c) with a diameter three times larger than that of the first column, the compressive stresses are only 28.3 kg/cm². Because the diameter of the column is three times larger, the compressive stresses produced by the same load become nine times smaller. The greatest reduction of compressive stresses will be attained by combining both possibilities:

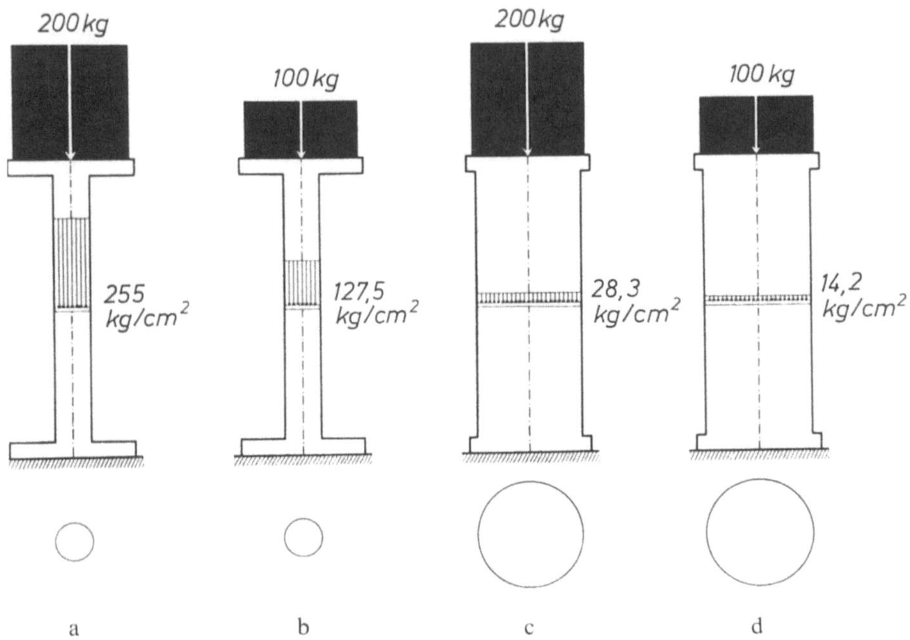

Fig. 137a–d. A reduction of the compressive stresses exerted in the column (a) is attained by decreasing the load (b), by enlarging the weight-bearing surface (c) or by combining these both possibilities (d). (From Pauwels, 1963)

diminution of the load and enlargement of the weight-bearing surface (Fig. 137d). The compressive stresses, which were 255 kg/cm^2 in the first column, are then reduced to 14.1 kg/cm^2. Let us now apply these general principles to the knee.

II. Biomechanical Treatment of Osteoarthritis of the Knee

In the knee, the articular compressive stresses can be decreased by diminishing the load or by enlarging the articular weight-bearing surfaces. But the ideal treatment combines both possibilities in order to reduce the mechanical stress acting on the joint as much as possible.

Diminishing the load and increasing the weight-bearing surfaces are essentially attained by correcting any flexion contracture, by displacing the tibial tuberosity anteriorly and by recentring the load. These three mechanical approaches to treatment can be used either separately or in combination. We shall successively study each of these, enunciating first the theoretical principles and then the techniques which permit the application of these principles. When necessary, graphic planning of these surgical procedures and the instruments which make them easier will be described. Clinical examples will illustrate the postoperative evolution of the disease. Patellectomy in the treatment of osteoarthritis of the knee will then be evaluated. Thinning of the patella by removal of a coronal slab will be briefly discussed. Finally, operative indications will be listed.

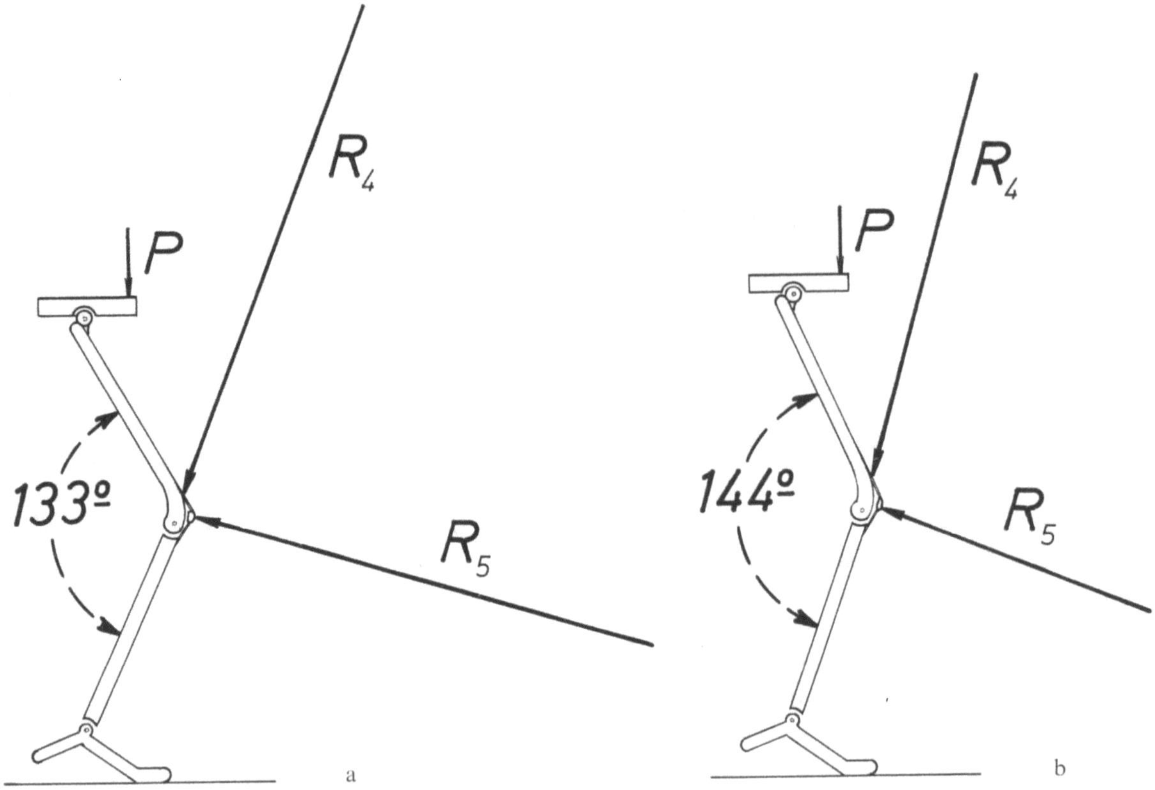

Fig. 138a and b. End result of a geometrical analysis. Extension decreases the forces R_4 pressing the femur on the tibia and R_5 pressing the patella against the femur

A. Correction of Flexion Contracture

1. Rationale

In Chapter V we have shown how a flexion contracture of the knee increases force R_4 compressing the femur against the tibia and force R_5 compressing the patella against the femur (Fig. 105). As shown by a geometrical analysis, extension diminishes forces R_4 and R_5 (Fig. 138) and displaces the support toward the anterior part of the femoral condyles, which has a larger radius of curvature than the posterior part (Fig. 139a and b). Consequently, full extension enlarges the femoro-tibial weight-bearing surfaces. Increasing the weight-bearing surfaces and diminishing the load decrease the articular compressive stresses.

Full extension can often be attained by simple physiotherapy or, if necessary, by a posterior capsulotomy.[12]

Extension produced by both an osteotomy or a posterior capsulotomy decreases the forces R_4 and R_5. However, full extension attained by release of soft tissue increases the weight-bearing area of the knee (Fig. 139b). This is in contrast to correction of flexion contractures by osteotomies, which usually do not achieve this.

After correction of a flexion contracture by an osteotomy of the lower end of the femur, the femoro-tibial weight bearing area remains restricted to the posterior aspect of the joint as it is in flexion (Fig. 139c).

Correcting a flexion contracture by an osteotomy of the upper end of the tibia provokes a recurvatum in the knee joint. This results in an uneven distribution of the articular stresses, which become concentrated and considerably increased in the anterior aspect of the joint (Figs. 108 and 115c, see also Fig. 234).

Capsulotomy can be performed alone to correct flexion contracture in patients with osteoarthritis involving the whole knee. For practical purposes, it is almost always combined with

[12] In some patients however, with massive joint destruction, full extension is anatomically impossible, even after capsulotomy. In such exceptional cases full extension can only be achieved by a distal femoral or a proximal tibial osteotomy which brings the osteophytes into the weight-bearing areas. In such cases a carefully planned osteotomy can both increase the weight-bearing surfaces and decrease the load

an anterior displacement of the tibial tuberosity and/or with a proximal tibial or distal femoral osteotomy, modifying the angle formed by the femur and the tibia in the coronal plane. Posterior capsulotomy can be performed at the same time as the main operation or separately.

2. Operative Procedure

a) Capsulotomy Alone

We use no tourniquet. With the patient lying prone, a *Z* incision with its central branch parallel to the posterior crease of the knee exposes the popliteal area. The neurovascular bundle is retracted laterally, the semimembranosus, semitendinosus and gracilis tendons are retracted medially. This gives access to the medial posterior capsule which is divided transversely on the femoral condyle. Incision of the capsule is done proximal to the joint space.

The external popliteal nerve is then found and retracted laterally along with the biceps tendon. The popliteal neurovascular bundle is retracted medially. The sural nerve must be found and protected. The lateral posterior capsule is divided proximal to the joint space. The knee can then be extended. Mobilization is started the next day and weight-bearing permitted. A posterior plaster splint is used at night for one or two months.

b) Capsulotomy Associated with Other Procedures

Since posterior capsulotomy is most often performed in association with an anterior displacement of the patella tendon or an osteotomy, in these cases it is usually carried out with the patient supine. We use no tourniquet. The knee is flexed at 90°. A short incision is made laterally between the vastus lateralis and the hamstrings. The lateral capsule is incised and then divided transversely with scissors behind the femoral condyle.

The medial capsule is incised in the same way through a posteromedial incision. The last fibres are stretched by a manual hyperextension of the joint. They give way with an audible snap and overextension is suddenly obtained. Care must be taken to retain it after surgery by daily exercises or by putting a sand back on the extended knee when the patient is seated since flexion contracture usually tends to recur.

3. Results

In most cases in which we performed a posterior capsulotomy, either an anterior displacement of the tibial tuberosity or an upper tibial or lower femoral osteotomy was carried out in addition. The results of the entire surgical intervention will be discussed in a later section.

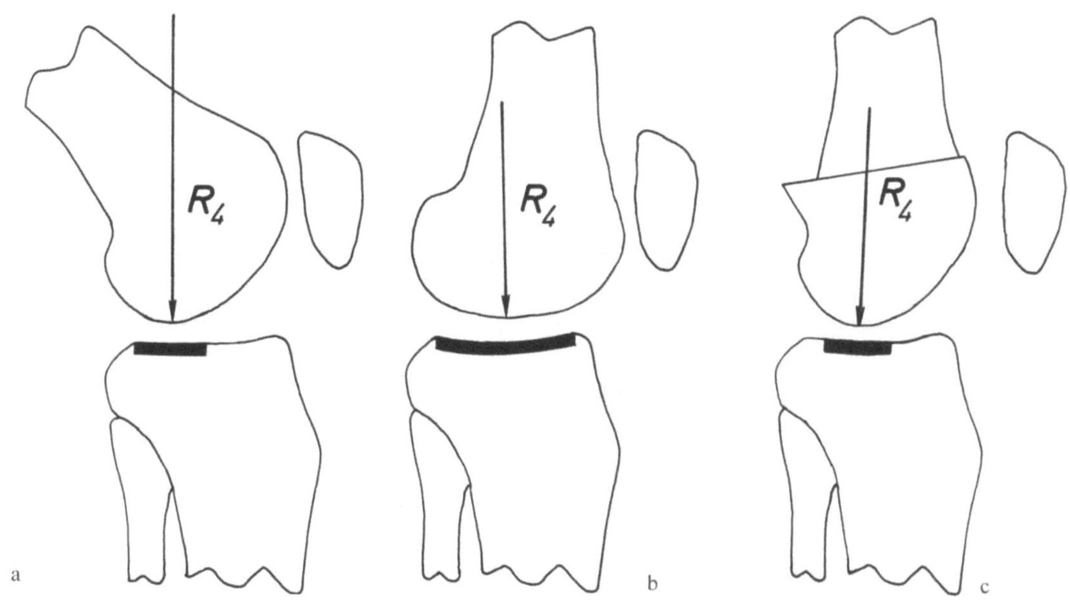

Fig. 139. (a) Flexed knee. (b) Extension. (c) Extension resulting from a femoral osteotomy

B. Anterior Displacement of the Tibial Tuberosity

1. Rationale

Displacing the tibial tuberosity anteriorly reduces force R_4 which compresses the femur against the tibia, diminishes force R_5 which compresses the patella against the femur and enlarges the patello-femoral weight-bearing surface. Displaced anteriorly the patella tendon P_a' acts with a longer lever arm c' (Fig. 140). Therefore, it can carry out the same work with a smaller force. The reduction of force P_a decreases the resultant R_4 transmitted from the femur to the tibia and the resultant R_5 transmitted from the knee-cap to the femur. The anterior displacement of the tibial tuberosity considerably opens the angle β formed by the lines of action of force P_a and force M_v, the quadriceps muscle (Fig. 141). Increasing this angle has the most dramatic effect in reducing the patello-femoral compressive force R_5. For example, when the knee is flexed at 45°, an anterior displacement of the tibial tuberosity by 2 cm reduces the force R_5 by about 50% (Fig. 142).

Anterior displacement of the tibial tuberosity by 2 cm lengthens the lever arm with which the force P_a rotates the tibia about the axis of the femoro-tibial joint, by 10% (Fig. 140):

$$P_a \cdot c = P_a' \cdot c',$$

$$P_a' = \frac{P_a \cdot c}{c'} = P_a \cdot \frac{10}{11}.$$

P_a' represents the force exerted by the patella tendon after anterior displacement of its tibial insertion and c' the new lever arm in relation to the axis G.

On the other hand, the lever arm k with which the force P_a acts on the patella (Fig. 55) is little changed by anterior displacement. k is the distance between the line of action of P_a and the centre of curvature of the patello-femoral weight-bearing surface. The change of k is slight, difficult to appreciate and can be neglected.

But the opening of the angle β, formed by the line of action of force P_a and that of force M_v (Fig. 141), is considerable and can be measured. We shall calculate how much the force R_5 compressing the knee-cap against the femur is reduced by the change of angle β produced

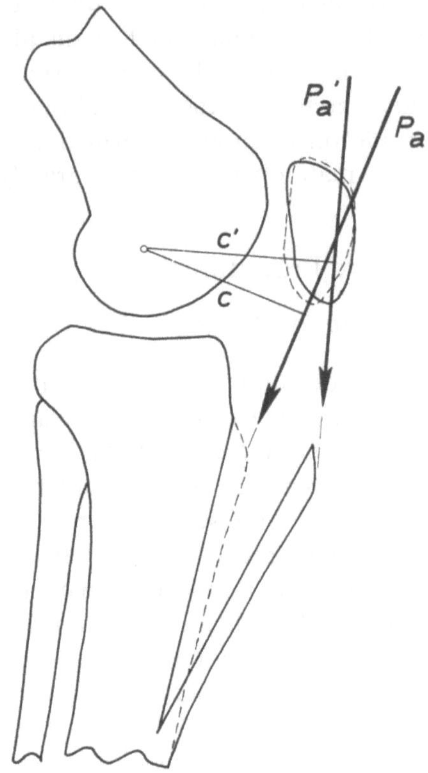

Fig. 140. Effect of an anterior displacement of the tibial tuberosity on force P_a. P_a: force normally exerted by the patella tendon. P_a': force exerted by the patella tendon when its distal insertion has been displaced forward. c: leverarm of P_a. c': leverarm of P_a'

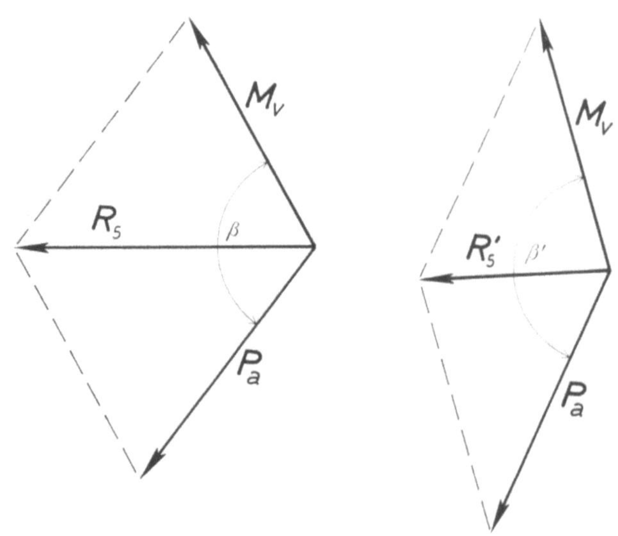

Fig. 141. Opening of the angle β formed by the lines of action of M_v and P_a decreases the magnitude of resultant force R_5

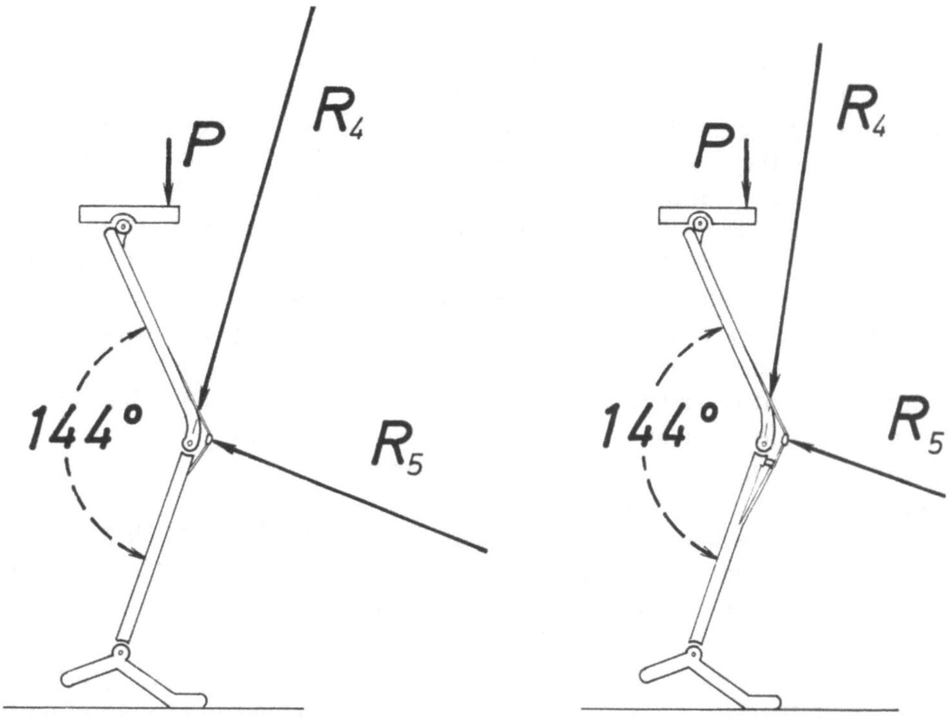

Fig. 142a and b. Modification of forces, produced by a 2 cm anterior displacement of the tibial tuberosity. (a) Normal knee. (b) After anterior displacement of the tibial tuberosity

by an anterior displacement of the tibial tuberosity by 2 cm.

Through the anterior displacement of the tendon, the angle β becomes β' and R_5 becomes R'_5.

$$R'_5 = \sqrt{P_a^2 + M_v^2 + 2P_a \cdot M_v \cdot \cos\beta'}.$$

The results are indicated in Table 17.

Table 17. Decrease of the patello-femoral compressive force caused by a 2 cm anterior displacement of the tibial tuberosity

Phases	Pull of the patella tendon		Patello-femoral compressive force	
	before P_a (kg)	after surgery P'_a (kg)	before R_5 (kg)	after surgery R'_5 (kg)
12	283.149	257.408	218.992	102.889
13	146.920	133.564	126.847	64.801
14	157.042	142.765	124.646	60.018
15	48.790	44.354	33.468	14.874

The comparison between the values of R_5 and R'_5 shows that the patello-femoral compressive force is decreased by about 50% during gait when the tibial tuberosity is displaced anteriorly by 2 cm.

Bandi (1972) has measured the force pressing the patella against the femur experimentally in a model and in anatomical specimens before and after anterior displacement of the tibial tuberosity by 1 cm and by 1.5 cm. He confirms the important diminution of the patello-femoral compressive force achieved by the anterior displacement – although small in his experiment – of the patella tendon.

The change in the direction of P_a causes the upper facet of the patella to be in contact with the femur earlier during flexion of the knee. The upper facet of the patella being larger than the middle facet, the weight-bearing area is enlarged. This assumption, which we formulated in 1963, has been confirmed by the experimental work of Tracy et al. (1982) and by Wagner et al. (1982b).[13]

[13] Enlargement of the patello-femoral contact area resulting from an anterior displacement of the tibial tuberosity (from Wagner et al., 1982b)

Magnitude of the anterior displacement	Flexion of the knee		
	30°	45°	60°
5 mm	+ 6.6%	+14%	+ 7%
10 mm	+14%	+25.3%	+26.5%
15 mm	+37.1%	+32.4%	+44.2%
20 mm	+10.7%	+21.1%	+ 3.5%
25 mm	+ 7.4%	+22.5%	+ 4.4%

By reducing the force exerted between the femur and the tibia, by decreasing the force compressing the patella against the femur and by enlarging the patello-femoral weight-bearing area, the anterior displacement of the tibial tuberosity decreases the femoro-tibial and patello-femoral compressive stresses.

Geometrical analysis and calculation show that the force exerted by the quadriceps is at its maximum during the first phases of the single support period of gait (phases 12 to 15). Electromyographic studies (Basmajian, 1967) confirm this fact. The phases of the single support period of gait when the quadriceps acts are the phases when the load R exerted on the femoro-tibial joint is at its maximum (Fig. 51). It is easy to show that this is the same for the load R_5 exerted on the patello-femoral joint.

On the other hand, anterior displacement of the tibial tuberosity is most efficient in extension or near extension. But during the phases when the quadriceps acts, the flexion of the knee is less than 30°. Therefore, the effects of the anterior displacement of the tibial tuberosity are exerted at the most favourable moments of gait, when the mechanical stress is at its greatest and during small amounts of flexion, giving the anterior displacement of the tibial tuberosity maximum efficiency.

Several authors (Bandi, 1972, 1980; Ferguson et al., 1979) have advocated anterior displacement of the tibial tuberosity by less than 2 cm. Ferguson even claims that beyond 2 cm, the articular stresses would be increased. Minimal anterior displacement may be sufficient in some cases. However, we had to operate again on patients in whom the tuberosity had been displaced anteriorly by 1 or 1.5 cm and who were not relieved of pain. They did well after more pronounced anterior displacement. On the other hand, none of the aforementioned authors could ever show in the postoperative X-rays any change which would demonstrate a decrease of the articular stresses. Burke and Ahmed (1980) have shown experimentally that the more considerable the anterior displacement of the tibial tuberosity (up to 3 cm in their experiment), the lower the articular stresses.

2. Operative Procedures

Anterior displacement of the patella tendon had been carried out previously by interposing an iliac graft between the tendon and the upper extremity of the tibia. Since 1968 we have elevated the tibial crest with its tibial tuberosity and maintained it in this position by an interposed graft. After upper tibial osteotomy, anterior displacement of the patella tendon is achieved by shifting the diaphysial fragment forward.

a) Anterior Displacement of the Tibial Tuberosity by Elevating the Tibial Crest (Fig. 143)

A full thickness graft 4 cm by 2 to 3 cm is first taken from the anterolateral aspect of the ilium. We retain the crest for the cosmetic appearance. No tourniquet is used. The incision on the anteromedial aspect of the leg is parallel and 0.5–1 cm posterior to the tibial crest, 10–13 cm long from the lower tip of the patella. The aponeurosis is incised medially along the patella tendon. The medial and lateral retinacula are divided subcutaneously with scissors in their distal part, on both sides of the patella, from the medial incision. The tibial periosteum is incised longitudinally. Holes are drilled transversely along a line parallel to and 0.7 cm behind the crest, through the anteromedial and anterolateral aspects of the tibia, which is then split with a thin chisel along the line joining the holes. The split separates the proximal part of the crest and the anterior tuberosity from the tibia. The anterior flap so created remains attached distally. The crest and anterior tuberosity are elevated, using a spreader which was specially designed for this purpose (Fig. 144). The iliac graft is inserted on edge behind the flap and maintains the elevation. The graft must be positioned at the proximal end of the elevated bone. If it is placed more distally, there is a danger of breakage of the tip of the bone due to the oblique pull of the patella tendon. Properly placed, the graft is maintained by the elasticity of the elevated bone and by the pull of the patella tendon. No screw should be used to fix it. Such a screw would be superfluous

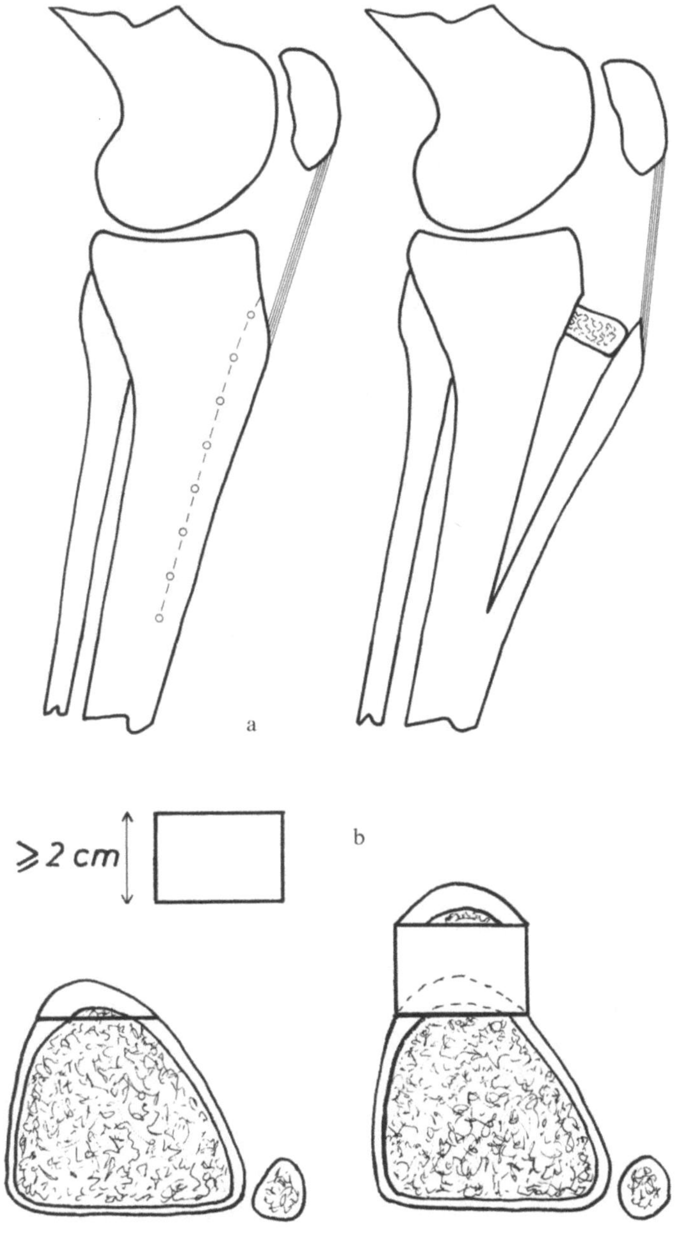

Fig. 143. Anterior displacement of the tibial tuberosity by splitting and moving the tibial crest forward. (a) Sagittal view. (b) Transverse cross section. The graft should be at least 2 cm wide. The two views a and b are on different scales for clarity's sake

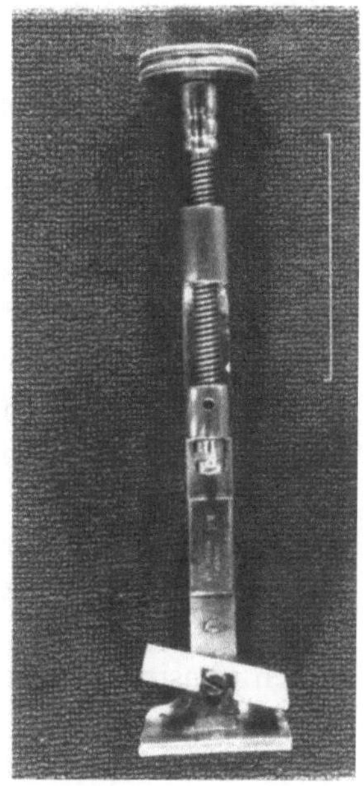

Fig. 144. Spreader used to separate the tibial crest and tuberosity from the remainder of the bone

and may lead to complications such as circumscribed necrosis of the skin. Supplementary smaller grafts can be added in the triangular space below the main graft. If the elevated bone accidentally breaks at its base, a screw may then be used as distally as possible, to fix it, but often it is not necessary. Subcutaneous fat and skin are carefully sutured on drainage without suction. Medial and lateral relieving incisions are sometimes called for to avoid tension of the skin. Mobilization and full weight-bearing are started immediately. The relieving incisions are skin-grafted after two weeks. Forceful muscular build-up must be avoided and particularly lifting weights hung from the ankle by extending the knee, the subject being seated. Such exercise has been proved to evoke articular compressive stresses far beyond the physiological range. This is exactly opposite to the aim of the procedure, which consists of decreasing the stresses in the joint.

This technique smooths the unaesthetic bulge distal to the knee-cap. It has not been followed by a partial resorption of the graft such as sometimes has been observed with the previous technique which moreover lowered the patella by the width of the graft. This inconvenience is avoided when the crest is elevated anteriorly. Lowering a patella alta or a normal patella usually results in patello-femoral pain. We avoid doing it.

The cosmetic appearance must be considered. We show the candidate for this operation a photograph of a knee which has been operated on (Fig. 145). The patient decides whether the pain he or she is complaining about is worth the bulge in front of the lower leg.

A 53-year-old female patient (Fig. 146) complained of pain in the right knee. She presented with the symptoms of osteoarthritis of the patello-femoral joint. In the X-ray a cup-shaped subchondral density indicated abnormally high compressive stresses (Fig. 146a). One year after an anterior displacement of the tibial tuberosity by about 2.5 cm the patient was painfree, had a full range of movement and lived normally. Eight years after the anterior displacement of the tibial tuberosity, the result remains excellent. The thin ribbon of subchondral dense bone in the patella means that the pressure is reduced and evenly distributed in the patello-femoral joint (Fig. 146b).

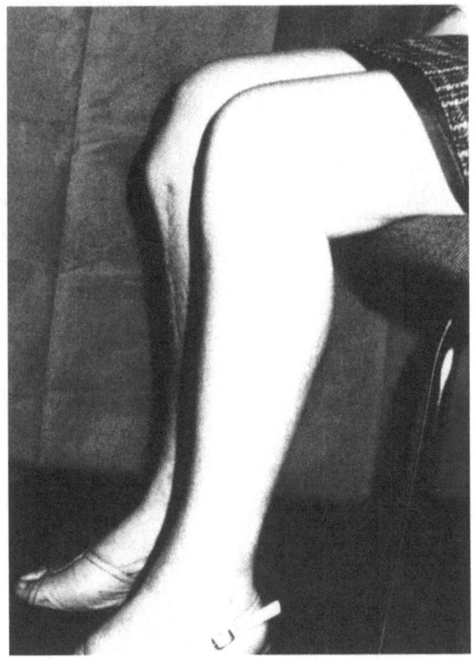

Fig. 145. Cosmetic appearance of the knee after a 3 cm anterior displacement of the tibial tuberosity

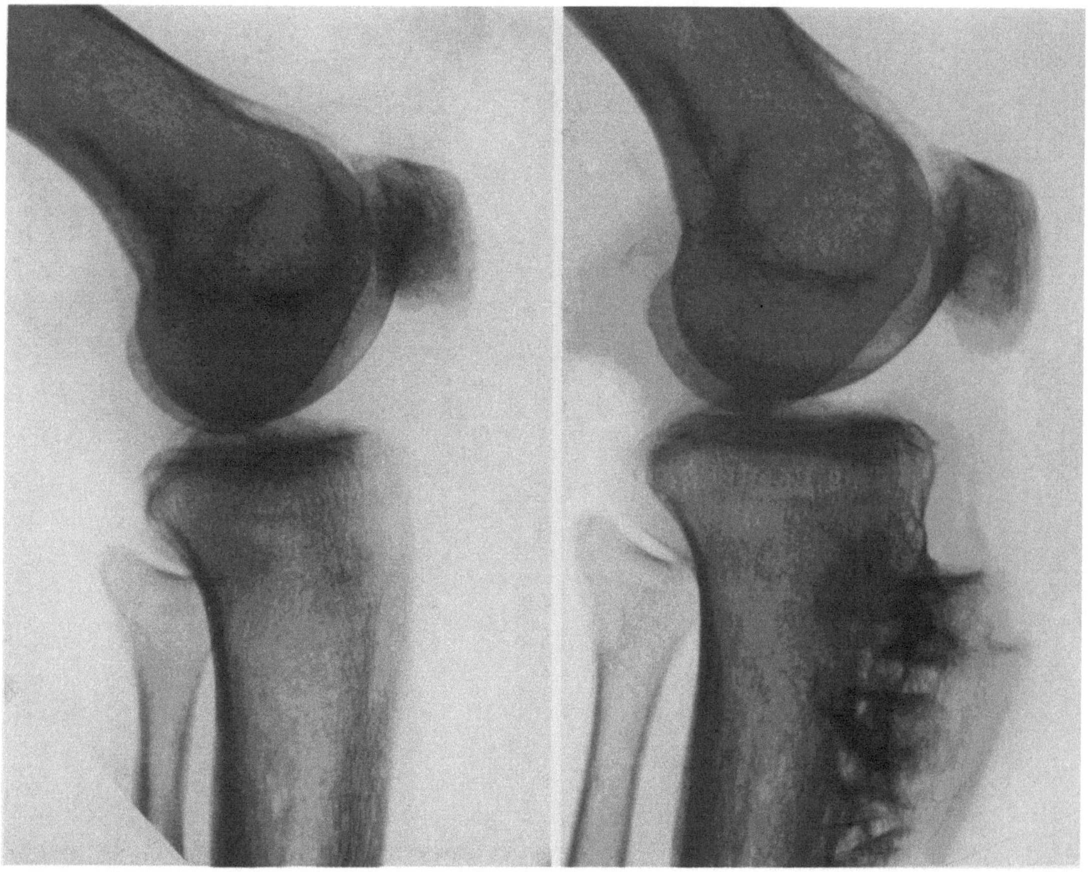

a b

Fig. 146a and b. A 53-year-old female patient before (a) and eight years after (b) an anterior displacement of the tibial tuberosity. The cup-shaped density in the patella has regressed

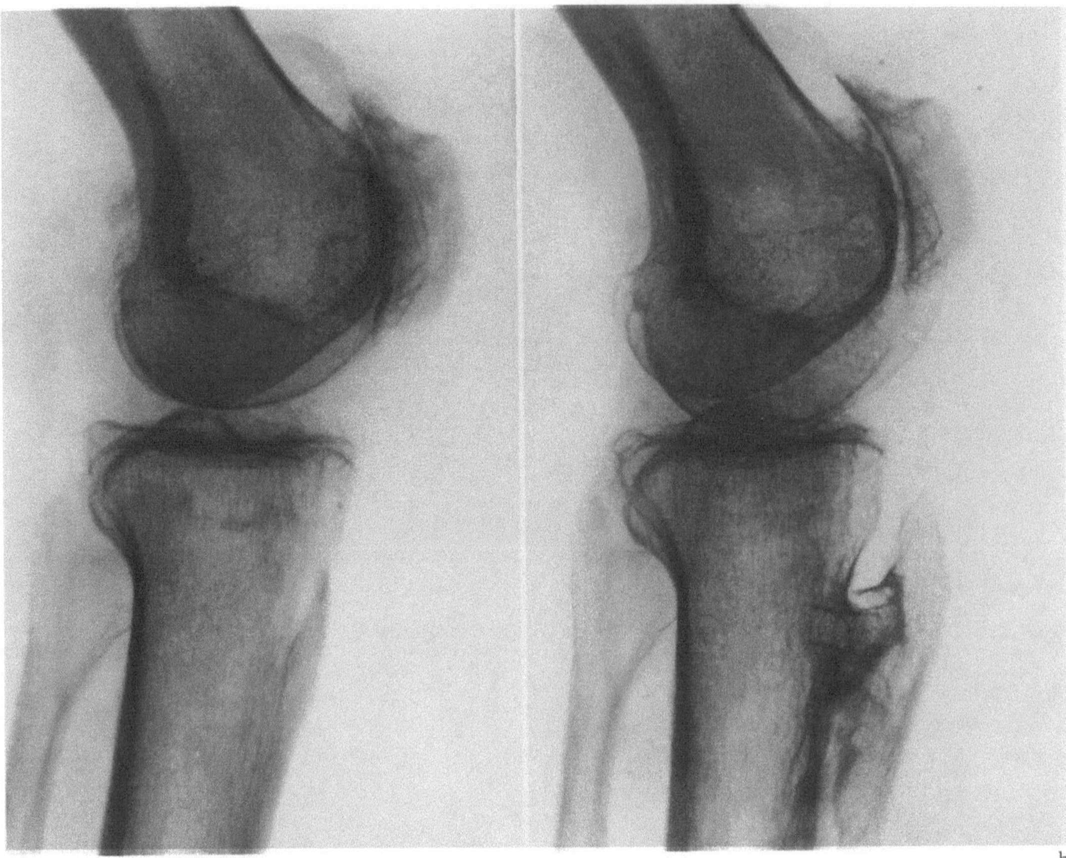

a b

Fig. 147a and b. A 60-year-old female patient before (a) and three years after (b) an anterior displacement of the tibial tuberosity. The graft should be inserted more proximally, at the level of the tip of the bony flap

A 60-year-old nun (Fig. 147a) presented with a painful patello-femoral osteoarthritis. Her tibial tuberosity has been displaced about 17 mm anteriorly. Three years later she is completely painfree and works normally. The subchondral sclerosis has regressed and a wide joint space has reappeared (Fig. 147b).

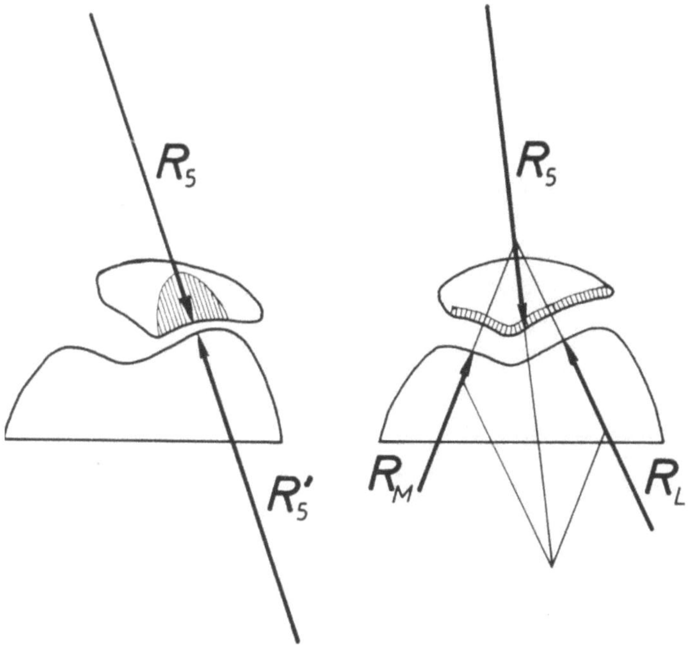

Fig. 148. Brought back into the intercondylar groove the patella transmits the load to the medial condyle as well as to the lateral

b) Anterior and Medial Displacement of the Tibial Tuberosity

If osteoarthritis is localized to the lateral part of the patella and to the corresponding femoral condyle or if the patella is subluxated laterally, the operative displacement must be medial as well as anterior. It will bring the kneecap back into the intercondylar groove, reduce the patello-femoral compressive force R_5 and distribute it on both sides of the groove. This increases the weight-bearing surface (Fig. 148).

An easy way of displacing the tibial tuberosity forward and medially consists of dividing the crest obliquely to the coronal plane (Beckers, 1982) (Fig. 149). The medial displacement depends on the degree of inclination of the osteotomy to the coronal plane and on the width g of the graft. The magnitude a of the anterior displacement and that m of the medial displacement can be calculated. If the angle formed by the coronal plane and the osteotomy cut is α, we have:

$$m = g \cdot \sin\alpha; \quad a = g \cdot \cos\alpha$$

Thus the inclination α of the osteotomy will be chosen and the graft tailored according to the desired magnitude of anterior and medial displacement.

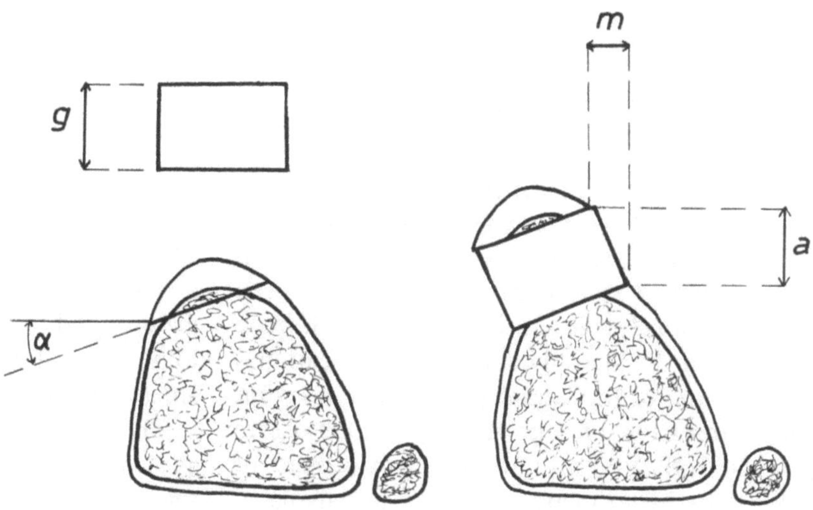

Fig. 149. Anterior and medial displacement of the tibial tuberosity. Cross section of the proximal extremity of the lower leg. α: angle formed by the osteotomy line and the coronal plane. g: width of the graft. a: anterior displacement. m: medial displacement

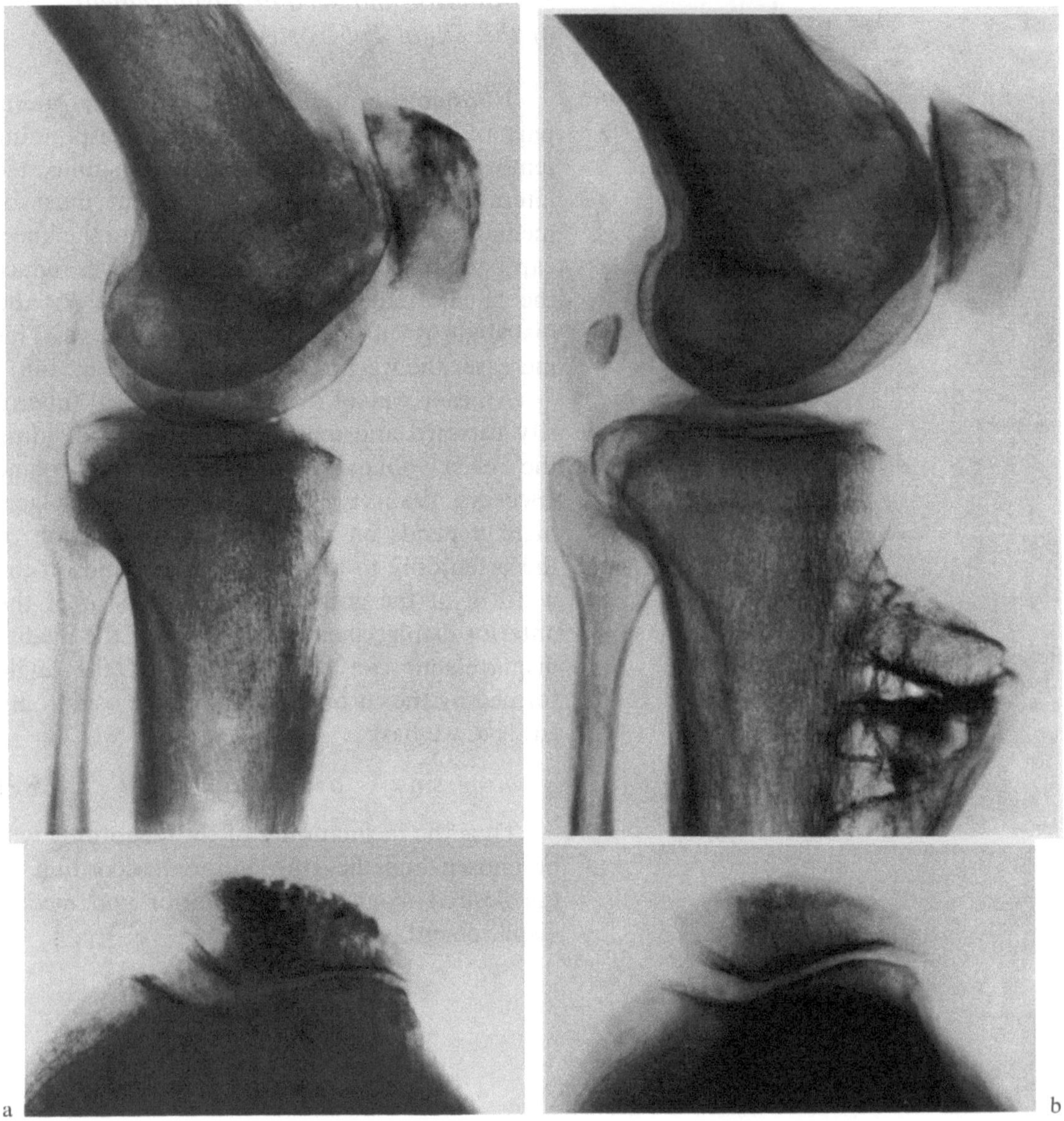

a b

Fig. 150a and b. A 68-year-old patient before (a) and six years after (b) an anterior and medial displacement of the tibial tuberosity

The 68-year-old male patient whose knee is shown in Fig. 150a presented with a very painful patello-femoral osteoarthritis with limitation of movement (from 0° to 30°). X-rays showed patello-femoral degenerative changes with bone densities in the laterally displaced knee-cap and a narrowing of the lateral patello-femoral joint space. Insertion of a graft as described above (Fig. 149) allowed an anterior displacement of more than 3 centimetres and a medial displacement of the tibial crest with the insertion of the patella tendon. Six years after surgery, the knee remains painfree. The range of movement has improved (from 5° hyperextension to 150° flexion). The patient lives normally and actively. The radiological examination (Fig. 150b) shows a patella well centred in the intercondylar groove and separated from the condyles by a joint space of even width. The intrapatellar densities have regressed.

The 43-year-old male patient whose knee is shown in Figure 151a complained of a painful knee. The patella was high and laterally subluxated. A subchondral cup-shaped density was a token of increased articular pressure in the patello-femoral joint. Nine years after a 3 cm

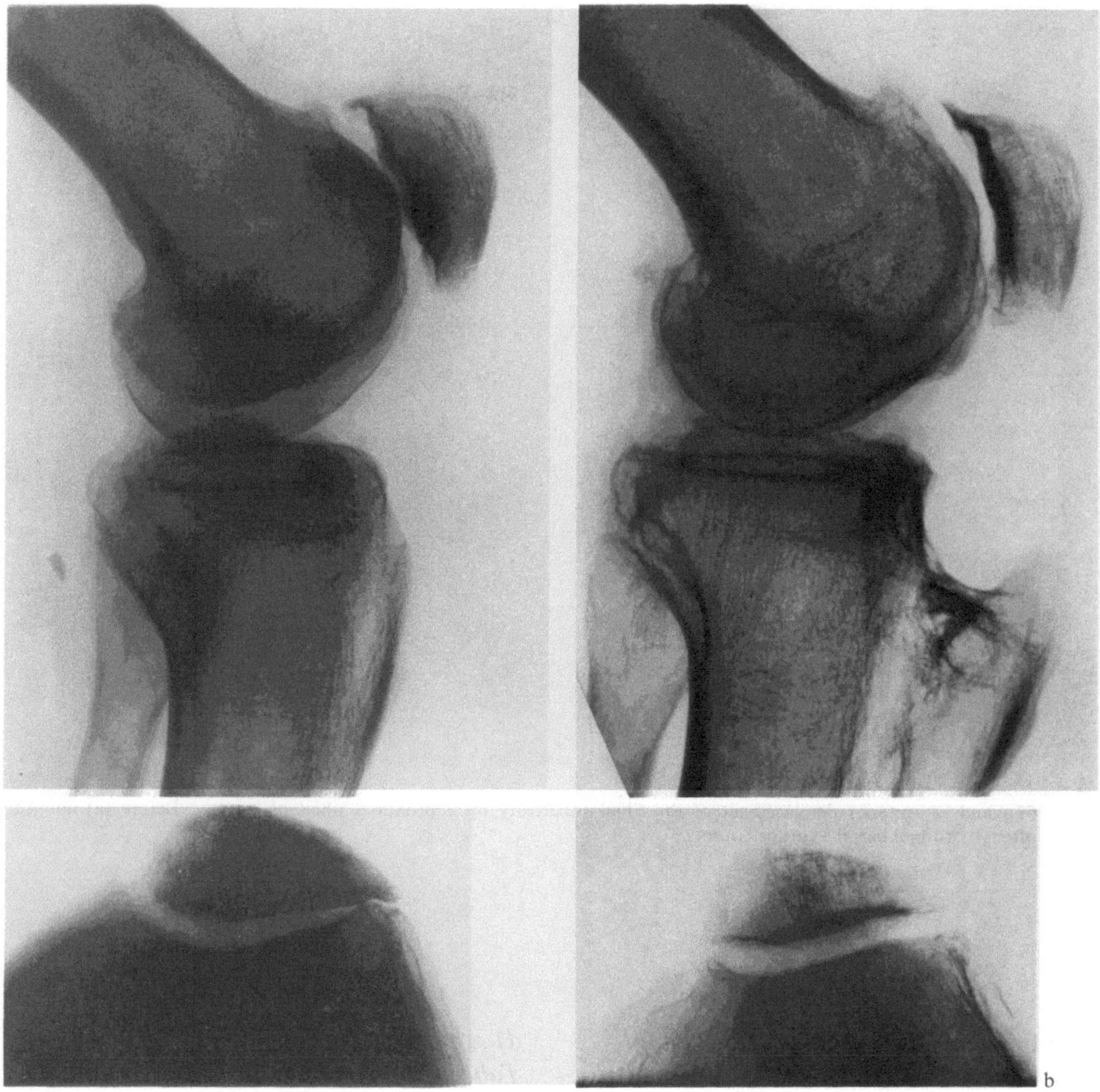

a b

Fig. 151a and b. A 43-year-old patient before (a) and nine years after (b) an anterior and medial displacement of the tibial tuberosity

anterior and medial displacement of the tibial tuberosity the patient remains painfree, has a full range of movement and is back to work. The patella is now well centred in the intercondylar groove. The subchondral sclerosis underlying the patello-femoral joint is of uniform width throughout. This indicates an evenly distributed joint pressure (Fig. 151b).

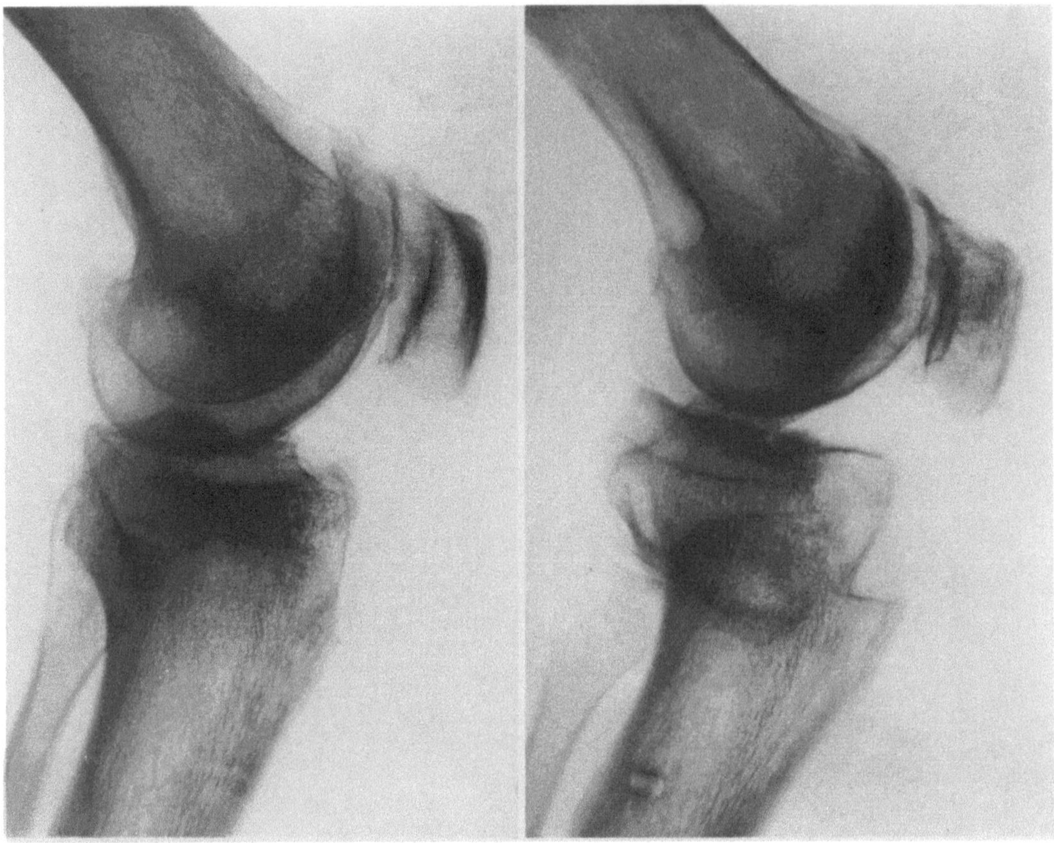

a b

Fig. 152a and b. Anterior displacement of the tibial tuberosity by a proximal tibial osteotomy. Before (a) and three years after (b) a tibial barrel-vault osteotomy

c) Anterior Displacement of the Tibial Tuberosity Combined with Upper Tibial Osteotomy

When an upper tibial osteotomy is carried out to correct a varus or a valgus deformity, anterior displacement of the tibial tuberosity is simply achieved by shifting the whole distal fragment anteriorly (Fig. 152). In this instance no graft is thus necessary (see pages 171 and 176). However, the degree of such an anterior displacement is restricted (1 to 1.5 cm) by the need not to impede stability of the tibial fragments.

If greater anterior displacement of the tuberosity is required, the tibial crest can be split and an iliac graft inserted, as described above, immediately after the upper tibial osteotomy or at a second operation.

d) Anterior Displacement of the Tibial Tuberosity Combined with Osteotomy of the Lower End of the Femur

In some cases we have combined an anterior displacement of the tibial tuberosity by splitting the tibial crest, as described above, with an osteotomy of the lower end of the femur. The anterior displacement of the tibial tuberosity is aimed at treating osteoarthritis in the patello-femoral joint and the femoral osteotomy at dealing with femoro-tibial osteoarthritis in a valgus knee (Fig. 228).

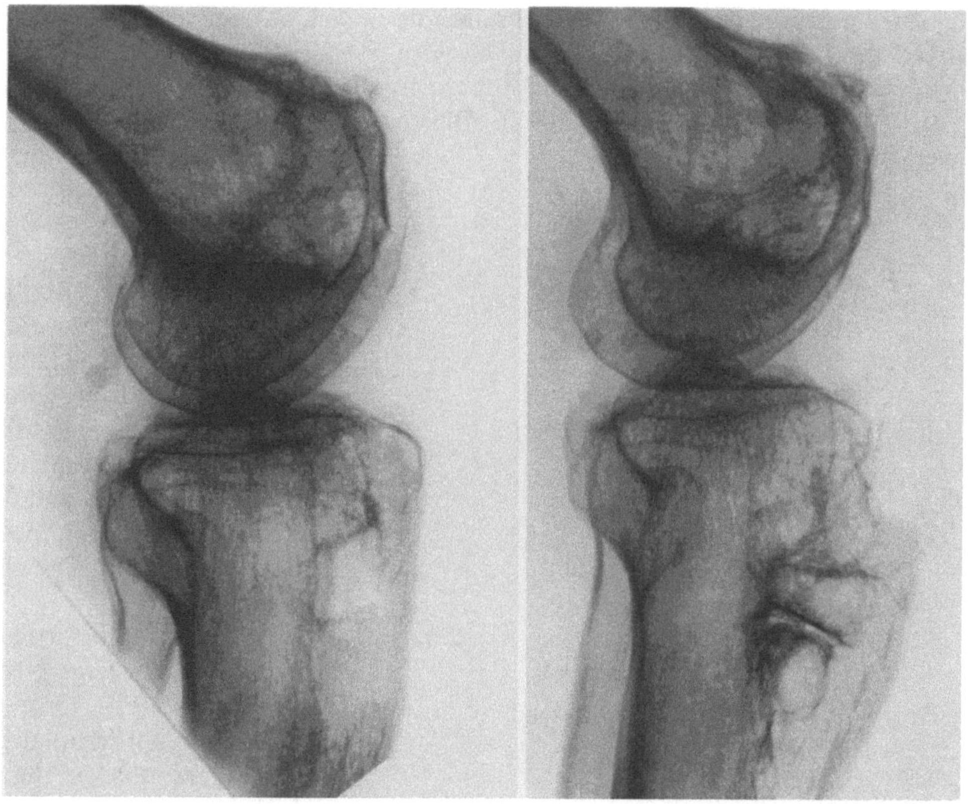

a b

Fig. 153 a and b. A 39-year-old patient after patellectomy (a), dramatically improved by an anterior displacement of the tibial tuberosity (b)

3. Anterior Displacement of the Tibial Tuberosity After Patellectomy

In our opinion, patellectomy should be carried out very rarely (see page 262). Even severely comminuted fractures of the patella can be treated without patellectomy (Pauwels, 1965a). Many patients continue to complain of knee pain after patellectomy, and often they can no longer fully extend the knee. In such cases, anterior displacement of the tibial tuberosity improves the efficiency of the quadriceps muscle by lengthening the lever arm of its tendon. It has a mechanical effect opposite to that of patellectomy. If the anterior displacement is sufficient, pain usually disappears and full active extension of the lower leg becomes possible (Fig. 153).

It is interesting to notice that nature has deprived the kangaroo of a patella but has provided it with a very prominent tibial tuberosity (Fig. 154). This gives the quadriceps an efficient leverage.

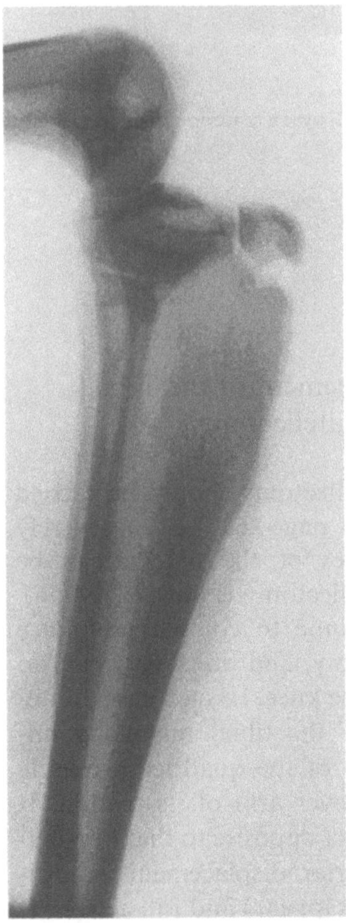

Fig. 154. The knee of a young female kangaroo. Nature has advanced her tibial tuberosity considerably!

4. Changes at Arthroscopy

Shiomi et al. (1978) have carried out arthroscopy in their patients systematically before and at regular intervals after displacing the tibial tuberosity anteriorly. Before surgery, arthroscopy in a 60-year-old female patient shows eburnated bone on the lateral facet of the patella and on the femoral condyle (Fig. 155a). Nine months after an anterior displacement of the tibial tuberosity by more than 2 cm, the two surfaces are being covered by patches of white tissue (Fig. 155b). The bone is still visible between the patches. Two years and ten months after the operation, the patella and femoral condyles are covered by tissue which looks more like normal cartilage (Fig. 155c). The clinical result is rated excellent.

When the anterior displacement of the tibial tuberosity is less, e.g. 1 cm in a 60-year-old female patient (Fig. 156), the covering tissue looks of much poorer quality. Correspondingly, the clinical result is less than good.

Arthroscopy confirms the X-ray changes following surgery. From the cases thus observed, it appears that at least 2 cm anterior displacement is necessary in order to ensure regrowth of cartilage-like tissue in the patellofemoral joint.

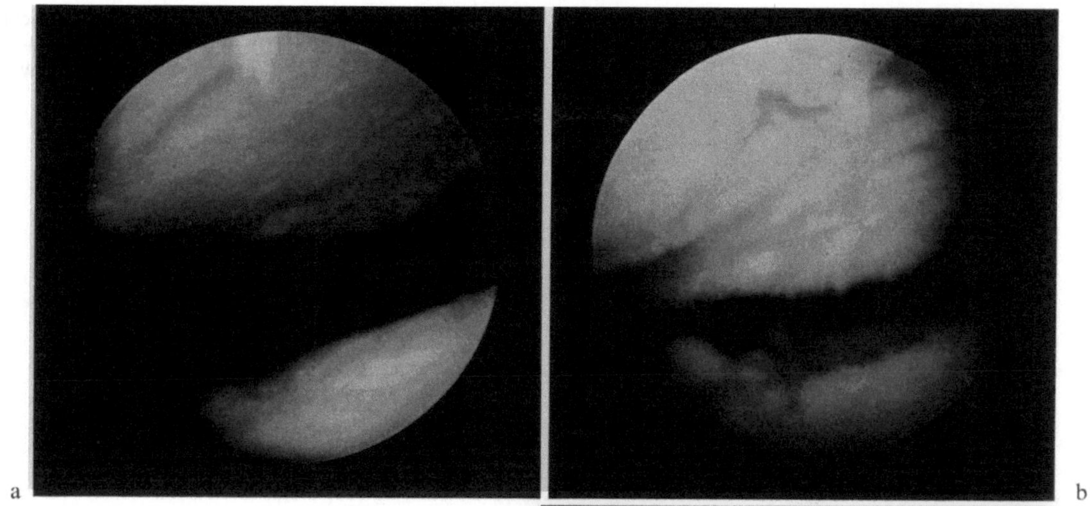

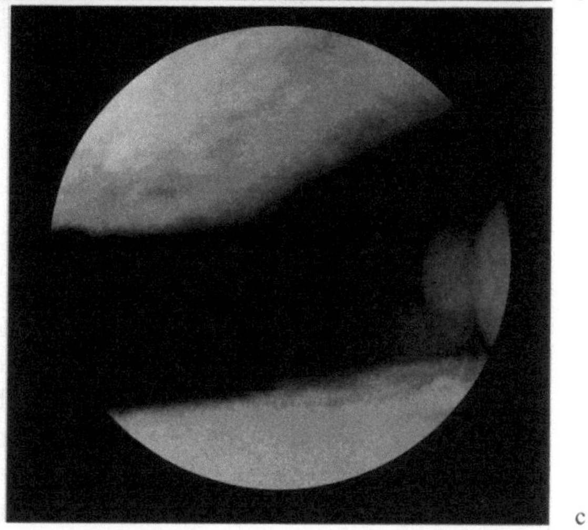

Fig. 155a–c. A 60-year-old female patient. Arthroscopy before (a), nine months after (b) and 34 months after (c) an anterior displacement of the tibial tuberosity. (From Shiomi et al. (1978). Permission to use these pictures has been granted by the authors and by the Japanese Arthroscopy Association)

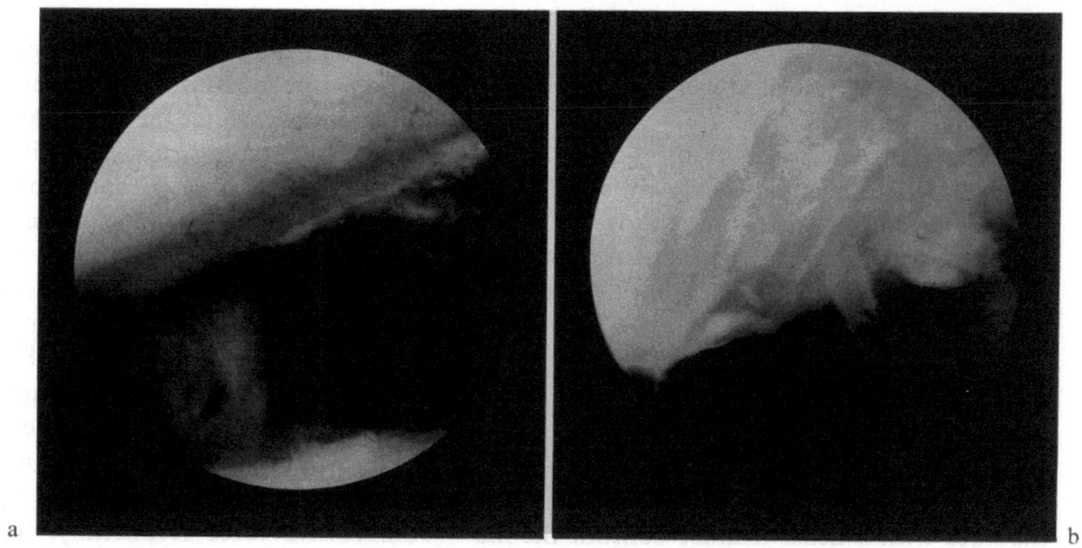

Fig. 156a and b. A 60-year-old female patient. Arthroscopy before (a) and one year and nine months after (b) a 1 cm anterior displacement of the tibial tuberosity. The replacement tissue looks obviously less good than in Fig. 155. The tuberosity has not been sufficiently displaced anteriorly. (From Shiomi et al. (1978). Permission to use these pictures has been granted by the authors and by the Japanese Arthroscopy Association)

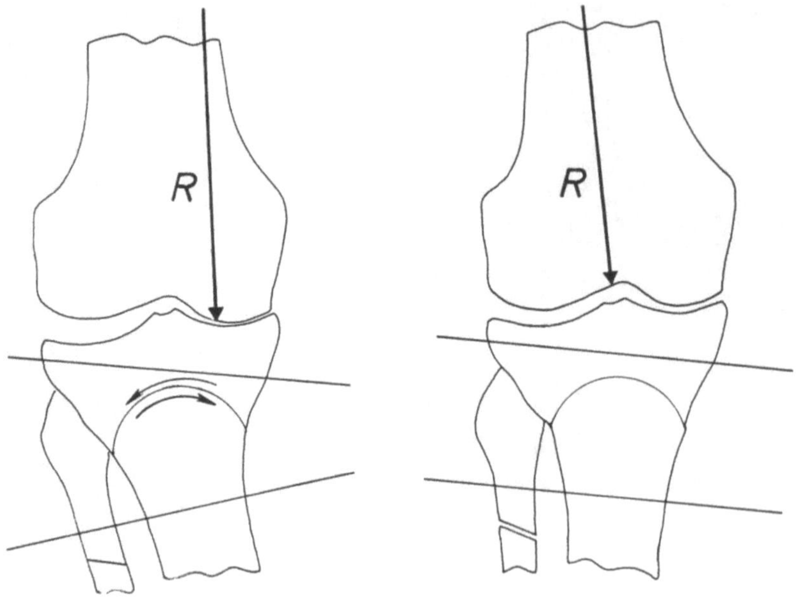

Fig. 157. The surgical treatment must bring back the load R to the centre of gravity of the weight-bearing surfaces

C. Recentring the Load

Rationale

As shown by theoretical analysis and X-rays, the compressive force R causes an asymmetrical distribution of the joint stresses if it is no longer exerted through the centre of gravity of the weight-bearing surfaces but displaced medially or laterally. The stresses are then abnormally high in the medial or in the lateral part of the knee. To obtain a balanced distribution of the compressive stresses over the weight-bearing surfaces, the load, force R, must be brought back between the tibial spines in the centre of the knee (Fig. 157).

Such a displacement of the line of action of force R is achieved usually by a valgus or a varus osteotomy performed above or below the knee, sometimes by the correction of a deformity located at a distance from the knee.

1. Osteoarthritis of the Knee with a Varus Deformity

Osteoarthritis prevailing in the medial part of the knee with a dense triangle beneath the medial tibial plateau and narrowing of the medial joint space results from a medial displacement of load R. Usually, but not always, it is accompanied by a varus deformity.

After correcting the flexion contracture, an upper tibial or a lower femoral osteotomy is planned to bring back the compressive force R to the centre of gravity of the weight-bearing surfaces in order to re-establish a symmetrical distribution of the compressive stresses in the joint.

a) Necessity of Overcorrecting the Varus Deformity

A return to the normal angle formed by the femur and the tibia in the coronal plane is not sufficient in most cases to recentre the load R. In patients in whom the varus deformity is secondary to osteoarthritis, returning the knee to neutral returns it to the mechanical conditions which provoked the appearance and development of the disease. Weakness of the muscles L which counterbalance force P seems to represent the most frequent origin of osteoarthritis with a varus deformity (Fig. 158a). Exact correction of the latter, i.e. re-alignment of the centres of the femoral head, the knee and the ankle, would shorten the lever arm a of force P and displace the line of action of the resultant compressive force R laterally, but not to the centre of gravity of the weight-bearing surfaces because of the weakness of the muscles L (Fig. 158b). In order to compensate for the lack of power of the muscles L, some valgum must

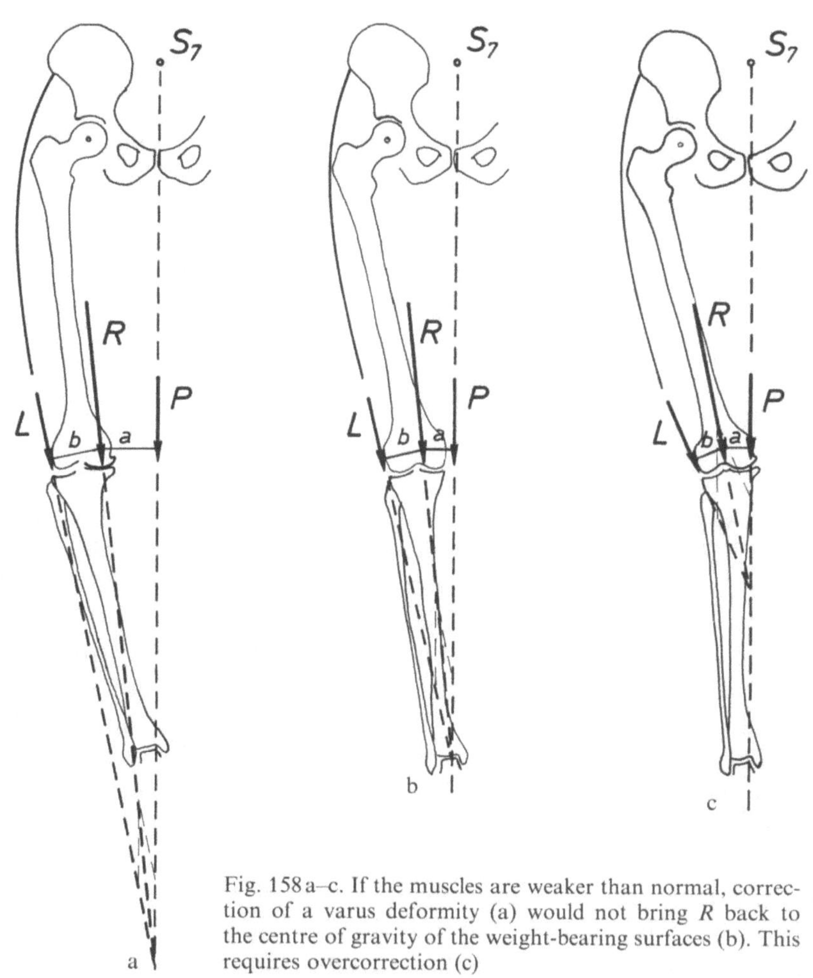

Fig. 158a–c. If the muscles are weaker than normal, correction of a varus deformity (a) would not bring *R* back to the centre of gravity of the weight-bearing surfaces (b). This requires overcorrection (c)

be created (Fig. 158c). Overcorrection makes the lever arm *a* of force *P* exerted by the partial mass of the body shorter than normal and thus decreases the moment $P \cdot a$. Force *P* can then be counterbalanced even by a weakened lateral tension band *L* and the line of action of the resultant force *R* is brought to the centre of the knee. This ensures an even distribution of the articular stresses over the largest possible weight-bearing surfaces. Since in many cases it is impossible to tell whether the varum is primary or secondary, overcorrection is the best course to take.

b) Accurate Estimation of Overcorrection

How does one accurately determine the necessary overcorrection? Several writers have proposed more or less complicated procedures for this. Kettelkamp and Chao (1972) attempt to calculate the actual distribution of the load between both tibial plateaux and deduce from it the angle which the long axes of the femur and of the tibia must form in the coronal plane in order to attain an ideal distribution. Their subjects are studied standing with symmetrical support on both feet. Each knee thus supports half the weight of the body above the knee, which is vertically exerted. This does not exactly correspond to the conditions of gait or even to that of the standing position on one foot, when the knee eccentrically supports not only the partial body weight but also the counterbalancing muscular forces. In fact the joint must be studied during gait or at least while standing on one foot. For these authors the distribution of the load on both tibial plateaux of a knee depends on the angle formed by the tibia and by the femur with a horizontal plane and on the anatomical form of the bones. Medial and lateral ligaments intervene as passive tension forces only in marked deformities with support on one pla-

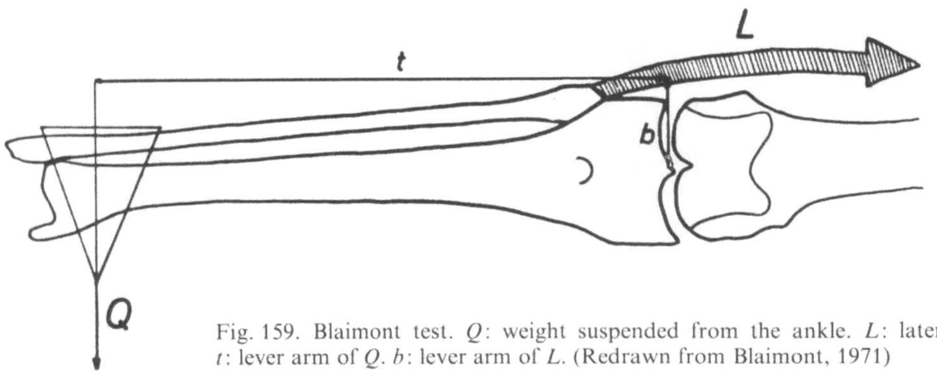

Fig. 159. Blaimont test. Q: weight suspended from the ankle. L: lateral muscular tension band. t: lever arm of Q. b: lever arm of L. (Redrawn from Blaimont, 1971)

teau. Applying their method to a normal subject of 60 kg they find a force of 24.5 kg exerted on the medial tibial plateau and 2.5 kg on the lateral plateau. "In this normal 60 kg subject, calculated compression force on the right medial tibial plateau was 24.5 kg; on the lateral 2.5 kg" (their Fig. 6). The X-ray of a normal knee is then shown with symmetrical subchondral scleroses underlying the tibial plateaux. The picture indicates a uniform distribution of the stresses and contradicts the results of the calculations. Indeed Kettelkamp and Chao use formulae which do not take the muscles (active forces of tension) into account. But, as we have seen, muscular action is essential to the equilibrium of the knee. It alone can explain the absence of osteoarthritis in many knee joints with valgus or varus deformity as well as the development of osteoarthritis in the medial part of the knee without any deformity and the development of osteoarthritis with a valgus deformity. Determination of the overcorrection to be done actually depends on the evaluation of the potential of the muscles which counterbalance force P. Neglect of this leads to conclusions which are not related to the facts.

Blaimont et al. (1971) describe a simple test which would permit measurement of the moment $L \cdot b$ of the lateral muscular tension band L in cases of osteoarthritis with varus deformity. From this measurement of the potential of the lateral muscles, one could deduce the degree of overcorrection to be obtained through an upper tibial osteotomy in order to recentre the load. The patient lies on his sound side, the osteoarthritic knee extended resting on a support. The other knee is flexed. The lower leg of the affected side extends beyond the examination table. The knee is observed antero-posteriorly by fluoroscopy. When the patient loosens his lateral muscles, a gap appears between the lateral tibial plateau and the femoral condyle. Voluntary contraction of the lateral muscles of the thigh closes the gap. From a ring fixed at the ankle, heavier and heavier weights are suspended. The minimum weight Q necessary to prevent closing of the joint space by the lateral muscles L of the thigh is recorded (Fig. 159). The distance t between the weight Q and the knee is measured. Blaimont et al. claim that the moment $Q \cdot t$, experimentally determined, measures the muscular potential for stabilizing the knee:

$$Q \cdot t = L \cdot b.$$

Since $L \cdot b = P \cdot a$, a simple calculation permits determination of the length of the lever arm a of the body weight P to be attained by osteotomy:

$$a = \frac{L \cdot b}{P}.$$

During surgery, after a curved osteotomy of the upper metaphysis of the tibia, a metal rod is placed on the patient using the image intensifier. This rod joins the centre of gravity of the body, conventionally located at the second sacral vertebra, and the centre of the ankle. The osteotomy fragments are displaced until the tibial spines are at a distance a from the rod. The correction is then fixed with compression clamps.

This test measures only the action of the tensor fasciae latae, the lower leg being ex-

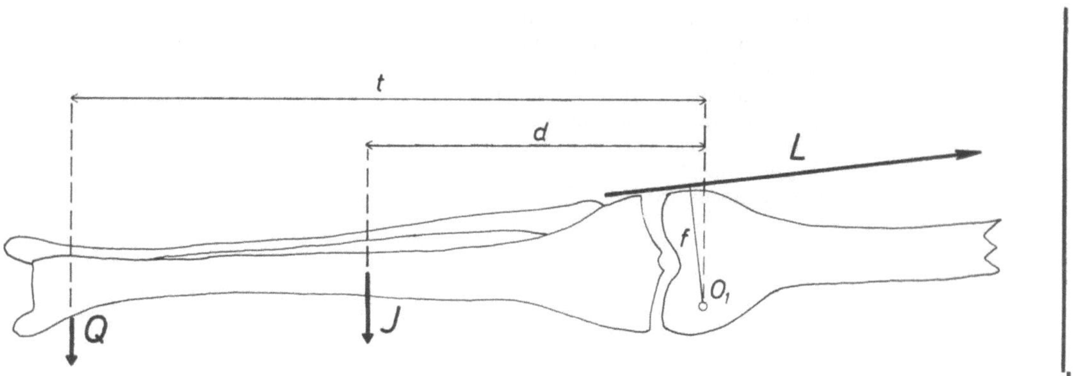

tended. But the moment $P \cdot a$ of force P, resultant of the weight of a part of the body and of forces of inertia, is at its maximum at the beginning of the single support period of gait when force P acting behind the knee is counterbalanced mainly by the quadriceps muscle and at the end of the single support period when force P acting in front of the knee is counterbalanced mainly by the hamstrings. The tensor fasciae latae constitutes the lateral muscular tension band only during a fraction of the step when the counterbalancing moment $P \cdot a$ is at its minimum (see Table 5, page 46).

The test of Blaimont et al. is not fully satisfactory because it measures only a part of the muscles which intervene to achieve equilibrium. Moreover, the moment $Q \cdot t$ is the product of the weight suspended from the ankle and the distance t between this weight and the knee. It would be equivalent to the muscular moment $L \cdot b$ (Fig. 159). However, the lateral muscles L must support not only the weight Q suspended from the ankle but also the weight J of the lower leg and foot maintained in space. This weight is not negligible. It acts with a lever arm which is the distance d between the knee and the centre of gravity of the mass of the lower leg+foot (Fig. 160). Consequently, the muscular moment is not equal to $Q \cdot t$ but to $Q \cdot t + J \cdot d$.

When standing on one foot the knee supports the body weight minus the loaded lower leg and foot. In the formula, P represents this partial weight. Because of that, the centre of gravity S_7 of the mass eccentrically supported by the knee does not lie on the median line.

Fig. 160. Critical analysis of the Blaimont test. J: weight of lower leg and foot. d: lever arm of J. f: lever arm of L in the position thus described. O_1: centre of curvature of the medial condyle. P: force exerted by the partial mass of the body. a: lever arm of P

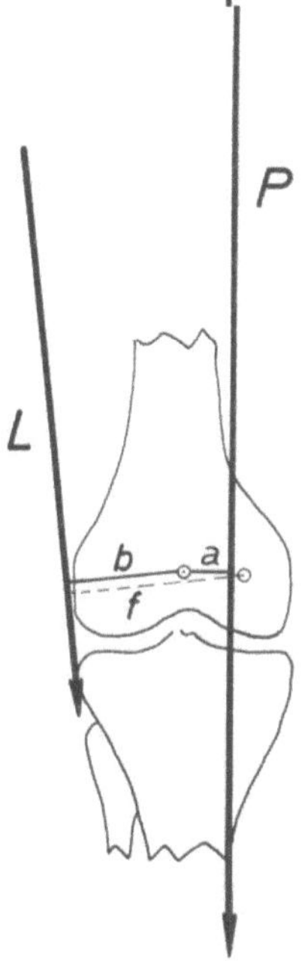

It does not correspond to the centre of gravity of the whole body.

When there is no lateral opening of the joint the resultant of the forces exerted on the knee can be anywhere between the centres of curvature O_1 and O_2 of the condyles. When a lateral gap appears, the tibia pivots in the coronal plane about the centre of curvature O_1 of the medial femoral condyle. The resultant of the forces exerted on the knee must therefore pass through the centre of curvature O_1 of the medial condyle, since there is then no direct contact remaining between the lateral tibial plateau and the femoral condyle. Consequently, the lever arm of the muscular force L is the distance f between the line of action of L and the centre of curvature O_1 (Fig. 160). But the distance at which the knee should be placed from force P is measured from the tibial spines. Lever arms a and f do not have the same origin and as

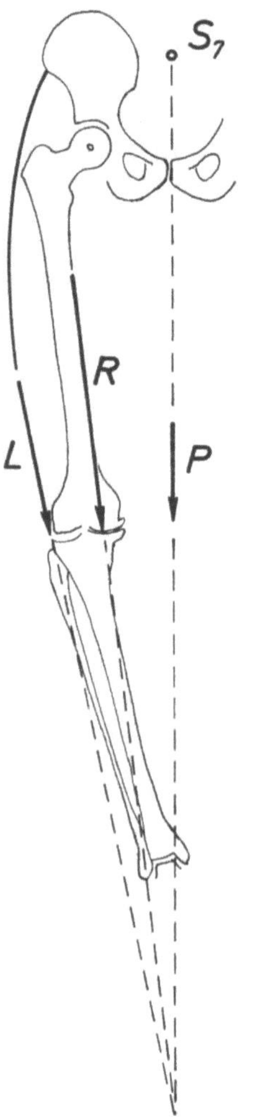

Fig. 161. In osteoarthritis with a varus deformity, the femur and the muscles *L* are less inclined to the line of action of force *P* than in a normal knee. Resultant force *R* is more vertical and acts at right angles to the plane tangential to the tibial plateaux. (See Fig. 73)

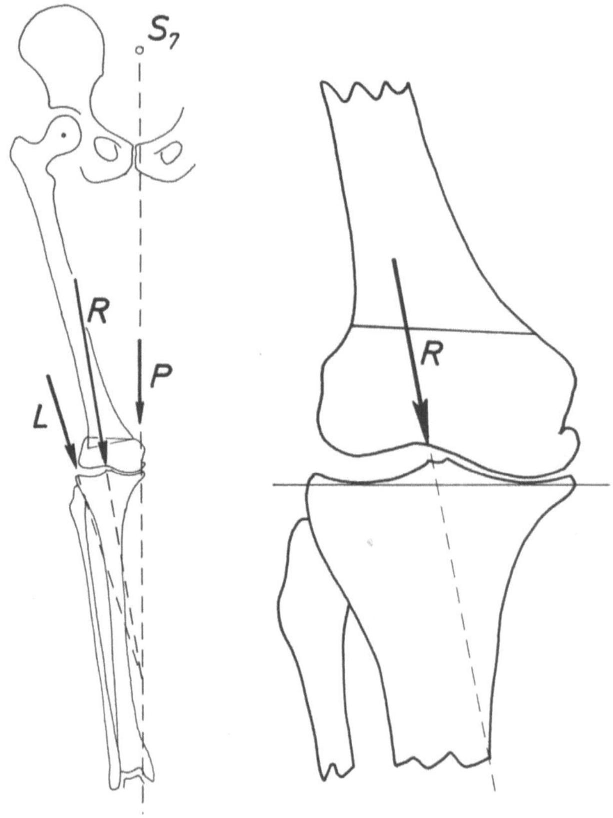

Fig. 162. Overcorrection of a varus deformity redirects the femur and the muscles, which become more inclined to the line of action of force *P*. If the overcorrection is arrived at by a femoral osteotomy, resultant force *R* becomes oblique to the plane tangential to the tibial plateaux

such cannot be used to establish the relation of equilibrium between *P* and *L* (Fig. 160).

On the other hand, forces of inertia intervene during gait and combine with the partial body weight. They vary with accelerations and continuously and considerably modify force *P* acting eccentrically on the knee.

Actually, even Blaimont (1982) himself no longer relies on his test routinely. He claims to use it only in considerable deformities with instability. Otherwise, he tries to achieve a "moderate" overcorrection, without any preliminary planning. In other words, he then works by the rule of thumb.

To measure accurately the necessary overcorrection, one ought to know the potential strength of the muscles which are involved in gait. At the present time no accurate method to determine this exists. The overcorrection to be attained must be chosen empirically for each case.

c) Choice of Procedure: Tibial or Femoral Osteotomy?

As mentioned above (page 80), in osteoarthritis with varus deformity, the line of action of the resultant force *R* is displaced medially and is less inclined to that of force *P* than in a normal knee. Because of the eventual erosion of the medial margin of the tibial plateau, *R* remains normal to the plane tangential to the plateaux (Fig. 161). After a supracondylar osteotomy of the femur overcorrecting the varus deformity, the femur and the muscles *L* are redirected more obliquely in relation to the line of action of force *P* (Fig. 162). The intersection of the lines of action of forces *P* and *L* is brought closer to the joint. Resultant force *R* becomes more inclined to the line of action of force *P*, rotating counterclockwise in our diagram. The whole lower leg and the distal fragment of the femur pivot about the heel, which remains on the line of action

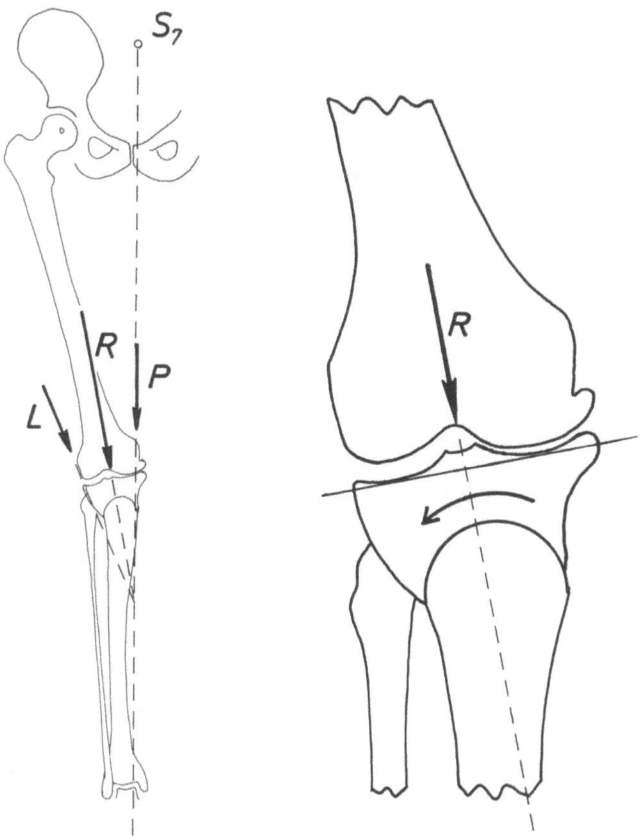

Fig. 163. If overcorrection of a varus deformity is arrived at by a tibial osteotomy, resultant force *R* remains normal to the plane tangential to the tibial plateaux

of force *P* (to be precise about 3 mm lateral to the line of action of *P*: see page 25). Because of this rotation of the lower leg, clockwise in our diagram, and the opposite rotation of the resultant *R*, force *R* is no longer normal to the plane tangential to the tibial plateaux. It acts then with a tangential component which tends to displace the femur medially on the tibia. Therefore, force *R* cannot be distributed evenly over the weight-bearing surfaces of the joint.

Supracondylar osteotomy of the femur with transverse displacement has also been proposed for correction of varus deformity (Izadpanah and Keönch-Fraknóy, 1977). It is, however, subject to the same criticism that applies to any supracondylar osteotomy. Moreover, the amount of correction, not to mention overcorrection, provided by such an osteotomy is much restricted by the very width of the femur.

Overcorrection of a varus deformity by an osteotomy of the upper end of the tibia will also open the angle formed by the femur and the lateral muscles *L* with the line of action of force *P* (Fig. 163). The resultant force *R* thus rotates counterclockwise in the diagram. But the proximal fragment of the tibia pivots in the same direction, i.e. in the opposite direction to the distal fragment, which rotates clockwise about the heel supported by the ground. Therefore, the resultant force *R* remains at right angles to the plane tangential to the tibial plateaux. If the overcorrection has been sufficient, force *R* is brought back to the centre of gravity of the weight-bearing surfaces and the stresses are then optimally redistributed over the joint. This has been demonstrated by Kummer (1977).

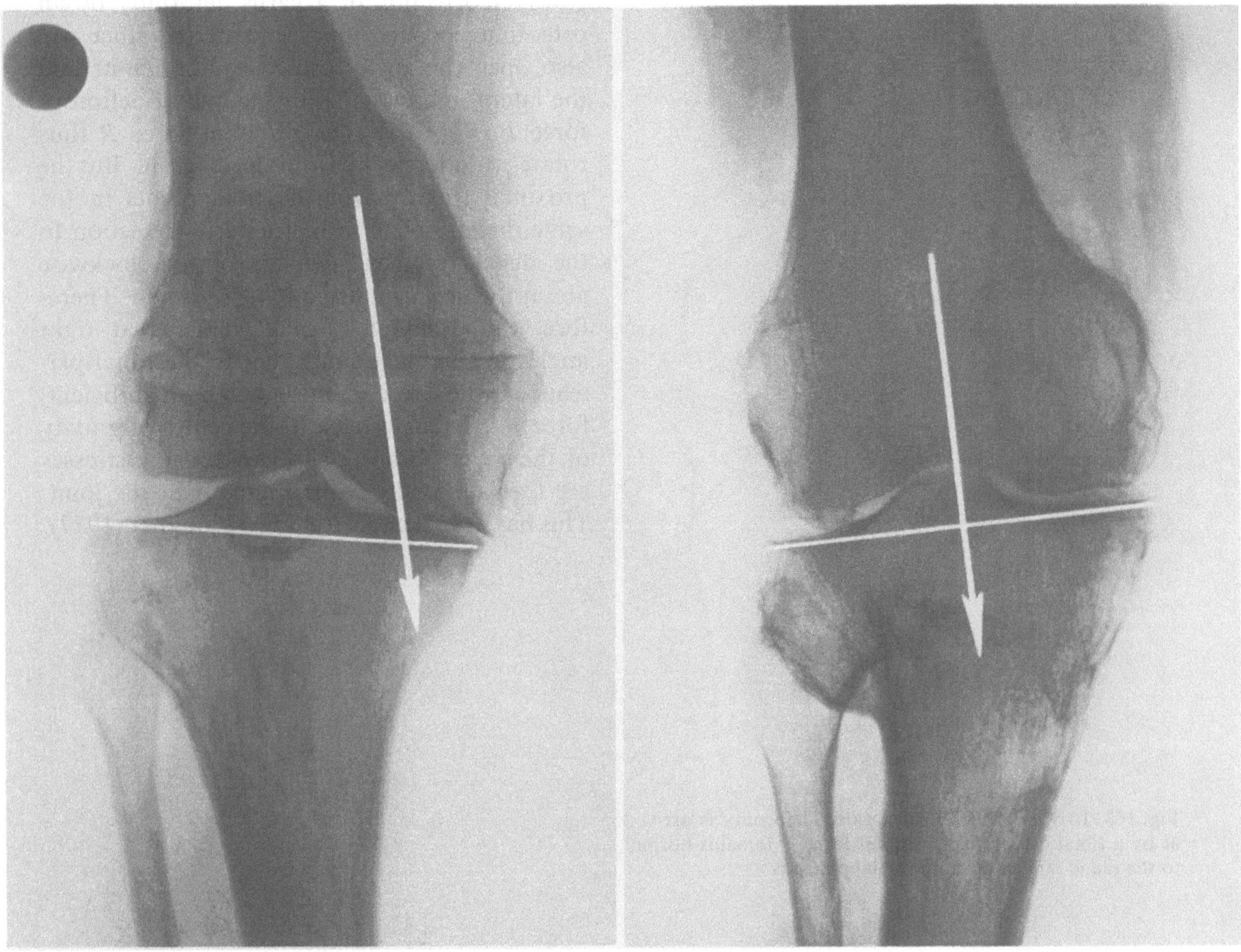

a b

Fig. 164a and b. Overcorrection of a varus deformity by a femoral osteotomy (a) and by a tibial osteotomy (b)

Comparison of X-rays after a distal femoral osteotomy (Fig. 164a) and after a proximal tibial osteotomy (Fig. 164b), both carried out for osteoarthritis with a varus deformity, shows the difference in the mechanical conditions. The resultant force is represented by a white arrow and the plane tangential to the tibial plateaux by a white line.

Osteoarthritis with a varus deformity thus should generally be dealt with by osteotomy of the upper end of the tibia.

d) Previous Operative Procedures

In osteoarthritis with varus deformity, a proximal tibial osteotomy is nearly always indicated. Many techniques have been proposed. We have used several of them before adopting the barrel-vault osteotomy. Simply to elevate the medial plateau with a triangular graft inserted beneath it has been disappointing in our hands. In both cases for which we used it, the graft was crushed and the correction was partially lost.

Lange (1951) advocated an inverted V osteotomy with the apex of the V one centimetre below the tibial spines. A wedge was resected along the lateral branch of the V and inserted in the medial branch after tilting the lower fragment about the apex of the V. The wedge consisting of cancellous epiphysial bone was not strong and was often crushed when inserted beneath the medial tibial plateau. Often the fragments got crushed when moved, making correction inaccurate. Fixation was seldom good enough to avoid plaster immobilization. Osteotomy with impaction of the fragments has the same drawbacks.

Osteotomy with resection of a lateral wedge between the joint and the insertion of the patella tendon (Conventry, 1965) can be carried out with good precision and makes anterior displacement of the patella tendon possible by moving the distal fragment forward. But it sacrifices the upper tibio-fibular joint. It cannot be used for the correction of large deformities because the space between the anterior tuberosity and the joint is seldom large enough to allow excision of a wide wedge. Fixation of the fragments with Blount staples often requires plaster immobilization. After such immobilization flexion of the knee may be limited. If performed below the anterior tuberosity of the tibia, the osteotomy heals much more slowly and does not permit the simultaneous anterior displacement of the patella tendon. It requires a longer immobilization or a heavy internal fixation.

After a wedge osteotomy Gariepy (1967) and Mac Intosh (1970) fix the fragments with compression clamps. Blaimont (1970) proposes the same fixation after a curved osteotomy proximal to the anterior tibial tuberosity and a rotation of the distal fragment inside the proximal fragment.

We have added to the technique described by Blaimont an anterior displacement of the patella tendon and a different preoperative planning. Greater accuracy can be obtained using this modified technique.

e) The Barrel-Vault-Osteotomy for Varus Deformity

α) *Planning*

An X-ray showing the whole lower limb under load makes it possible to determine the angle α formed by the tibia and the femur in the coronal plane (Fig. 165). A line is drawn from the centre of the femoral head to the midpoint of the cross section of the tibia at the level of the planned osteotomy. Another line is drawn from this point in the tibia to the centre of the ankle. The angle α is formed by the femoral line and the proximal prolongation of the tibial line. The outlines of the knee to be operated on are drawn on transparent paper from an A.-P. view of the loaded knee (Fig. 166a). The osteotomy curve is drawn. Its radius approximates 2.5 cm. The insertion of the patella tendon lies in its concavity. Two transverse lines are added to the drawing, one above and the other below the osteotomy line. The transverse line distal to the tuberosity is at right angles to the long axis of the tibia, the one proximal is oblique to this axis. They form an angle open laterally of $\alpha + 3°$ to 6°. The lower fragment is now traced on a second transparent sheet of paper with the line which intersects it (Fig. 166b). The second sheet is rotated on the first and the convexity of the distal fragment moved inside the concavity of the proximal until the straight line on the second sheet is parallel to the straight line intersecting the proximal fragment on the first sheet (Fig. 166c). The upper tibial fragment and the femur are then traced on the second sheet (Fig. 166d). In this way the surgical procedure is carried out graphically preoperatively (Fig. 166e).

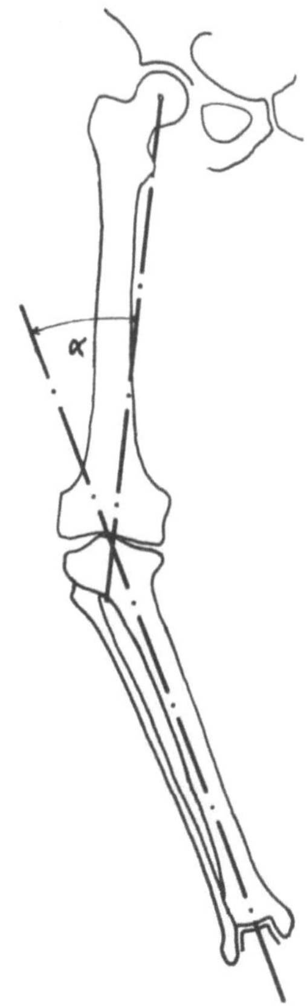

Fig. 165. Tracing from X-rays of the whole limb. The axes of the femur and the tibia delineate the angle α

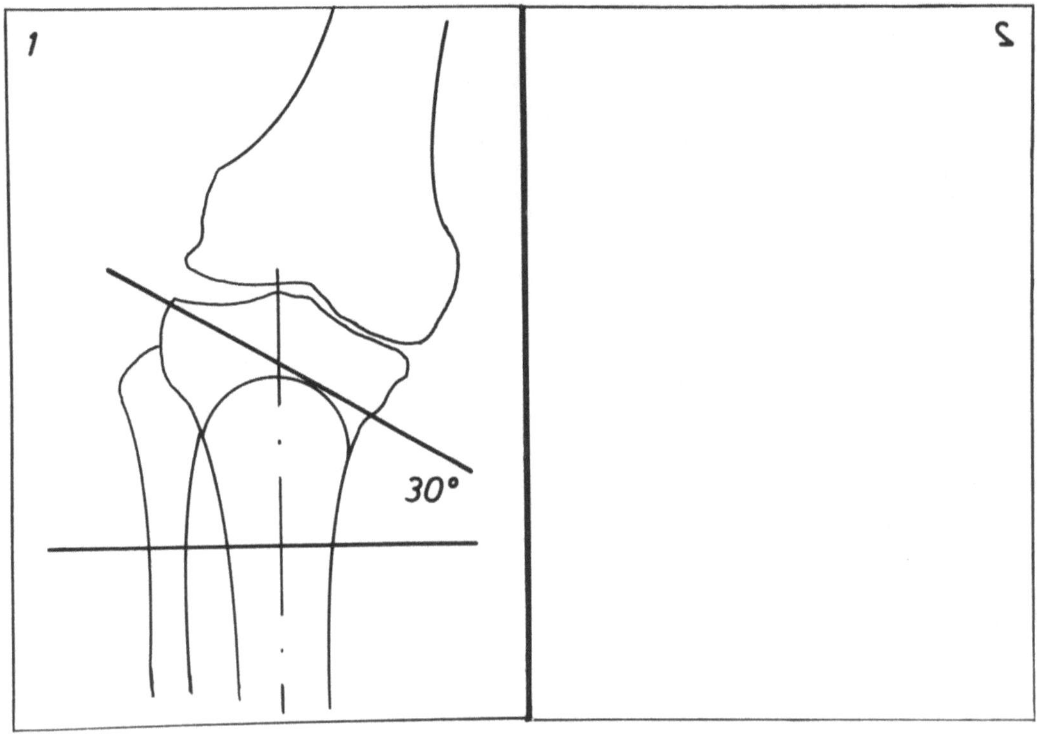

a

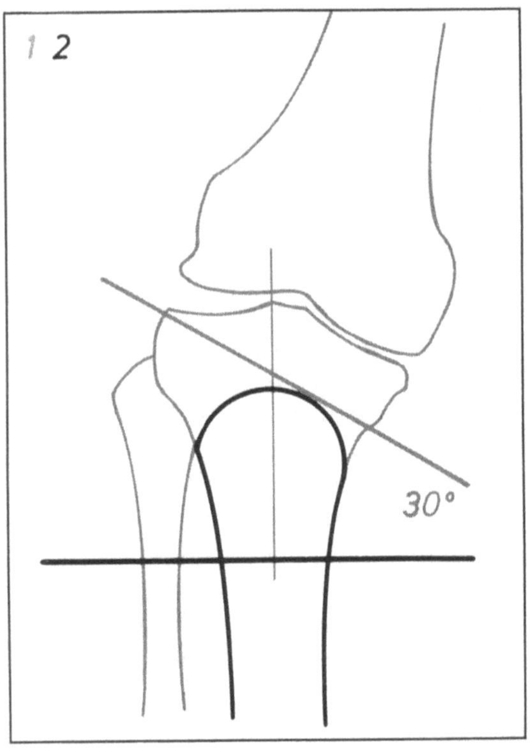

b

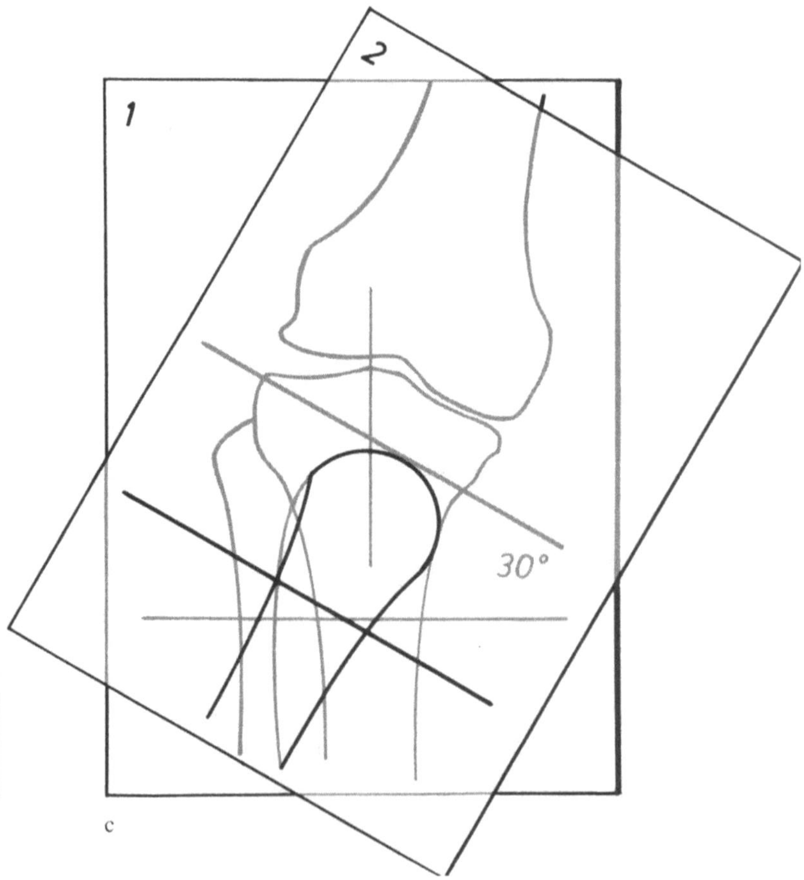

c

Fig. 166a–e. Planning the barrel-vault osteotomy of the tibia

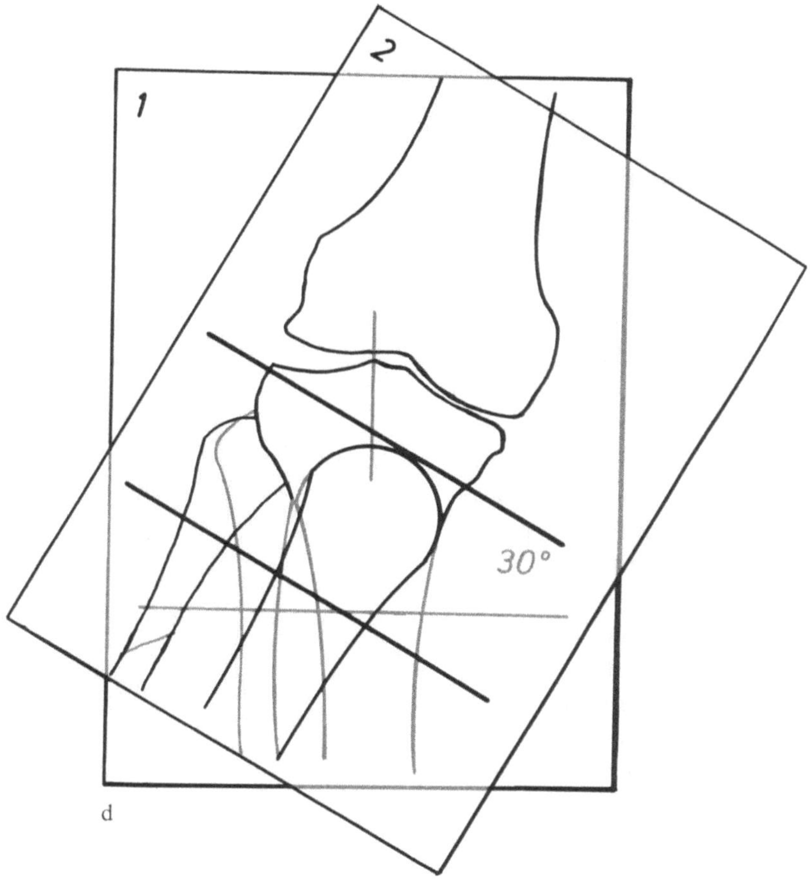

d

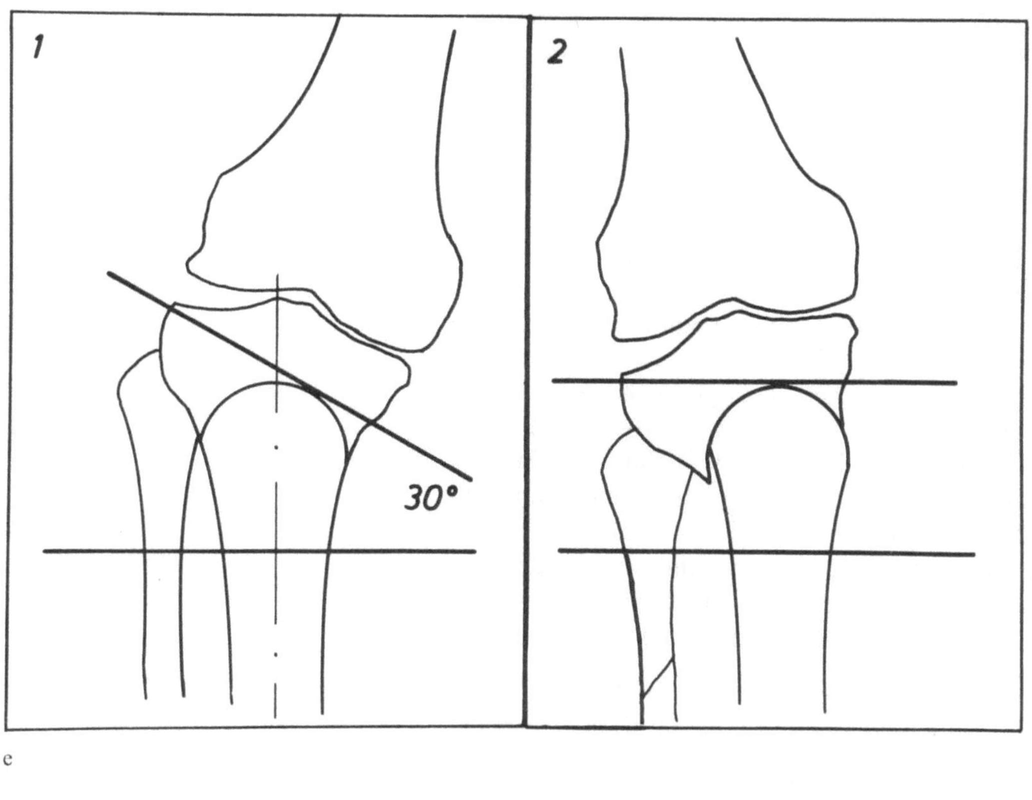

e

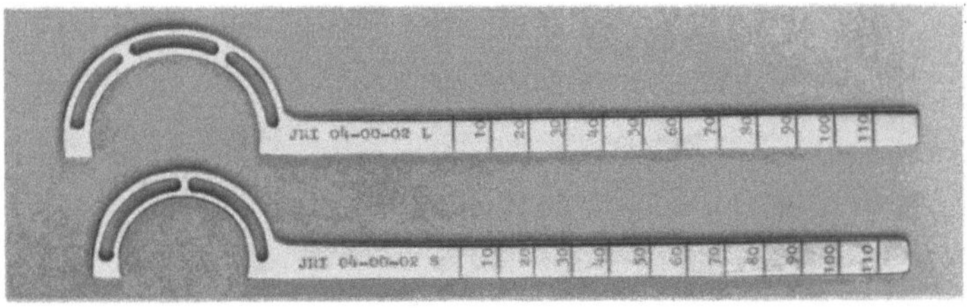

Fig. 167a. Osteotomy guide for the barrel-vault osteotomy. Two sizes are available

Fig. 167b. X-ray during operation

β) Instruments. We use special instruments for this procedure: an osteotomy guide, a pin insertion guide and a metal sleeve with a blunt-ended Steinmann pin.

Our osteotomy guide (Fig. 167a) is made up of a straight handle ending in a slotted curve. When the curve is inserted behind the patella tendon (Fig. 167b), it guides the insertion of a Kirschner wire driven by an air-powered drill. The osteotomy line is traced by multiple insertions of the Kirschner wire through the slot. The holes form a curve corresponding to the shape of the osteotomy guide. The guide is supplied in two sizes each with a different radius of curvature.

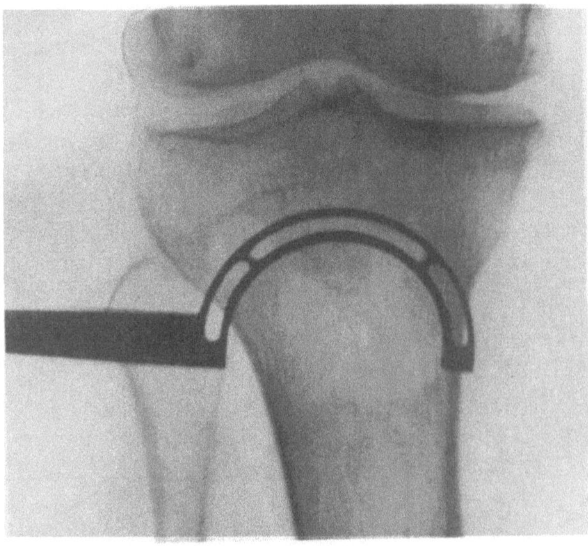

b

The pin insertion guide[14] consists of a bar on which two mobile parts can move (Fig. 168). One of these shows a graduation in degrees and can rotate about an axli fixed in the bar. The other mobile piece can slide along the bar and be fixed at any level. A Kirschner wire or a Steinmann pin is passed through a tunnel in each of the mobile parts. The two wires or pins form the angle indicated by a pointer on the graduated scale. The sliding part allows modification of the distance between the wires or the pins without changing the angle which they form.

A third mobile piece sliding along the bar is added for the femoral osteotomy. The Steinmann pins inserted through both pieces which can slide along the bar are strictly parallel.

The sleeve (Fig. 169) is a thin tube exactly adapted to the diameter of the Steinmann pins (5 mm). A blunt-ended Steinmann pin is introduced in the sleeve. Sleeve and blunt-ended pin are pushed to the bone, tibia or femur, through the muscles. The pin is then removed and a

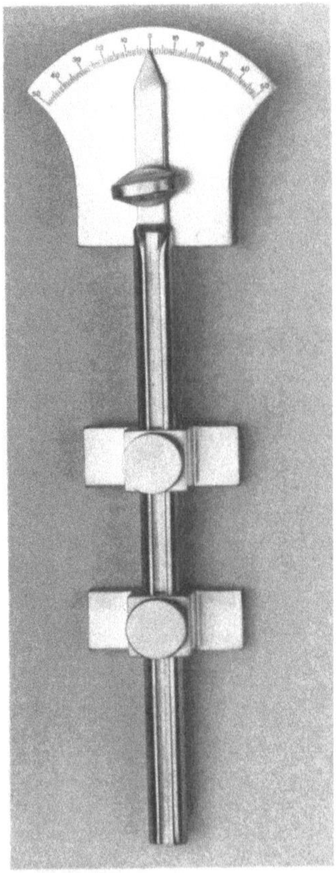

Fig. 168. Pin guide for proximal tibial and distal femoral osteotomy

[14] The pin insertion guide has been designed by Joint Replacement Instrumentation Ltd, London, as an improvement of a first instrument made by the author

Fig. 169a and b. Sleeve and blunt-ended pin used to protect the muscles

a

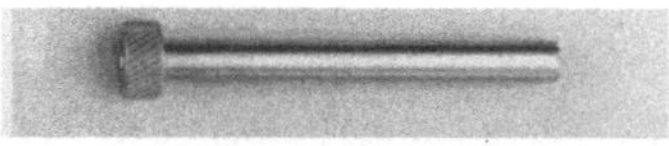

b

pointed Steinmann pin is introduced through the sleeve to be drilled through the bone without damage to the soft tissues.

γ) Surgical Procedure. The operation is performed without a tourniquet in order to avoid venous stasis and to diminish the risks of thrombophlebitis and subsequent pulmonary embolism. The surgeon operates more comfortably if seated.

1. With the patient supine under general anaesthesia the knee is flexed. A short posterolateral incision gives a view of the middle third of the fibula between the soleus posteriorly and the peroneal muscles anteriorly. The periosteum is incised and cautiously elevated. The fibula is divided obliquely from below upward and medially, with an oscillating saw. The fragments must be able to slide on each other when the tibial fragments are rotated. One must avoid tearing the fibular vein, which is very close to the medial periosteum. This vein can bleed badly and is difficult to ligate. Only the subcutaneous tissue and the skin are sutured using suction drainage.

2. The knee is now extended. A longitudinal incision 5 cm long centred on the anterior tibial tuberosity is made (Fig. 170). The aponeurosis is split on both sides of the patella tendon. A curved periosteal elevator is used to clear the bone behind the tendon and on both sides of the tibial tuberosity. The osteotomy guide is inserted behind the patella tendon (Fig. 167b). Its position is checked by the image intensifier. A series of holes is drilled in the tibial epiphysis with a Kirschner wire positioned by the guide. These holes delineate a curve around the tibial tuberosity.

The sleeve and blunt-ended pin are pushed to the tibia through a 5 mm incision on the lateral aspect of the lower leg, 5 to 7 cm below the tibial tuberosity. The assistant holds the foot. If he perceives a movement of the latter, the sleeve and pin are changed and inserted again more anteriorly or posteriorly. Actually, they should abut against the bone as posteriorly as possible. The blunt-ended pin is then replaced by a pointed one which we use to drill through the bone by hand, with a handle, to avoid heating of the tissues. The distal pin must be at right angles to the long axis of the tibia. The pin insertion guide is then adjusted to the distal pin. Its pointer indicates the desired angle. The proximal pin is passed through the appropriate tunnel and, through a 5 mm incision, to the lateral aspect of the tibia, immediately beneath the joint space. Its position is checked by the image intensifier. For a good hold in the bone, the proximal pin should pass through the subchondral sclerosis if at all possible.

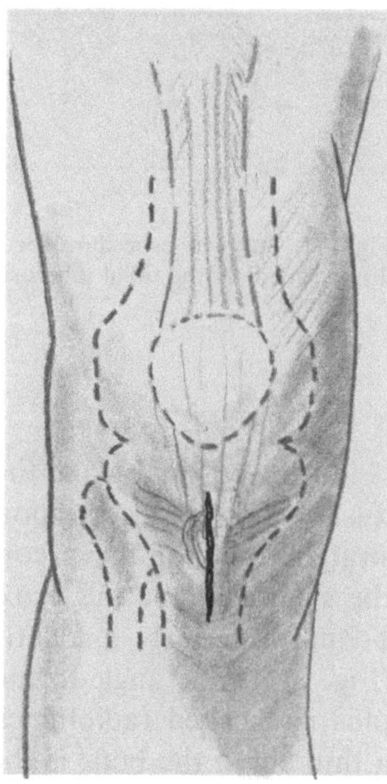

Fig. 170. Skin incision for the barrel-vault osteotomy of the tibia

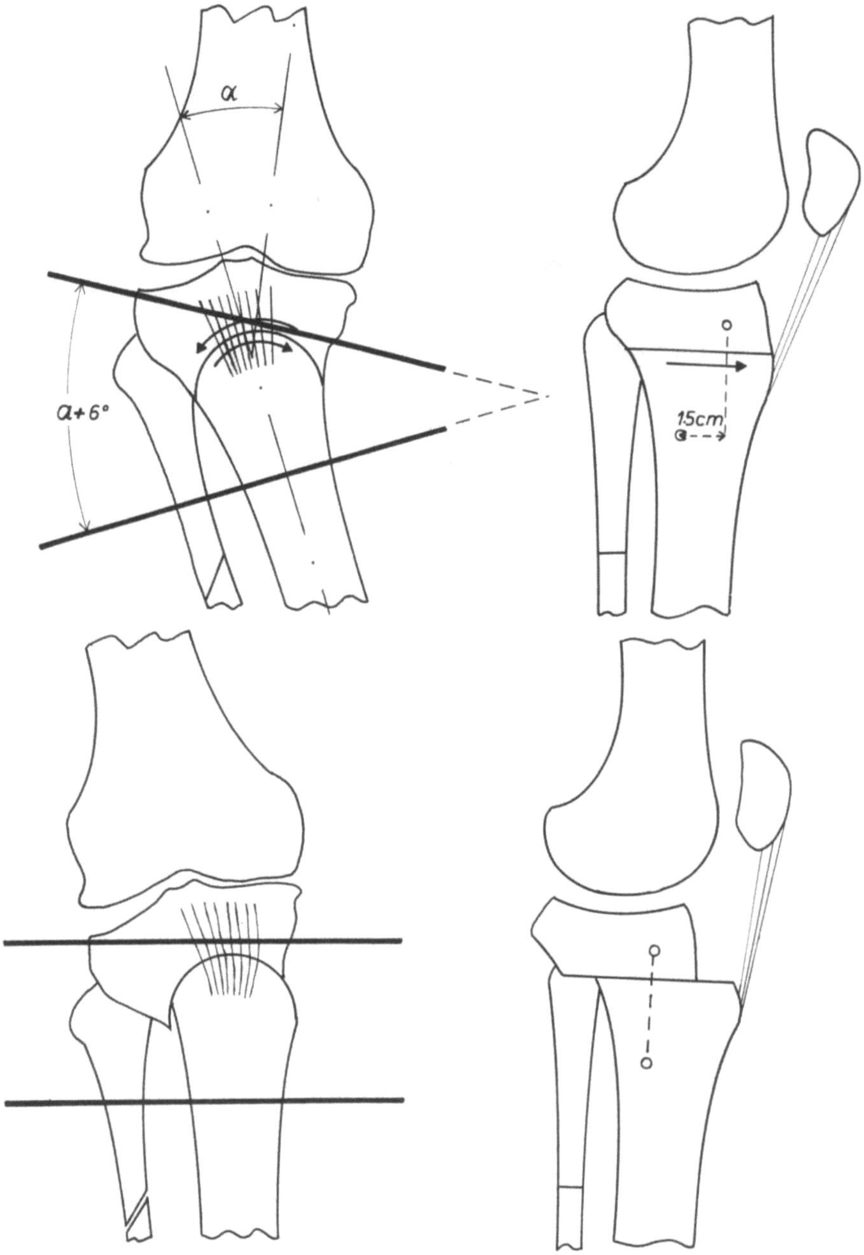

Fig. 171. Surgical procedure for the barrel-vault osteotomy combining overcorrection of varus deformity and anterior displacement of the tibial tuberosity

In the coronal plane the pins form an angle $\alpha + 3°$ to 6° and correspond to the transverse straight lines of the preoperative drawing. In the sagittal plane the proximal pin is inserted 1–2 cm anteriorly in relation to the distal pin (Fig. 171). The angle formed by the Steinmann pins is checked radiologically (Fig. 172). With a thin chisel the bone is then divided along the curved line delineated by the Kirschner wire holes. The tibial fragments are rotated until both Steinmann pins are parallel and the distal fragment is displaced forward until the pins lie in the same coronal plane. Two clamps, placed over the pins, fix the fragments under compression (Fig. 173). The wound is sutured using suction drainage.

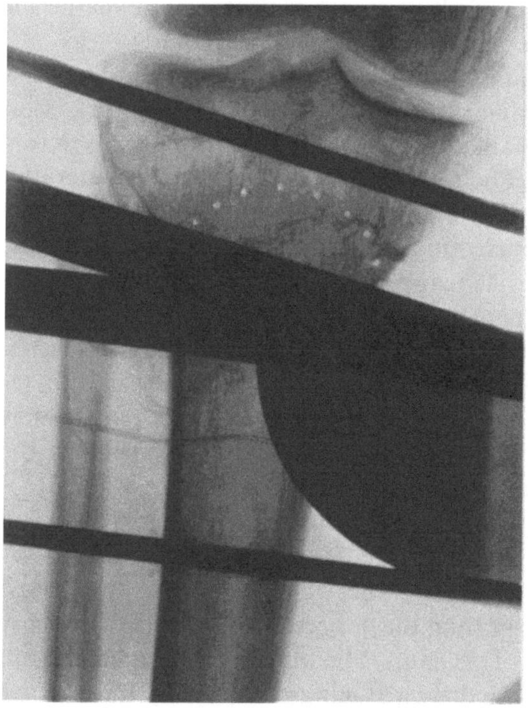

Fig. 172. X-ray control of the angle formed by the Steinmann pins. Notice the holes delineating the osteotomy curve

δ) Postoperative Care. The knee is passively and actively flexed from the day after surgery. On the second day the patient stands and walks with two crutches. He puts as much weight as he can on the operated leg. A long X-ray showing the whole leg is taken after four or five days. If the angle formed by the femur and the tibia does not appear satisfactory, overcorrection of the deformity can be modified to some extent by tightening the lateral or the medial compression clamp. After eight weeks the Steinmann pins are removed after an X-ray has shown that the osteotomy has healed. The use of crutches is progressively diminished as the patient feels stable on the operated leg, which occurs between two and four months after surgery.

ε) Comments and Examples

The upper tibial osteotomy as described here allows an accurate overcorrection of the varus deformity and an anterior displacement of the patella tendon (Fig. 171) without any troublesome bulge below the knee. It avoids plaster immobilization and permits early weight-bearing. The pain relief is usually immediate. The X-ray changes appear during the first year, although here only the last follow-up will be compared with the preoperative picture.

The overcorrection of the varus deformity is to compensate for the weakness of the lateral tension band *L*. After surgery, the plane of the tibial plateaux should be at right angles to the line of action of the resultant force *R* (pages 162, 163). The orientation of the tibial plateaux in relation to the horizontal plane is irrelevant. The plateaux must not tilt backward more than normally, and above all not forward. If necessary, a third Steinmann pin parallel to and in the same transverse plane as the proximal one can be added to avoid such tilting. This requires special mobile units for the compression clamps.

At surgery the collateral ligaments are not tightened. They spontaneously contract postoperatively. Laxity disappears completely if the deformity has been sufficiently overcorrected.

The results of the barrel-vault osteotomy for varus deformity are good if a sufficient overcorrection has been achieved. The valgum shortens the lever arm *a* of force *P* exerted on the knee by the mass S_7 of the body. That reduces the force required from the lateral muscular tension

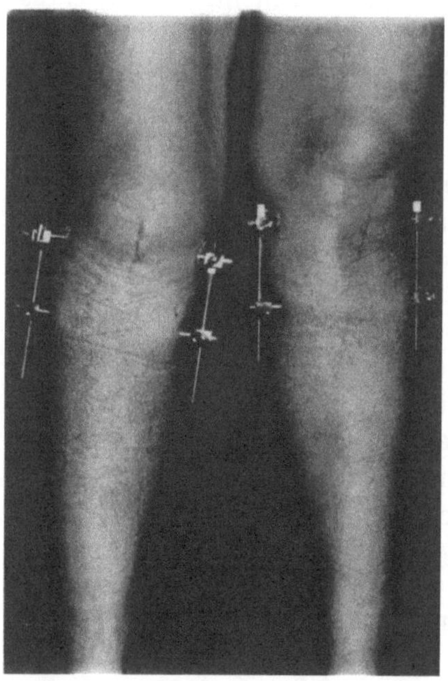

Fig. 173. Fixation by compression clamps. The 4 to 5 cm scar is clearly visible as is the overcorrection. The two knees are rarely operated on at the same session

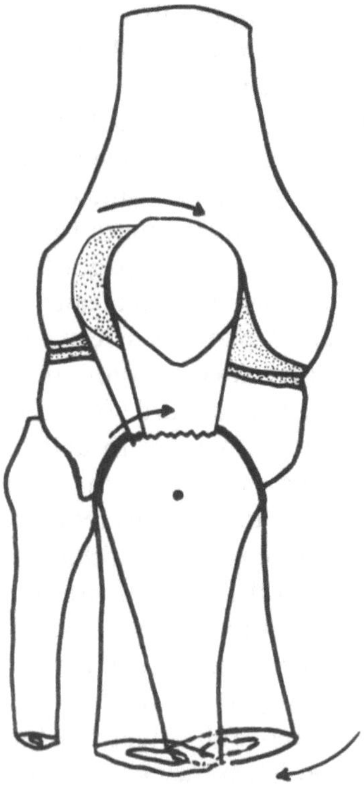

Fig. 174. In the barrel-vault osteotomy, turning the convexity of the distal fragment in the concavity of the proximal fragment displaces the tibial tuberosity medially (drawn by Y. Andrianne)

band L and thus diminishes the compressive force R acting on the knee. But essentially, because of its lateral displacement, force R is distributed over the largest possible weight-bearing surfaces of the knee. Consequently, the valgus proximal tibial osteotomy reduces the articular compressive stresses in the femoro-tibial joint.

The objection has been raised that correcting the varus deformity would subluxate the patella laterally. This is theoretically true with most techniques of tibial osteotomy, but in fact just the opposite occurs with the barrel-vault osteotomy. The patella tendon is inserted at the apex of the convexity of the distal fragment. When the latter rotates inside the concavity of the proximal fragment, clockwise in our diagram, the insertion of the patella tendon is displaced medially (Fig. 174). This results in a recentring of the patella into the intercondylar groove of the femur rather than in a lateral subluxation (see page 151). Consequently, the barrel-vault osteotomy of the tibia, such as described above, displaces the tibial tuberosity anteriorly and medially. This provokes a decrease and a redistribution of the compressive stresses over larger surfaces in the patello-femoral joint.

This is illustrated by the evolution in a 60-year-old female patient (Fig. 175a) who complained of painful knees with a varus deformity. The right patella was subluxated laterally. One year after a barrel-vault osteotomy (15°) overcorrecting the deformity, the medial joint space has reappeared and the subchondral scleroses have become symmetrical beneath the tibial plateaux (Fig. 175b). This means an even distribution of the compressive stresses in the femoro-tibial joint. Moreover, the patella has been displaced medially and is recentred in the intercondylar groove (Fig. 175b).

Some examples illustrate the results achieved by this procedure. The final angle formed by the femur and the tibia in the coronal plane is larger than the desired overcorrection in many cases. This cannot be otherwise. Overcorrection is planned on a tracing of an X-ray of the knee which is often deprived of cartilage in its medial compartment. Immediately after surgery, the desired angle between the femur and the tibia may have been achieved, but as soon as cartilage or cartilage-like tissue regrows, it separates the articular surfaces from each other, increasing the overcorrection.

A 57-year-old patient (Fig. 176a) complained of a permanently painful knee. There was a flexion contracture and 7° of varus deformity (Fig. 176b). In the A.-P. X-ray the medial joint space is narrowed and underlined by sclerotic bone. The lateral view shows a dense triangle under the posterior aspect of the joint. A 10° barrel-vault osteotomy was planned (Fig. 176c) and carried out (Fig. 176d and f). The patient walked after two days, putting some weight on the operated leg. She was relieved of pain immediately. After two months, when the pins were removed, the range of movement was normal with full extension. The crutches were discarded after three months. Eight years later the clinical result remains excellent: no pain, full range of movement and normal function. In the X-rays (Fig. 176e) the medial joint space is about as wide as the lateral, the subchondral scleroses are symmetrical and the dense triangle under the posterior aspect of the knee has disappeared. Note the anterior displacement of the tibial tuberosity.

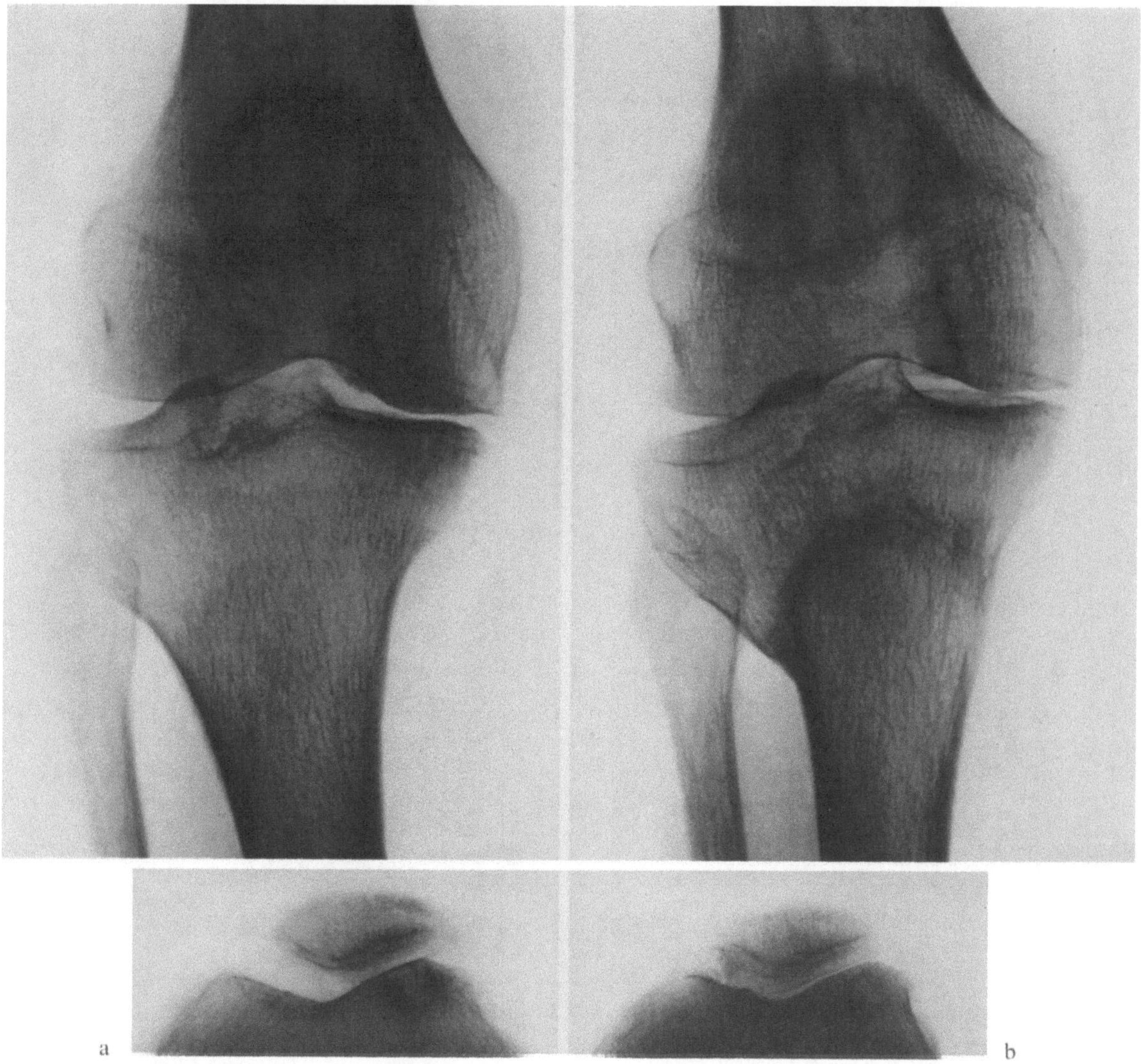
a b

Fig. 175a and b. A 60-year-old female patient before (a) and one year after (b) a barrel-vault osteotomy overcorrecting a varus deformity. After this operation, the patella is recentred into the intercondylar groove

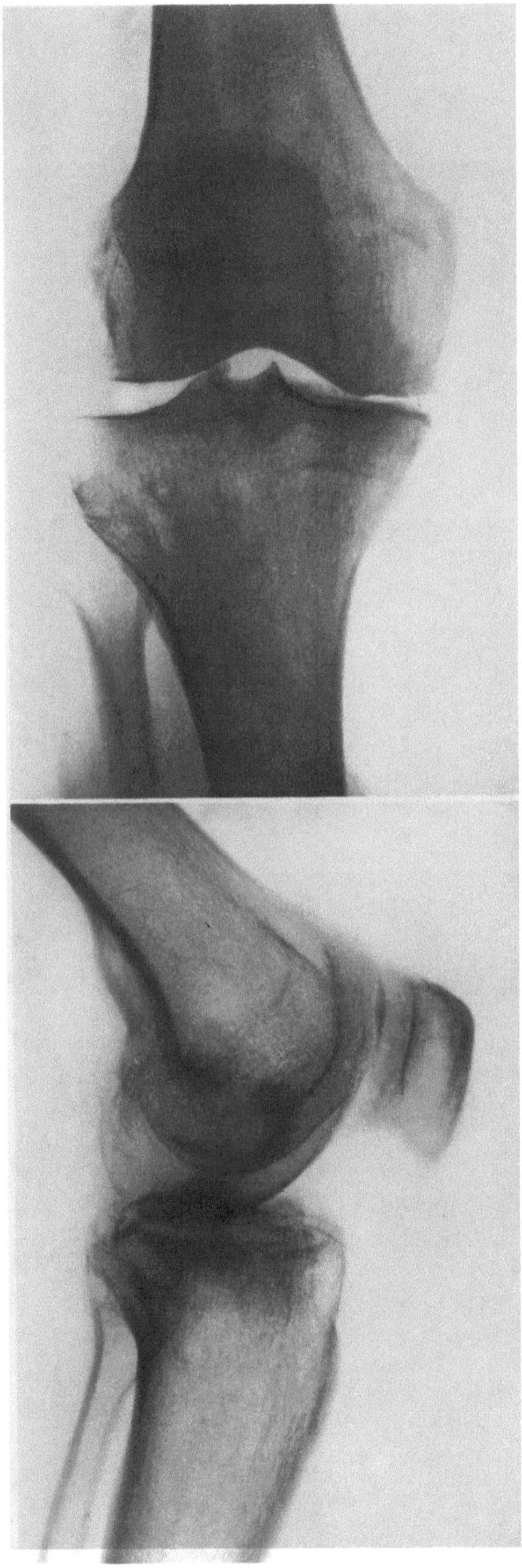

Fig. 176a

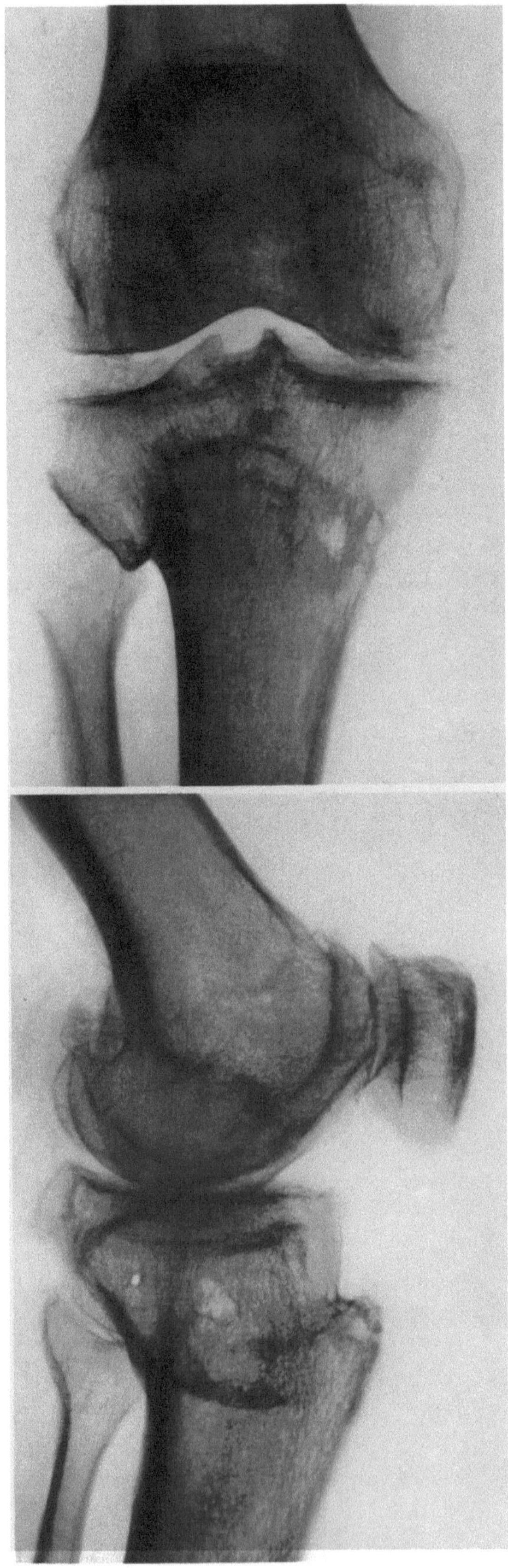

Fig. 176e

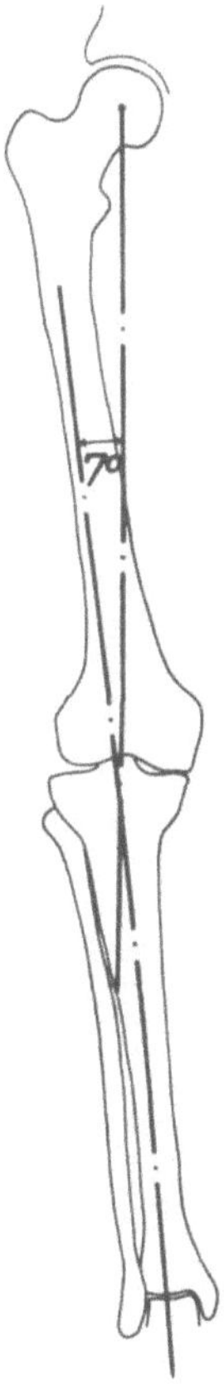

b

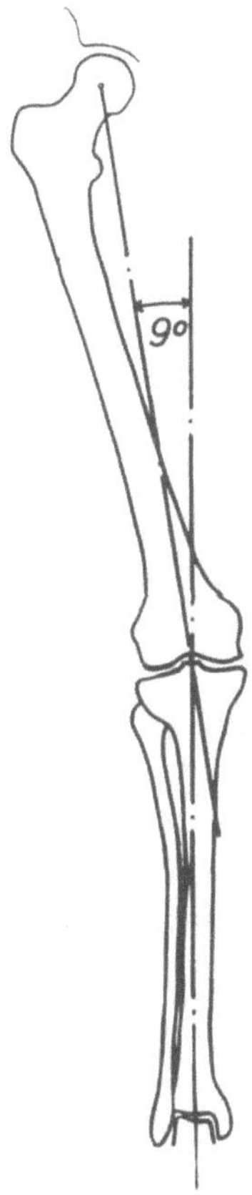

f

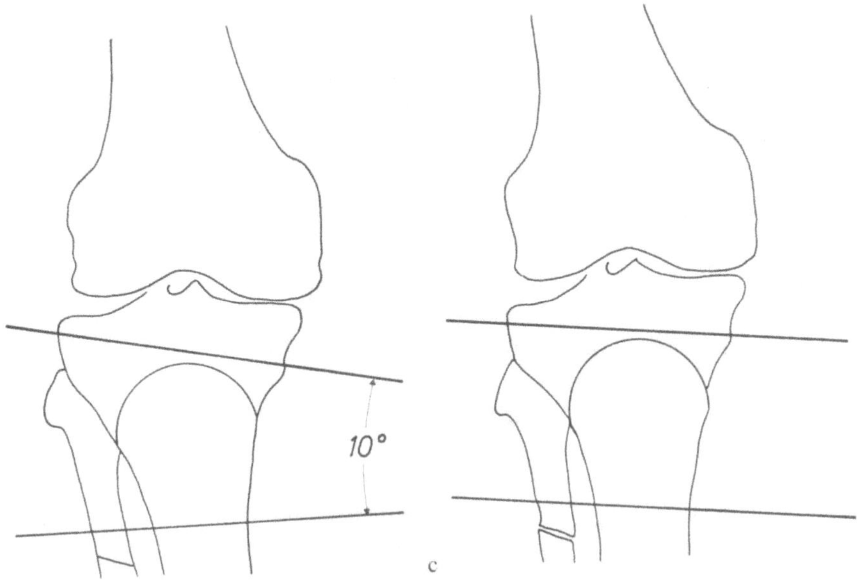

c

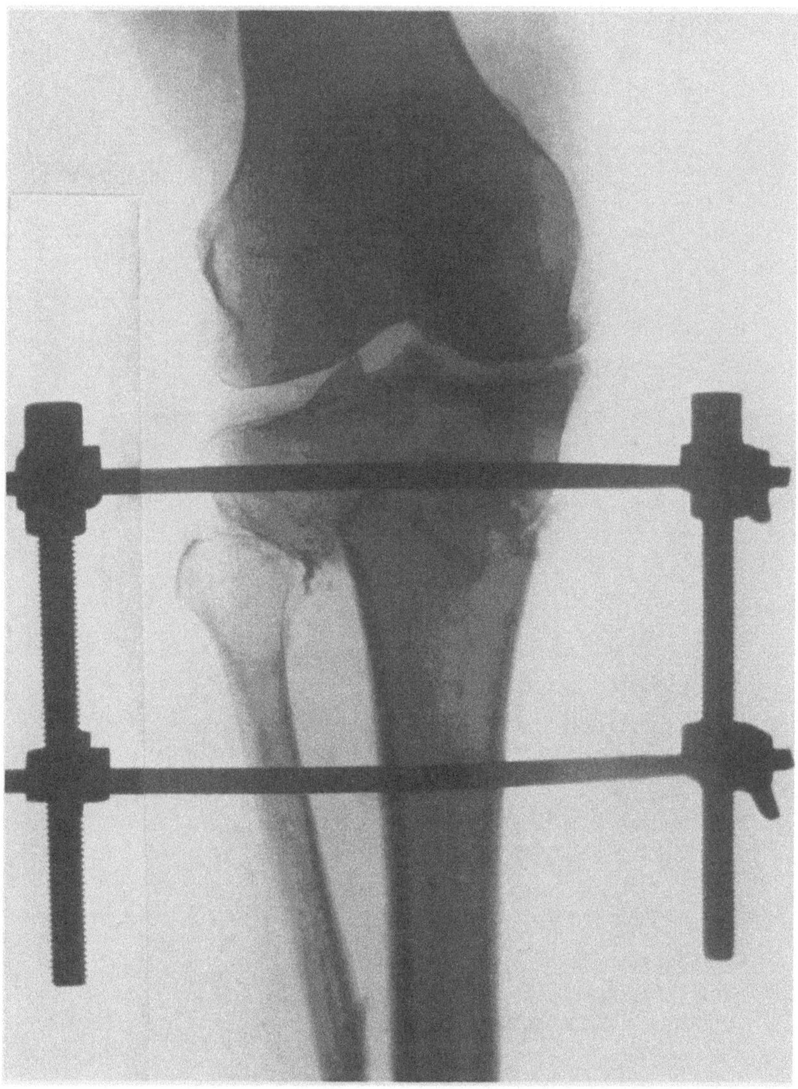

Fig. 176a–f. A 57-year-old patient with medial osteoarthritis of the knee (a). Tracing of the preoperative full length X-ray (b). Preoperative drawing (c). Immediately after surgery (d). Eight years later (e and f)

d

The next case (Fig. 177) also illustrates the regression of the dense triangle under the medial plateau and the appropriate redistribution of the joint pressure after surgery. The 55-year-old patient had his medial meniscus removed 3 years previously. He developed osteoarthritis and a varus deformity. A dense triangle betokens the concentration of joint pressure beneath the medial plateau (Fig. 177a). Four years after a barrel-vault osteotomy has slightly overcorrected the deformity, the patient remains pain-free, has a full range of movement and works normally. The medial plateau is now underlined by a cup-shaped subchondral dense bone symmetrical with the lateral subchondral sclerosis (Fig. 177b). The pressure is again evenly distributed over large weight-bearing surfaces.

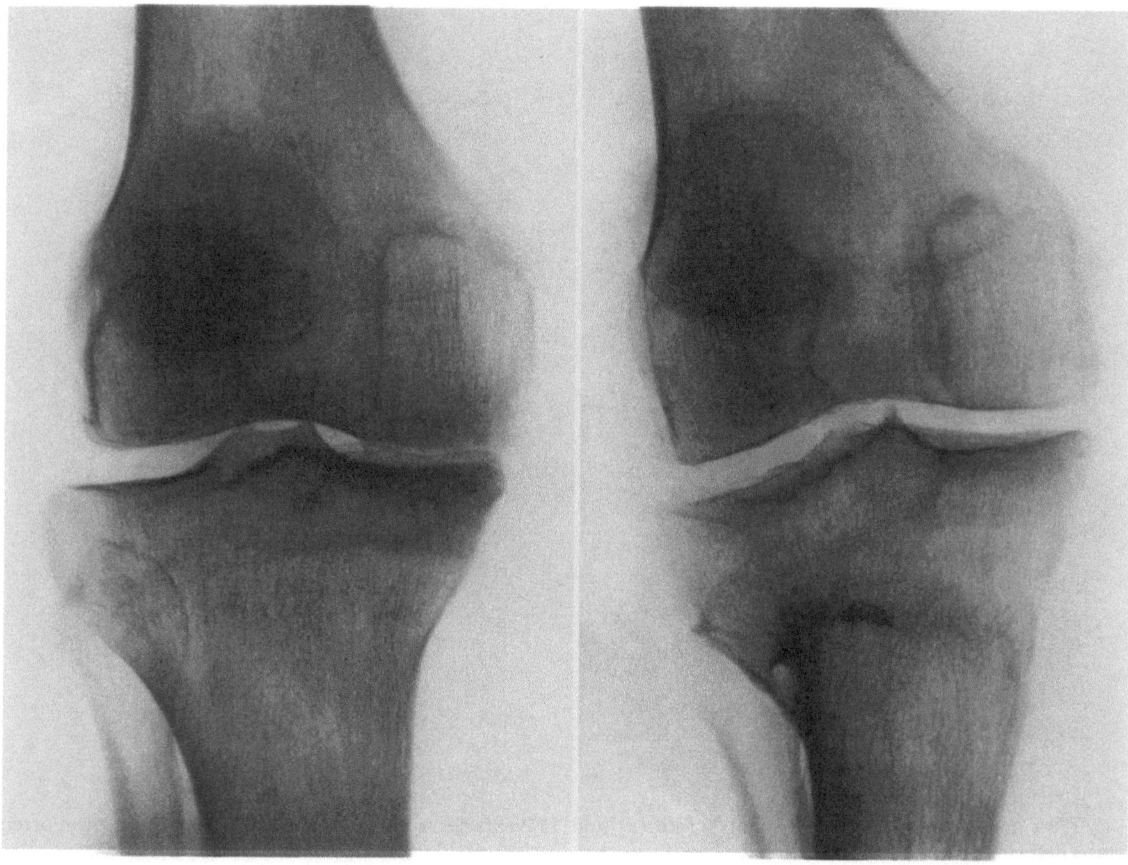

a b

Fig. 177a and b. A 55-year-old patient before (a) and 4 years after (b) a barrel-vault osteotomy slightly overcorrecting the deformity

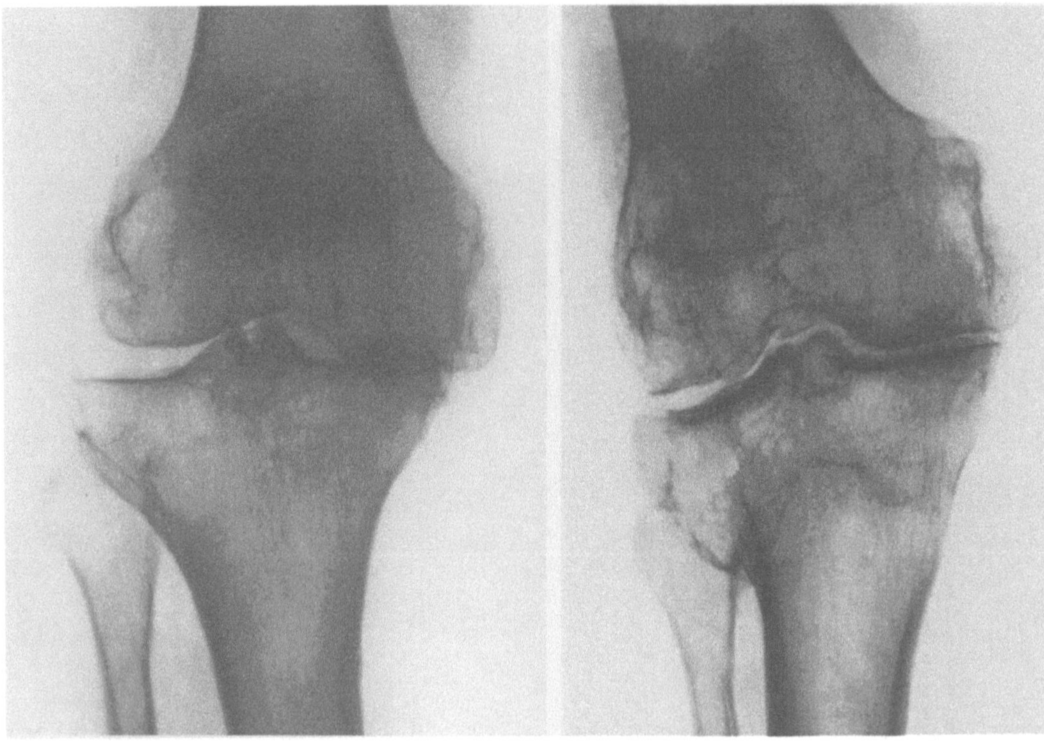

a b

Fig. 178a and b. A 65-year-old patient before (a) and 11 years after (b) a barrel-vault osteotomy overcorrecting the varus deformity

Another example is given by a 65-year-old female patient (Fig. 178). The varus deformity of the right knee was 21° (Fig. 180a). The pain occurred when standing and when walking. There was a lateral laxity of the joint. The X-rays showed a triangle of more pronounced trabeculae beneather the medial plateau. The medial joint space was narrowed (Fig. 178a). Eleven years after a barrel-vault osteotomy of 25°, the pain has gone, the patient has a full range of movement and the knee is stable. The X-ray while standing shows a uniform bone density beneath both plateaux, indicating a regular distribution of the compressive stresses. A medial joint space has reappeared (Fig. 178b). The left knee (Fig. 179a) was operated on two years after the right with a similar good result. The deviation of this left knee reached 14° (Fig. 180a). The fragments have been rotated by 18°. The follow-up here is nine years (Fig. 179b). Both knees have been sufficiently overcorrected to achieve satisfactory distribution of the articular compressive stresses and consequent healing of the condition (Fig. 180b).

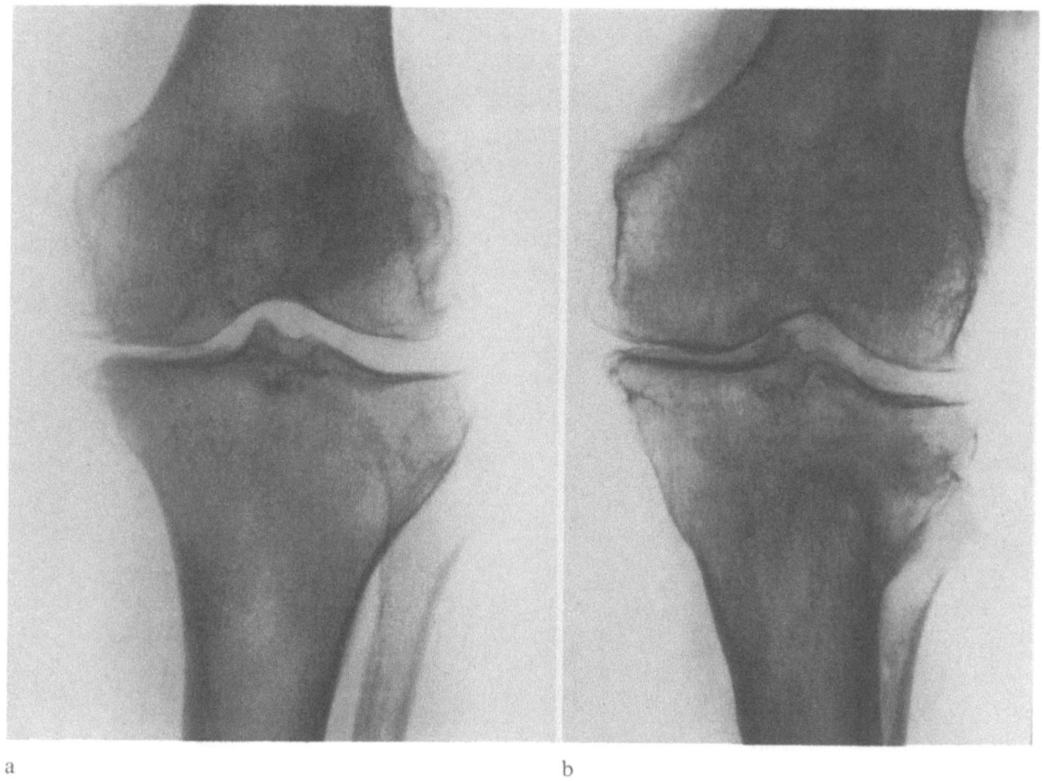

Fig. 179a and b. Same patient as Fig. 178. Left knee operated on at age 67 (a) and the 9-year follow-up (b)

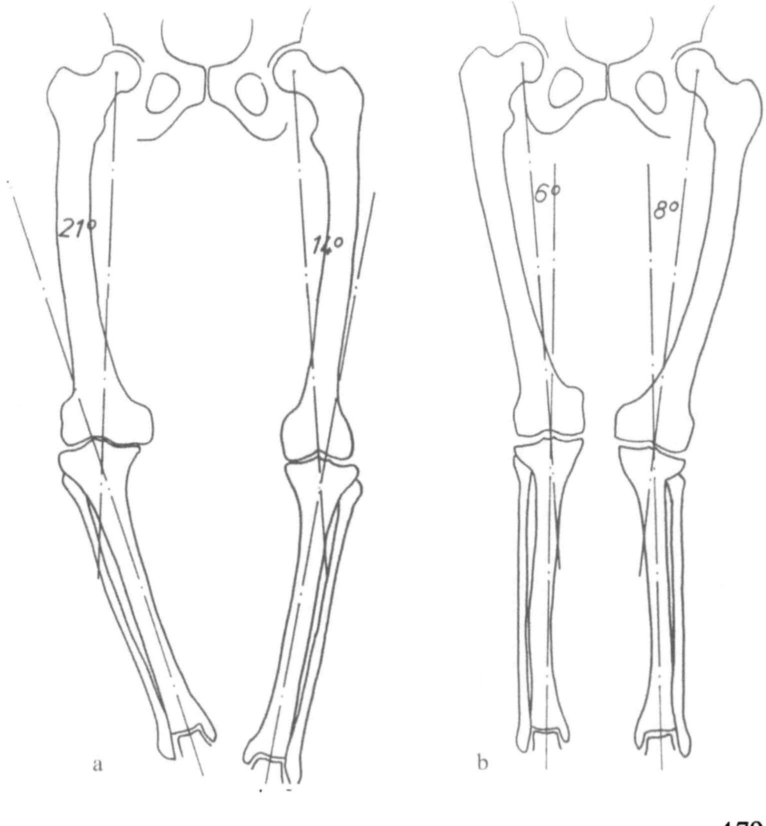

Fig. 180a and b. Patient of Fig. 178 and 179 before (a) and after (b) a bilateral barrel-vault osteotomy

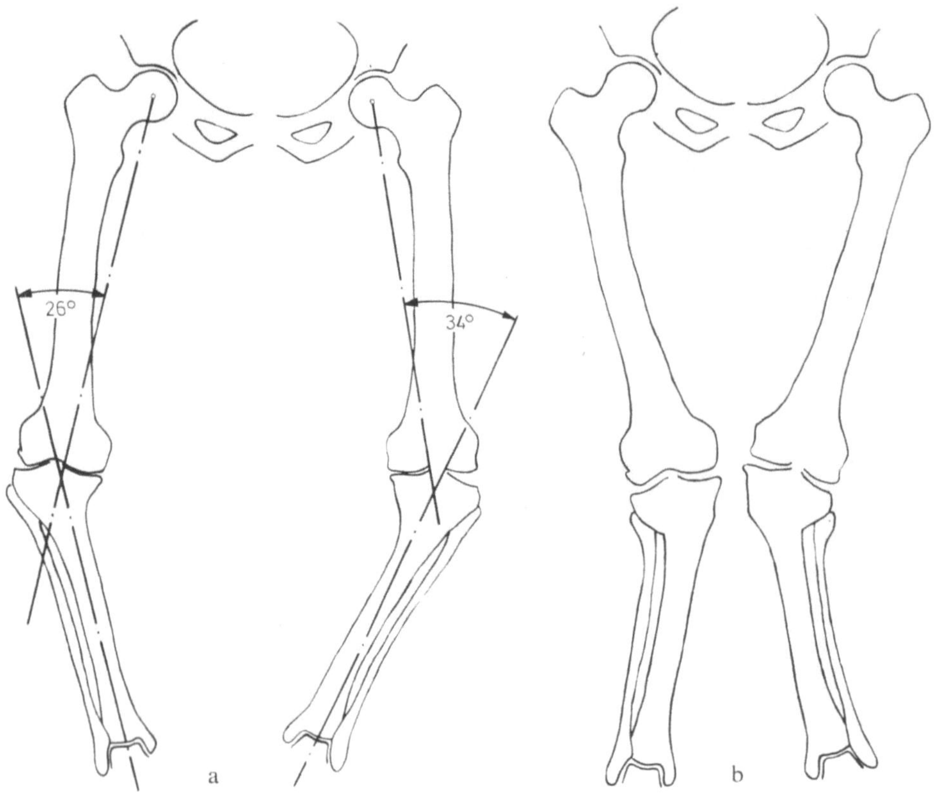

Fig. 181 a and b. A 65-year-old patient before (a) and 5 years after (b) surgical treatment. See Figures 182 and 183

In some patients the deformity to be corrected can be very severe. A 65-year-old obese female patient complained of constant pain, day and night, in both knees. These were deformed in varum, 15 cm from each other when the medial malleoli were together (Fig. 181 a). A lateral laxity existed. The X-rays showed bilateral osteoarthritis with narrowed medial joint spaces and flattening of the medial plateaux underlined by more pronounced trabeculae (Figs. 182a and 183a). A barrel-vault osteotomy of 39° was performed on the left tibia. Ten months later a 30° osteotomy was carried out on the right tibia. The deformity has been largely overcorrected (Fig. 181 b). The pain has immediately disappeared. Five years later the patient walks 2–3 km at one time every day. The standing A.-P. view shows regression of the signs of osteoarthritis and a medial joint space in both knees (Figs. 182b and 183b). In the lateral X-ray the dense triangle beneath the posterior part of the plateaux (Figs. 182a and 183a) is replaced by a cup-shaped sclerosis beneath the plateaux throughout (Figs. 182b and 183b).

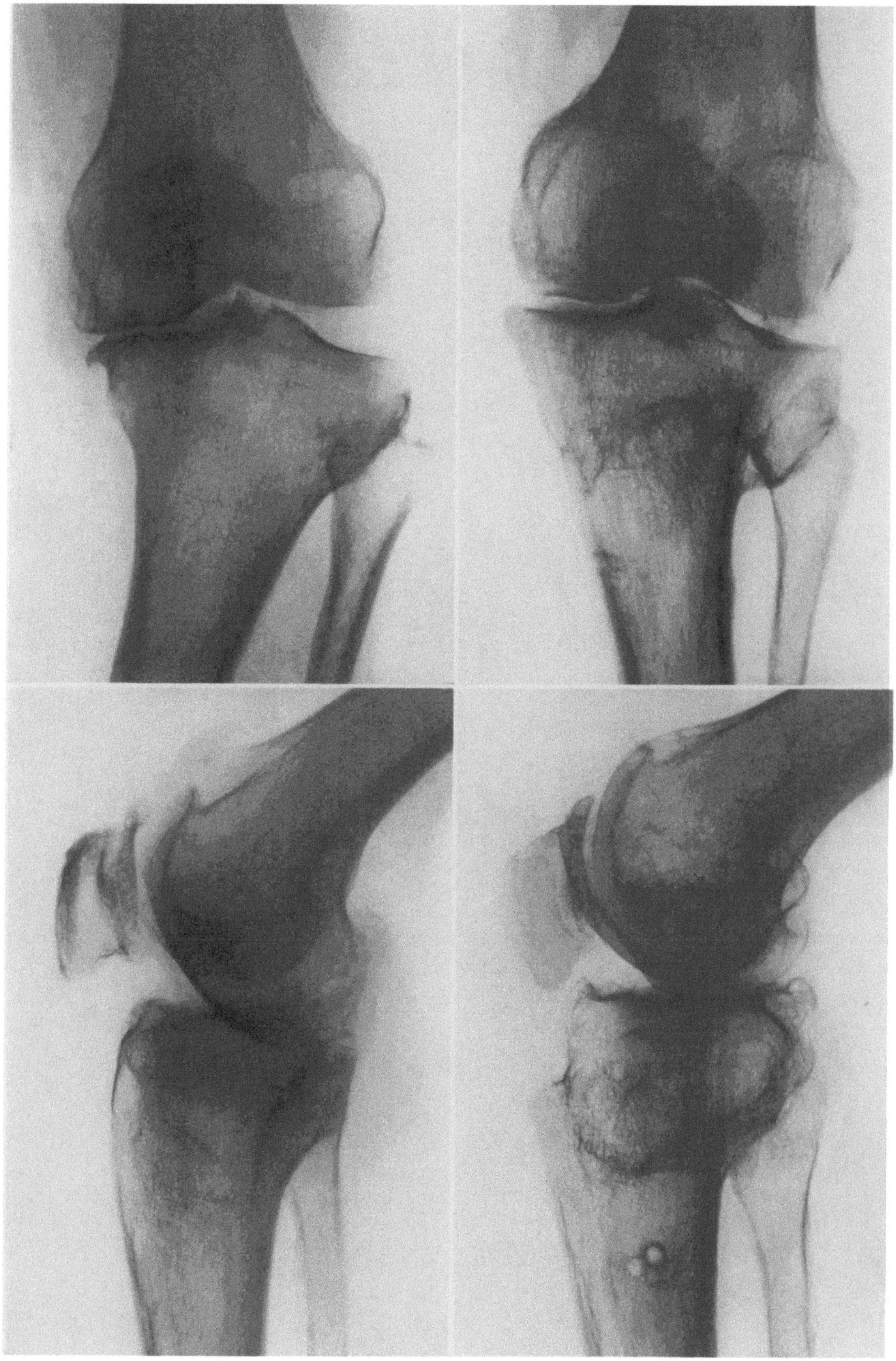

a b

Fig. 182a and b. A 65-year-old patient (left knee) before (a) and 6 years after (b) a barrel-vault osteotomy overcorrecting the varus deformity. See Figure 181

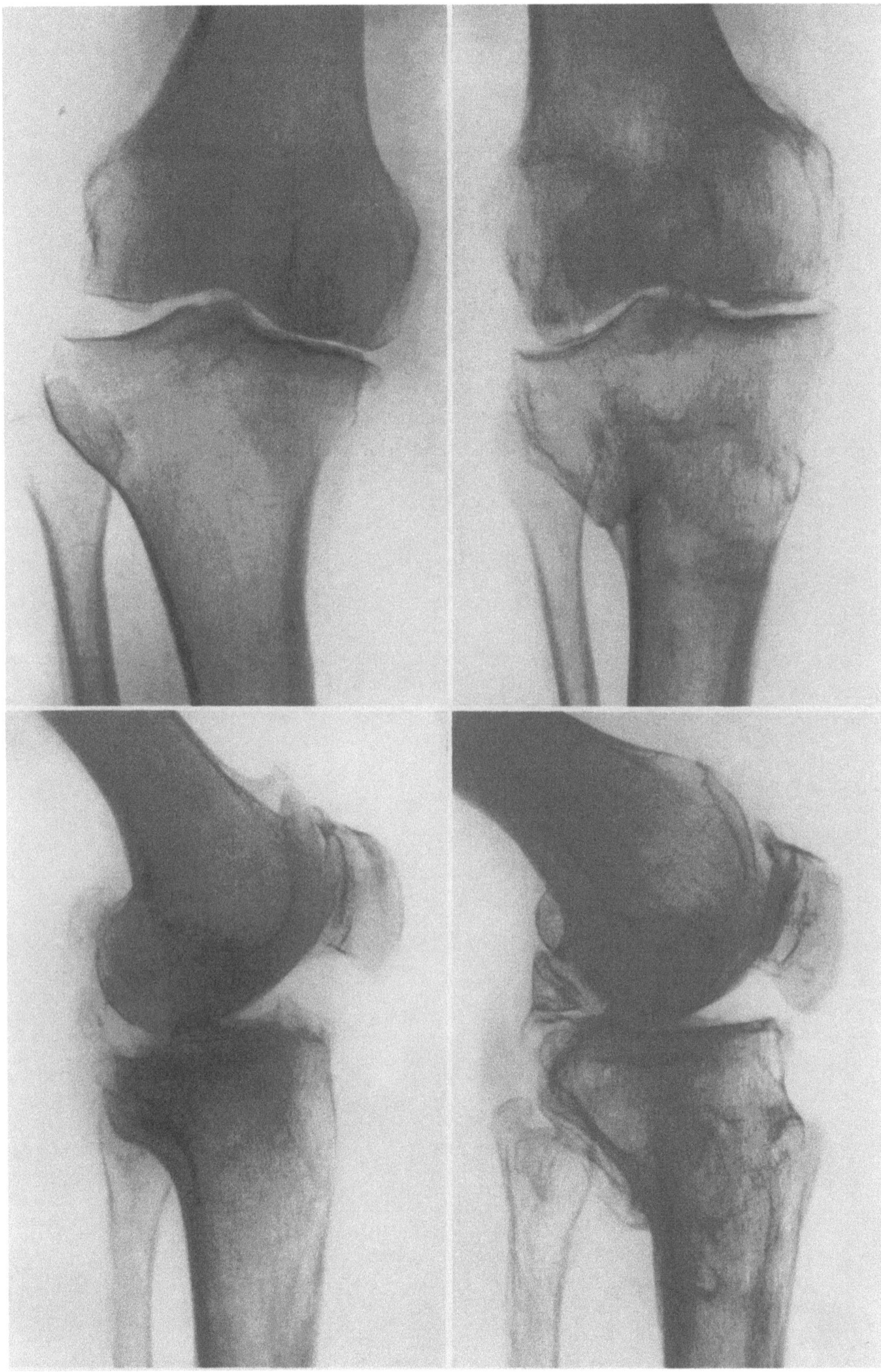

a b

Fig. 183 a and b. A 66-year-old patient, right knee, before (a) and 5 years after a barrel-vault osteotomy overcorrecting the varus deformity (b). See Figure 181

A 69-year-old female patient (Fig. 184a) developed osteoarthritis in both knees due to a varus deformity, the consequence of rickets in infancy. Before surgery the knees were separated by 18 cm while standing. Gait was painful and ugly. A dense triangle underlined the medial tibial plateaux (Fig. 184e) indicating locally increased pressure. In the lateral view it appeared posteriorly (Fig. 184e). Twelve years after barrel-vault osteotomies combined with anterior displacement of the tibial tuberosities (Fig 184c and d), the dense triangles have disappeared and the cancellous trabeculae beneath the lateral plateaux are more pronounced than before (Fig. 184f). The operations have decreased the articular pressure by diminishing the load and by distributing it over larger weight-bearing surfaces. The clinical result is excellent from the points of view of relief of pain, function and cosmetic appearance (Fig. 184b). The preoperative laxity of the ligaments has subsided completely. The knees are stable.

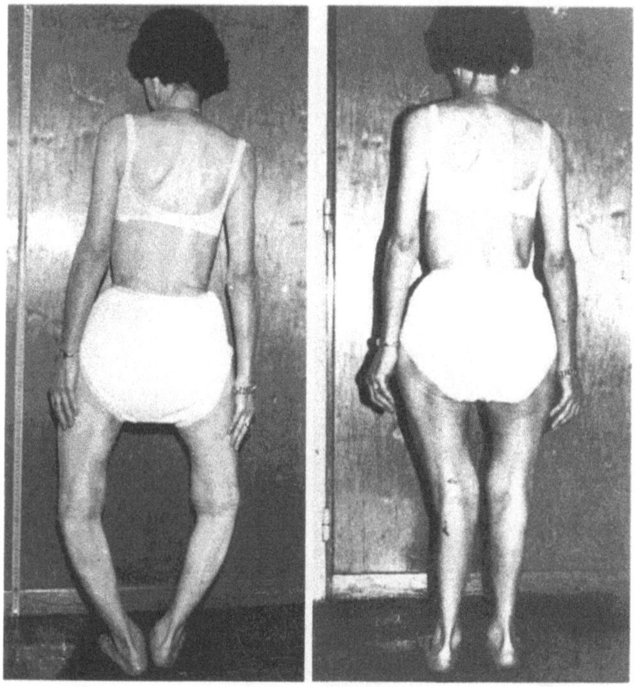

a b

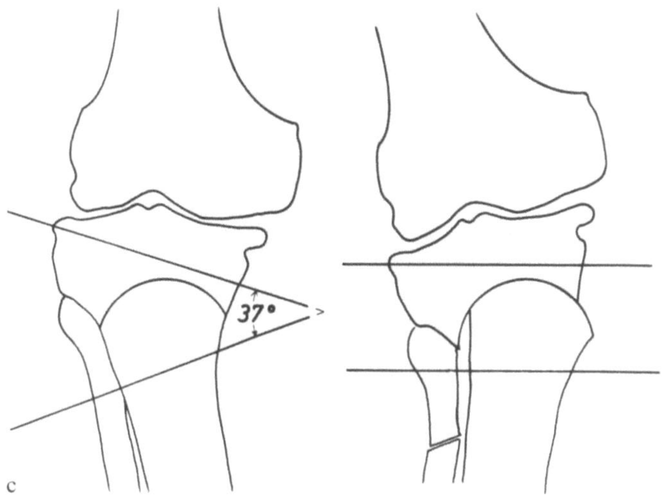

c

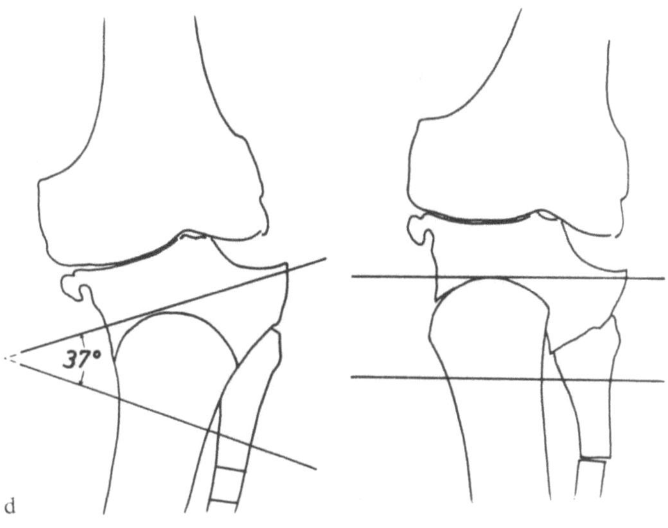

d

Fig. 184a–f. A 69-year-old patient before (a) and after (b) a bilateral barrel vault osteotomy of the tibia with anterior displacement of the tibial tuberosity. Planning for the right knee (c) and left knee (d). Standing A.-P. and lateral X-rays before operation (e) and at the 12-year follow-up (f)

Fig. 184e–f, see page 184–185

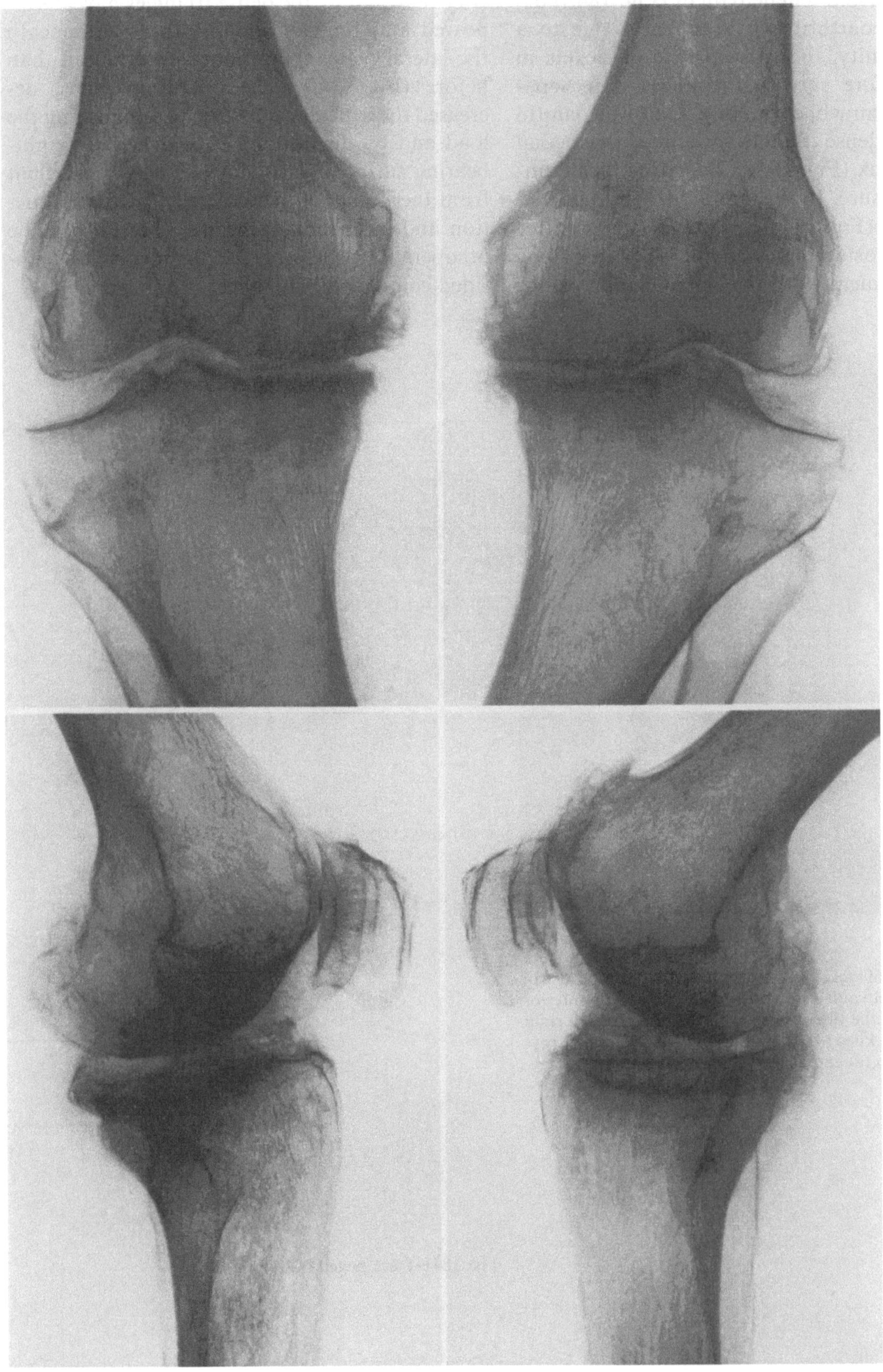

Fig. 184e

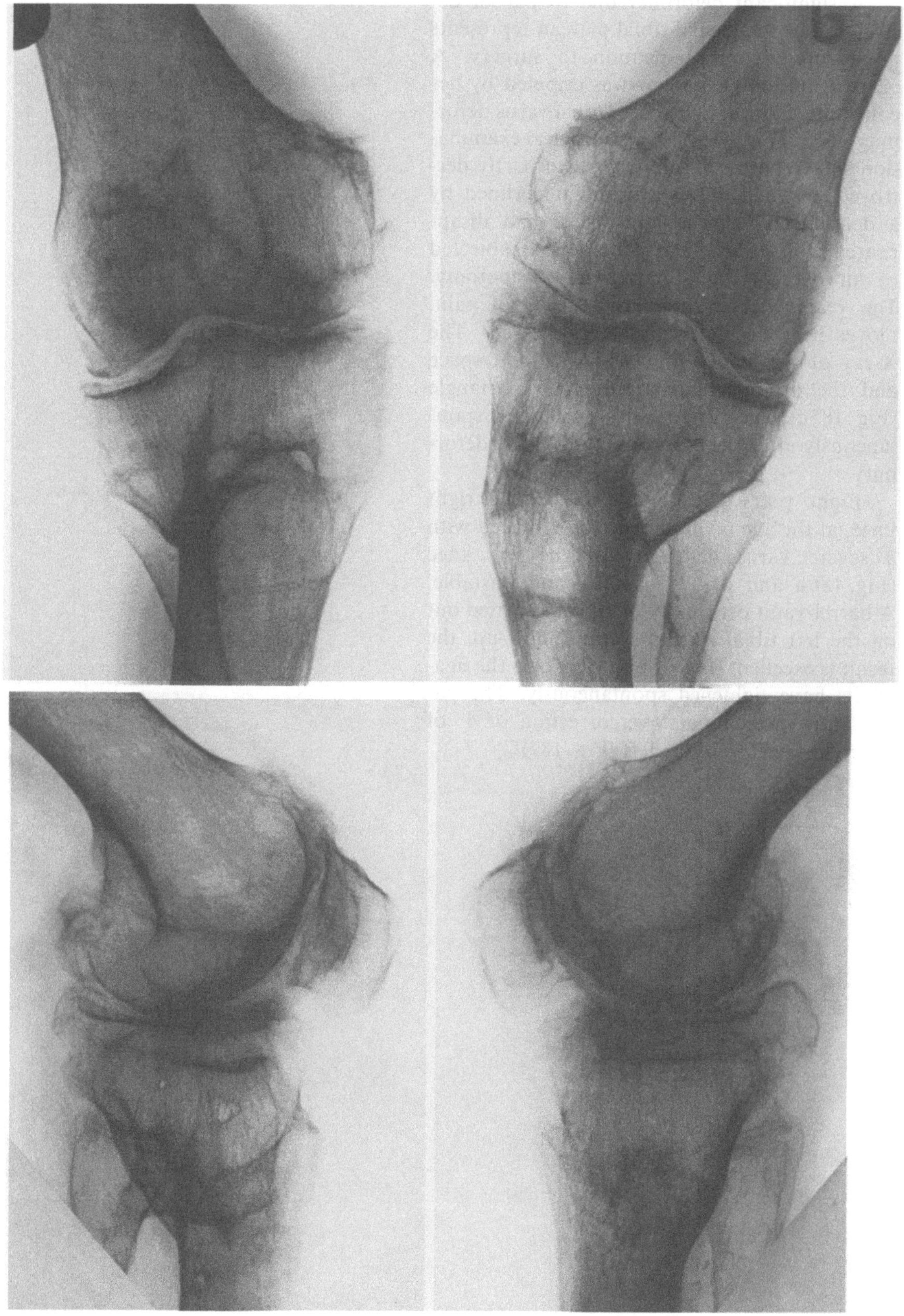

A significant deformity due to partial destruction of the medial tibial plateau represents no absolute contra-indication to surgery. A 62-year-old female patient was crippled by her constantly painful right knee with a varus deformity of 25° (Fig. 185a). A radiological examination showed a medial tibial plateau partly destroyed, inclined downward and underlined by a dense triangle. The joint space had disappeared (Fig. 185b). The patient was subjected to an overcorrecting barrel-vault osteotomy. Ten years later she remains relieved of pain, moves her knee freely and lives actively. The X-ray of the loaded joint shows a joint space and the disappearance of the dense triangle (Fig. 185c). The ligaments have tightened spontaneously after the overcorrection of the deformity.

Eight years after operation on the right knee, at the age of 70, the patient returned with a severe varus deformity of the left knee (Fig. 186a and b), also painful and unstable. A barrel-vault osteotomy of 29° was carried out on the left tibia. At the 2-year follow-up, the result is excellent (Fig. 186c). Here also the ligaments have tightened spontaneously. The patient now presents an overcorrection of 4° of the right knee, 6° of the left (Fig. 187).

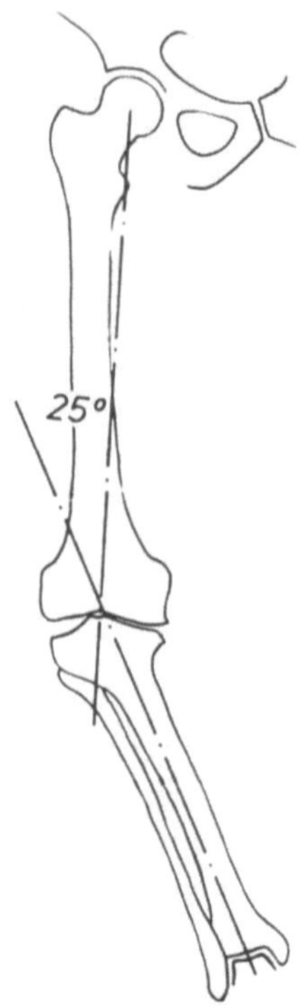

Fig. 185a

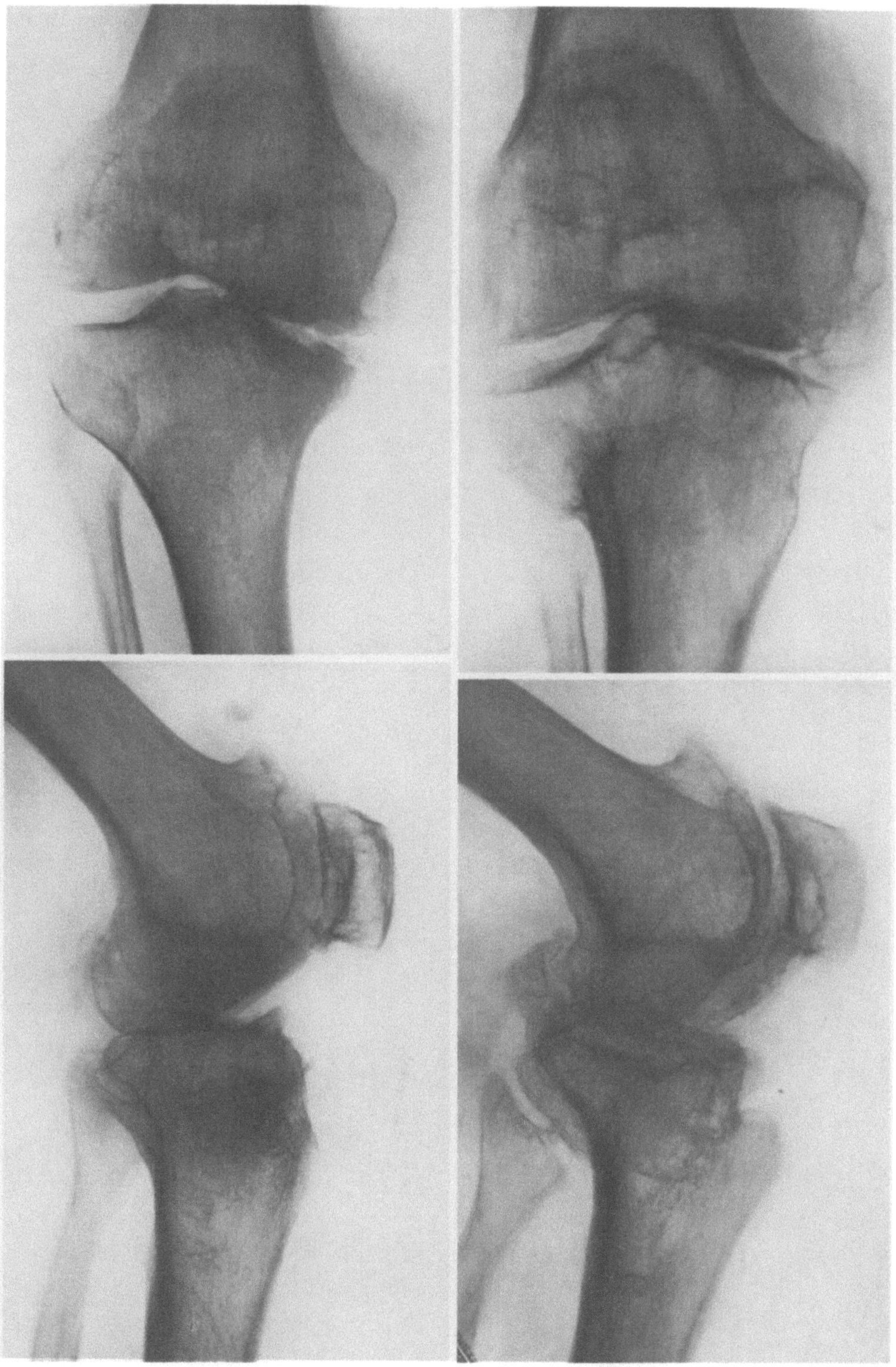

b c

Fig. 185a–c. A 62-year-old patient before (b) and 10 years after (c) a barrel-vault osteotomy overcorrecting the varus deformity of the right knee. Preoperative deformity (a)

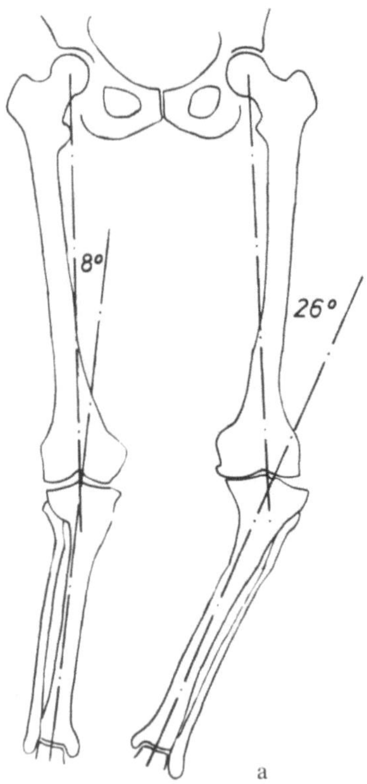

a

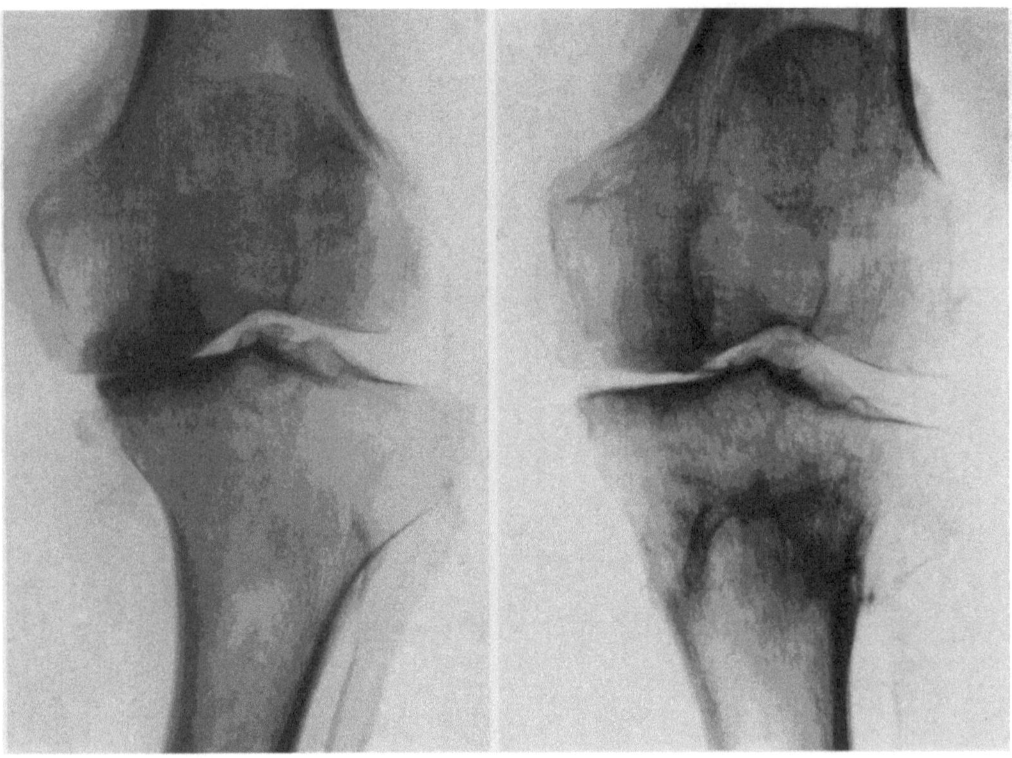

b

c

Fig. 186a–c. Same patient as Figure 185. Left knee (a) operated on at age 70. Deformity of the knee (b). Two-year follow-up (c)

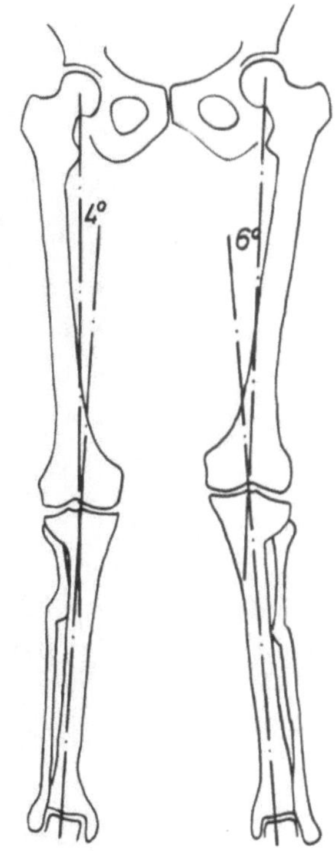

Fig. 187. Same patient as Figures 185 and 186. Overcorrection of the deformity

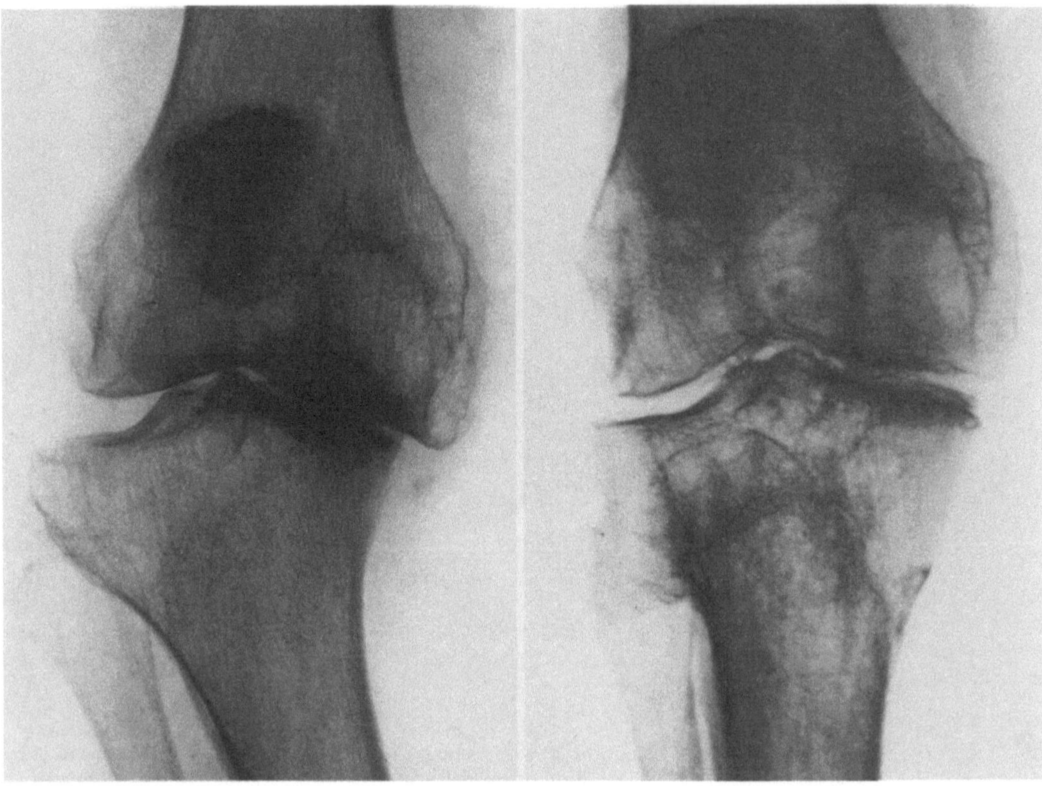

a c

Fig. 188a–e. A 75-year-old patient with unstable knees (a). Preoperative drawing (b). Six years after a barrel-vault osteotomy overcorrecting the varus deformity (c). The other knee before operation (d) and 5 years after surgery (e)

Even subluxated knees can heal when properly overcorrected. The ligaments seem to tighten spontaneously when the mechanical conditions of the joint are improved by surgery. A 75-year-old female patient was crippled by a bilateral painful osteoarthritis with subluxation of the knees and instability because of ligamentous laxity (Fig. 188a and d). The deformities were overcorrected by barrel-vault osteotomies combined with anterior displacement of the tibial tuberosities (Fig. 188b). When reviewed, at age 81, the patient walks painfree. The knees are stable again. A joint space has reappeared. The X-rays indicate a proper distribution of the compressive stresses in the joints (Fig. 188c and e).

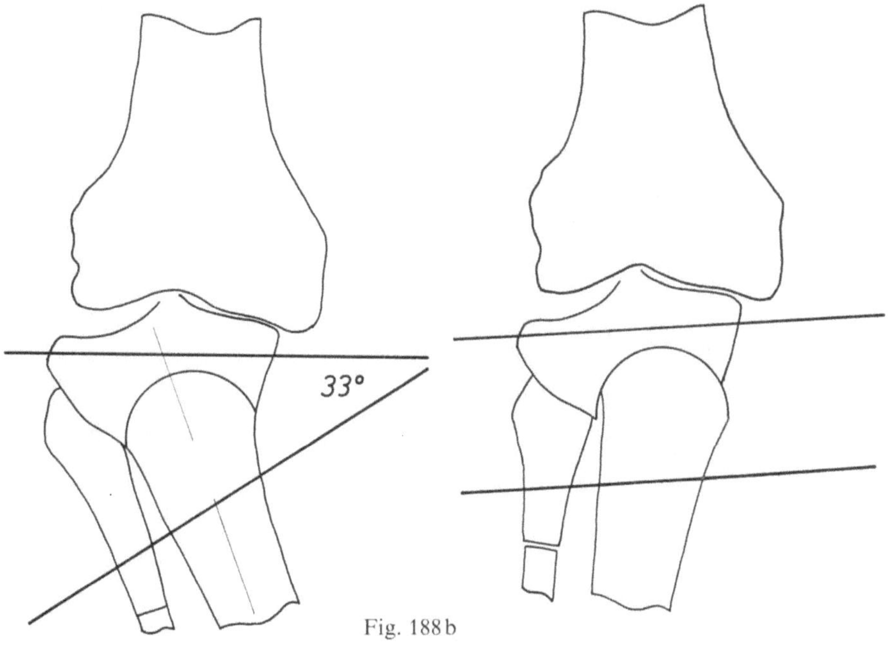

Fig. 188b

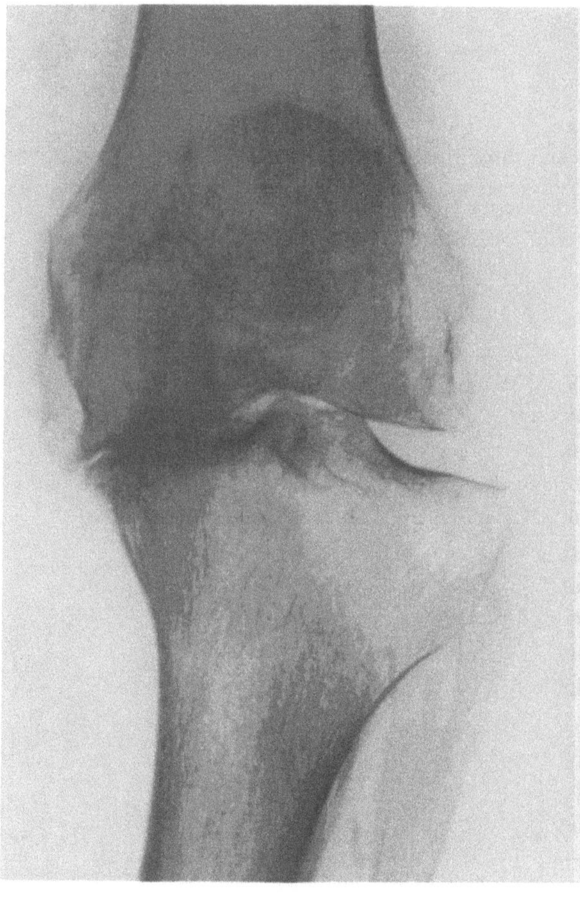

Fig. 188d

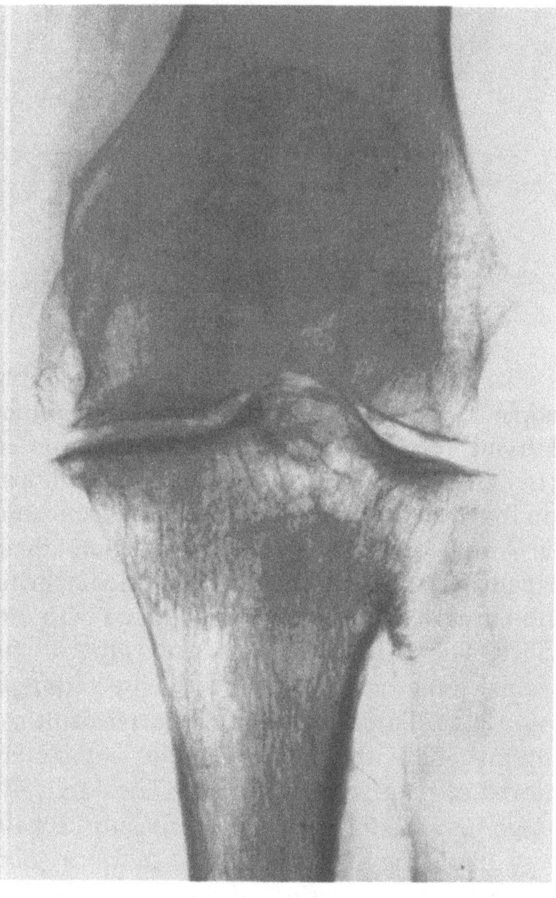

Fig. 188e

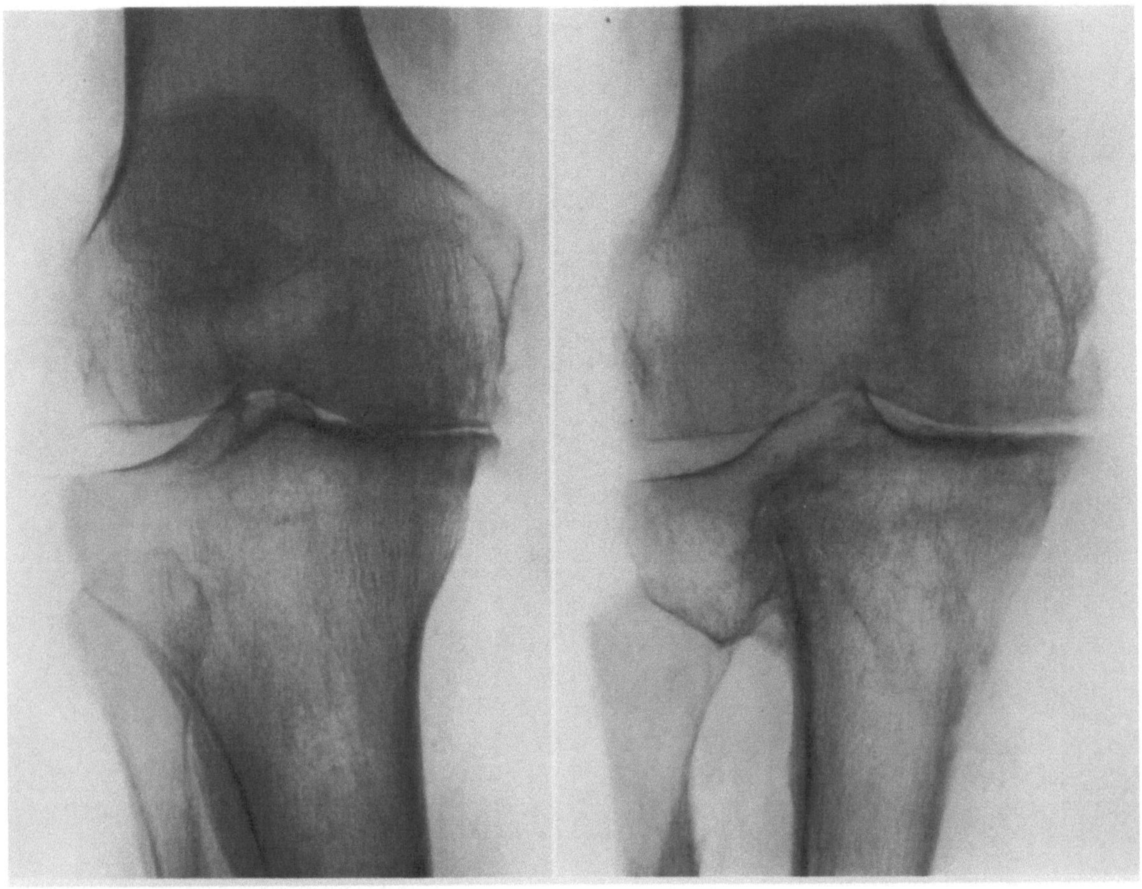

a b

Fig. 189a and b. A 80-year-old female patient before (a) and three years after (b) overcorrection of a varus deformity. Redistribution of the stresses in the joint

From the results attained in these elderly people it can be concluded that the possibilities of tissue regeneration are maintained even in extreme old age. Age is thus no contra-indication for this type of surgery. Here is a last example. An 80-year-old female patient complained of pain day and night from her osteoarthritis with varus deformity (12°) and walked with difficulty. The X-ray showed a narrowing of the medial joint space with a dense triangle (Fig. 189a). Three years after a barrel-vault osteotomy of 15° the patient remains completely relieved of pain, works normally, lives actively, can go up and down stairs and goes for a walk every day. The standing X-ray shows a joint space and the regression of the signs of osteoarthritis (Fig. 189b).

f) Cases Requiring a Derotation of the Leg

Because of the curved shape of the osteotomy, the procedure described above does not allow for a derotation of the tibia. This can be necessary if the ankle and the distal tibia and fibula are externally rotated in relation to the knee. In such exceptional cases the operative technique must be modified. We shall describe it by reporting a clinical example. A 14-year-old patient (Fig. 190a) with rheumatoid arthritis had bilateral dislocation of the patella. In each knee a synovectomy was performed with transfer of the insertion of the patella tendon to the antero-medial aspect of the tibia. The dislocation of the patellae was corrected. Four years later the sedimentation rate was back to normal. One of the knees was deformed in varum ($\alpha = 17°$) (Fig. 190b) and the ankle was externally rotated, 80° in relation to the knee (normally 30°, according to Fick). The cancellous trabeculae were more pronounced beneath the medial tibial plateau.

A second operation was carried out (Fig. 190c). A fragment of the upper third of the fibula was resected. The skin was then incised below the tibial tuberosity. One Steinmann pin was inserted beneath the tibial plateau and a second through the tibial diaphysis to form an angle of 19° open laterally in the coronal projection. Projected on a transverse plane, they formed an angle of 50° (Fig. 190c). A periosteal elevator was used to free the posterior aspect of the tibia at the level of the osteotomy. It was kept in place to protect the popliteal artery. The bone was divided transversely distal to the insertion of the patella tendon with an oscillating saw and a thin chisel. The fragments were moved and the lower one impacted into the upper until the pins were parallel. They were then fixed with two compression clamps. Overcorrection of the varus deformity and derotation were thus achieved without any supplementary and undesirable medial displacement of the insertion of the patella tendon.

After two months the Steinmann pins were removed. The correction was followed by an excellent functional result. The X-ray taken nine years later shows an even distribution of the articular compressive stresses (Fig. 190d).

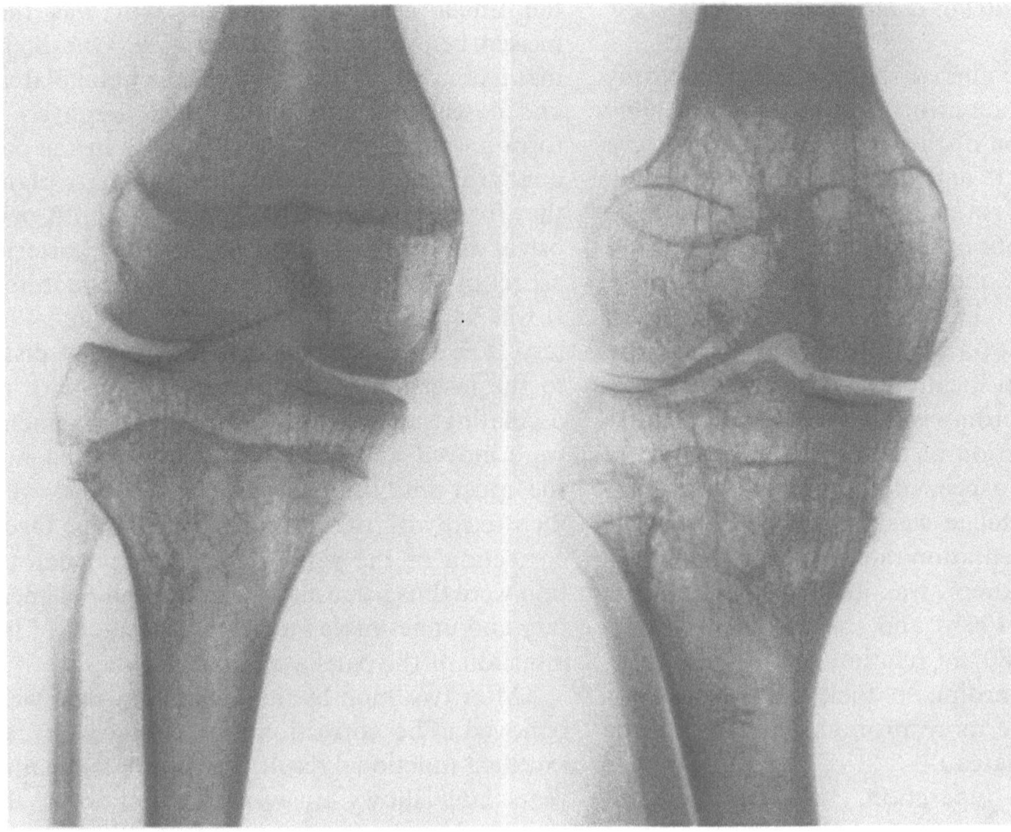

b

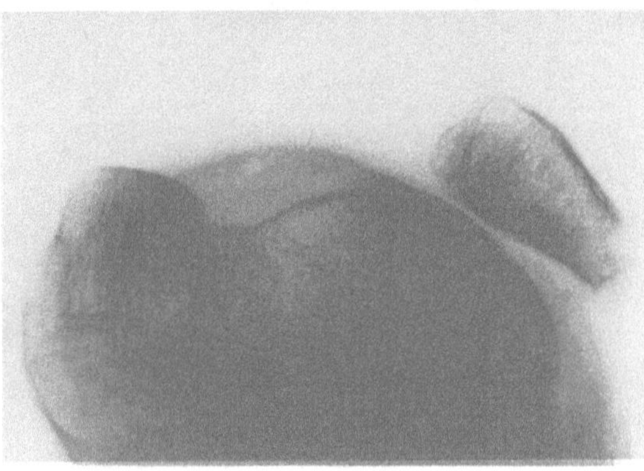

a

Fig. 190. (a) A 14-year-old patient with rheumatoid arthritis and luxated patella. (b) Four years after medial transplantation of the tibial tuberosity and synovectomy. (c) Overcorrection (19°) of the varus deformity combined with a derotation of the lower leg skeleton (50°). Projections on the coronal and on the horizontal plane. (d) Nine years after the second operation

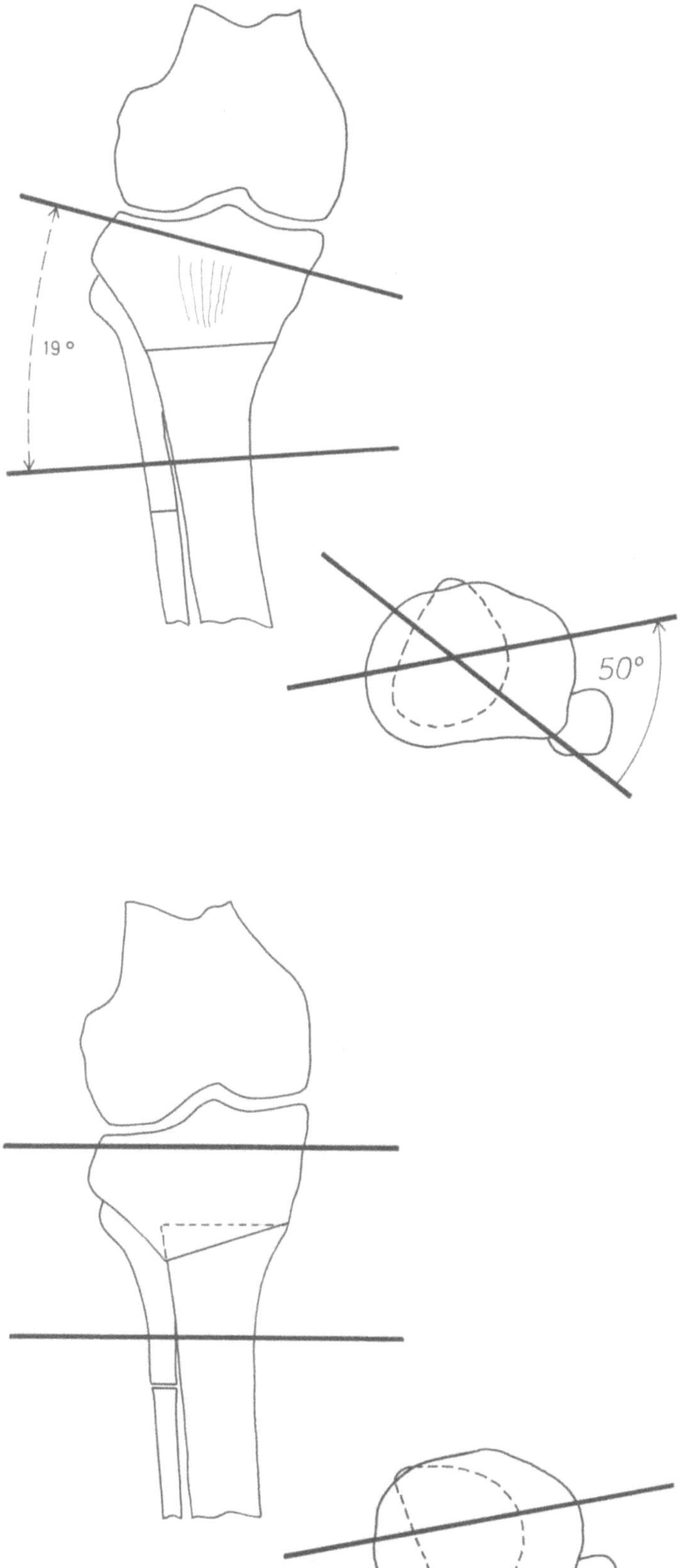

c

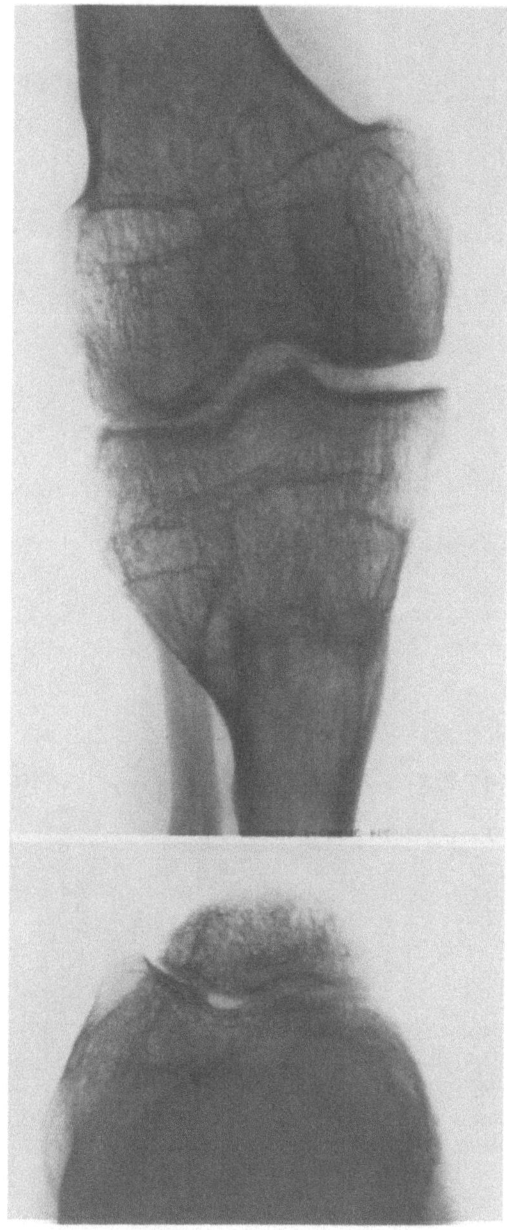

d

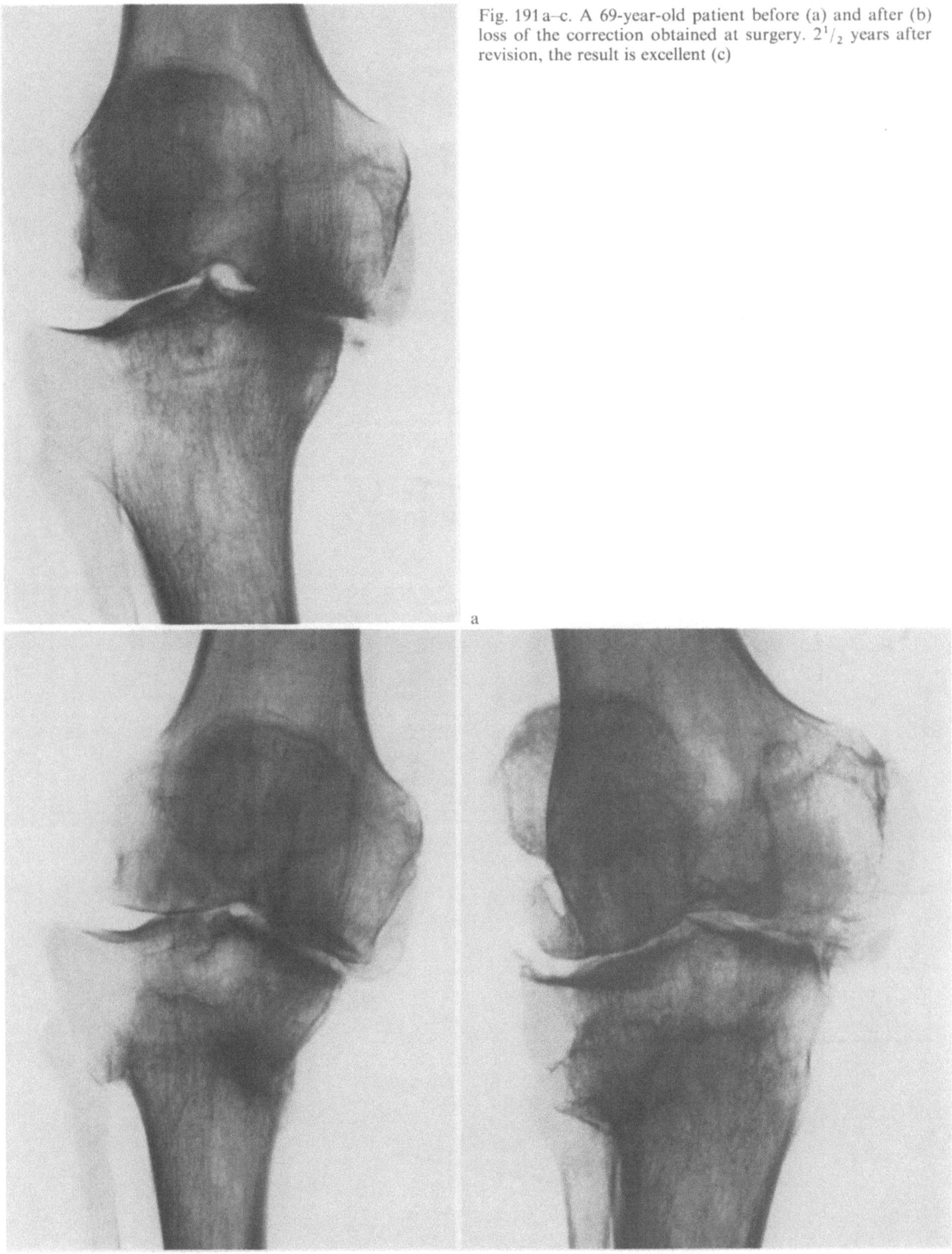

Fig. 191 a–c. A 69-year-old patient before (a) and after (b) loss of the correction obtained at surgery. $2^1/_2$ years after revision, the result is excellent (c)

Fig. 192a–c. Same patient as Figure 190, before (a) and after (b) the first osteotomy. End result after revision (c)

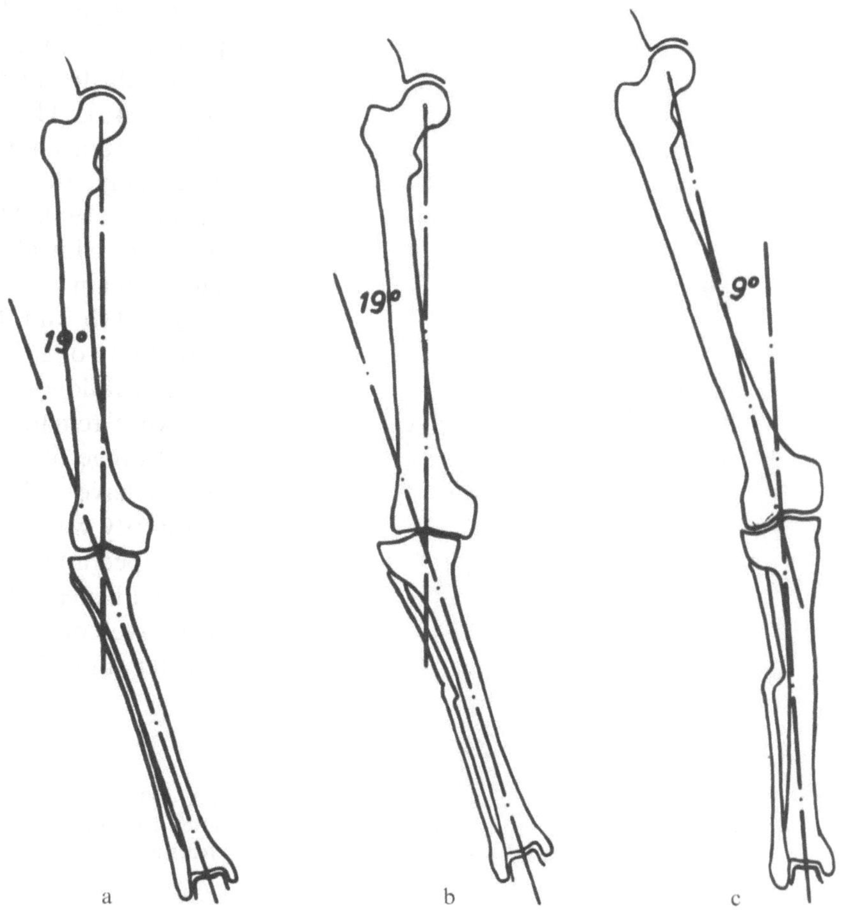

g) Revisions

Some operations had to be revised because the varus deformity had been either undercorrected or exaggeratedly overcorrected.

α) Revision After Undercorrection

Clinical improvement and dramatic changes in the X-ray pictures are not usually observed when the varus deformity has not been overcorrected sufficiently. This means that the biological alterations provoked by dividing the bone are not sufficient to lead to healing. The mechanical effect of the procedure thus appears to be essential.

A 69-year-old female patient (Fig. 191a) presented with severe osteoarthritis of the knee with a varus deformity of 19° (Fig. 192a). A barrel-vault osteotomy of 23° was carried out and the pins were removed after two months as usual. However, the osteotomy was probably not yet consolidated. The deformity recurred (Figs. 191b and 192b) and a new barrel-vault osteotomy of 24° was carried out. At the $2^1/_2$-year follow-up, the overcorrection is 9° (Fig. 192c). Subchondral scleroses of uniform width underline both tibial plateaux, illustrating an even distribution of the compressive stresses in the joint (Fig. 191c). The clinical result is outstanding. The knee is stable, pain-free and has a full range of movement.

Age should not hinder the prospect of such a revision. In elderly patients, inability to walk painlessly may constitute a greater risk to life than a second barrel-vault osteotomy.

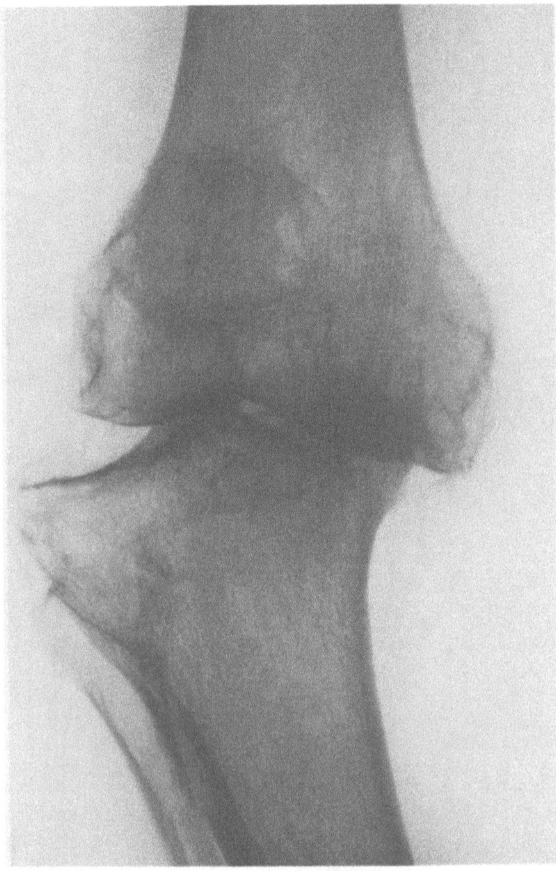

a

A 78-year-old female patient (Fig. 193a) presented with a painful flail knee with severe osteoarthritis and a varus deformity of 37° (Fig. 194a). The laxity of the ligaments was taken into account in planning the barrel-vault osteotomy; the fragments were rotated by 30° only (Fig. 195). After a short period of improvement, the clinical and the X-ray picture deteriorated again and the deformity recurred (Figs. 193b and 194b). A second barrel-vault osteotomy of 27° was carried out (Fig. 196). At the last follow-up, $7^1/_2$ years after the revision, the knee remains painfree and stable and has a full range of movement. Here again the ligaments have tightened spontaneously after the deformity was overcorrected. A joint space has developed and the subchondral sclerosis has regressed beneath the medial plateau (Fig. 193c). The overcorrection could have been a little greater (Fig. 194c).

Fig. 193a–c. A 78-year-old patient before (a) and 2 years after (b) a first barrel-vault osteotomy. Excellent result $7^1/_2$ years after revision (c)

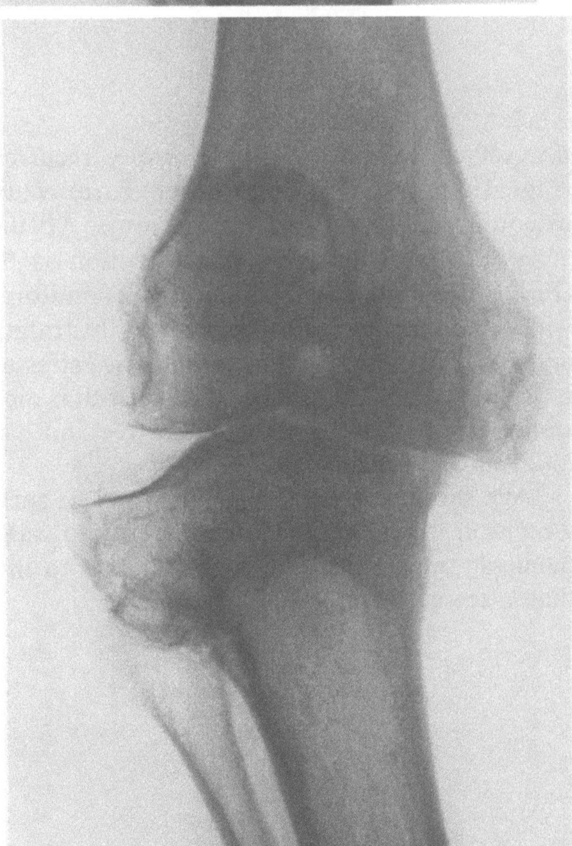

b

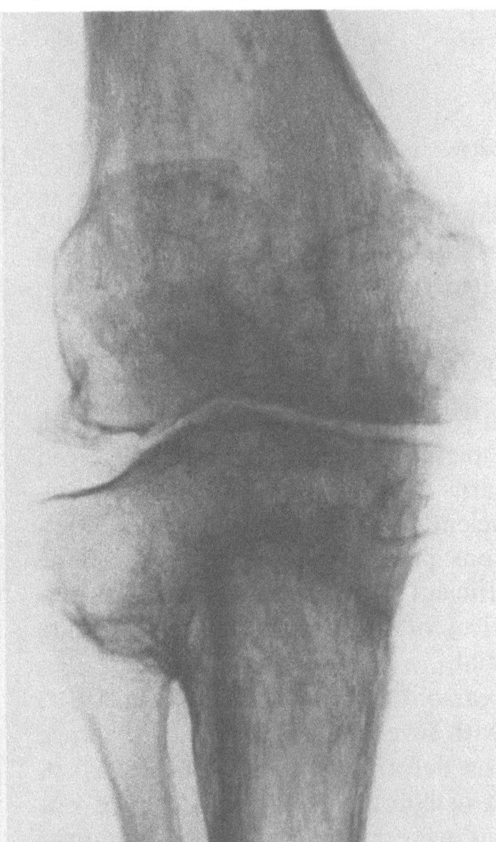

c

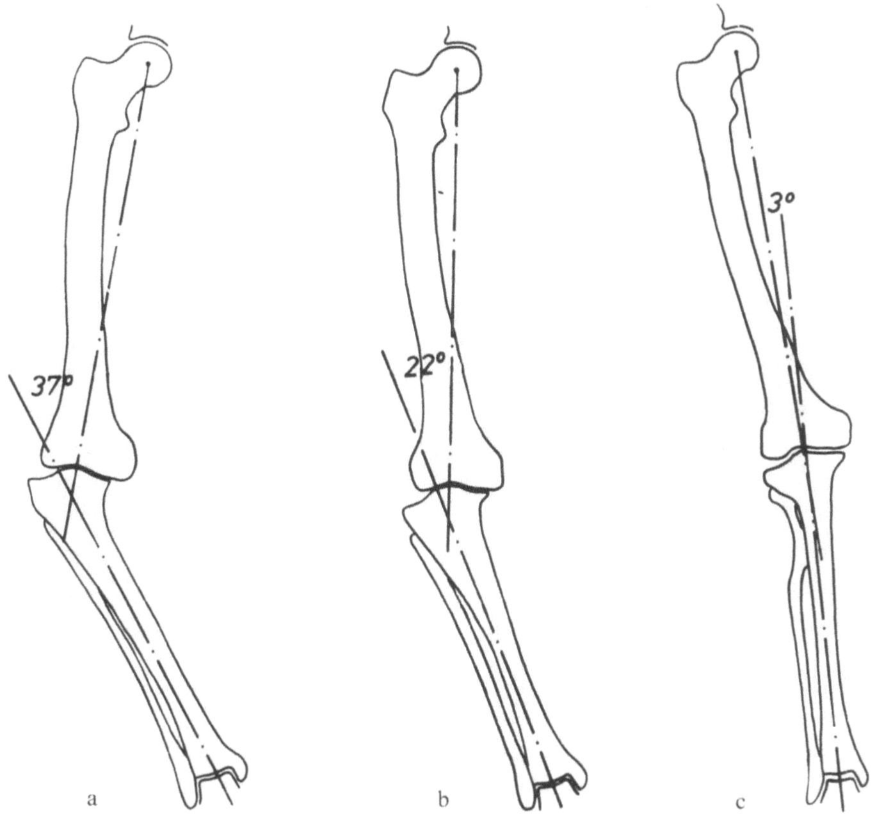

Fig. 194a–c. Same patient as Figure 193, before (a), after (b) the first osteotomy and after revision (c)

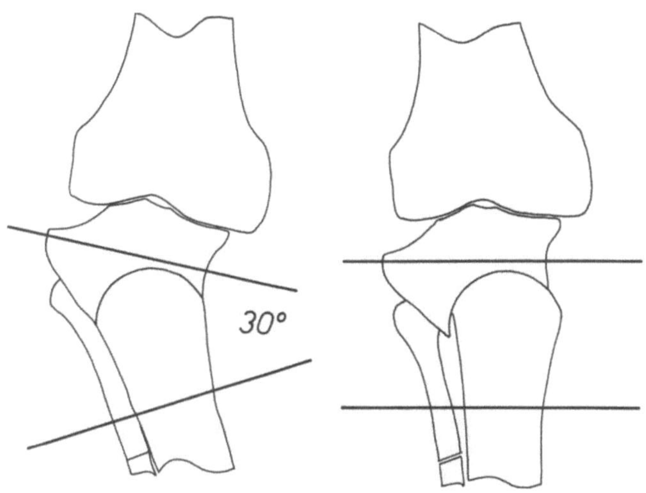

Fig. 195. Same patient as Figure 193. Planning of the first osteotomy

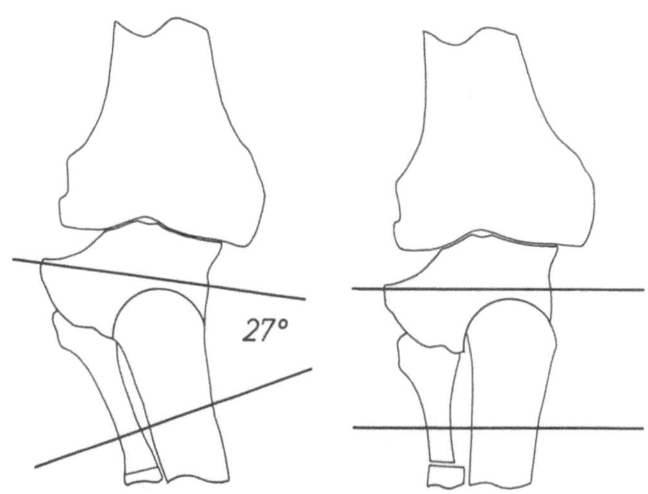

Fig. 196. Same patient as Figure 193. Planning of the revision

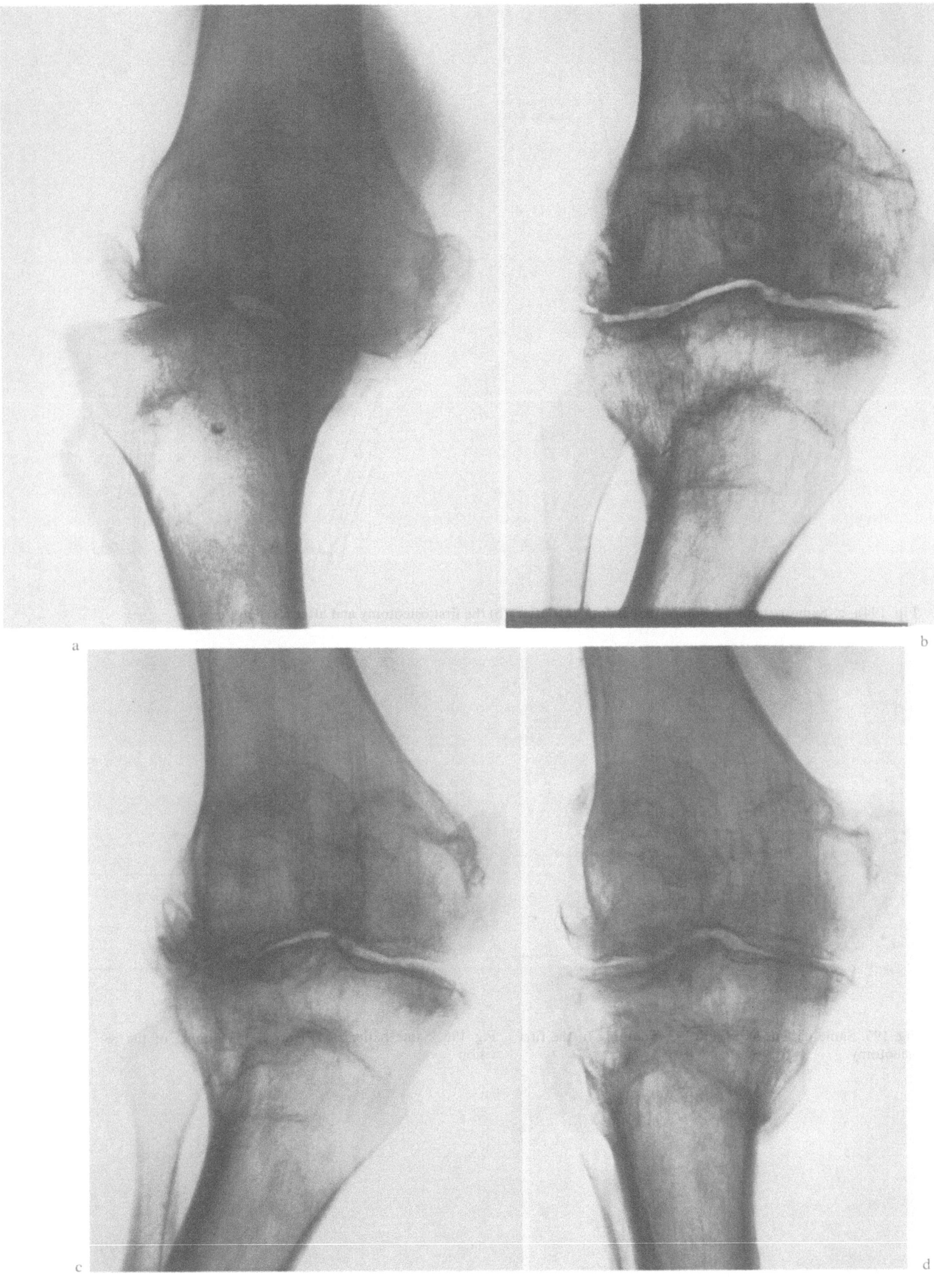

a b c d

Fig. 198a–c. Same patient as Figure 197, before (a) and after (b) exaggerated overcorrection of the varus deformity, after revision (c)

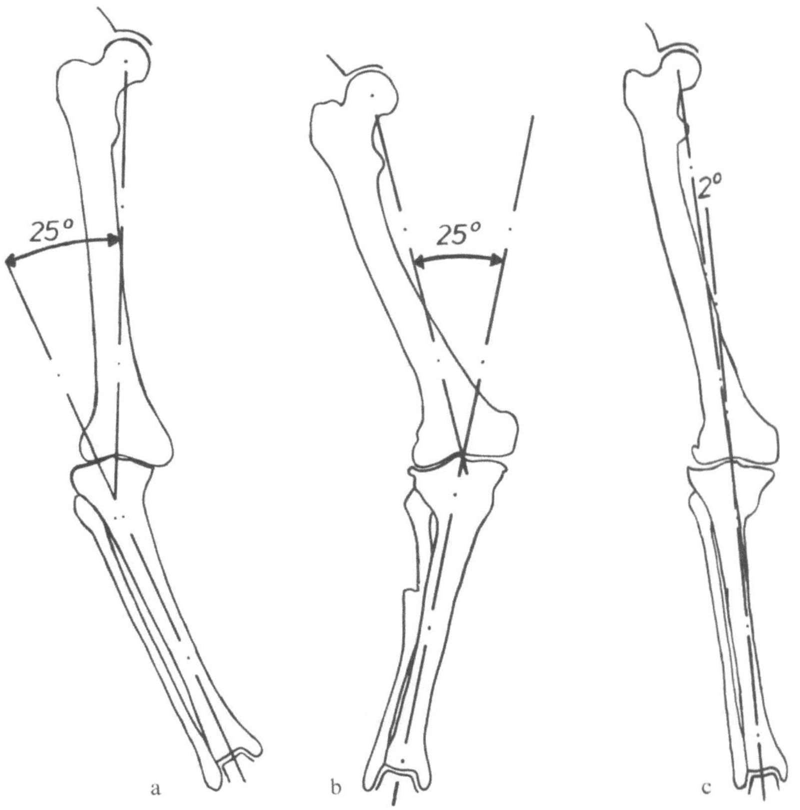

β) Revision after Exaggerated Overcorrection

In our experience, exaggerated overcorrection of a varus deformity leads to regression of the signs of osteoarthritis very quickly but may impede easy walking and, after a time, even provoke lateral osteoarthritis.

The knee (Fig. 197a) of a 57-year-old female patient was completely unstable and its varus deformity reached 25° (Fig. 198a). Here the laxity of the collateral ligaments was disregarded and a barrel-vault osteotomy of 28° was planned and carried out (Fig. 199). Clinical and X-ray improvement was quick. After a year the knee looked healed (Fig. 197b) but the gait was impeded by the knock-knee. We suggested a revision, which the patient postponed. By the time she agreed, $6^1/_2$ years after the first operation, she had developed lateral osteoarthritis (Fig. 197c) and the valgus deformity of the knee had reached 25° (Fig. 198b). A fresh barrel-vault osteotomy (25°) was carried out to diminish the valgum (Fig. 200). Pain disappeared and gait soon became normal. At the one-year follow-up, the scleroses have become symmetrical again beneath the tibial plateaux and a joint space has developed (Fig. 197d). The patient remains with a valgum of 2° (Fig. 198c).

◄ Fig. 197a–d. A 57-year-old patient with flail knee (a). One year (b) and $6^1/_2$ years (c) after exaggerated overcorrection of the varus deformity (b). One year after revision (d)

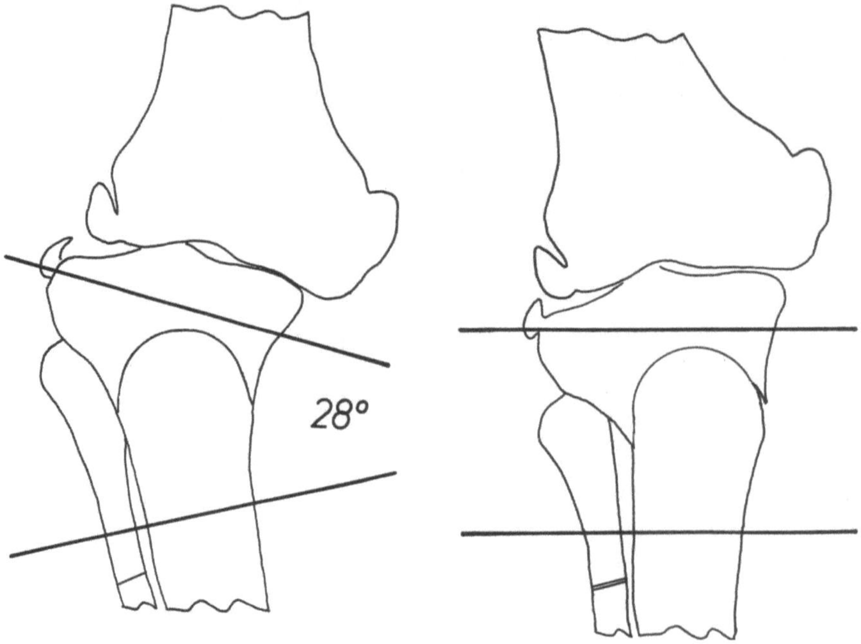

Fig. 199. Same patient as Figure 197. Planning of the first osteotomy

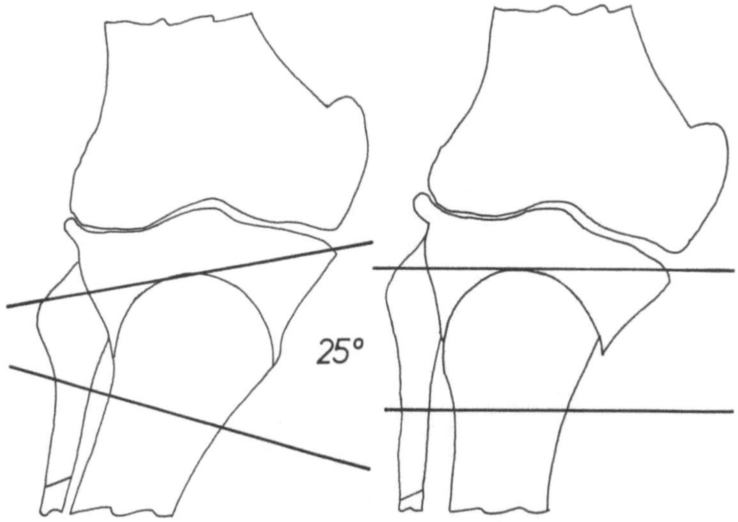

Fig. 200. Same patient as Figure 197. Planning of the revision

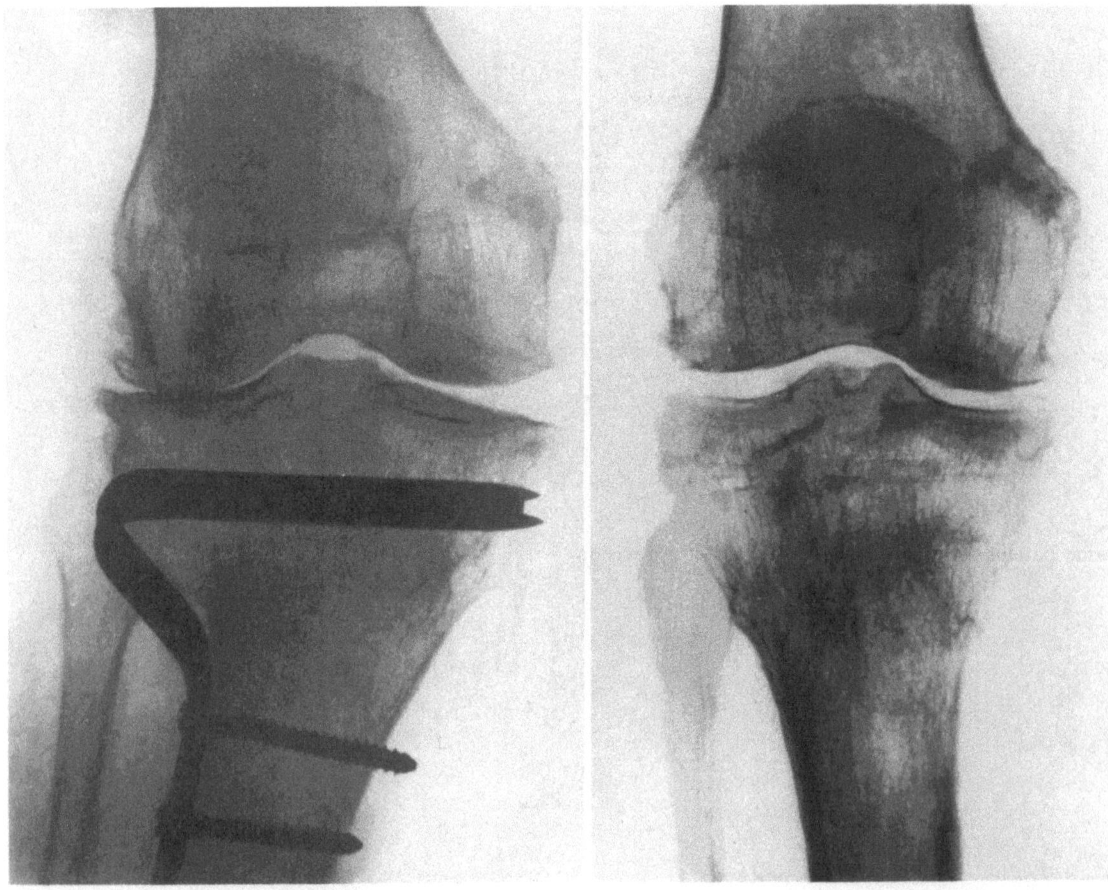

a e

Fig. 201. (a) A 69-year-old patient after an exaggerated overcorrection of a varus knee. (b) The surgically created valgus deformity. (c) Preoperative drawing for a revision. (d) The leg after revision. (e) Two years after surgery

The knee of a 69-year-old female patient (Fig. 201) was in varum. The operation carried out elsewhere created a valgus deformity which soon reached 23° (Fig. 201 a and b). One year later the pain was sufficient for the patient to accept a revision. A barrel-vault osteotomy was planned to reduce the deformity by 21° (Fig. 201 c). The operation was carried out permitting a slight valgum (2°) to remain (Fig. 201 d). Two years after the revision, the knee is painfree and a joint space has developed (Fig. 201 e).

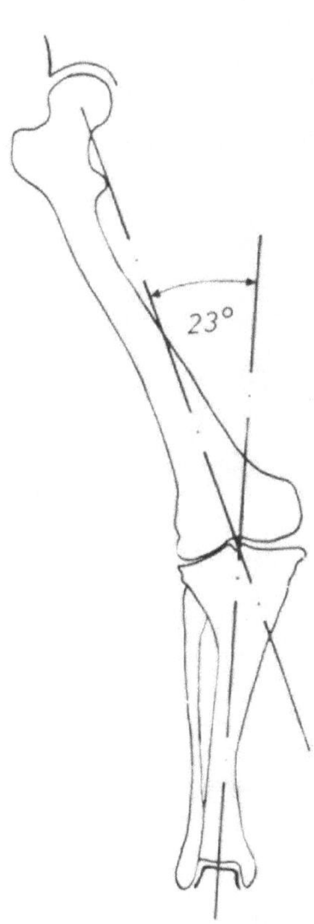

Fig. 201 b

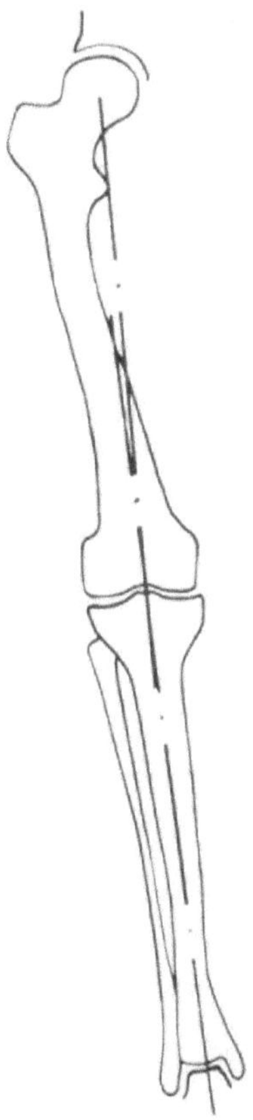

Fig. 201 d

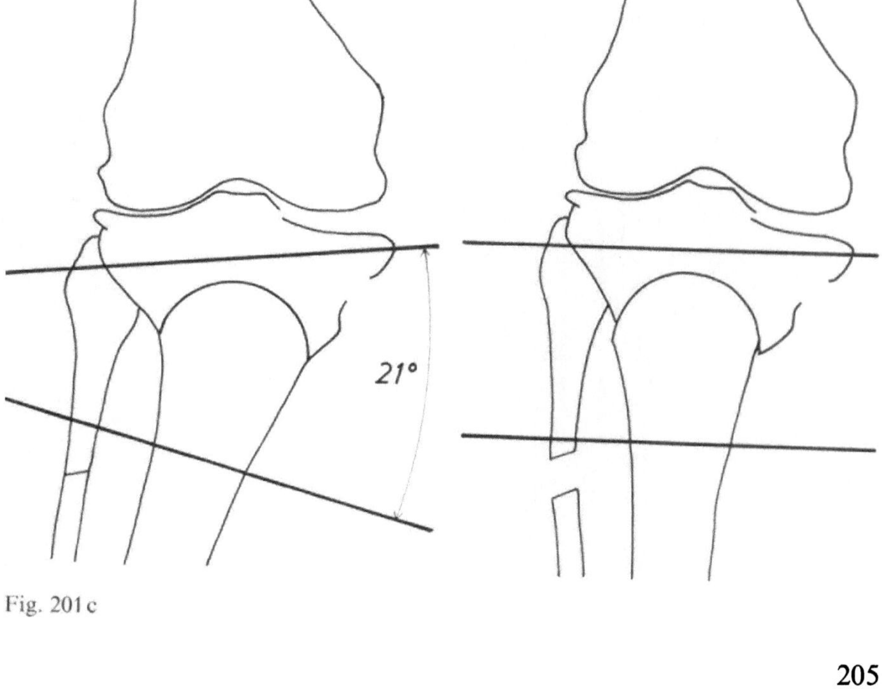

Fig. 201 c

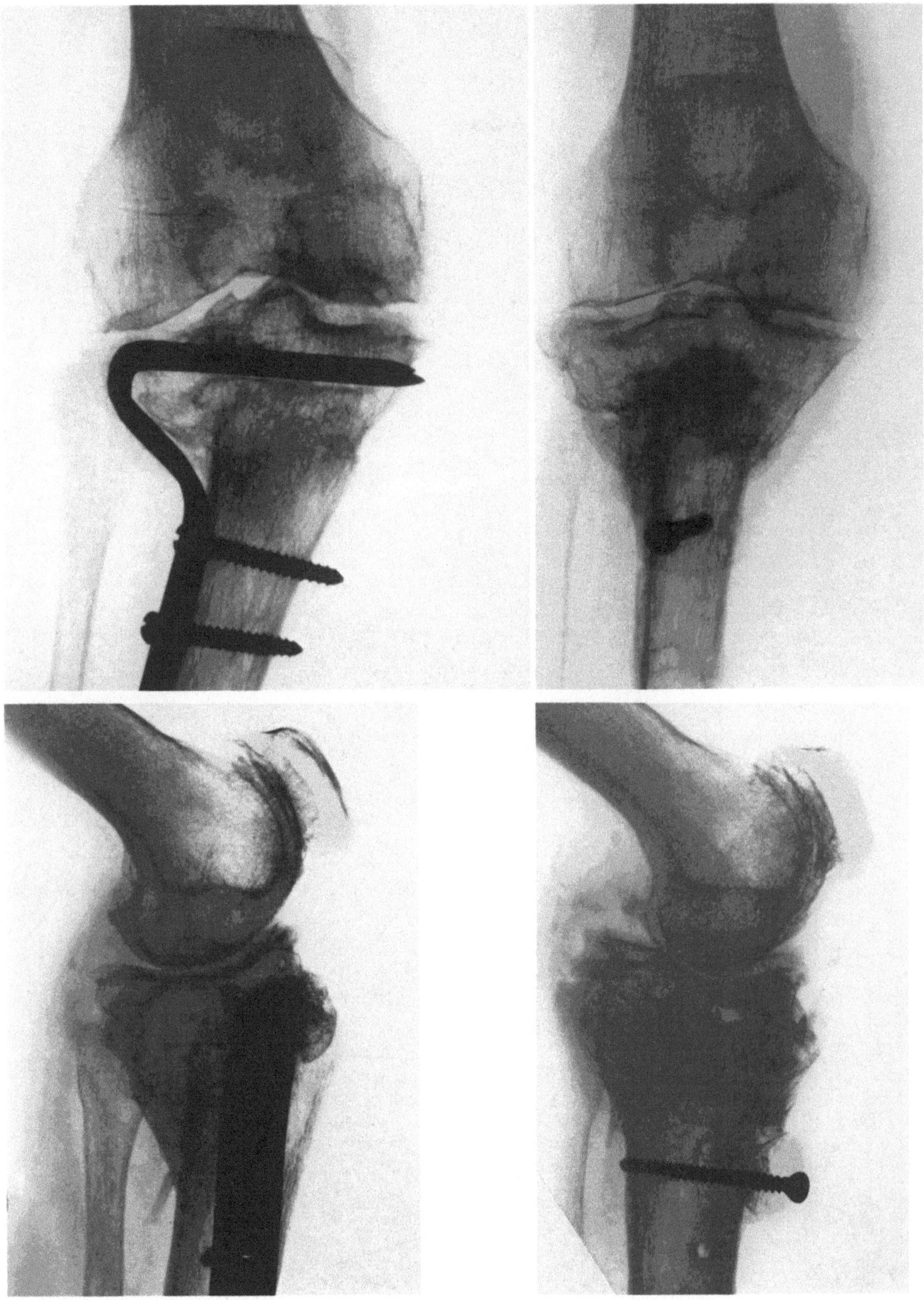

a b

Exaggerated overcorrection combined with flexion deformity can make the patient chair-bound.

A 76-year-old patient (Fig. 202a) was operated on elsewhere for osteoarthritis with a varus deformity. When we saw her, she presented with a valgum of 24° (Fig. 203a) and a flexion deformity of more than 20°. She could no longer walk. In the coronal plane, a barrel-vault osteotomy was planned as usual in order to retain a slight valgum (3°). In the sagittal plane, we planned to remove a wedge with the base anterior (Fig. 204). To avoid raising the patella, the tibial tuberosity had to be excised and replaced on the tibial crest below the osteotomy site. The operation was carried out with the utmost precision. Two Steinmann pins forming an angle of 21° open medially were inserted in the coronal plane, made parallel after the barrel-vault osteotomy and fixed by two compression clamps. Two more pins were placed from before backward through the two fragments of the tibia. The distal one, normal to the long axis of the bone, formed an angle of 20° open anteriorly with the proximal one. After removal of the wedge, these pins were made parallel and fixed by an anterior compression clamp. The tibial tuberosity was screwed to the tibial crest in such a way that the distance between the patella and the tibial plateaux remained unchanged. The patient was encouraged to move and to walk with partial weight-bearing. One year after the revision, the clinical result is good. The knee is painfree, stable and mobile. The patient now walks. She retains a slight valgum of 3° (Fig. 203b). The X-ray picture is satisfactory (Fig. 202b).

Overcorrection seems to be tolerable to about 15° of valgum. Beyond that, it seems advisable to diminish the overcorrection by a new osteotomy before osteoarthritis has developed in the lateral aspect of the knee. Then a valgum of 3° to 9° should be maintained. Flexion deformity soon becomes intolerable. More than 10° slope of the tibial plateaux backward usually requires correction.

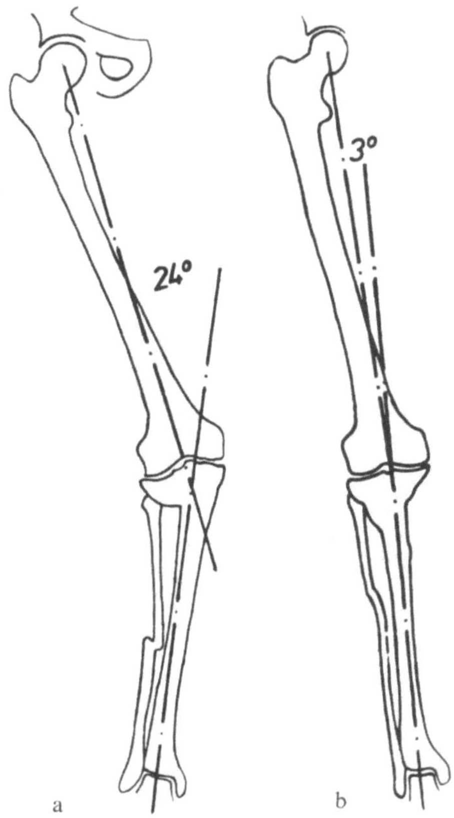

Fig. 203. Same patient as Figure 202, before revision (a) and after revision (b)

◀ Fig. 202a and b. A 76-year-old patient with iatrogenic valgus and flexion deformity, before revision (a) and 1 year after revision (b)

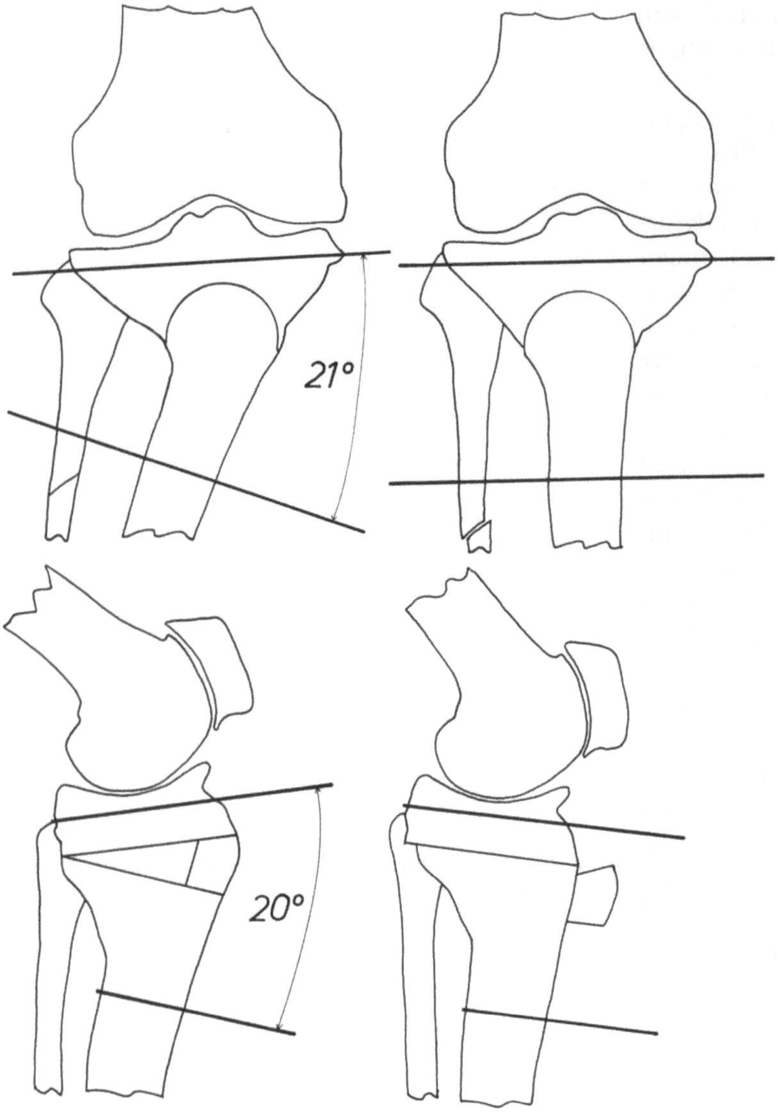

Fig. 204. Same patient as Figure 202. Planning of the revision

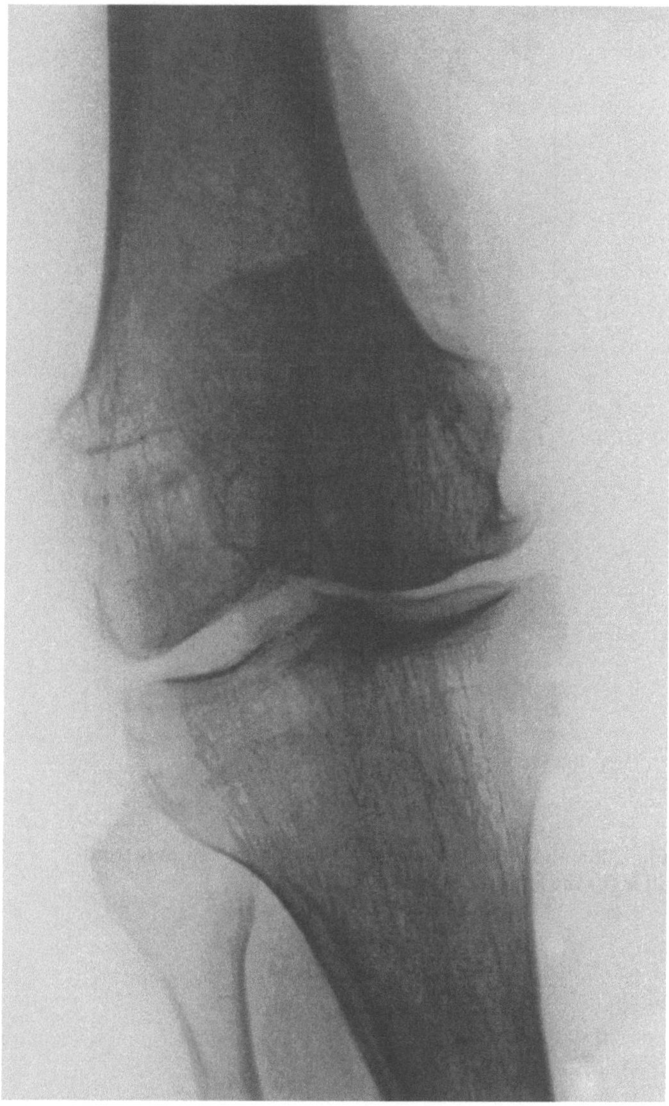

Fig. 205. Varus deformity with sclerotic triangle in the vicinity of the intercondylar eminence

h) The Exceptions. Femoral Osteotomy for a Varus Deformity of the Knee

When the varus deformity results from curvature of the femur convex laterally and the tibia convex medially, the plane of the tibial plateaux is oblique to the resultant force acting on the knee. This occurred in a 67-year-old female patient (Fig. 205). A sclerotic triangle underlies the medial plateau in the vicinity of the intercondylar eminence, reminiscent of the X-ray picture of a knee with lateral osteoarthritis. The trabeculae in the femoral metaphysis tend to form gothic arches, the typical structure of bending stress combined with shearing (see page 123). After correction of the deformity by a barrel-vault osteotomy of the upper end of the tibia, the tibial plateaux would remain oblique to the resultant force acting on the knee. In such a case, only an osteotomy of the lower end of the femur can improve the mechanical conditions, by rotating the tibial plateaux in the opposite direction to the rotation of the line of action of the resultant force.

α) Planning

The degree of deformity is measured on a long X-ray showing the whole leg (Fig. 206a). A line is drawn from the centre of the femoral head to the midpoint of the cross section just above the femoral condyles. From this point another line is drawn to the centre of the ankle. The angle formed by the two lines (here 22°) is measured. The outlines of the knee are traced on transparent paper from an A.-P. view of the loaded knee (Fig. 207). Two parallel transverse lines representing Steinmann pins are drawn through the femoral condyles and two through the diaphysis of the femur. The latter form an angle of 25° open laterally with the former. In this case we thus aim at an overcorrection of 3°. The osteotomy line is drawn just proximal to the condyles. The distal fragment with its two transverse lines is now traced on a second sheet of transparent paper. The second sheet is rotated over the first about the medial end of the supracondylar osteotomy line in such a way that the lateral aspect of the proximal fragment penetrates into the distal until the four transverse lines corresponding to Steinmann pins are all parallel. The outline of the femoral diaphysis and the two proximal transverse lines are then traced on the second transparent sheet. The femoral osteotomy has thus been carried out graphically.

β) Surgical Procedure

The patient lies supine. No tourniquet is used. The knee is slightly flexed, resting on a cushion placed in the popliteal area. A 6 cm longitudinal lateral incision lying between the vastus lateralis and the biceps femoris gives access to the lateral aspect of the distal metaphysis of the femur. The periosteum is incised and elevated in front of and behind the femur, just above the condyles, without opening the joint. A Steinmann pin is passed transversely through the femoral condyles, proximal to the intercondylar notch. Using the pin insertion guide, another Steinmann pin is inserted through the diaphysis, forming with the first one the desired angle, open laterally. A third pin is inserted parallel and proximal to the first one through the condyles and a fourth parallel and distal to the second through the diaphysis. The pin insertion guide with its two parallel mobile units makes

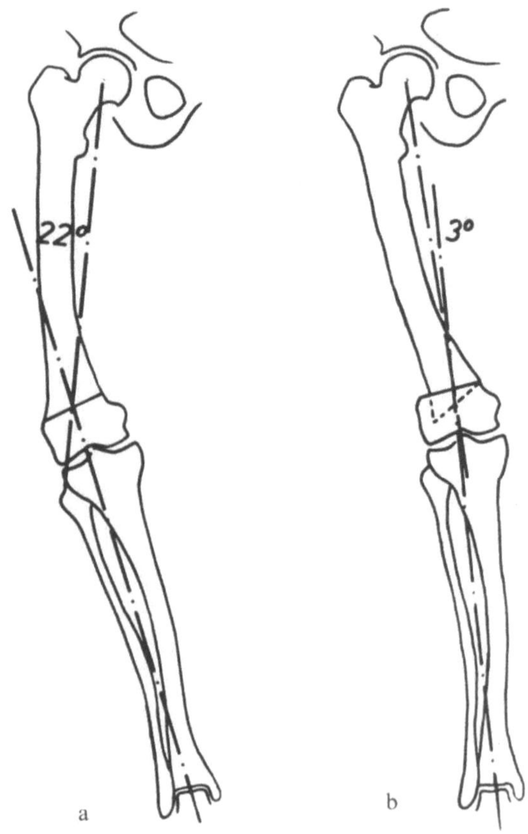

Fig. 206a and b. Same patient as Figure 205, before (a) and after (b) the femoral osteotomy

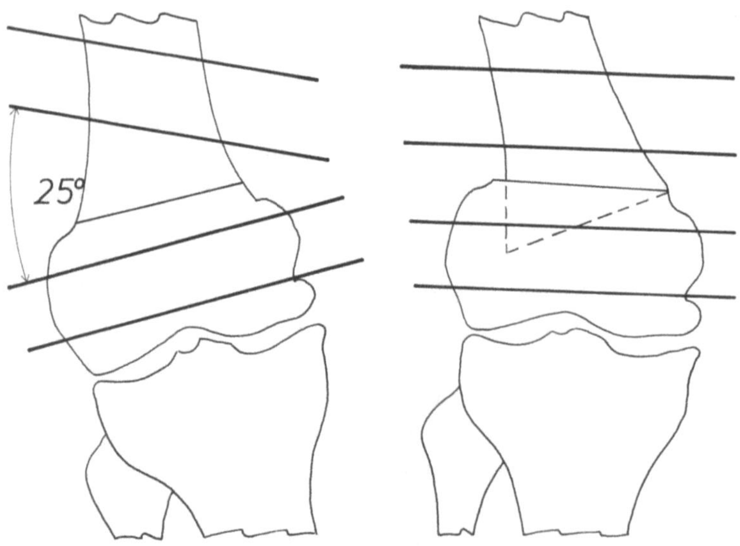

Fig. 207. Same patient as Figure 205. Planning of the femoral osteotomy

this step easy. The Steinmann pins are inserted *from the medial side* and must not emerge through the lateral incision. It is better if they are not involved at all in this incision. For the diaphysial pins, the sleeve and its blunt-ended pin are first driven to the bone, then the pin is removed and a pointed pin is inserted with an air-powered drill. The angle formed by the two pairs of Steinmann pins is checked by X-ray. Then the bone is divided transversely proximal to the femoral condyles with an oscillating saw and a thin chisel. During this step, a large periosteal elevator is maintained behind the femur to protect the popliteal nerves and vessels. The medial cortex is incompletely divided. The lower end of the proximal fragment is bevelled off in its anterior, lateral and posterior aspect. The fragments are then impacted, using the medial cortex as a hinge, until all the Steinmann pins are parallel in the same coronal plane. They are fixed by two compression clamps each equipped with four mobile units. The aponeurosis, the subcutaneous fat and the skin are sutured using suction drainage.

γ) Postoperative Care

The surgeon completely flexes and extends the knee on the operation table. The patient is encouraged to move immediately. He stands up and starts walking with crutches the second day after surgery, as soon as the suction drainage has been removed. The pins are removed two months later after clinical examination and X-ray have shown consolidation of the osteotomy. The patient can discard the crutches as soon as he wishes.

In the present case (Fig. 206) the desired overcorrection of 3° has been achieved. The result is excellent.

δ) Second Example

This operation was carried out bilaterally on a 24-year-old female patient (Fig. 208a). The varus deformity of her legs resulted from abnormal curvature of her femora and (to a lesser degree) of her tibiae. The clinical result is excellent and the cosmetic appearance has improved dramatically (Fig. 208b). Before the operation the medial plateau of each knee was underlined by an exaggerated cup-shaped subchondral sclerosis. The subchondral sclerosis under the lateral plateau was less than normal (Fig. 209a). Two years after a slight overcorrection of the deformity through bilateral femoral osteotomy, the subchondral scleroses are less pronounced beneath the medial plateaux (Fig. 209b). This means a better distribution of the compressive stresses.

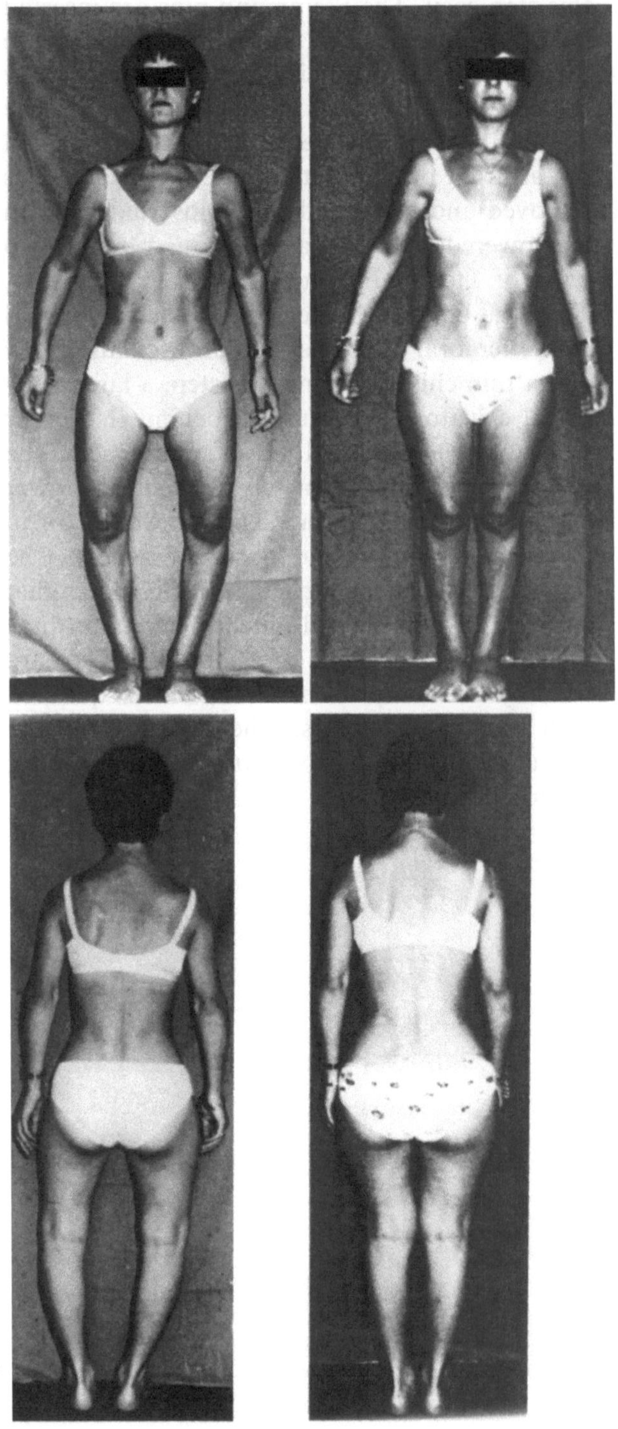

a b

Fig. 208a and b. A 24-year-old patient with varus knees before (a) and after (b) femoral osteotomies

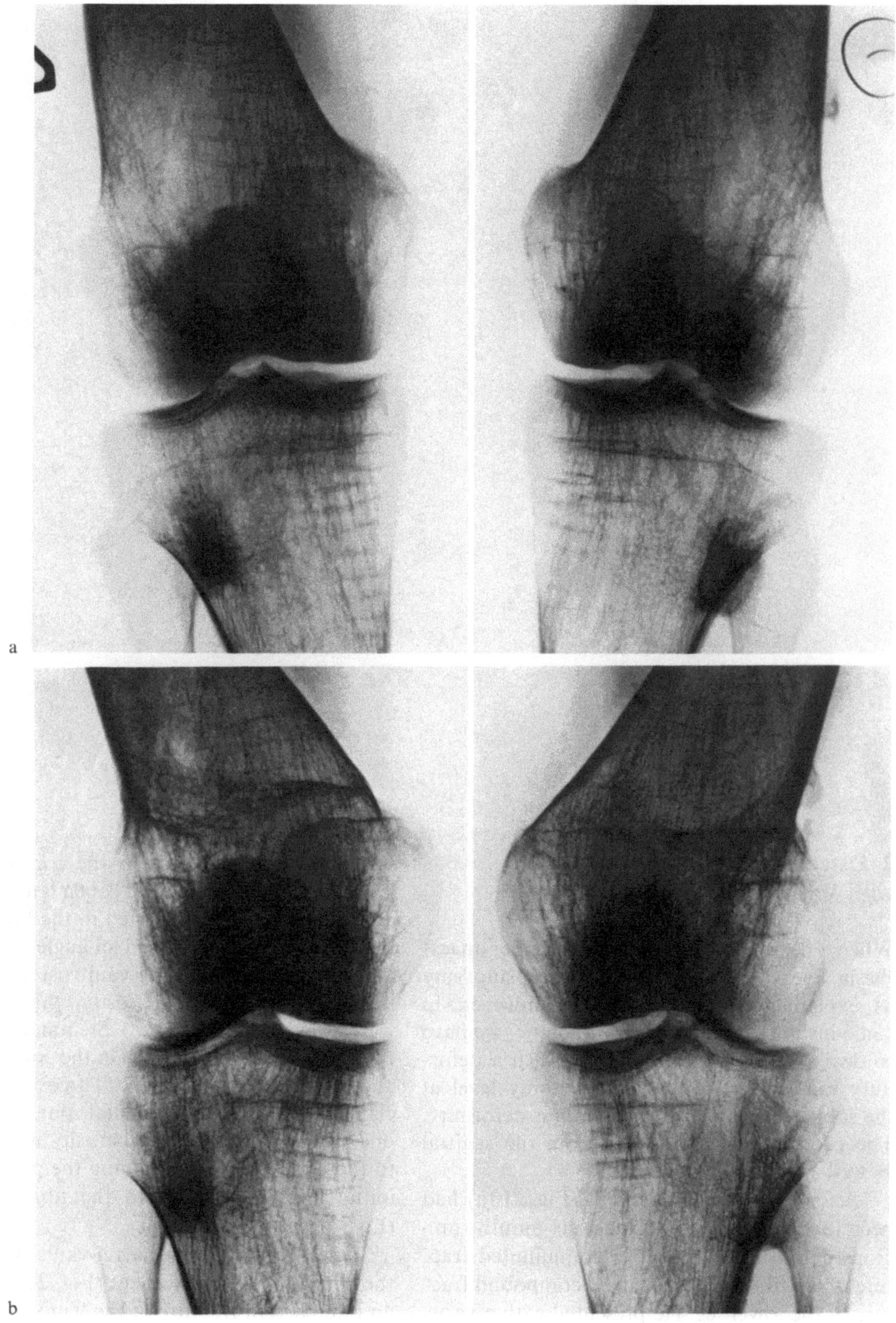

Fig. 209 a and b. Same patient as Figure 208 before (a) and $1^1/_2$ years after (b) femoral osteotomies

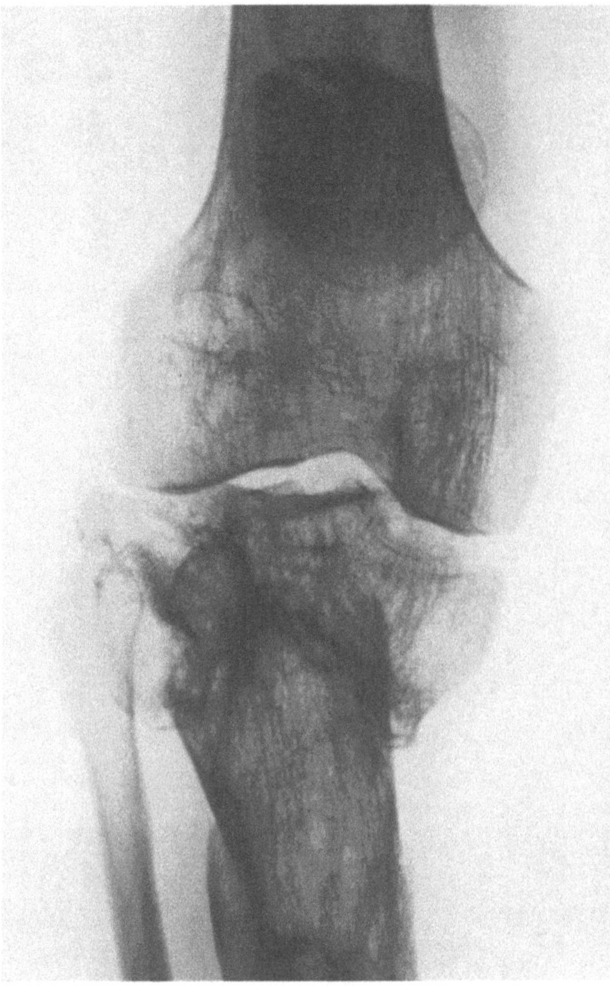

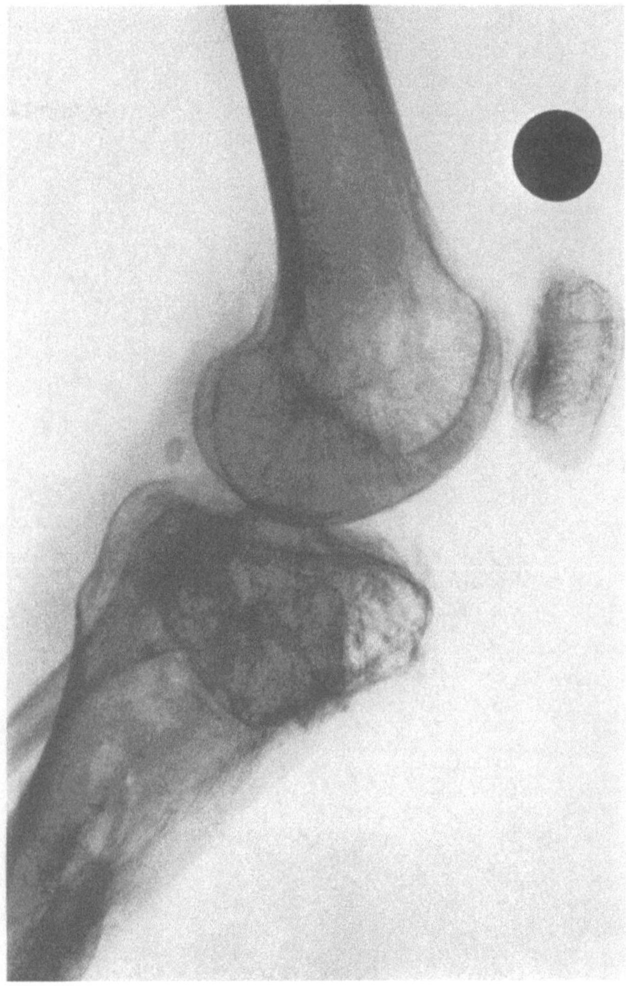

2. Osteoarthritis of the Knee with Varus and Flexion Deformity

When a flexion *contracture* persists under anaesthesia, we carry out a posterior capsulotomy as explained above (page 143). However, in some instances, mostly post-traumatic, we have to deal with a flexion *deformity*. Such a deformity must be corrected at osteotomy level at the same time as the varus or valgus deformity. The procedure is then planned in the sagittal as well as in the coronal plane.

A 26-year-old male patient (Fig. 210a) had been involved in a car accident six months previously and had sustained a comminuted fracture of the tibial plateaux and a compound fracture of the lower leg. He presented with a varus deformity of 10° (Fig. 210e) and a flexion deformity of 16°. Following the preoperative planning (Fig. 210b), two Steinmann pins were inserted through the tibia in the coronal plane. They formed an angle of 12° open laterally. Two additional pins were inserted in the bone in the sagittal plane. They formed an angle of 16° open anteriorly. After a barrel-vault osteotomy and lifting of the tibial tuberosity, the fragments were moved until the two Steinmann pins of each pair were parallel, two in the coronal plane and two in the sagittal plane. Three compression clamps were used, one medial, one lateral and one anterior. The tibial tuberosity was screwed to the tibial crest, maintaining the patella at its initial level in relation to the tibial plateaux (Fig. 210b and c).

One year after the barrel-vault osteotomy, the clinical result is excellent (Fig. 210e) despite an intercurrent fracture of the femoral condyles in another accident. The X-ray is satisfactory (Fig. 210d).

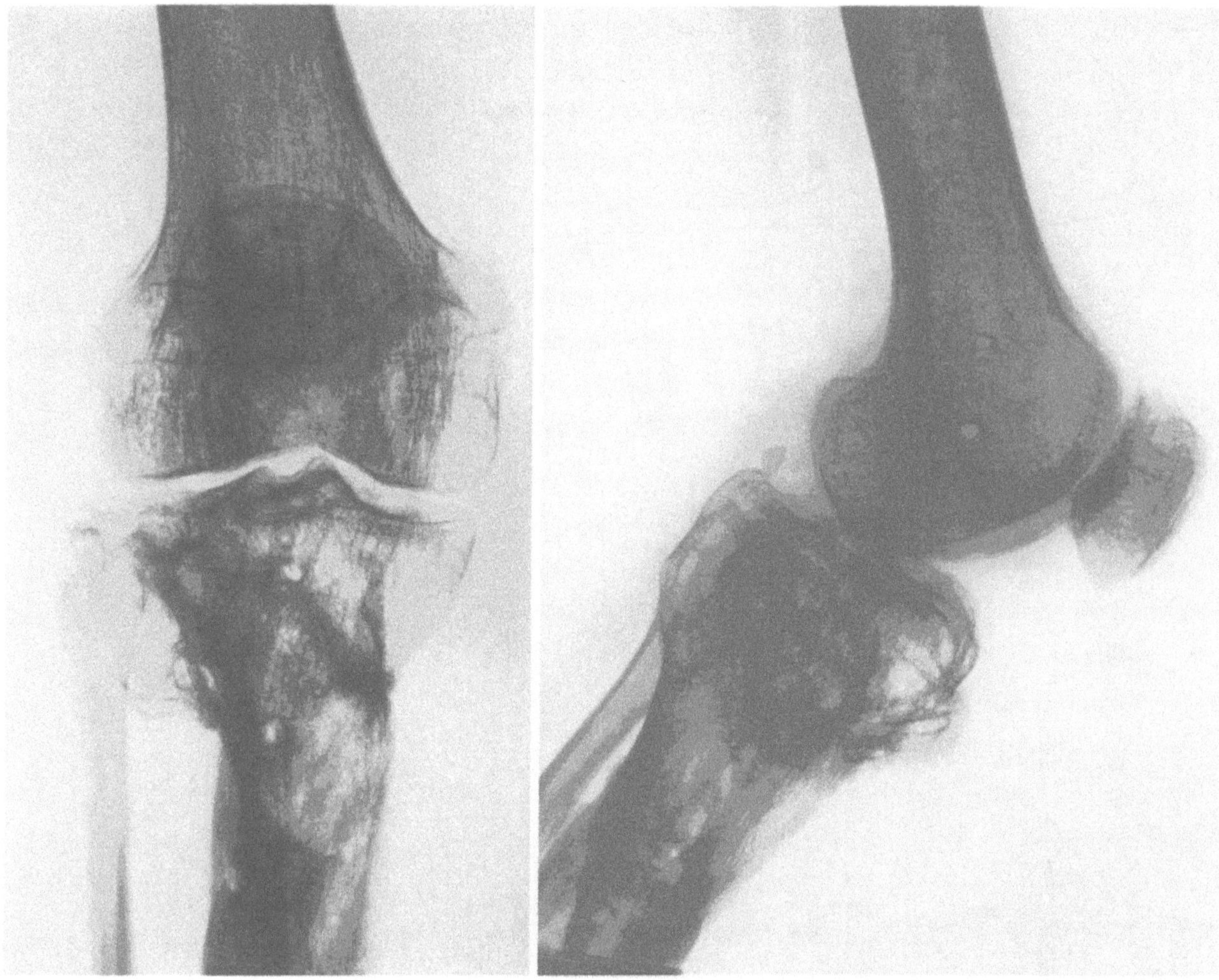

d

Fig. 210a–g. A 26-year-old male patient with post-traumatic varus and flexion deformity (a). Planning of the correction (b). After the operation (c). Result at one-year follow-up (d). Before and after the correction (e). Range of movement (f, g)

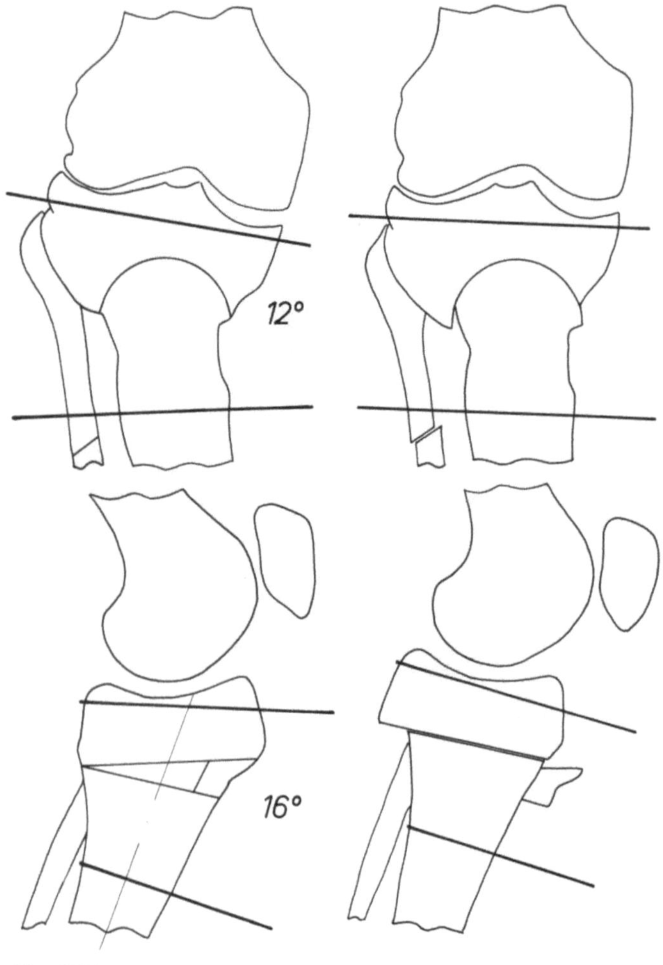

Fig. 210b

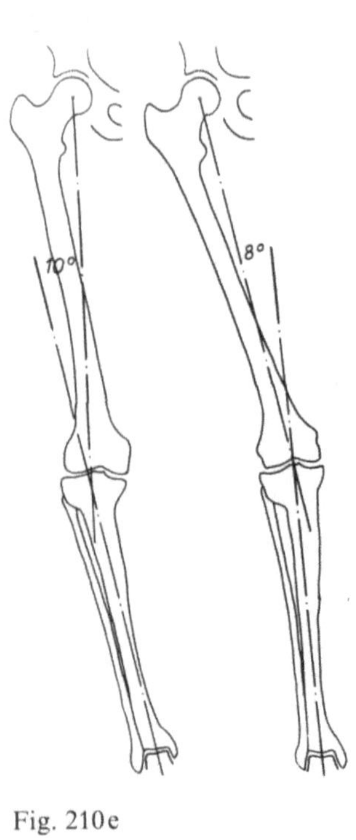

Fig. 210e

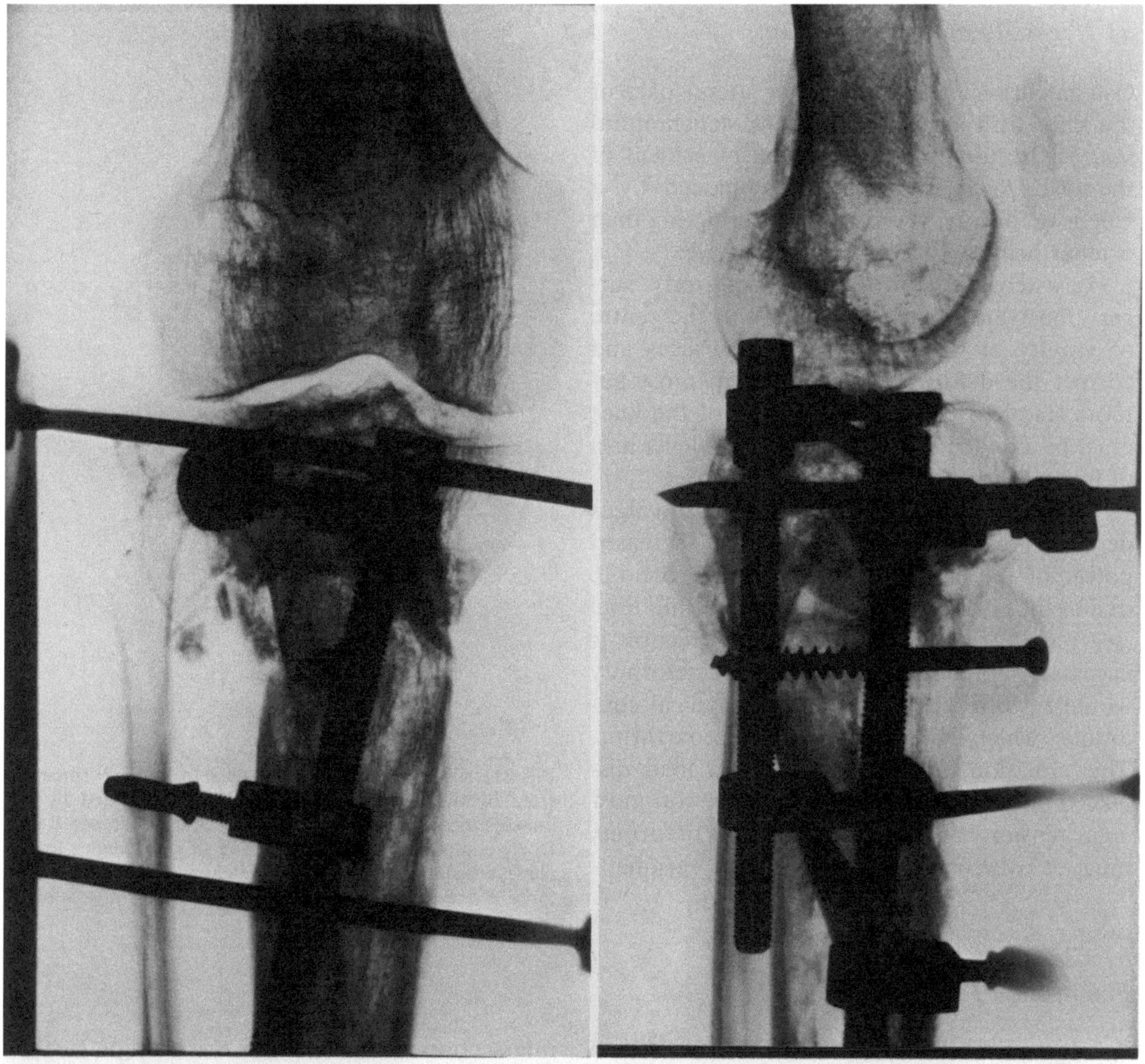

Fig. 210c

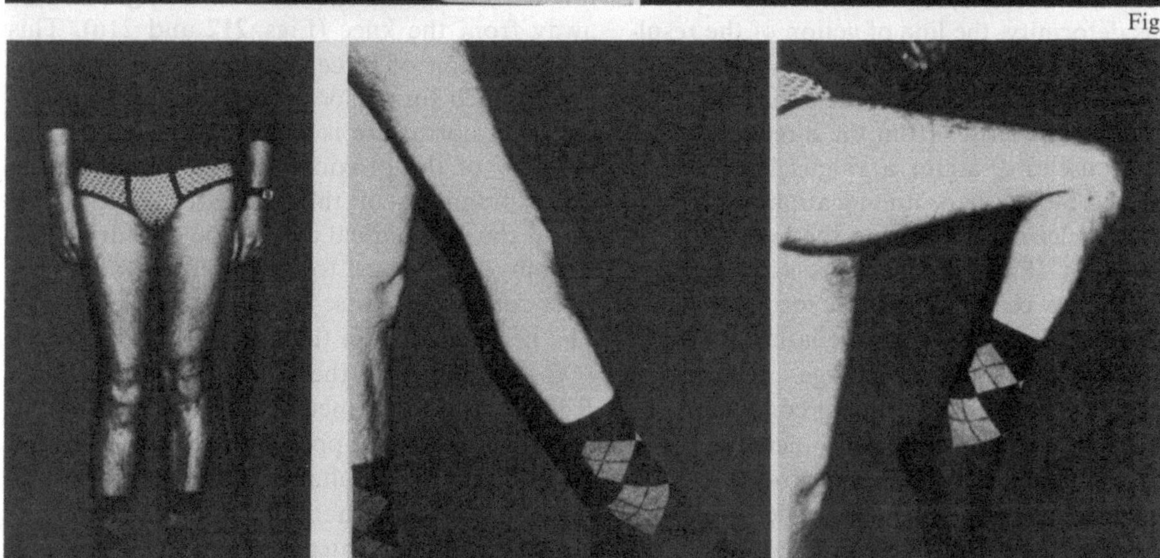

3. Osteoarthritis with a Valgus Deformity

a) Necessity of Overcorrection

Osteoarthritis prevailing in the lateral part of the knee with an increased lateral subchondral sclerosis results from a lateral displacement of the load *R*. It is usually accompanied by a valgus deformity. It is often more crippling than osteoarthritis with a varus deformity.

In osteoarthritis with valgus deformity, surgery must bring back the load *R* to the centre of gravity of the weight-bearing surfaces and correct the deformity. According to most authors (Herbert et al., 1967), varum of the knee must be avoided since it lengthens the lever arm of force *P*.

But in most cases, osteoarthritis with valgus deformity can only be explained by increased power of the lateral muscular tension band *L* due to the conditions of equilibrium at hip level (see pages 24 and 81). Then overcorrection is necessary. An exact correction of the deformity would restore the pre-existing mechanical conditions which have caused the osteoarthritis. This situation is characterized by a load displaced laterally because of a too powerful muscular tension band *L*. In principle the deformity must be overcorrected into moderate varum.

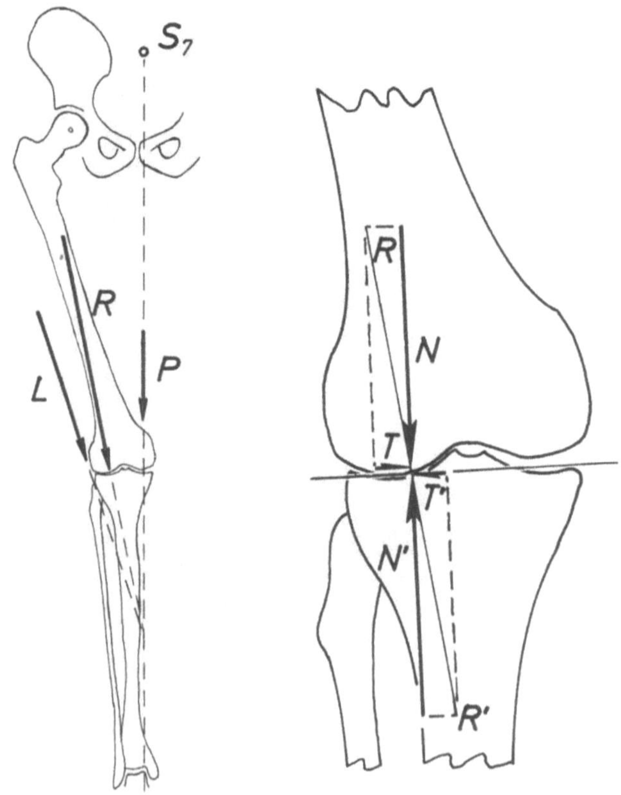

Fig. 211. Osteoarthritis with valgus deformity. *P*: force exerted by the partial body mass. *L*: force exerted by the muscles to counterbalance *P*. *R*: resultant of forces *P* and *L*. *N*: component of force R acting at right angles to the plane tangential to the tibial plateaux. *T*: component of force *R* acting in the direction of the plane tangential to the tibial plateaux

b) Choice of Procedure: Femoral or Tibial Osteotomy?

As described above, in osteoarthritis with a valgus deformity, the line of action of the resultant force *R* is displaced laterally and is oblique to the plane tangential to the tibial plateaux (Fig. 211). This arises from the direction of the lateral muscles *L* about a valgus knee and is confirmed by the shape and location of the subchondral sclerosis underlying the lateral aspect of the joint (Fig. 114). Overpressure, concentrated in the centre or in the medial part of the lateral plateau, deepens the latter after destroying the articular cartilage. Resorption reduces the height of the lateral condyle of the femur. The valgus deformity increases. The picture is completely different from that of medial osteoarthritis with a varus deformity.

Correcting or even overcorrecting the deformity redirects the femur and the muscles *L*. The intersection of forces *L* and *P* is brought further away from the knee (Figs. 212 and 216). This re-orientation of force *L* changes the direction of *R*, which turns clockwise in our drawing. If the (over)correction is achieved through an osteotomy of the proximal end of the tibia, the upper fragment of this bone turns clockwise about the geometrical axis of the cylindrical osteotomy, in contrast to the lower leg, which rotates counterclockwise about the heel (Fig. 212). The heel remains on the line of action of force *P* for equilibrium (the line of action of force *P* is exactly 3 mm medial to the centre of the heel). Turning the upper fragment in the same direction as the resultant force does not allow force *R* to be made normal to the plane tangential to the tibial plateaux. Force *R* is oblique to this plane and continues to act with a normal component *N* and a tangential component *T* (Fig. 211). The femur tends to slide medially on the tibia, pushed by the tangential component *T*. This hinders optimal distribution of the stresses, which are concentrated in the vicinity lateral to the tibial spines. The medial plateau, is oblique to the line of action of force *R* as a consequence of the deformity, and therefore can take hardly any part in the load transmission.

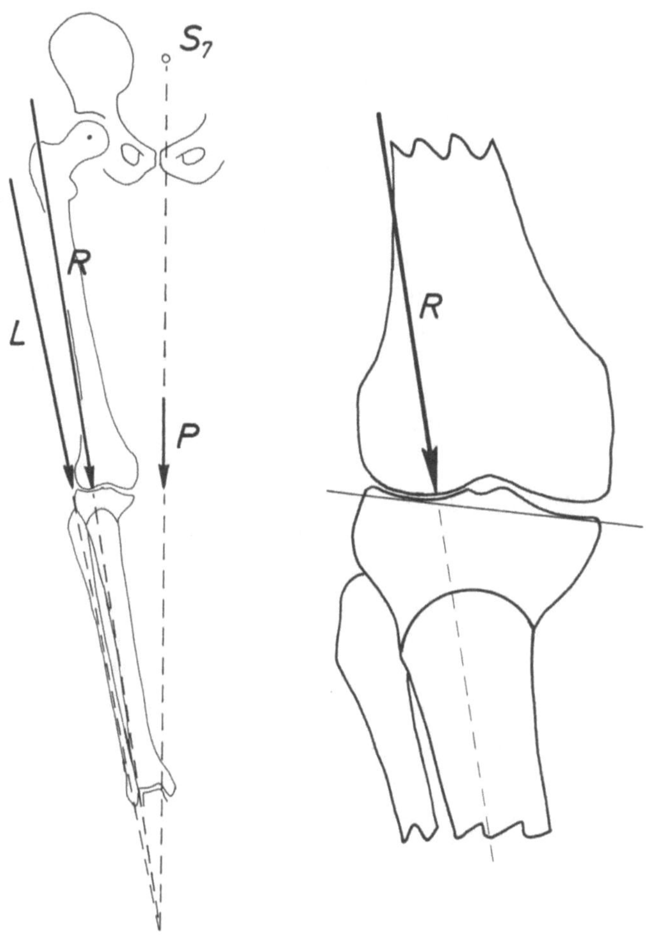

Fig. 212. Correcting the valgus deformity by a tibial osteotomy results in a force *R* acting obliquely to the plane tangential to the tibial plateaux

Medial sliding of the femur on the tibia is impeded only by the lateral condyle abutting against the intercondylar eminence. The subchondral density developed at this place demonstrates the increase of compressive stresses (Fig. 214b and 217a).

Furthermore, (over)correction of the valgus deformity removes the knee away from the line of action of force *P*. Lengthening of the lever arm *a* of force *P* increases the moment $P \cdot a$ (Fig. 213). Force *L* must increase to maintain equilibrium since its lever arm *b* is only insignificantly modified, if at all. Consequently, the load *R* exerted on the knee is greater and the weight-bearing surfaces are not enlarged. Surgery has worsened the mechanical conditions.

These facts are illustrated by the example of a 67-year-old female patient (Fig. 214) in whom a proximal tibial osteotomy has precisely corrected the valgus deformity. Only the lateral part of the joint is transmitting from the femur to the tibia a load increased by the lengthening of the lever arm *a* of force *P*. The X-ray one year later confirms the clinical deterioration (Fig. 214b).

Fig. 213. Proximal tibial osteotomy for accurate correction of a valgus deformity, insufficient to move the line of action of resultant *R* toward the medial plateau

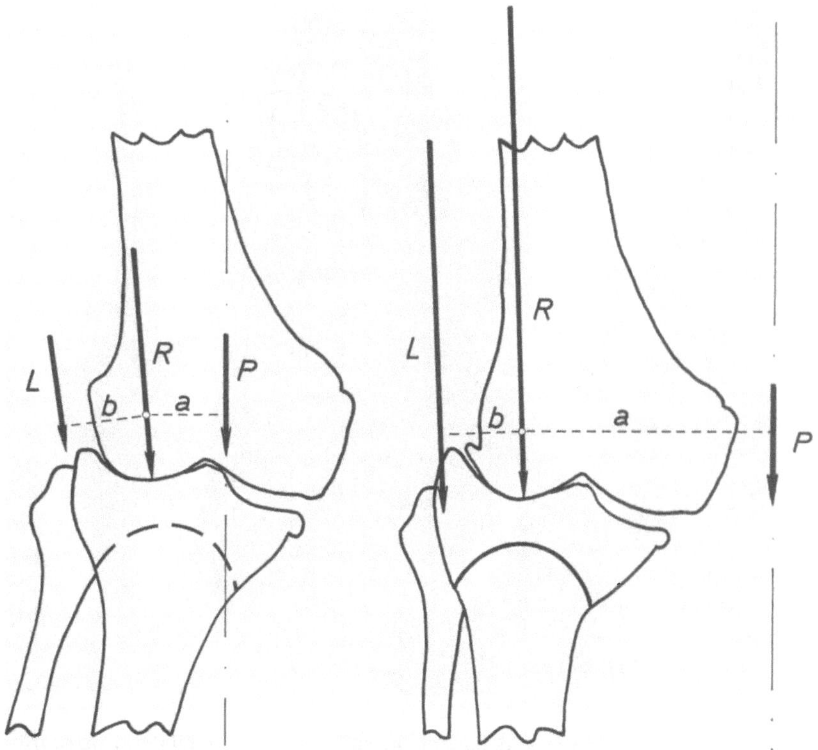

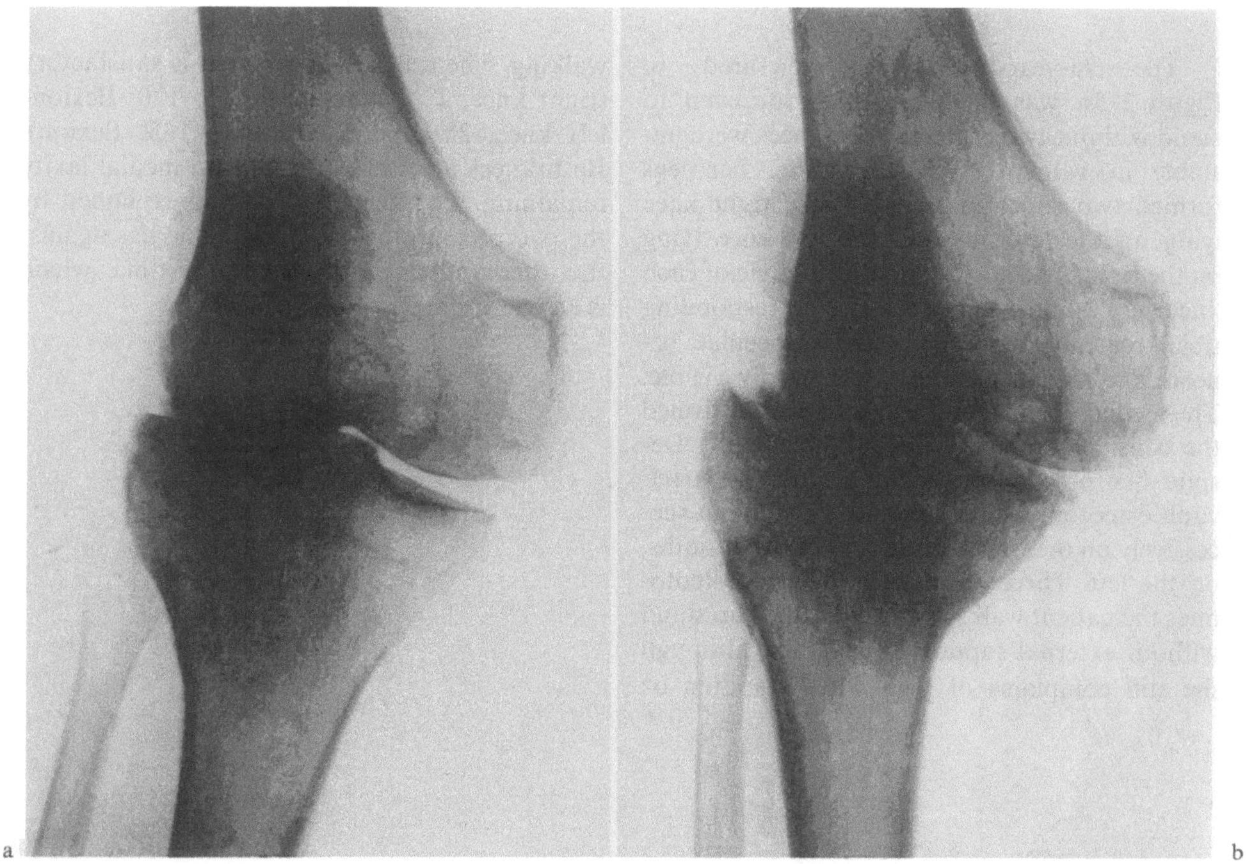

Fig. 214a and b. A 67-year-old patient (a) after an exact correction of her valgus osteoarthritic knee. One year later her knee is worse (b)

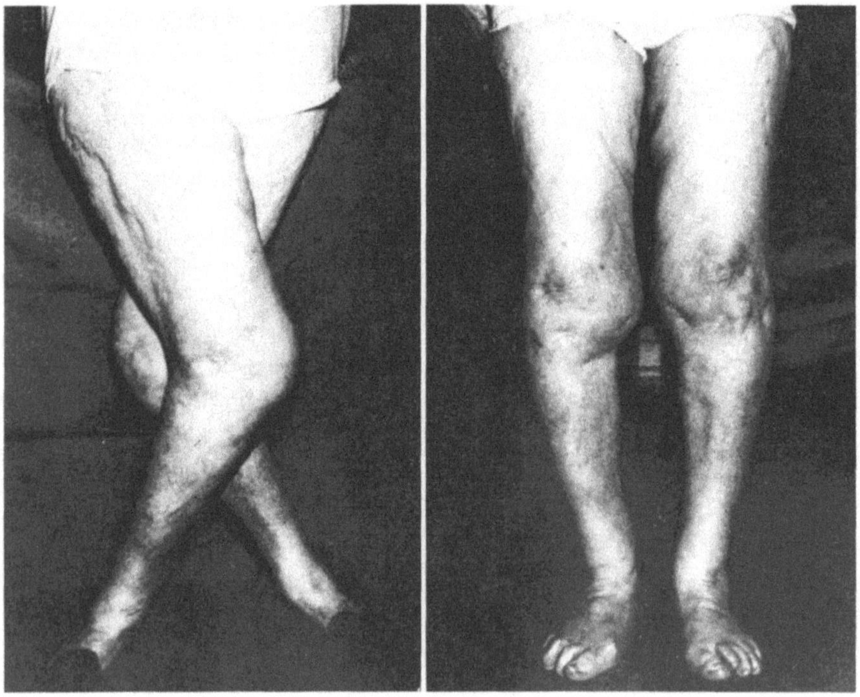

a c

Fig. 215a–d. A 75-year-old patient (a) suffering from a bilateral osteoarthritis with considerable valgus deformity (b) and unstable knees. The right knee is on the left and the left knee on the right (a and b). Three years after a bilateral overcorrecting osteotomy the cosmetic appearance has improved (c), but the knees remain painful (d)

The 75-year-old patient pictured in Figure 215a was unable to walk or even to stand without two crutches. Her knees were unstable in valgum. When standing, her legs formed two superimposed X's, the right knee lying on the left side and the left knee lying on the right. The lateral femoral condyle of each knee had sunk deeply into the corresponding tibial plateau (Fig. 215b). The trabeculae beneath the medial plateaux were hardly visible. The medial opening of the joint space confirmed the considerable laxity observed clinically. Despite her precarious general health, a barrel-vault osteotomy of the tibia was carried out successively on the right side and, after five months, on the left. Three years after the two osteotomies the patient walks with a stick and can stand without external support (Fig. 215c), although she still complains of pain when standing or walking. The range of movement is satisfactory (right knee, 2° hyperextension to 120° flexion; left knee, 2° hyperextension to 135° flexion). Both knees are stable: there is no medial laxity remaining. The persistent pain is explained by the overpressure of the lateral condyles against the intercondylar eminence of the tibia, which is shown in the X-rays (Fig. 215d).

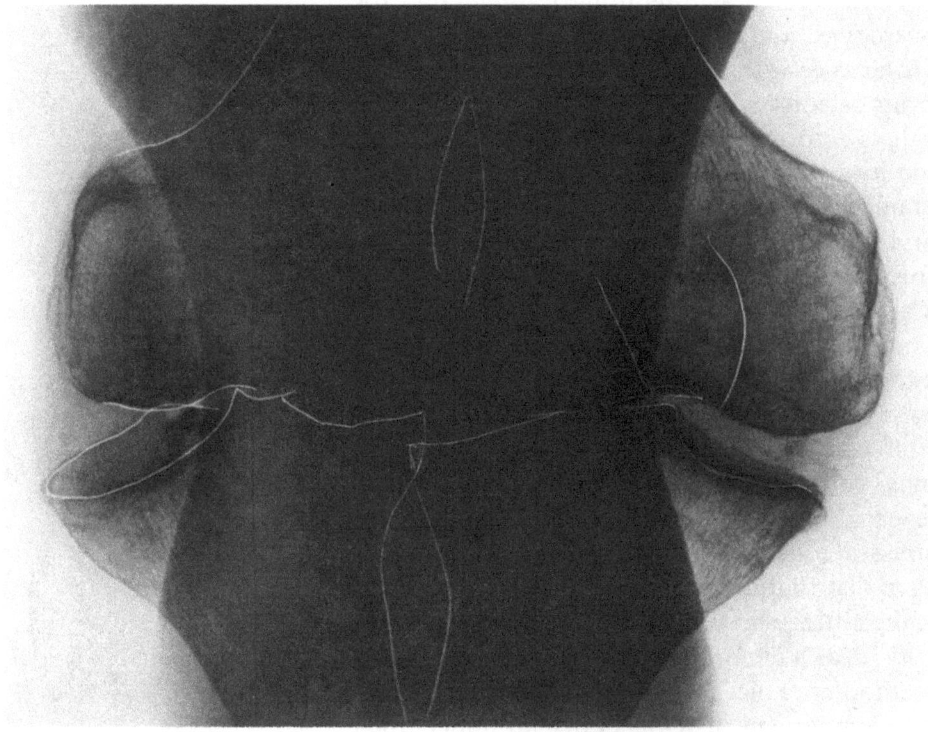

b

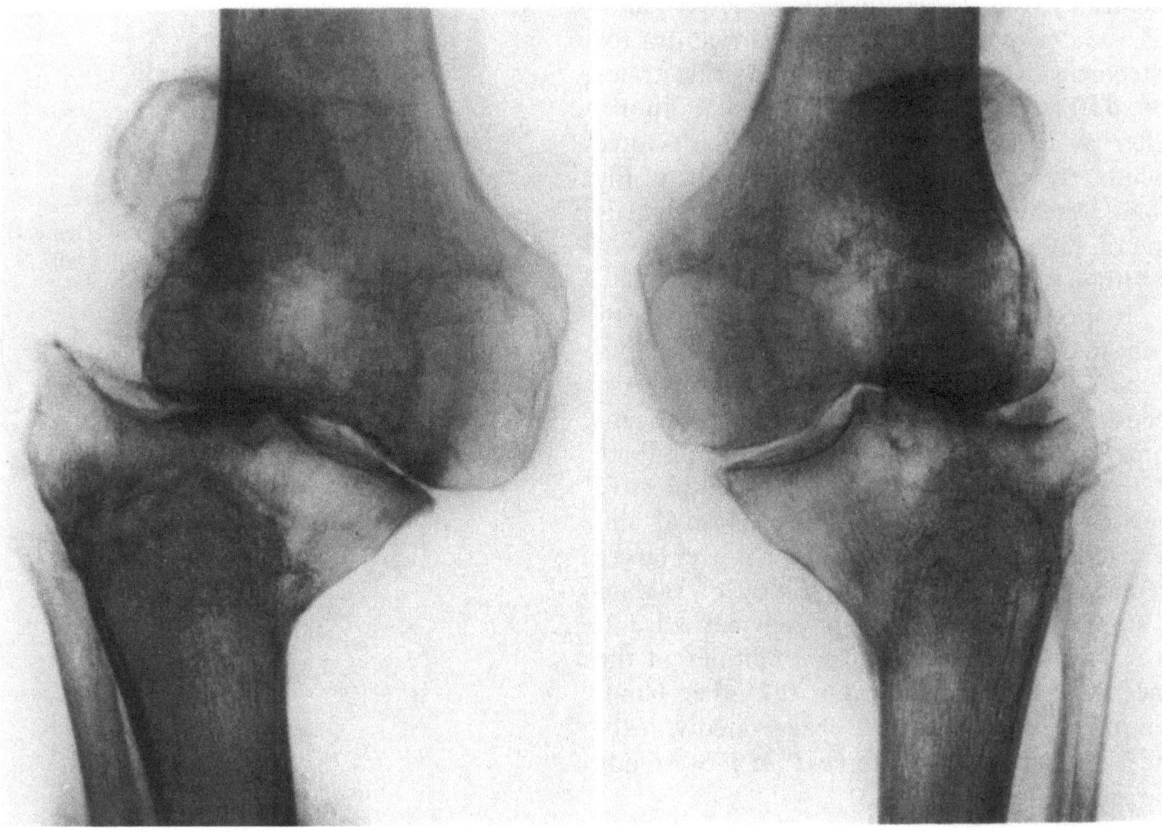

d

Distal femoral osteotomy is preferable for overcorrection of most cases of osteoarthritis of the knee with valgus deformity. A supracondylar osteotomy of the femur overcorrecting the valgus deformity rotates the tibial plateaux with the lower leg about the heel, counterclockwise in our diagram (Fig. 216). Redirecting the femur and the muscular force *L* turns the resultant force *R* in the opposite direction, clockwise. This reduces or even neutralizes the component tangential to the plane tangential to the tibial plateaux. Resultant *R* then acts on this plane as a pure compressive force. If the overcorrection has been sufficient, force *R*, although it may be increased by some lenghtening of the lever arm *a* of force *P*, evokes compressive stresses considerably reduced because of their even distribution over the largest possible load-transmitting surfaces of the joint. This explains why distal femoral osteotomy represents the method of choice in most instances.

Comparison of X-rays after an upper tibial osteotomy (Fig. 217a) and after an lower femoral osteotomy (Fig. 217b), both carried out for osteoarthritis with valgus deformity, illustrates the difference in the mechanical conditions. After a tibial osteotomy, the resultant force (white arrow) acts obliquely to the plane of the tibial plateaux. After a femoral osteotomy, its line of action is at right angles to the plane of the tibial plateaux.

It appears more difficult to give proper treatment to osteoarthritis associated with valgus deformity than to that with varus deformity. Correction of varum, even if not complete, automatically shortens the lever arm *a* of force *P*. It always decreases the articular compressive stresses, at least by diminishing the load and at best, in addition, by distributing it over larger weight-bearing surfaces. Correction of valgum is more delicate. If inappropriately carried out, it aggravates the mechanical conditions of the knee (Fig. 213 and 214). On the other hand, too much overcorrection exaggeratedly lengthens the lever arm of force *P* and overloads the joint.

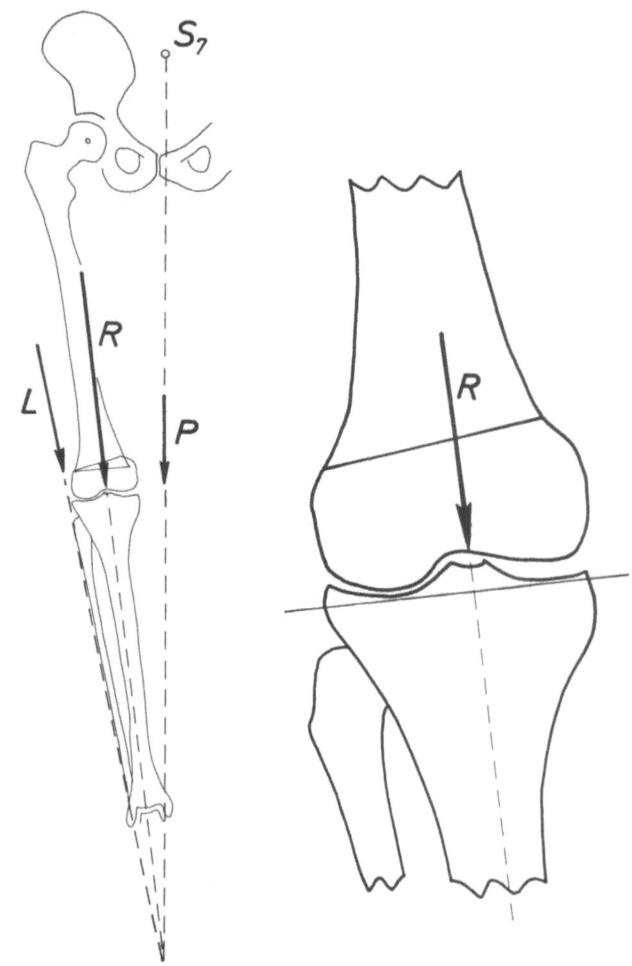

Fig. 216. Correcting the valgus deformity by a femoral osteotomy results in a force *R* acting at right angles to the plane tangential to the tibial plateaux

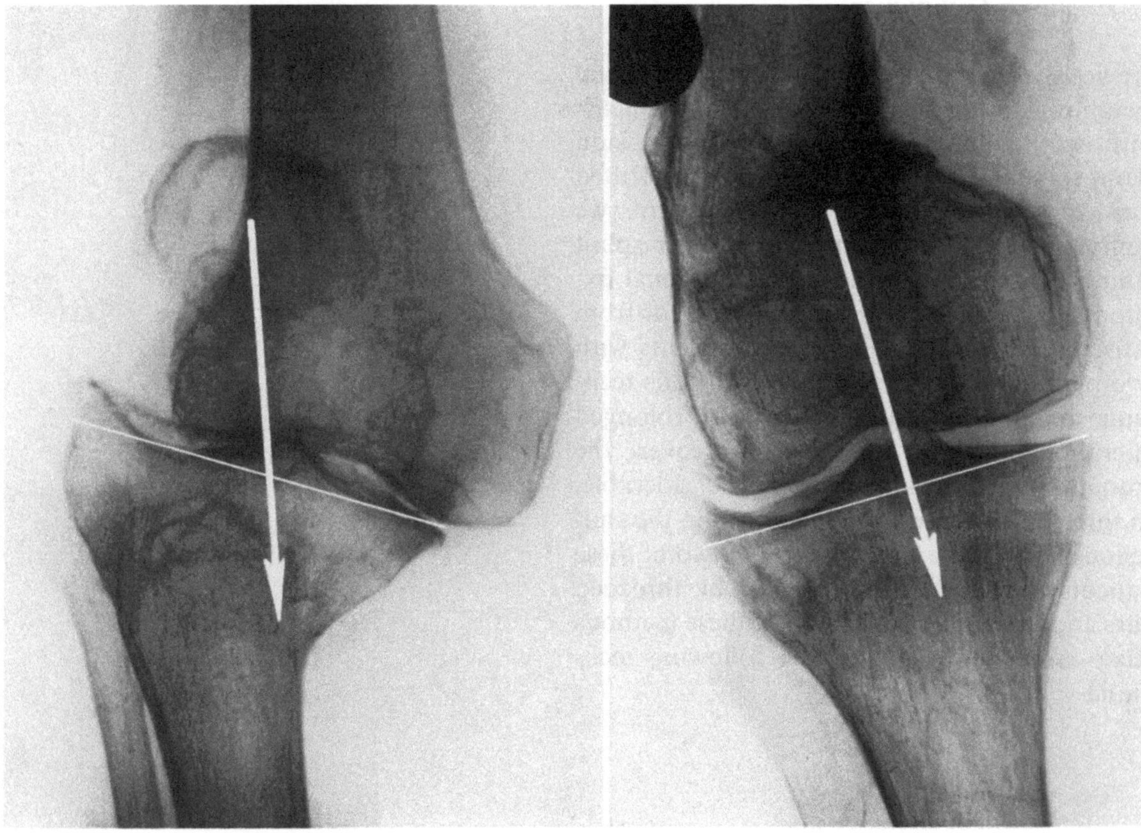

a b

Fig. 217a and b. Correction of a valgus deformity by tibial osteotomy (a) and by femoral osteotomy (b). The white arrow illustrates the resultant force *R* and the transverse white line the plane tangential to the tibial plateaux

c) Previous Techniques

Previously, after osteotomy through the distal metaphysis of the femur we fixed the fragments with two Steinmann pins and two compression clamps. In all the cases the lower fragment tilted forward after some days. In some patients we immobilized the lower limb in a plaster splint after the osteotomy, without any additional fixation. The accuracy of the method was far from satisfactory. We then fixed the fragments with two plates and screws (Fig. 233). But this technique required two incisions and a prolonged operation. Healing was slow. Moreover, the amount of material implanted was considerable. In three patients the screws loosened in the cancellous bone of the epiphysis and in two of these patients they had to be replaced by threaded bars and bolts. The drawbacks of these methods have caused us to adopt the following technique.

d) Distal Femoral Osteotomy with Fixation by Four Steinmann Pins and Two Compression Clamps

We use a technique which combines accuracy, simplicity and stability.

α) Planning

Here also the deformity is measured on a long X-ray showing the whole leg. A line is drawn from the centre of the femoral head to a point at the midpoint of the cross section just above the femoral condyles. Another line is drawn from this point to the centre of the ankle. The angle formed by these two lines measures the deformity at the level of the planned osteotomy (Fig. 218). The outlines of the knee and of the adjacent bones are traced on a sheet of transparent paper from an A.-P. view of the joint taken with the subject standing (Fig. 219a). Two parallel transverse lines are drawn through the condyles and two other transverse lines are drawn through the diaphysis of the femur, all representing Steinmann pins. The latter make with the former an angle $\alpha+$ from 1° to 3° open medially. This means that we aim at an overcorrection of from 1° to 3°. The osteotomy line is drawn just above the condyles. The joint and the distal fragment of the femur with its two transverse lines are then traced on a second sheet of transparent paper (Fig. 219b). This sheet is rotated on the first about the lateral end of the osteotomy line, the proximal fragment of the femur penetrating into the distal, until all the transverse lines are parallel (Fig. 219c). Then the proximal fragment and its Steinmann pins are traced on the second sheet (Fig. 219d). The operation has thus been carried out on paper (Fig. 219e).

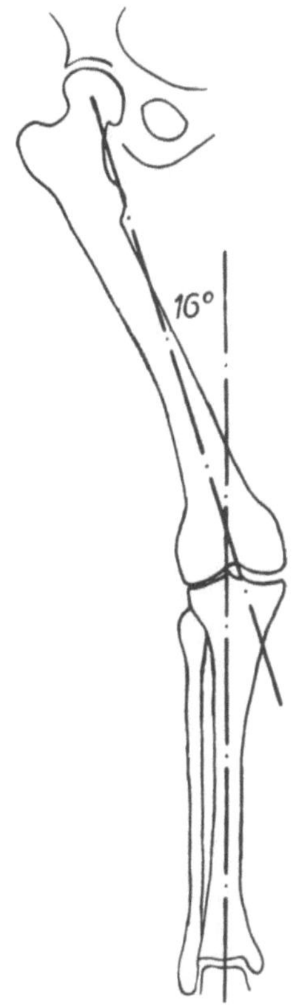

Fig. 218. The angle formed by the axis of the femur and that of the tibia measures the valgus deformity

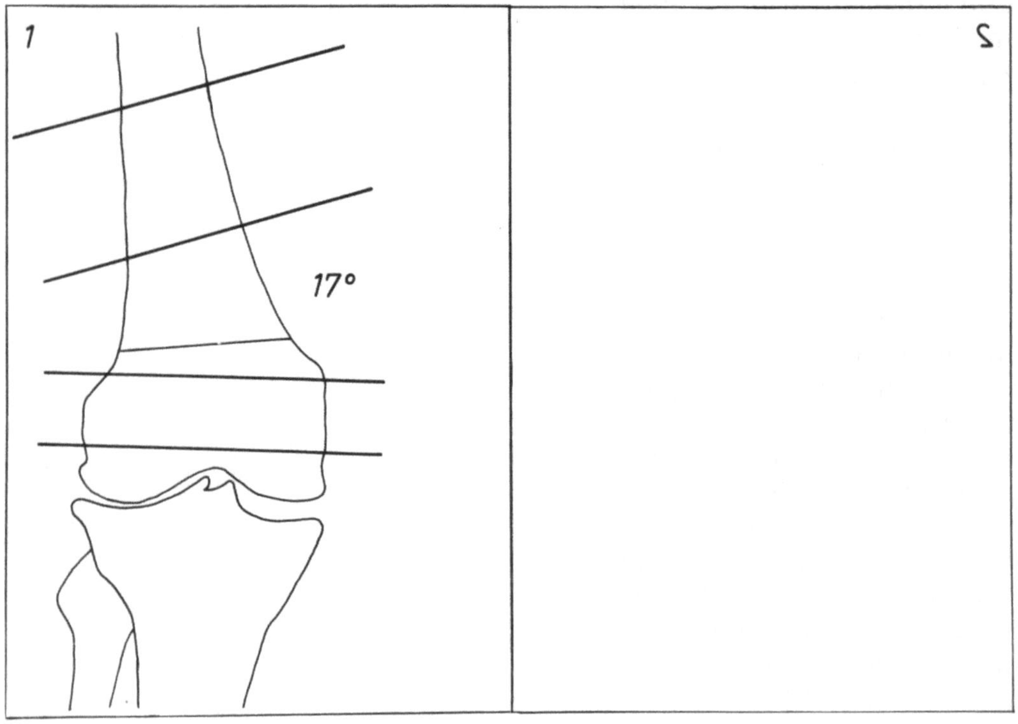

a

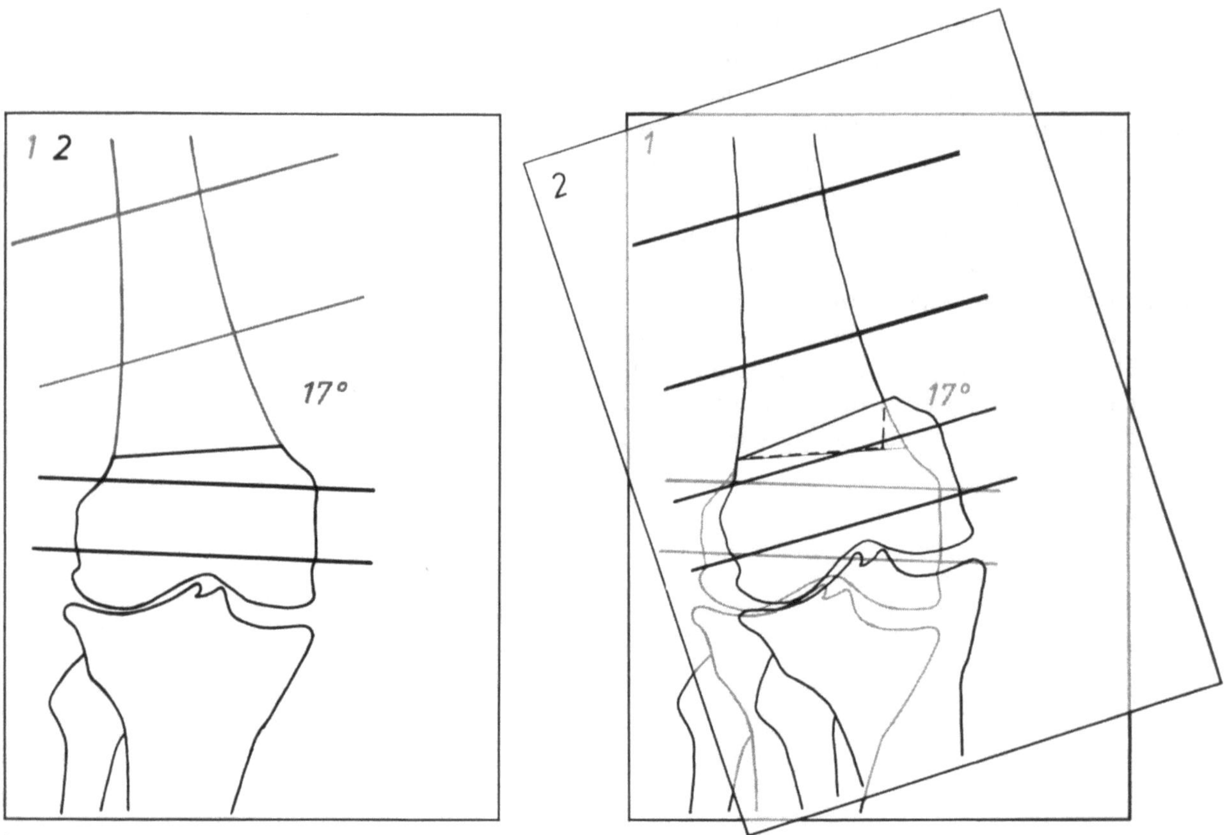

b

d

Fig. 219a–e. Planning a supracondylar osteotomy of the femur

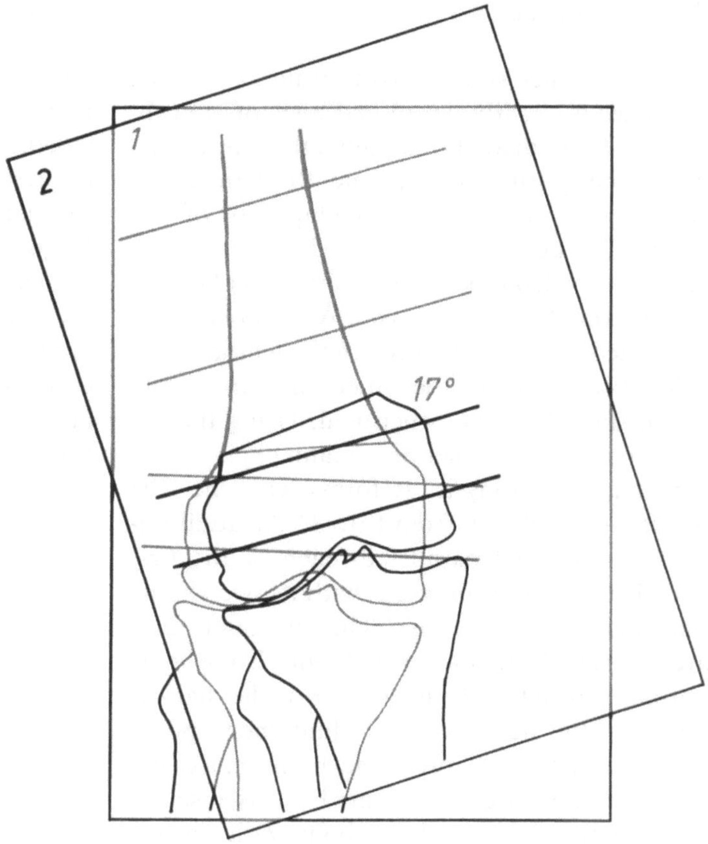

c

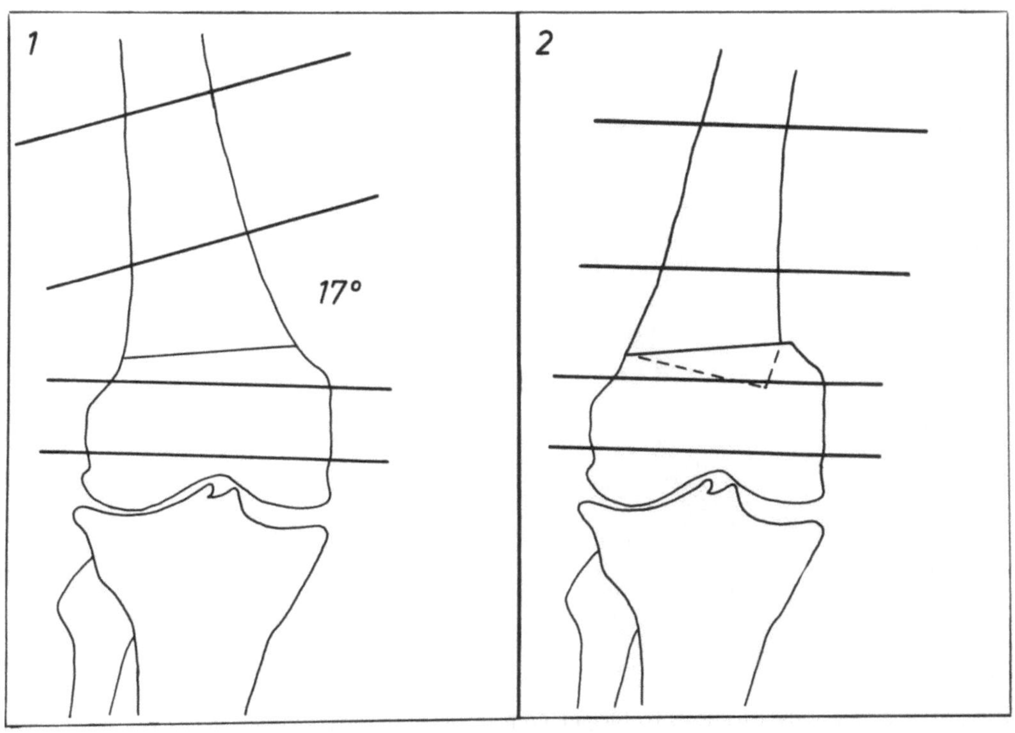

e

β) Surgical Procedure

The patient lies supine. No tourniquet is used. The knee is slightly flexed and rests on a cushion. A 6 cm medial longitudinal incision between the vastus medialis and the hamstrings gives access to the distal metaphysis of the femur. The periosteum is incised longitudinally and elevated from the bone in front and behind, just proximal to the condyles. A Steinmann pin is inserted transversely through the femoral condyles proximal to the intercondylar notch, which should not be penetrated. Using the pin insertion guide, another Steinmann pin is passed through the diaphysis. It forms with the first the desired angle $\alpha+$ from 1° to 3°. An additional pin is inserted parallel to each of the first two. The pin insertion guide with its two sliding units makes this step easy. The Steinmann pins are inserted from the medial side. They must not be involved in the incision. For the diaphysial pins, the sleeve with its blunt-ended pin is first driven to the bone. The blunt-ended pin is then removed and a pointed pin is inserted through the sleeve and the bone with an air-powered drill. The angle formed by the two pairs of Steinmann pins is checked by X-ray. The bone is then divided transversely proximal to the femoral condyles with an oscillating saw and a thin chisel. While the bone is divided, a large periosteal elevator is maintained behind the femur, protecting the popliteal vessels and nerves. The lateral cortex is incompletely divided. The lower end of the proximal fragment is bevelled off in its anterior, medial and posterior aspects and is then impacted into the distal fragment, using the incompletely divided lateral cortex as a hinge, until all the Steinmann pins are parallel in the same coronal plane. The pins are then fixed by two compression clamps each equipped with four mobile units (Fig. 220). The aponeurosis, subcutaneous fat and skin are sutured using suction drainage.

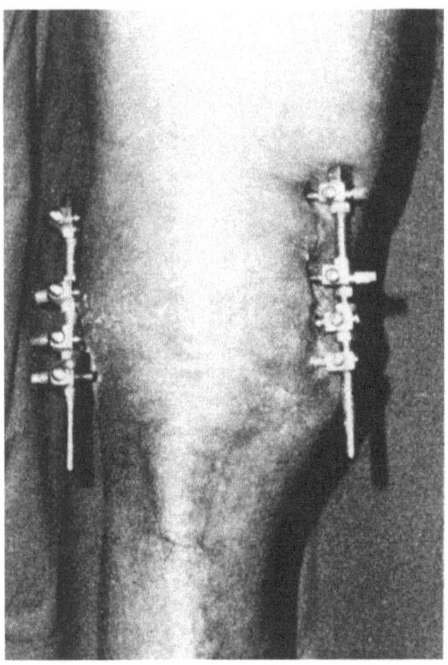

Fig. 220. Fixation of the fragments by Steinmann pins and compression clamps

γ) Postoperative Care

The surgeon completely flexes and extends the knee on the operation table. The patient is encouraged to move immediately. On the second day the suction drainage is removed and the patient is helped to stand up and walk with crutches. The pins are removed two months later, after clinical examination and X-ray have shown consolidation of the osteotomy. The patient discards the crutches as soon as he feels able to do so.

δ) Comments and Examples

When the overcorrection is sufficient to utilize the greatest weight-transmitting surface of the joint, the clinical result is generally excellent. The X-ray shows regeneration of the knee.

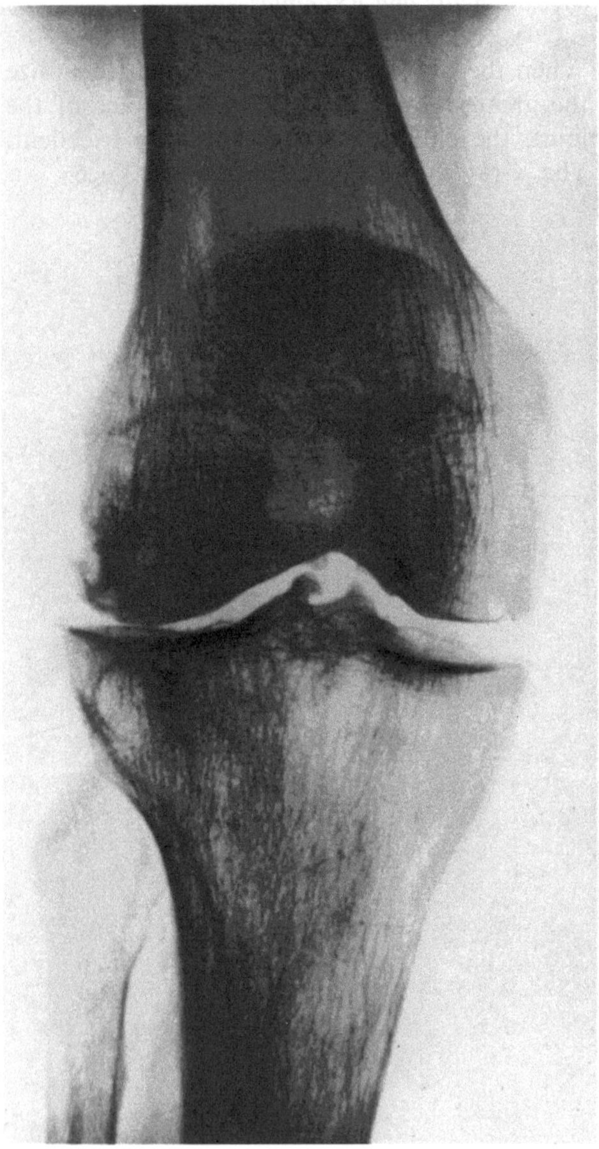

Fig. 221 a

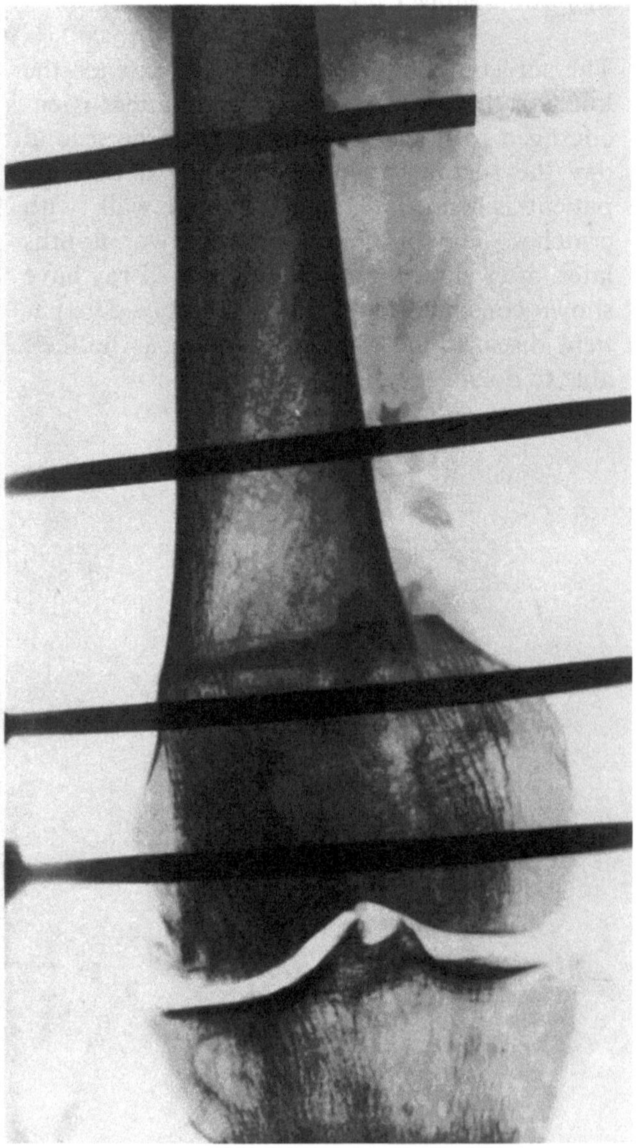

Fig. 221 b

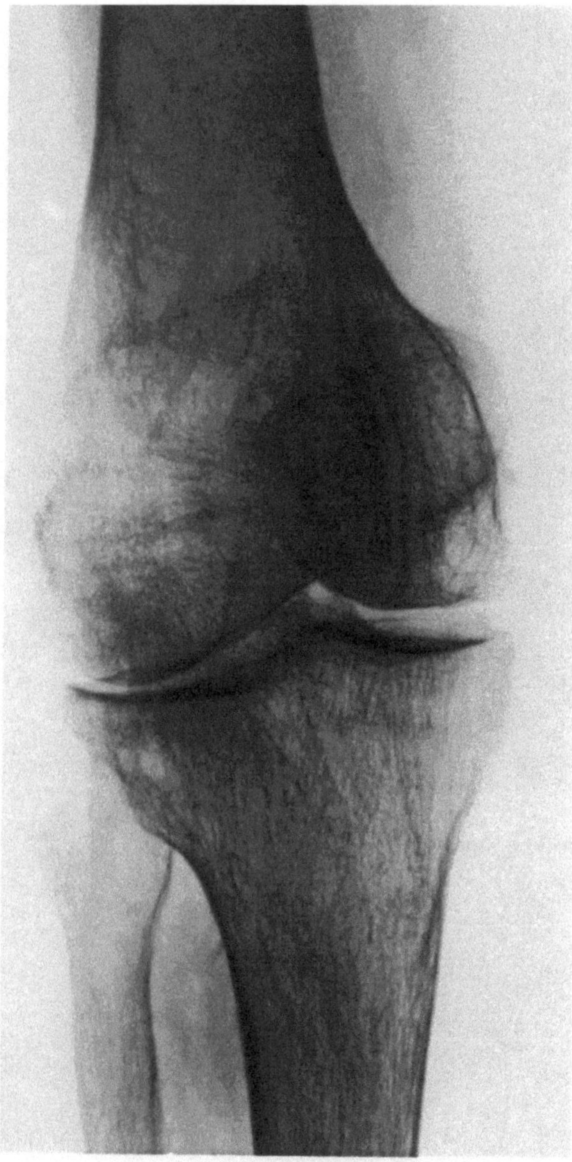

c

A 74-year-old female patient (Fig. 221 a) complained of a painful knee. The leg presented with a 16° valgus deformity (Fig. 221 d). In the X-ray, the subchondral sclerosis was increased beneath the lateral tibial plateau and the cancellous bone had faded away beneath the medial plateau. A varus osteotomy of 17° of the distal femur was carried out (Fig. 219). The postoperative X-ray shows the penetration of the proximal fragment into the distal and the Steinmann pins appear parallel (Fig. 221 b). Pain was relieved immediately and six years later the knee remains painfree. Its range of movement is satisfactory despite some loss of flexion. In the X-ray, the two tibial plateaux are underlined by symmetrical scleroses. The cancellous structure also appears symmetrical in the medial and lateral aspects of the tibial epiphysis (Fig. 221 c). This means an even distribution of the compressive stresses in the joint. The deformity has been a little overcorrected (Fig. 221 d).

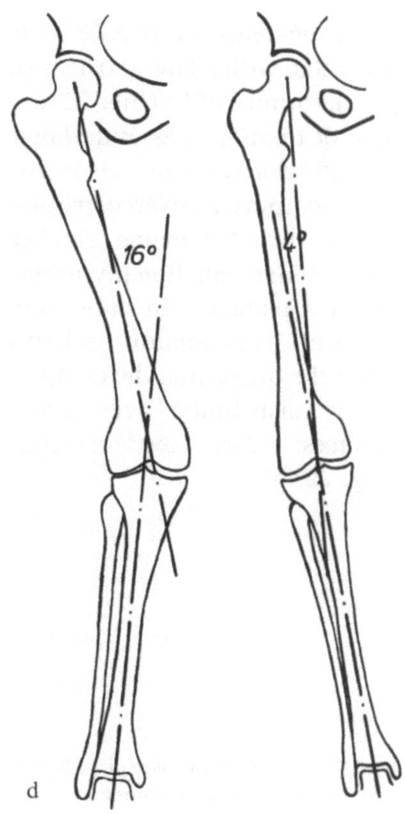

d

Fig. 221 a–d. Osteotomy and fixation of the fragments by four Steinmann pins and two compression clamps. A 74-year-old patient before (a), after surgery (b) and six years later (c). Pre-operative planning: see figure 219. The leg before surgery and at the latest follow-up (d)

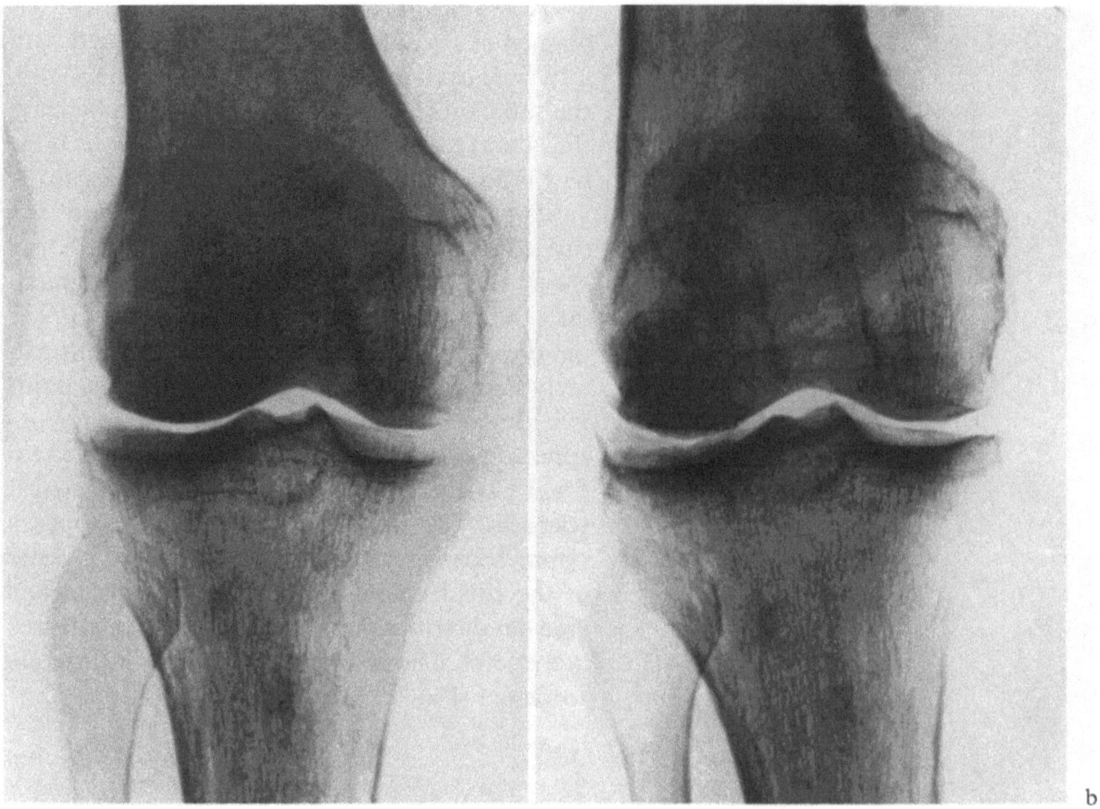

a b

Fig. 222a and b. A 66-year-old patient before (a) and 4 years after (b) a supracondylar femoral osteotomy

A 66-year-old female patient (Fig. 222a) presented with a lateral osteoarthritis with a valgus deformity of 8° (Fig. 223a). The lateral plateau of the tibia was underlined by a thick cup-shaped sclerosis. Four years after a distal femoral osteotomy to overcorrect the deformity, the clinical result remains excellent; the knee is painfree and functionally normal. In the X-ray, the two plateaux are now underlined by symmetrical subchondral scleroses (Fig. 222b). Since the operation, the compressive stresses are evenly distributed over large weight-bearing surfaces. The overcorrection reaches 4° (Fig. 223b).

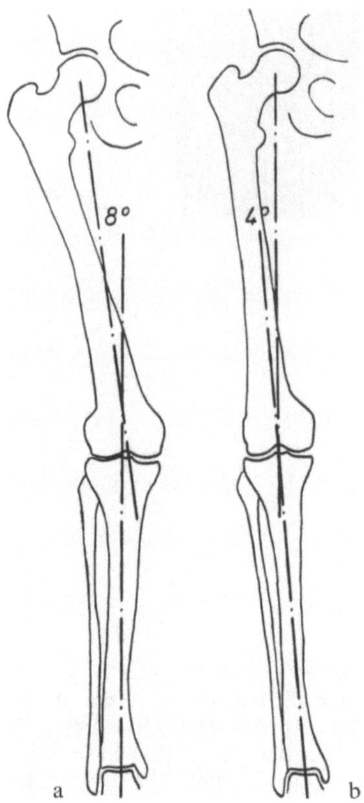

a b

Fig. 223a and b. Same patient as Figure 222, before (a) and after (b) the femoral osteotomy

a 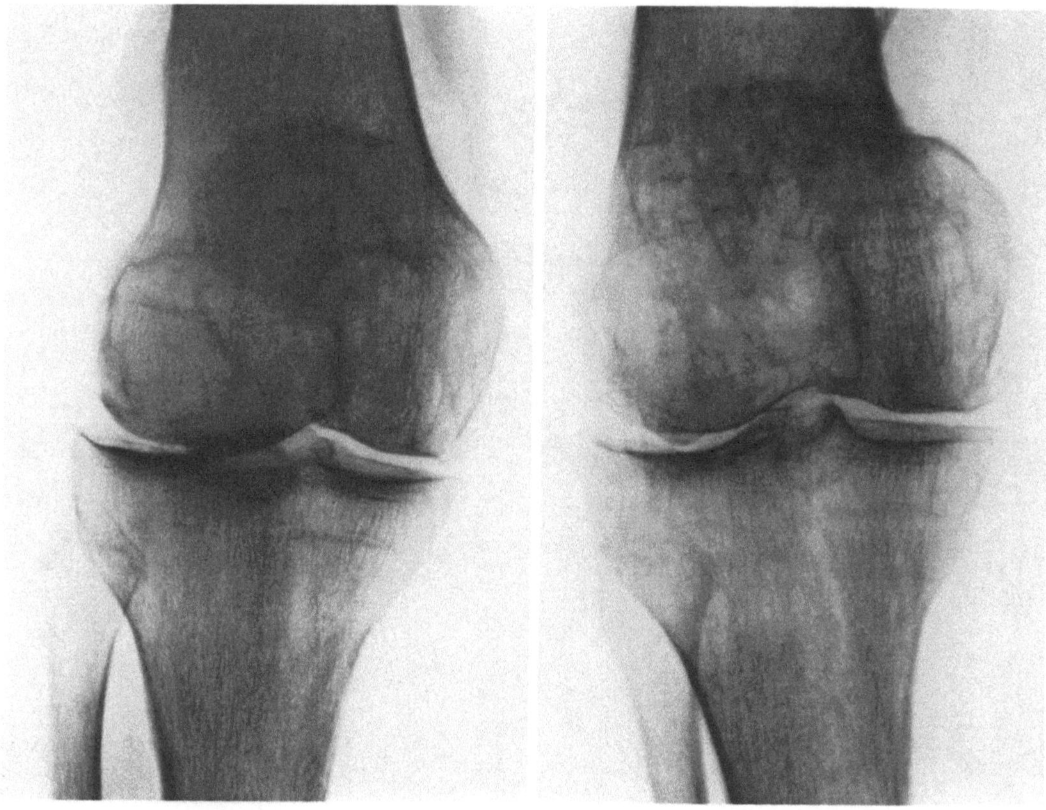 b

Fig. 224a and b. A 57-year-old patient before (a) and 1 year after (b) a supracondylar femoral osteotomy

The change in the subchondral sclerosis appears clearly in the case of a 57-year-old female patient (Fig. 224a). Before the operation, the sclerosis was accentuated close to the tibial spines. One year later the two plateaux are underlined by symmetrical subchondral scleroses (Fig. 224b).

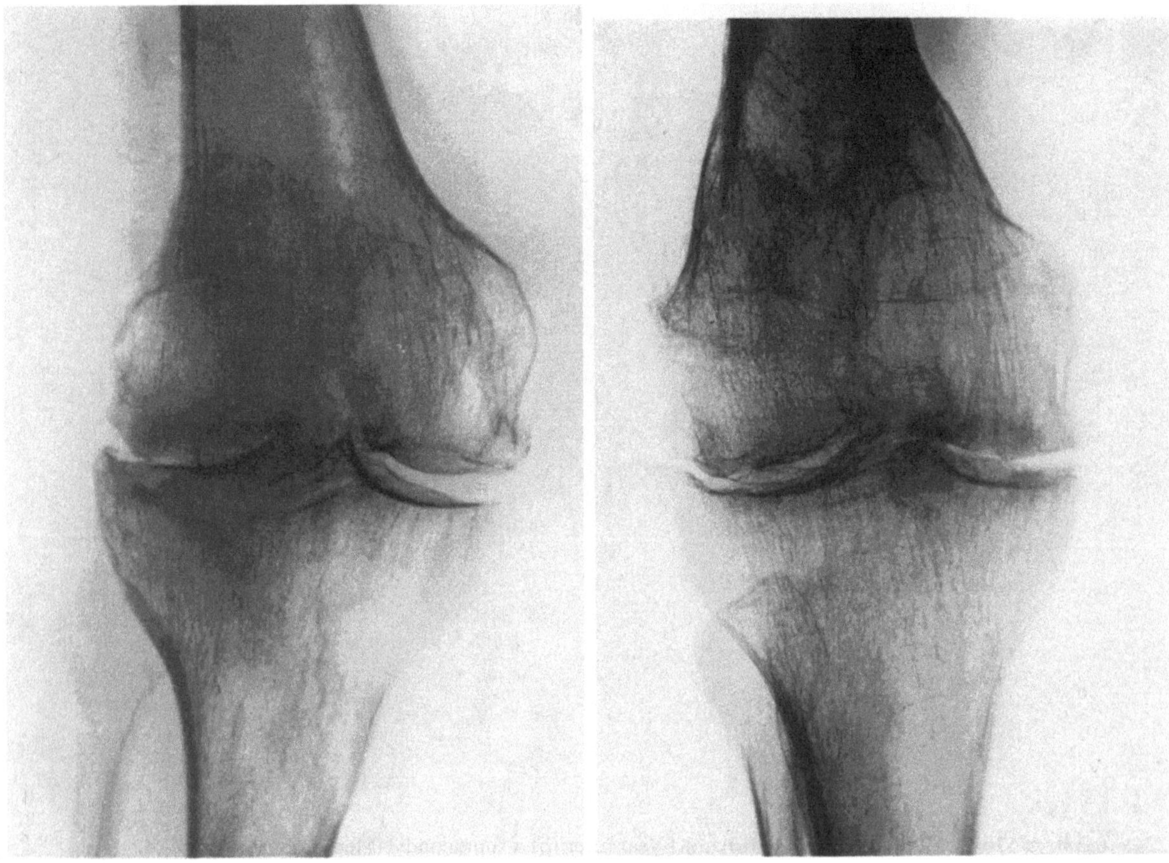

a b

Fig. 225a and b. A 66-year-old patient before (a) and twelve years after (b) a distal femoral osteotomy overcorrecting the valgus deformity

The improvement seems to be lasting, as seen in the knee of a patient (Fig. 225) operated on at the age of 66. Here again the subchondral sclerosis and the cancellous structure are more pronounced beneath the lateral plateau before surgery, whereas the cancellous structure is hardly visible beneath the medial plateau (Fig. 225a). At the twelve-year follow-up, the subchondral sclerosis and cancellous structure appear symmetrical beneath both plateaux (Fig. 225b).

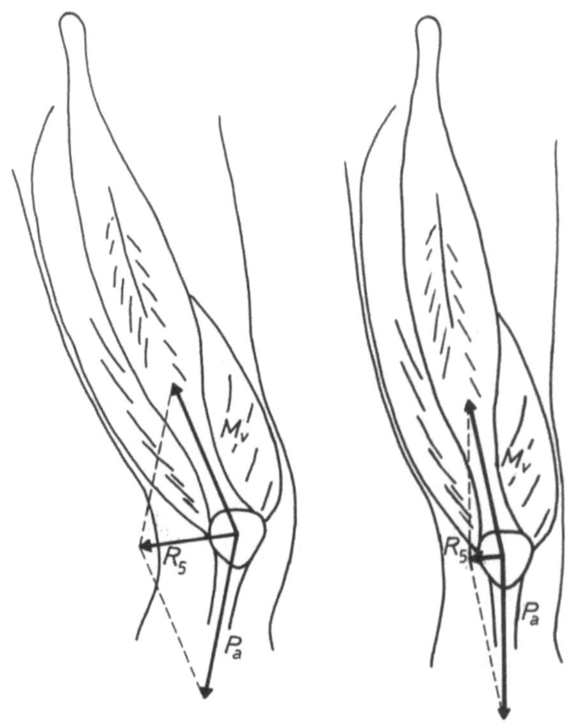

Fig. 226. Opening of the angle formed by the femur and the tibia and thus by the quadriceps tendon M_v and the patella tendon P_a decreases the projection of the resultant R_5 on the coronal plane

e) Distal Femoral Osteotomy Combined with Anterior Displacement of the Tibial Tuberosity

A varus osteotomy of the distal end of the femur opens the angle formed by forces M_v exerted by the quadriceps and P_a exerted by the patella tendon in the coronal projection (Fig. 226). This reduces the magnitude of the resultant force R_5 in this projection and should tend to reduce a lateral subluxation of the patella and to recentre the patella into the intercondylar groove.

However, when lateral osteoarthritis is accompanied by patello-femoral osteoarthritis, a supracondylar osteotomy can be combined with an anterior displacement of the tibial tuberosity. The two operations can be carried out separately or in the same session.

A 79-year-old female patient (Fig. 227) had complained essentially of patello-femoral pain six years previously and had undergone an anterior displacement of the tibial tuberosity. The operation was followed by complete relief of pain but, after a hip arthroplasty, she progressively developed lateral osteoarthritis with increasing valgus deformity (Fig. 227a). There was augmentation of the lateral subchondral sclerosis and accentuation of the cancellous structure beneath the lateral plateau, whereas beneath the medial plateau the subchondral sclerosis decreased and the cancellous structure faded away. A distal femoral osteotomy was carried out and resulted in a dramatic clinical improvement: relief of pain, restored stability and full range of movement of the knee. Subchondral sclerosis and cancellous structure now appear symmetrical beneath both plateaux. A wide joint space has developed. Follow-up is at one year (Fig. 227b).

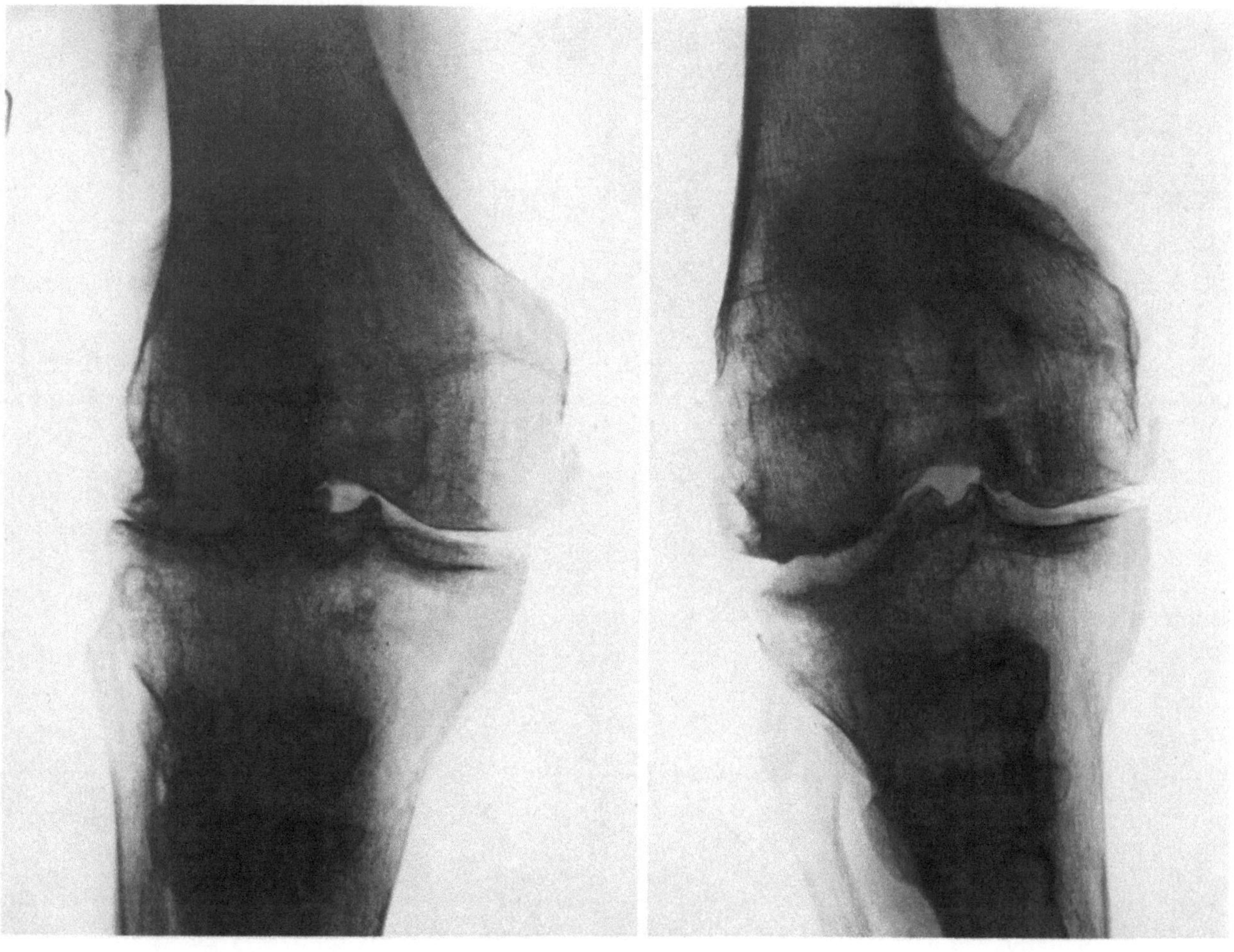

a b

Fig. 227a and b. A 79-year-old patient before (a) and 1 year after (b) a supracondylar femoral osteotomy carried out 6 years after an anterior displacement of the tibial tuberosity

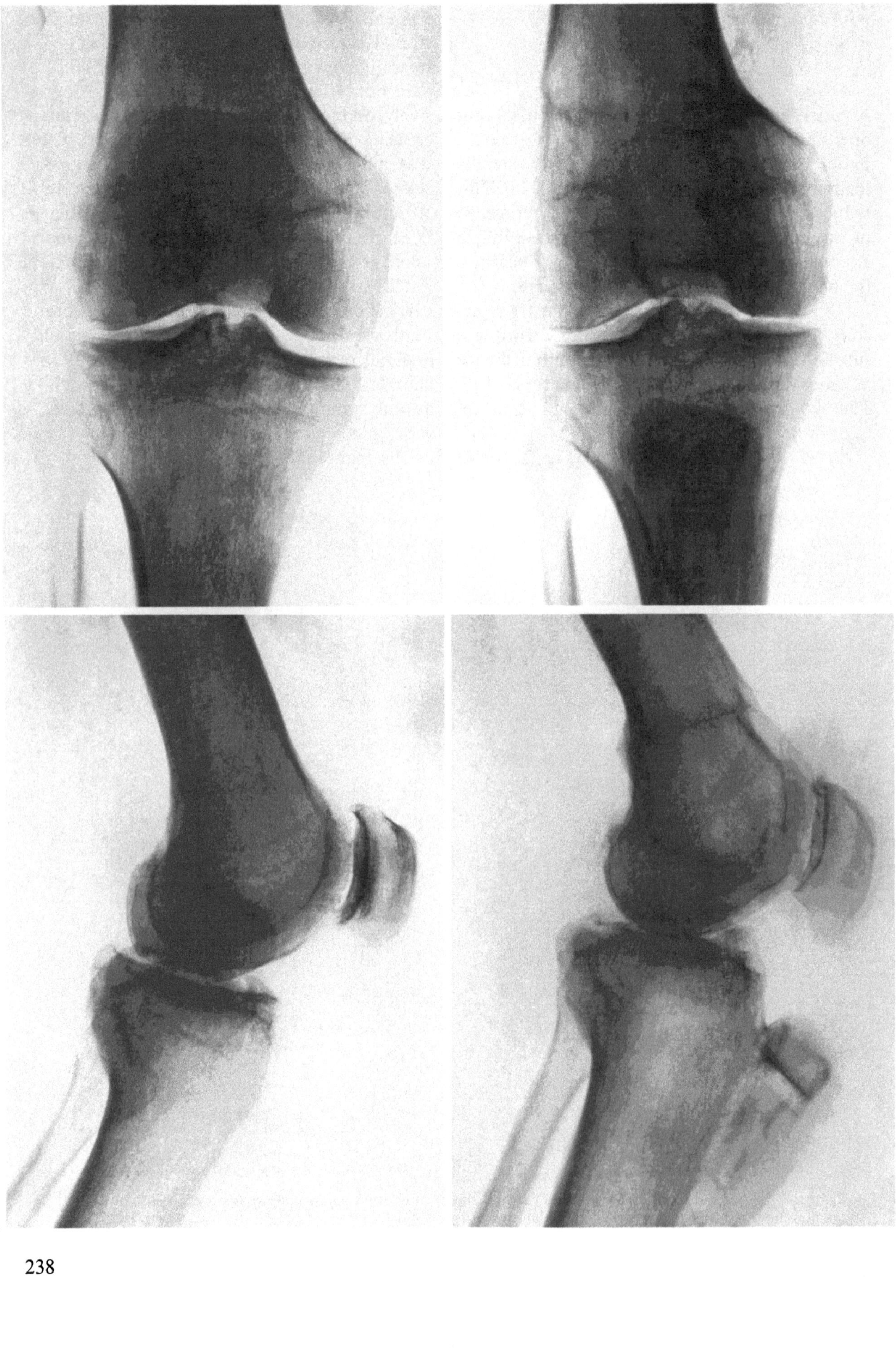

a

t

In a 38-year-old male patient (Fig. 228a), distal femoral osteotomy and anterior displacement of the tibial tuberosity were carried out at the same time. One year after surgery the X-ray changes are striking in the A.-P. view as well as in the lateral. In the initial A.-P. view, the subchondral sclerosis was prominent beneath the lateral plateau, but now both plateaux are symmetrically underlined. In the lateral view, a dense triangle was observed beneath the posterior aspect of the tibial plateaux and the subchondral sclerosis was increased in the patella, but now the dense triangle has disappeared and the patellar subchondral sclerosis looks much thinner and of even width throughout (Fig. 228b).

◄ Fig. 228a and b. 38-year-old male patient before (a) and one year after (b) a supracondylar femoral osteotomy combined with an anterior displacement of the tibial tuberosity

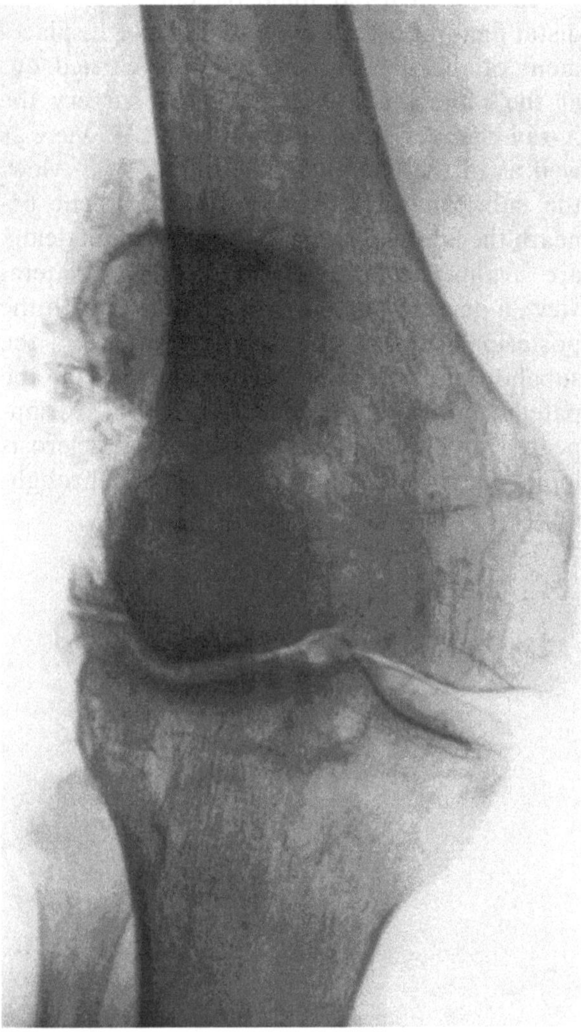

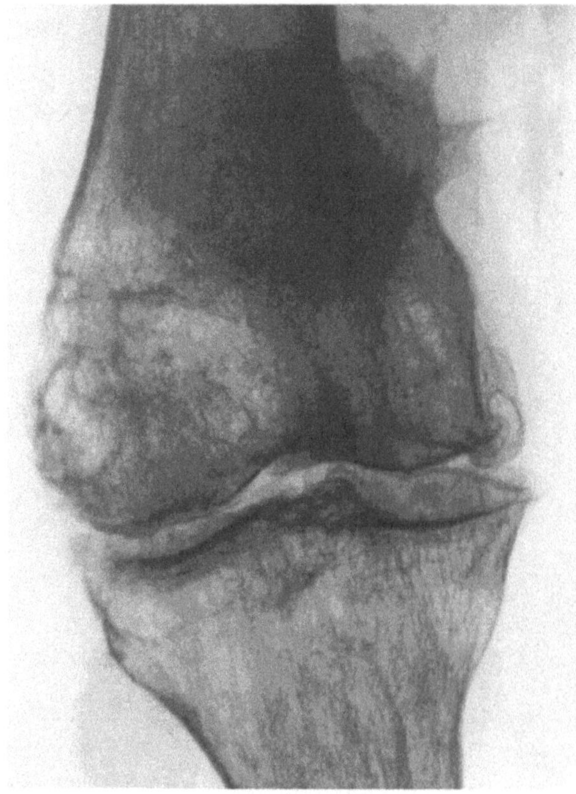

Fig. 229 b

Fig. 229 a

A 66-year-old female patient presented with an osteoarthritic, painful and unstable knee (Fig. 229a) with a valgus deformity of 22° (Fig. 230a). After a distal femoral osteotomy, the fragments were immobilized in a plaster splint, which cannot ensure accuracy of correction. The knee became painfree and stable with a sufficient range of movement: 2° hyperextension to 115° flexion. The X-ray showed a symmetrical distribution of the compressive stresses in the joint (Fig. 229b). After six years the patient asked for a correction of her bow leg; by then the varum had attained 22° (Fig. 230b). A fresh distal femoral osteotomy brought the knee to 1° valgum (Fig. 230c), and the patient was fully satisfied. However, this revision may have been too much, as three years later the subchondral sclerosis has again increased beneath the lateral plateau (Fig. 229c). At the six-year follow-up the result remains excellent.

f) Revisions

Here also some operations were revised because the valgus deformity had been either exaggeratedly overcorrected or miscorrected.

α) Revision After Exaggerated Overcorrection

Exaggerated overcorrection of the valgus deformity seems to be followed by quick clinical and radiological improvement. However, it gives an unaesthetic gait and certainly increases the overall load on the knee.

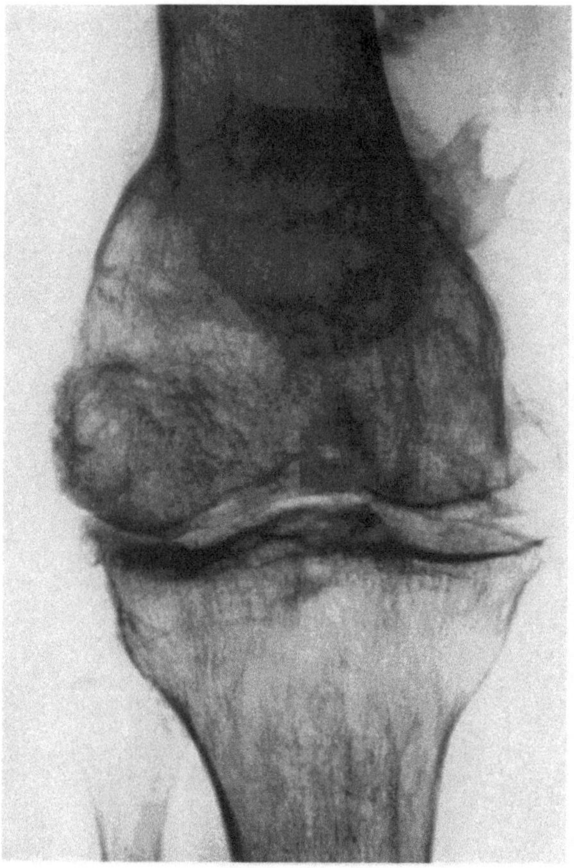

c

Fig. 229a–c. A 66-year-old patient before (a) and 6 years after (b) a supracondylar femoral osteotomy exaggeratedly overcorrecting a valgus deformity. Three years after revision (c)

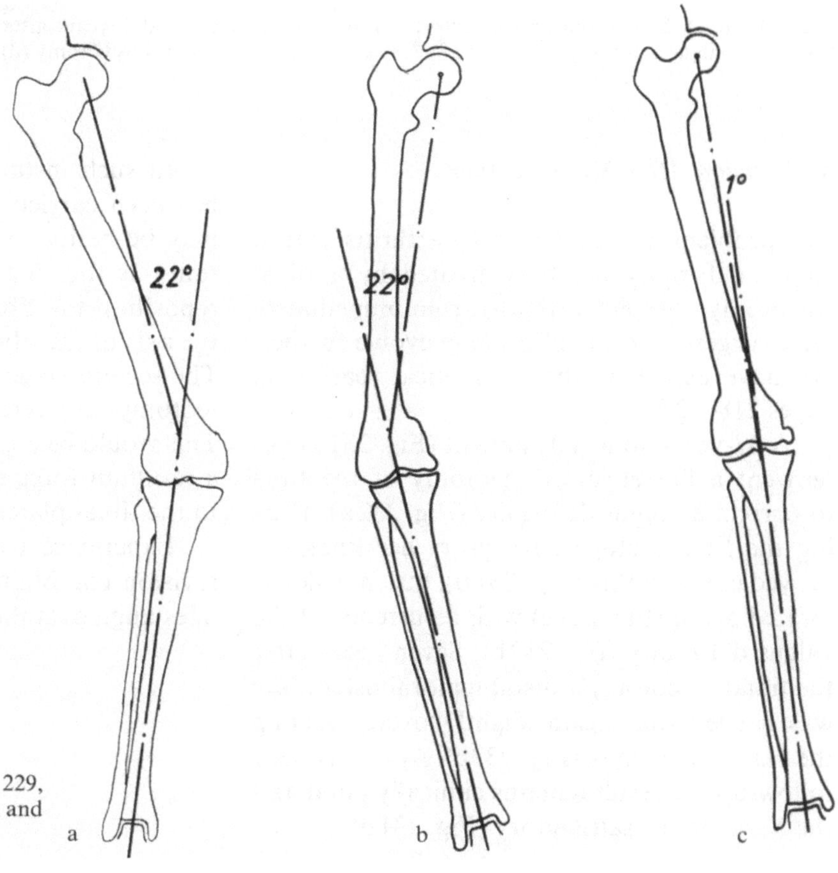

Fig. 230a–c. Same patient as Figure 229, before (a), after (b) the first operation and after the revision (c)

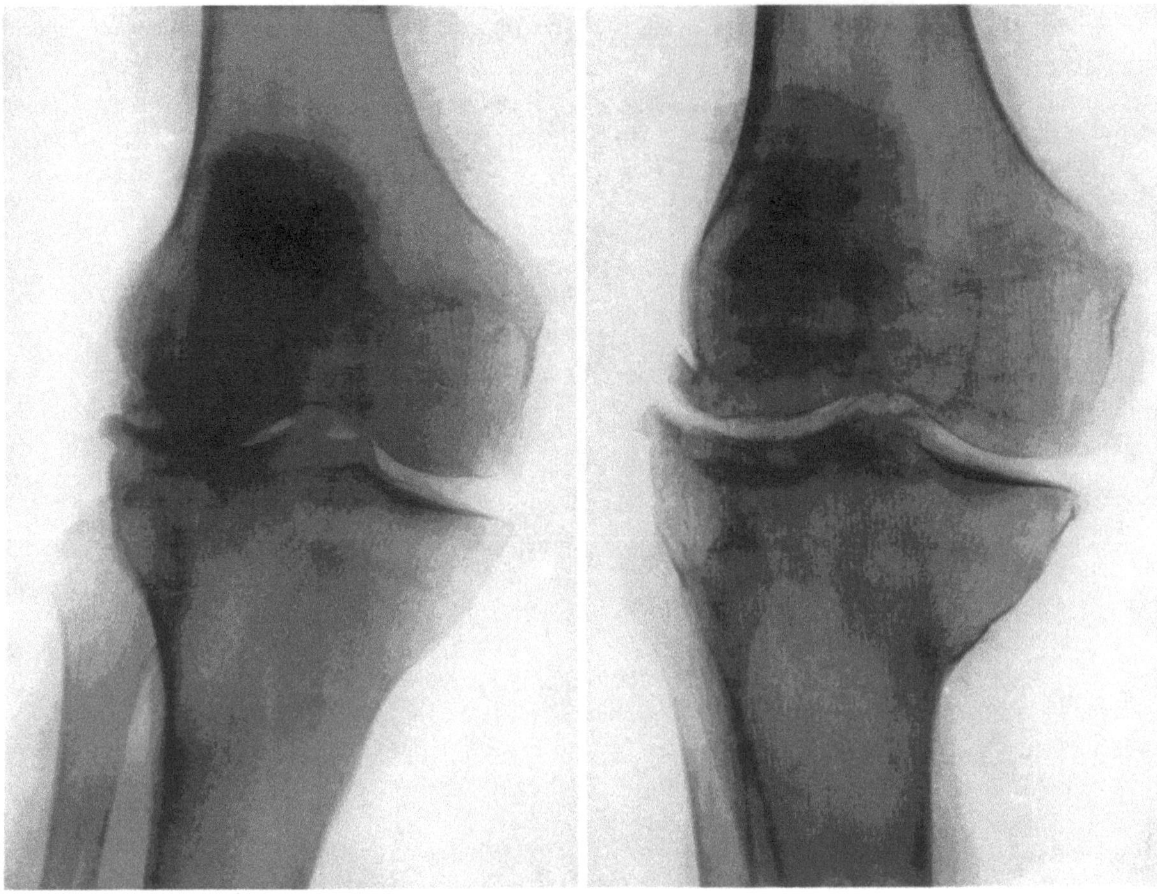

a b

Fig. 231 a–d. A 57-year-old patient before (a), 4 years after (b) and 7 years after (c) a barrel-vault osteotomy of the tibia for a valgus deformity. Six years after revision through a femoral osteotomy (d)

β) Revision After Miscorrection

The problem arises when osteoarthritis with a valgus deformity has been treated by a tibial osteotomy. Despite a slight varum immediately after surgery, the condition may evolve further. We have explained the mechanical reason on pages 218–225.

A 57-year-old female patient (Fig. 231 a) underwent a barrel-vault osteotomy of the tibia to correct a valgus deformity (Fig. 232a). During the first postoperative years her knee improved considerably (Fig. 231 b), but then deteriorated again (Fig. 231 c) with recurrence of the valgus deformity (Fig. 232b). Seven years after the tibial osteotomy a distal femoral osteotomy was carried out, again slightly overcorrecting the valgus deformity (Fig. 232c). At the six-year follow-up the result remains clinically good and the X-ray looks satisfactory (Fig. 231 d).

In such instances, when a tibial osteotomy has been carried out for a valgus deformity, it may be better to perform another tibial osteotomy as the first stage of revision in order to reposition the tibial plateaux at right angles to the axis of the tibia, thus increasing the valgum. The second stage would be a distal femoral osteotomy to overcorrect the deformity slightly. This would be a more reliable way of redirecting a resultant force normal to the plane tangential to the tibial plateaux (see page 224).

Experience teaches that performing such a revision combining femoral and tibial osteotomies augments the risks of inaccuracy.

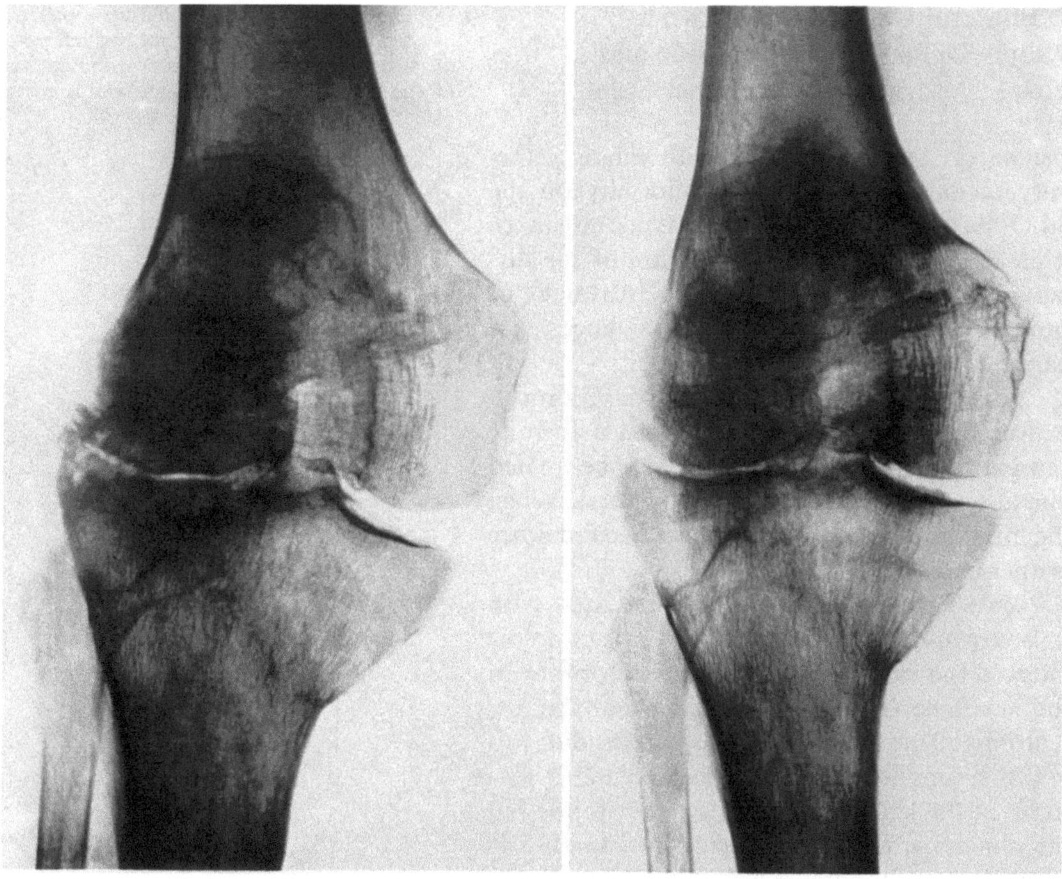

c d

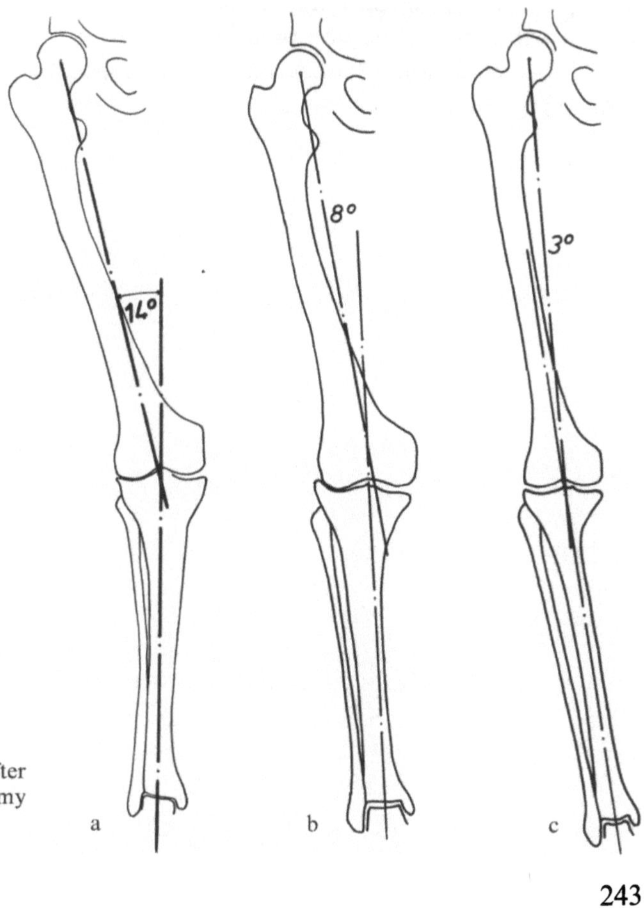

a b c

Fig. 232a–c. Same patient as Figure 231, before (a), after (b) the tibial osteotomy and after the femoral osteotomy (c)

4. Bilateral Osteoarthritis with a Valgus Deformity on One Side and a Varus Deformity on the Other Side

Osteoarthritis of both knees with valgus deformity on one side and varus deformity on the other may result from a deformity at a distance from the affected knees. Correction of the distant deformity may then be the treatment of choice, leading to healing of both knees (see page 250).

When no distant cause of the condition can be found or when an existing distant deformity cannot be corrected, the knees should be treated one at a time, the more painful one first. When the first osteotomy is stable, the second knee is operated on.

A 75-year-old male patient presented with severe bilateral osteoarthritis with valgus deformity of the right knee and varus deformity of the left knee (Fig. 223a). A supracondylar osteotomy of the right femur was carried out, followed six months later by a barrel-vault osteotomy of the left tibia. At the six-year follow-up the clinical result remains excellent. In the X-rays, the subchondral sclerosis underlining the tibial plateaux of both knees demonstrates an even distribution of the compressive stresses in the joints. Wide joint spaces have reappeared (Fig. 233b).

Fig. 233a and b. A 75-year-old male patient with bilateral osteoarthritis, right valgus knee and left varus knee, before (a) and 6 years after (b) a supracondylar varus osteotomy of the femur on the right side and a barrel-vault valgus osteotomy of the tibia on the left side

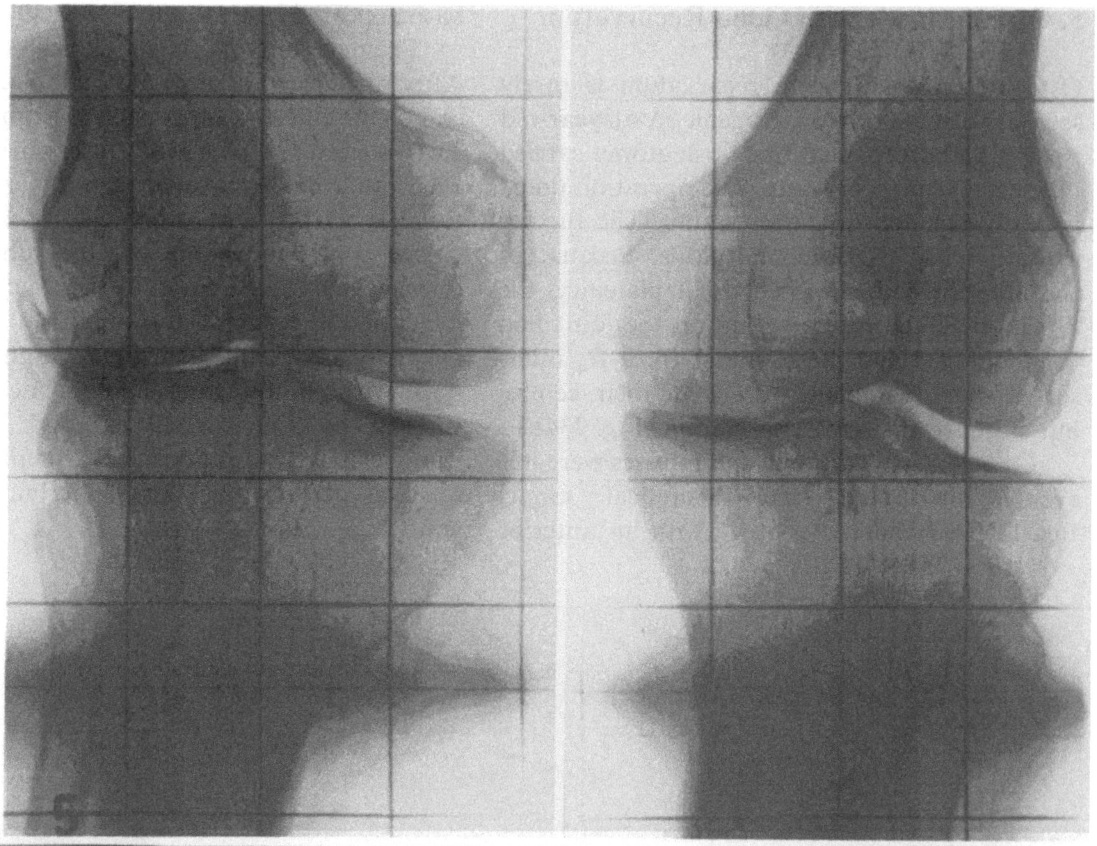

a

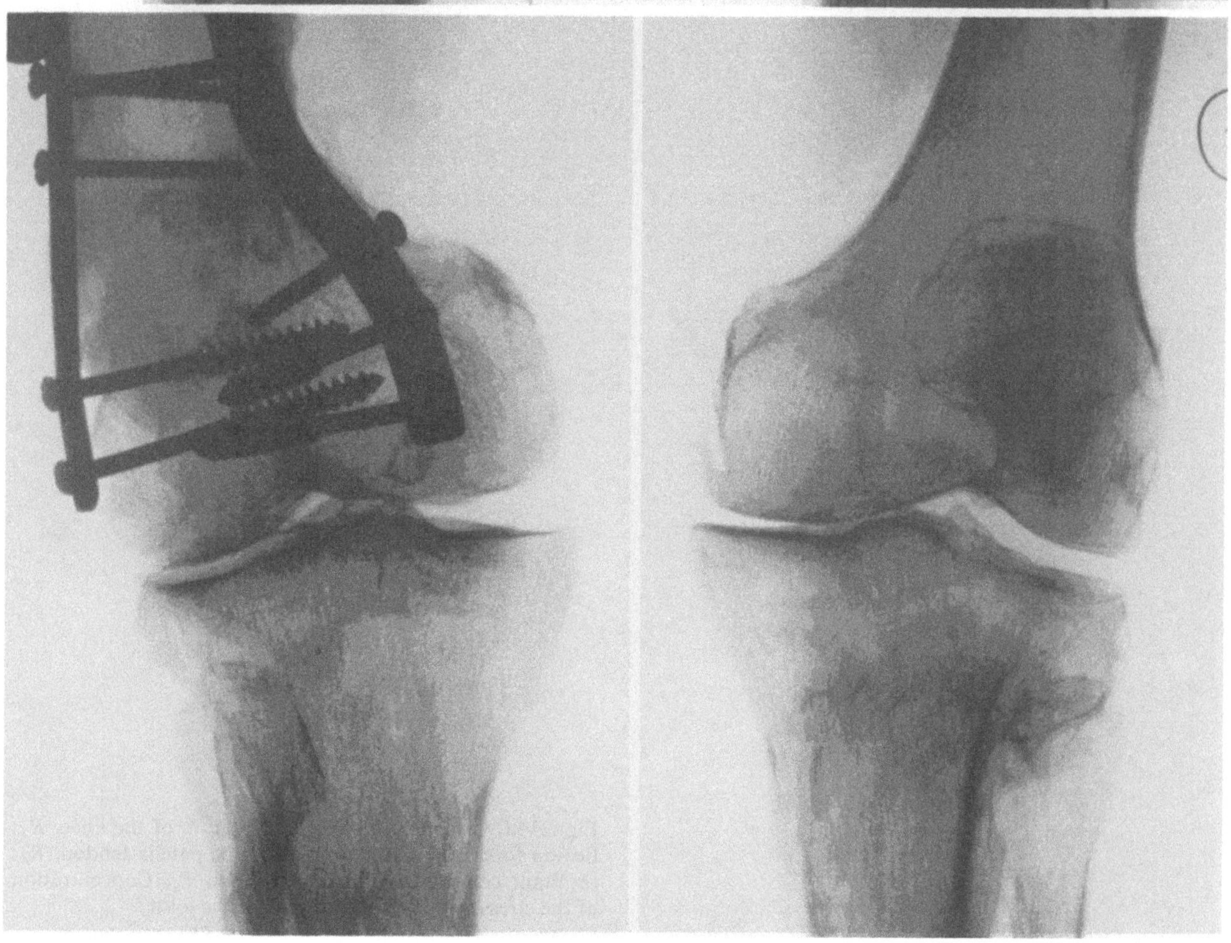

b

5. Osteoarthritis with Genu Recurvatum

Osteoarthritis with genu recurvatum is rarely seen. Its origin is often iatrogenic. A 61-year-old patient had a proximal tibial osteotomy carried out seven years previously. The operation unexpectedly resulted in a recurvatum. The further evolution had experimental value. Because of the anterior tilting of the tibial plateaux, the direction of the forces acting on the joint had been changed and the resultant force R_4 intersected the tibial plateaux not in their centre, but closer to their anterior margin (Fig. 234a).

Consequently, the articular stresses were unevenly distributed (see theoretical model Fig. 13) and a dense triangle with an anterior base appeared beneath the plateaux (Fig. 234b).

The deformity has been surgically corrected. After resection of a piece of the fibula, the upper end of the tibia was approached from the front. Two Steinmann pins were inserted parallel to each other in the coronal plane, one above the tibial tuberosity, the other below. A posterior wedge was removed below the tuberosity. It formed a 20° angle equal to the exact measurement of the recurvatum (Fig. 234d). After removal of the wedge, the tibial fragments were fixed with compression clamps. One year after surgery the clinical result is good. The dense triangle has disappeared and is replaced by a sclerosis of even thickness underlying the plateaux (Fig. 234c).

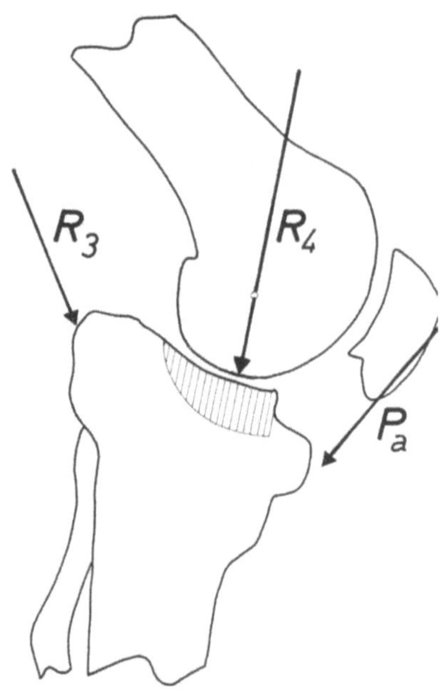

Fig. 234a. Osteoarthritis with recurvatum of the knee. R_3: flexion forces. P_a: force exerted by the patella tendon. R_4: resultant compressive force of R_3 and P_a. Concentration of the stresses in the anterior part of the joint

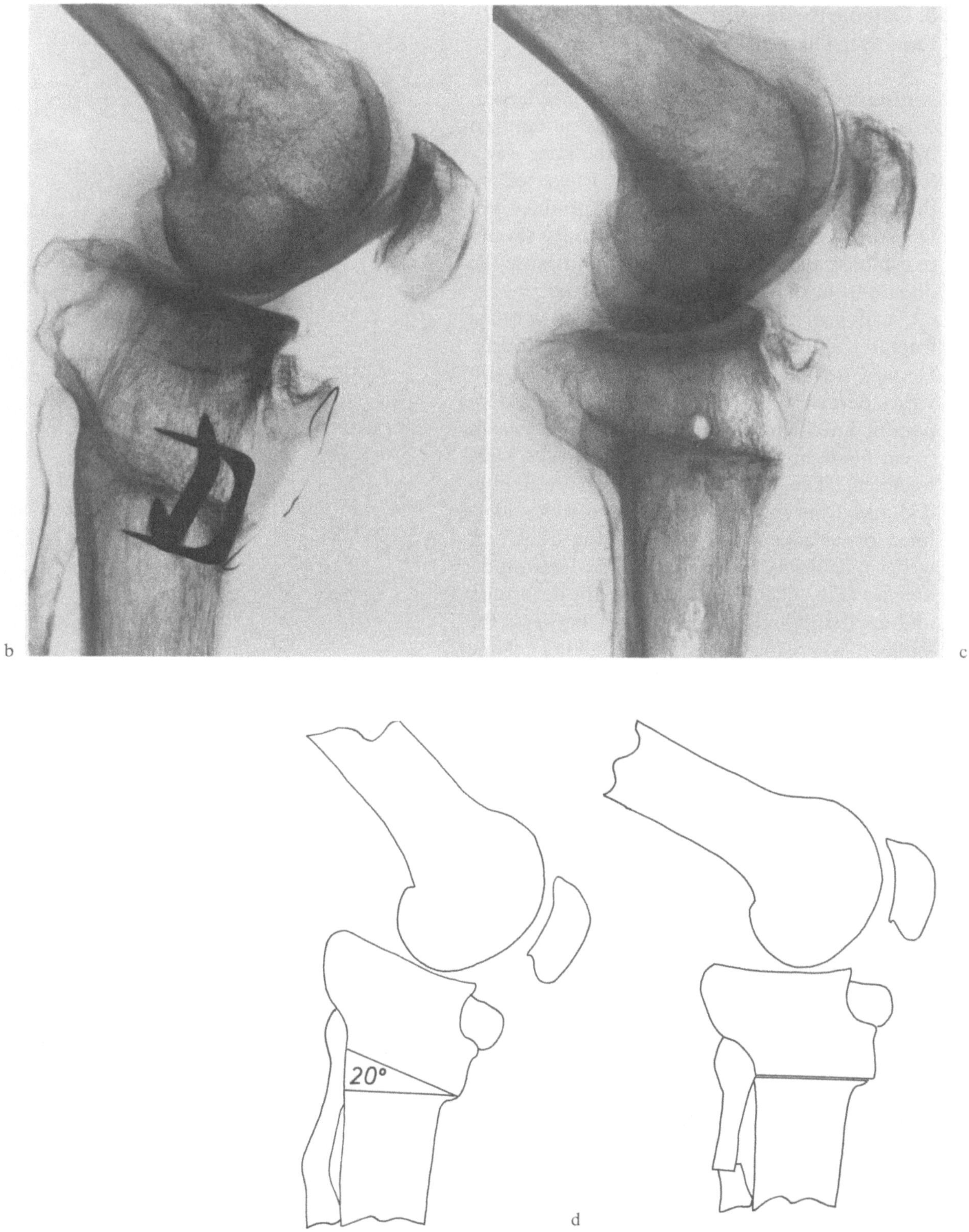

Fig. 234b–d. A 61-year-old patient before (a and b) and one year after (c) correction of a genu recurvatum. Planning of the operation (d)

6. Osteoarthritis of the Knee Due to a Distant Deformity

A distant deformity can disturb the mechanical conditions of the knee. Obviously a femoral fracture healed in varum brings the knee away from the line of action of force *P* exerted by the mass of the body. This can produce osteoarthritis. Correction of the deformity should re-establish the normal conditions and cause the disease to heal. This is what is observed.

A 49-year-old male patient had sustained a fracture of the femur twenty years previously. He was treated by traction and healed with a varus deformity of 15°. He complained of a painful knee which was deformed into varum (6 cm between the knees when the ankles were together). The joint space was narrowed medially and the medial tibial plateau was underlined by a large dense triangle (Fig. 235a). A wedge osteotomy (17° wedge) of the femoral diaphysis (Fig. 235b) overcorrected the deformity and created a mild valgum (1 cm between the malleoli when the knees were together). Nine years later the trabeculae beneath the medial plateau are less pronounced (Fig. 235c). The dense triangle has disappeared and a joint space has developed. The range of movement is satisfactory. The patient remains painfree and lives an active life.

a 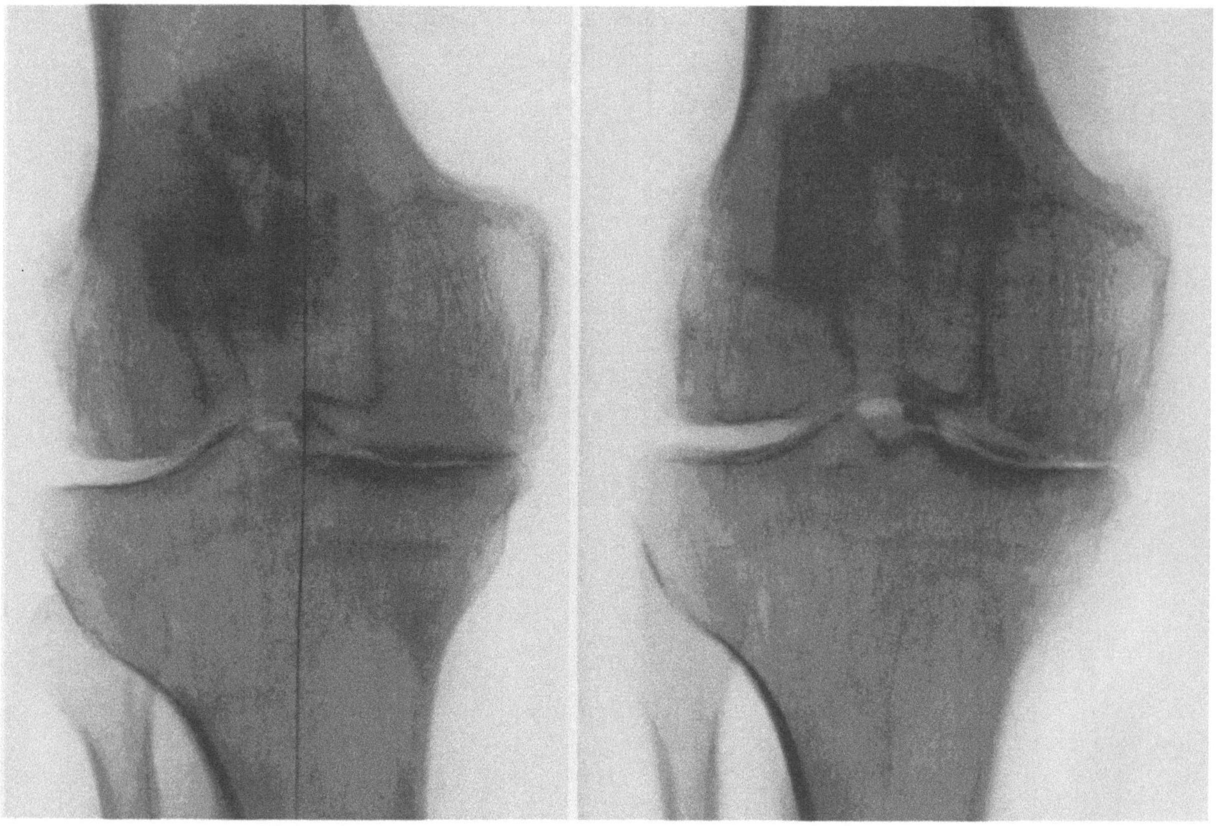 c

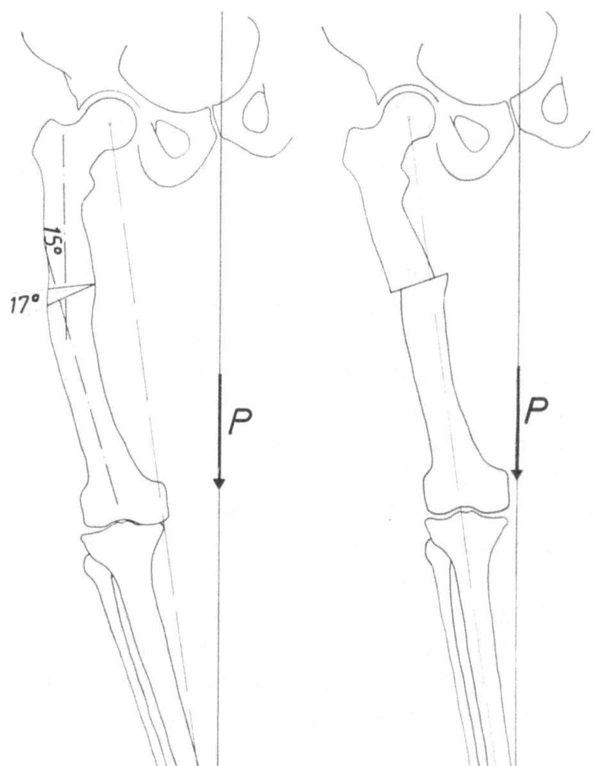

b

Fig. 235a–c. A 49-year-old patient 20 years after a fracture of the femoral shaft (a). Overcorrecting osteotomy at the fracture site (b). Result nine years after surgery (c)

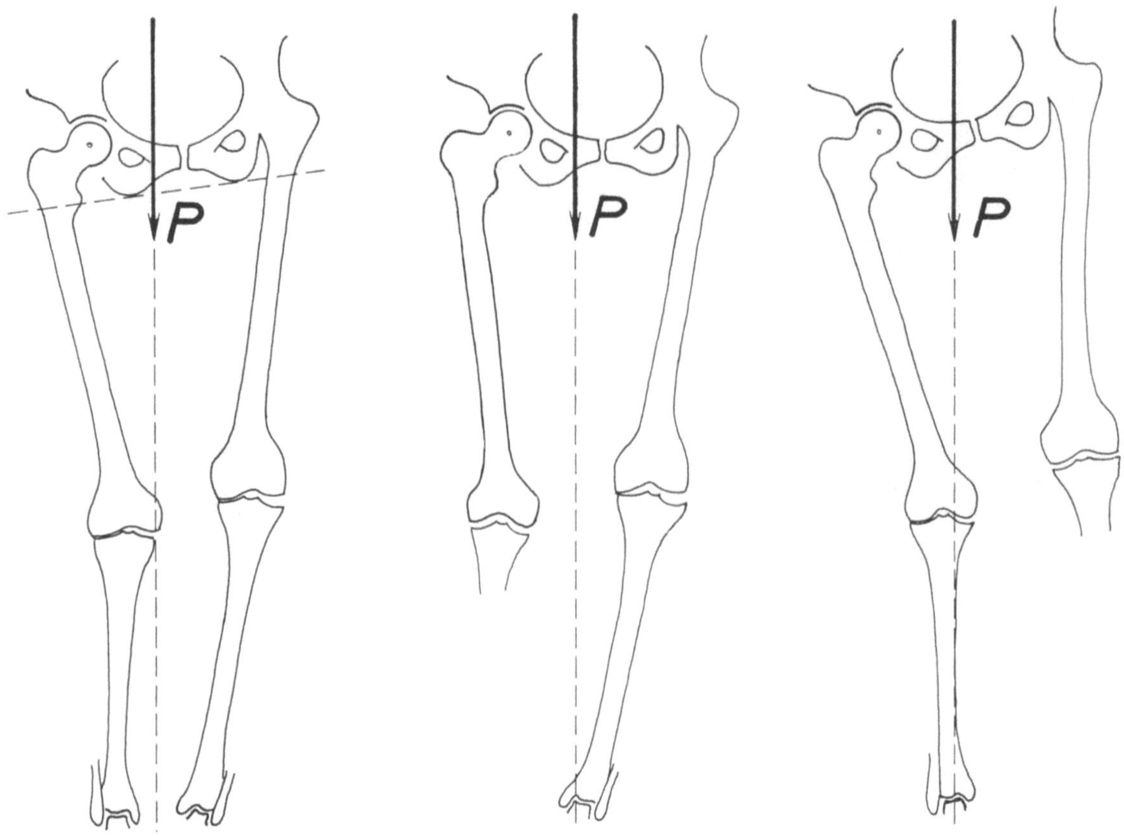

Fig. 236a. A 55-year-old patient. Left hip arthrodesed with adduction of the thigh

A much more distant deformity can cause limping, displace the partial centre of gravity S_7 and modify the mechanical conditions at knee level. These changes can move the compression force R medially or laterally and produce osteoarthritis (see Chapter V). A distant deformity can cause a fixed flexion contracture of one or both knees, as for example in patients with flexion contracture of a hip or fixed pelvic obliquity. In some cases, correction of the deformity can reduce the compressive stresses in the knee sufficiently and heal the osteoarthritis. Every case must be analysed individually, taking the mechanical information given in the preceding chapters into account.

As an example, a 55-year-old female patient had a left hip ankylosed in adduction for 27 years (Fig. 236a). The trunk was tilted to the opposite side. The centre of gravity of the body was displaced to the right. The right knee was held flexed to make up for the apparent leg lengthening. The patient had developed osteoarthritis with a valgum and a flexion contracture of the right knee (Fig. 237a). The left knee had developed a varus osteoarthritic deformity (Fig. 238a). Revision of the arthrodesis through an intertrochanteric osteotomy made the pelvis horizontal and brought the centre of gravity back to its normal position (Fig. 236b). A posterior capsulotomy of the right knee was also performed. Eight years later the patient walks comfortably, relieved of pain in her knees. The radiological signs of osteoarthritis have regressed. A large joint space can be seen in the standing X-rays (Figs. 237b and 238b). Such a result can be explained only by the change in the mechanical stress in the knees. Biological modifications, vascular or otherwise, cannot be invoked since surgery was performed on the hip.

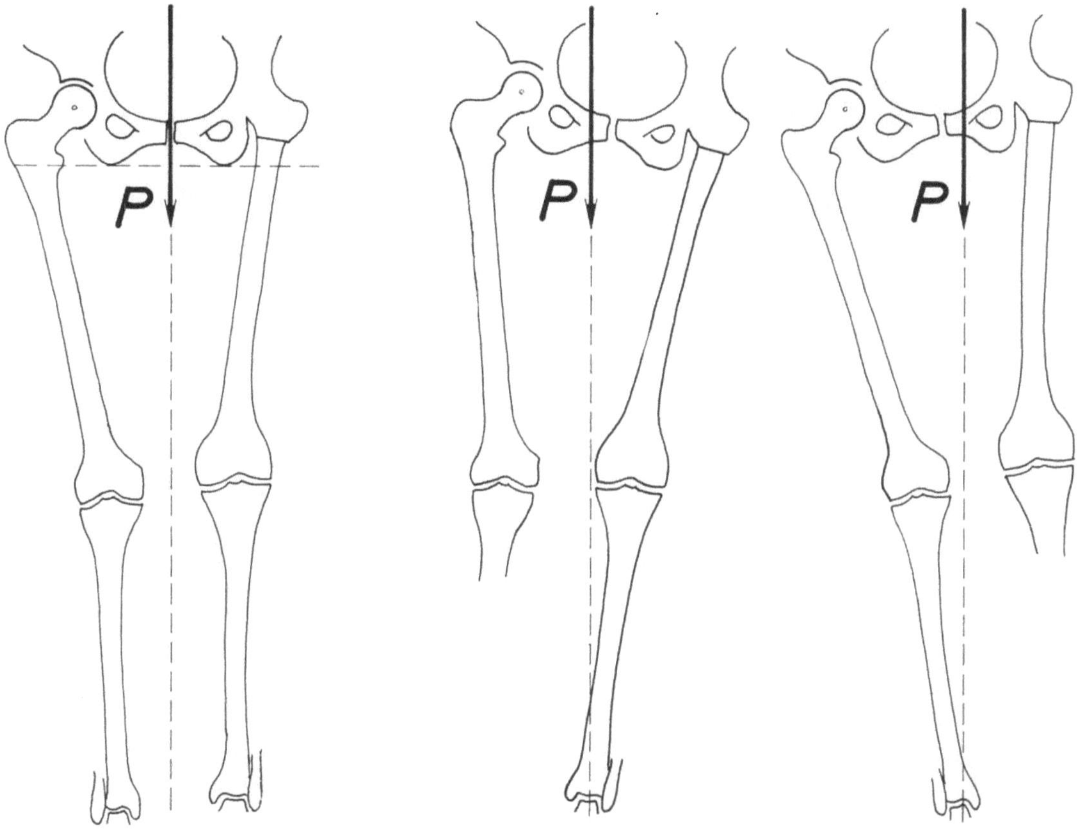

Fig. 236b. Correction of the hip arthrodesis

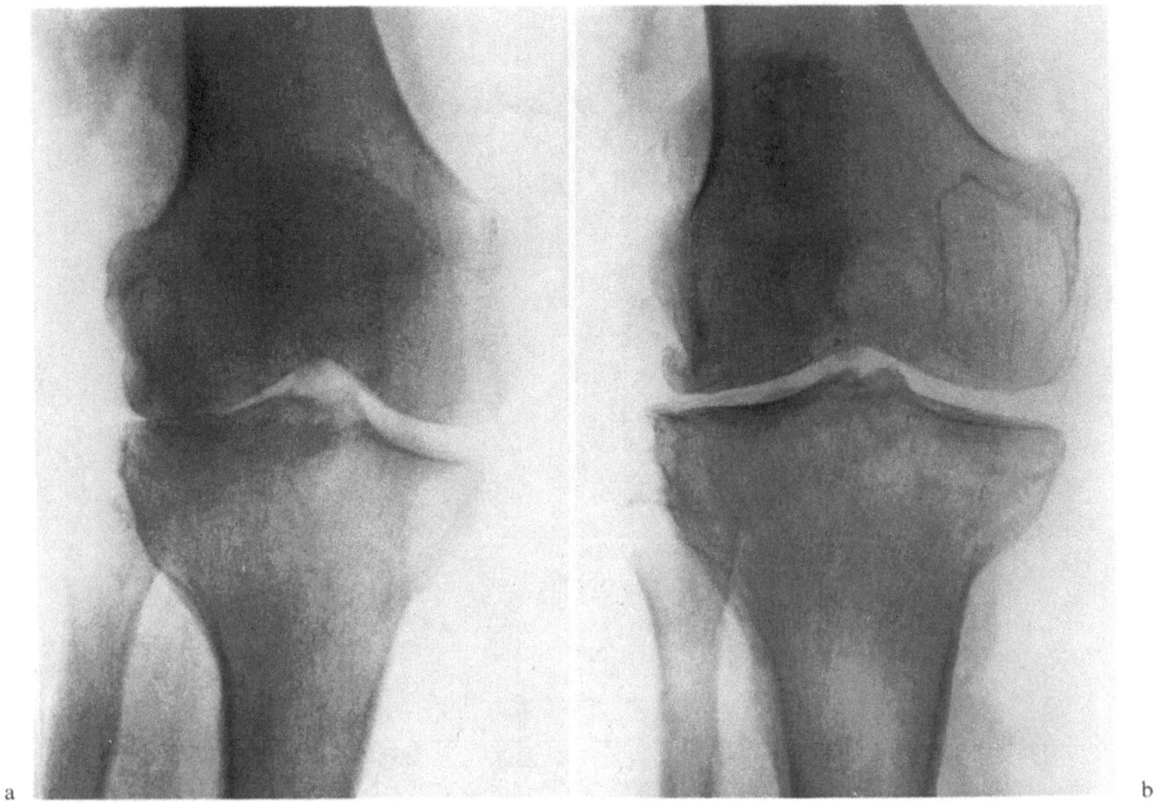

Fig. 237a and b. Right knee of the patient with an arthrodesed hip, before (a) and 8 years after (b) revision of the arthrodesis

a 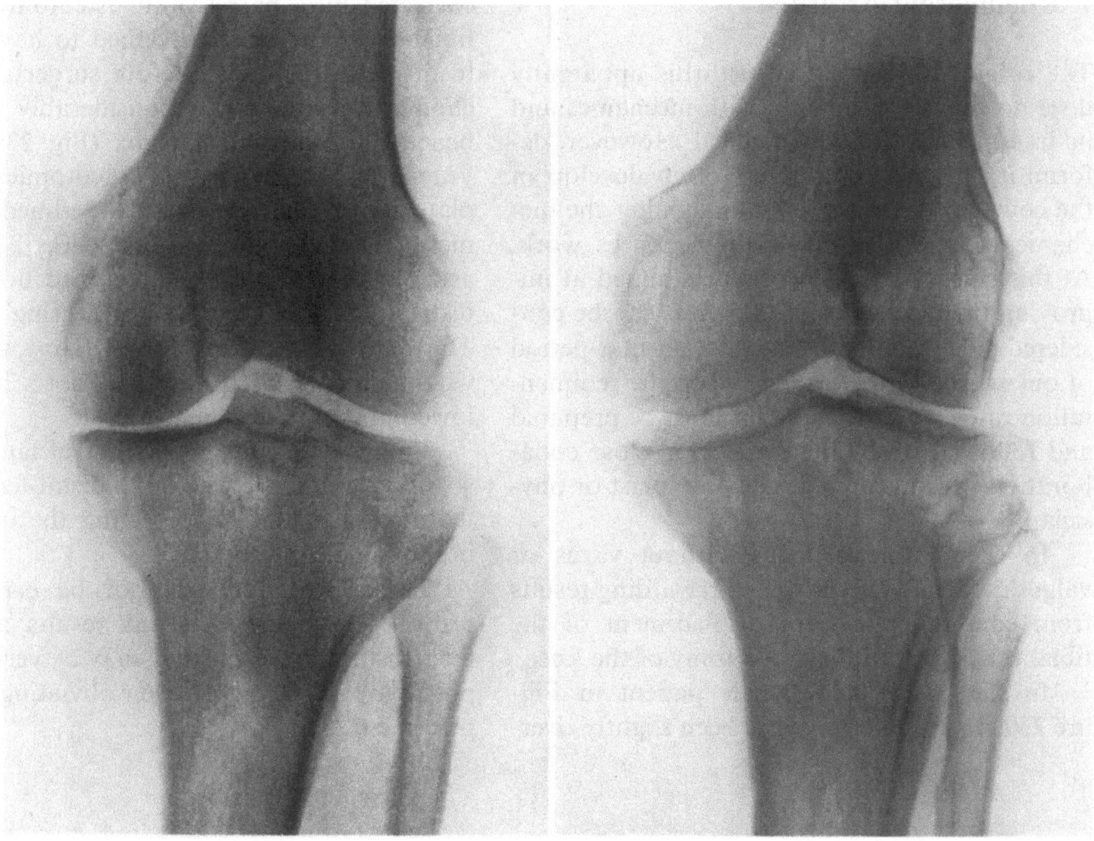 b

Fig. 238a and b. Left knee of the patient with an arthrodesed hip, before (a) and 8 years after (b) revision of the arthrodesis

7. Rheumatoid Arthritis

The origin of rheumatoid arthritis apparently does not have much to do with mechanics and its treatment is basically medical. However, deformities of the lower limbs often develop in the course of the disease and may alter the mechanical conditions in which the knees work. At this stage, surgical procedures aimed at improving the mechanics of the joints can be considered. They should be carried out in a period of quiescence of the disease, when the sedimentation rate is low. The patient must be prepared and followed up by the surgeon in close collaboration with his or her rheumatologist or physician.

In rheumatoid arthritis without varus or valgus deformity, we have had rewarding results from combining anterior displacement of the tibial tuberosity with synovectomy of the knee.

In the 52-year-old female patient in Figure 239, varus deformity has been slightly overcorrected by a barrel-vault osteotomy of both tibiae after the disease seemed to have yielded to medical treatment. Before surgery, the subchondral sclerosis was considerably increased beneath the medial plateaux (Fig. 239a). Two years after the barrel-vault osteotomies, the two plateaux of each knee are underlined by symmetrical subchondral scleroses (Fig. 239b). The articular compressive stresses are now evenly distributed over the weight-bearing surfaces. The patient was chair-bound. Now she walks with neither limp nor stick. She is being followed up by her rheumatologist.

Valgus deformity with signs of lateral overstressing was dealt with by a distal femoral osteotomy slightly overcorrecting the deformity, in the same conditions.

These procedures cannot be expected to provide the same excellent results as in osteoarthritis; however, they may be very helpful, possibly postponing or even obviating total replacement.

Fig. 239a and b. A 52-year-old patient with rheumatoid arthritis, before (a) and 2 years after (b) a barrel-vault osteotomy of the tibiae to overcorrect a varus deformity

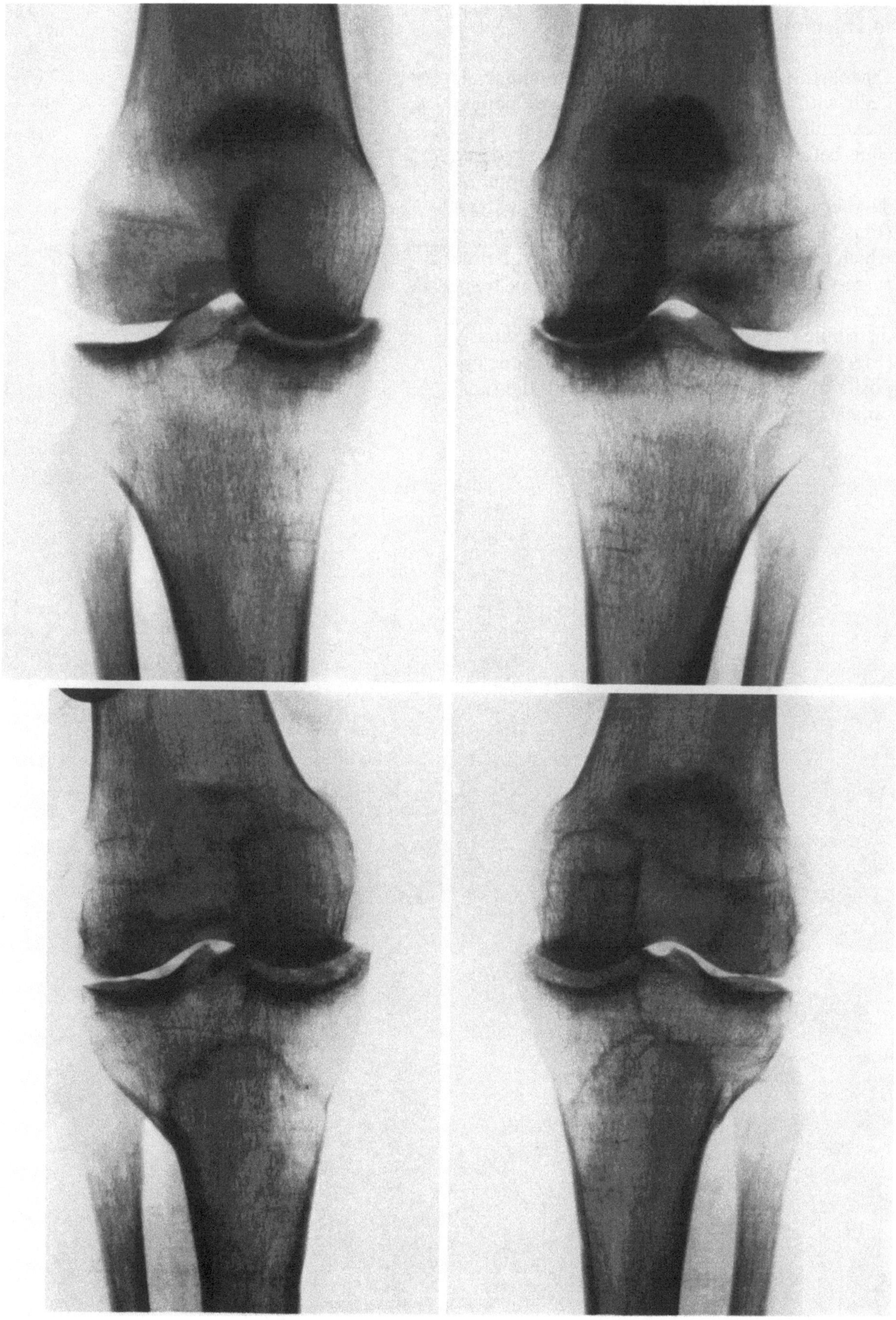

a

b

8. Osteoarthritis of the Knee in Haemophiliacs

Osteoarthritis of the knee in haemophiliacs is dealt with as rheumatoid arthritis. The patient is carefully examined and prepared by the physician before surgery. Operation is considered only after transfusion of cryoglobulin. A 42-year-old male haemophiliac patient (Fig. 240a) underwent a barrel-vault osteotomy which overcorrected his varus deformity. At the three-year follow-up, a joint space has reappeared and the stresses seem to be more evenly distributed in the joint than previously (Fig. 240b). A barrel-vault osteotomy of the opposite tibia was also carried out, with the same satisfactory result.

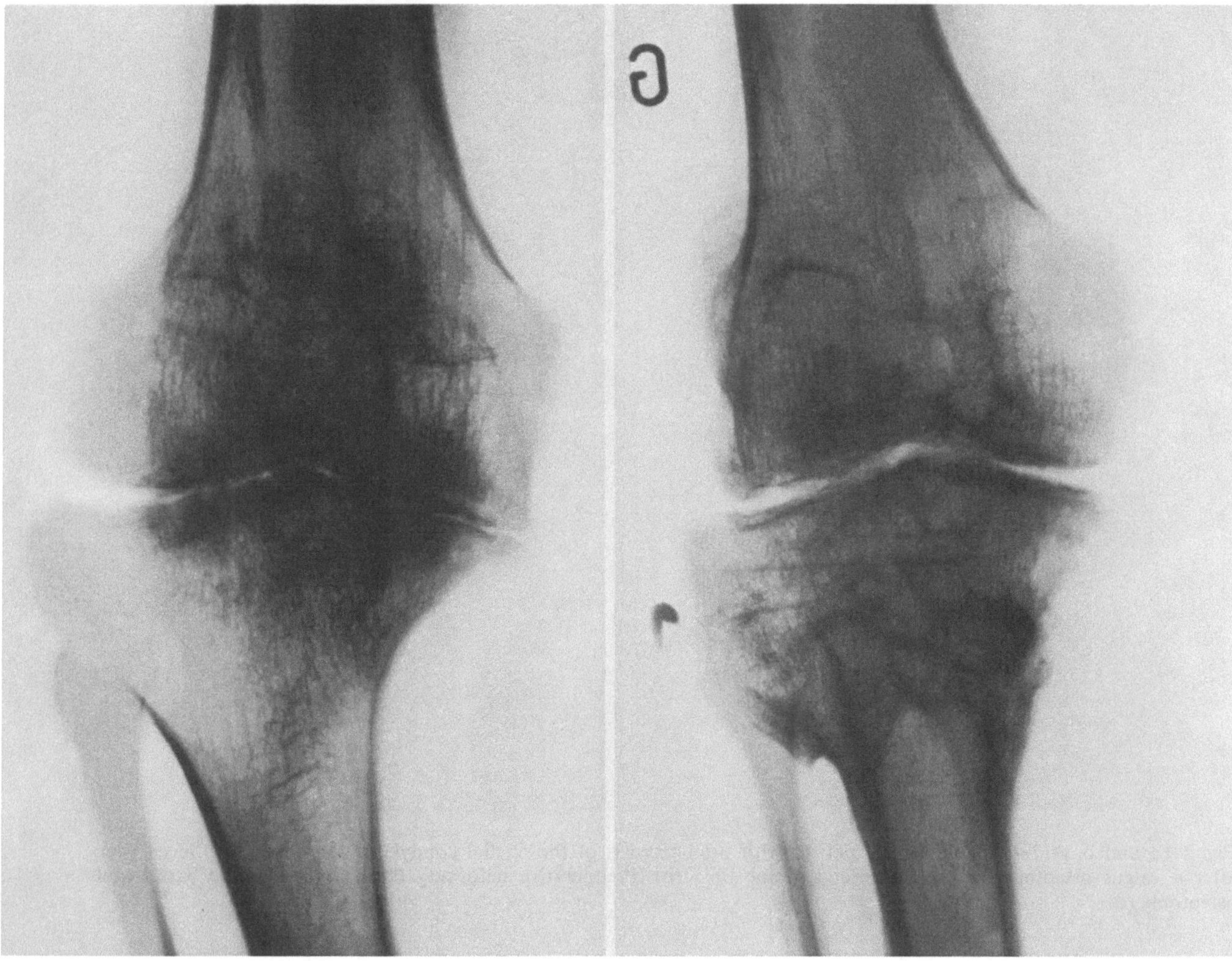

Fig. 240a and b. A 42-year-old haemophiliac patient before (a) and 3 years after (b) a barrel-vault osteotomy of the tibia to overcorrect a varus deformity

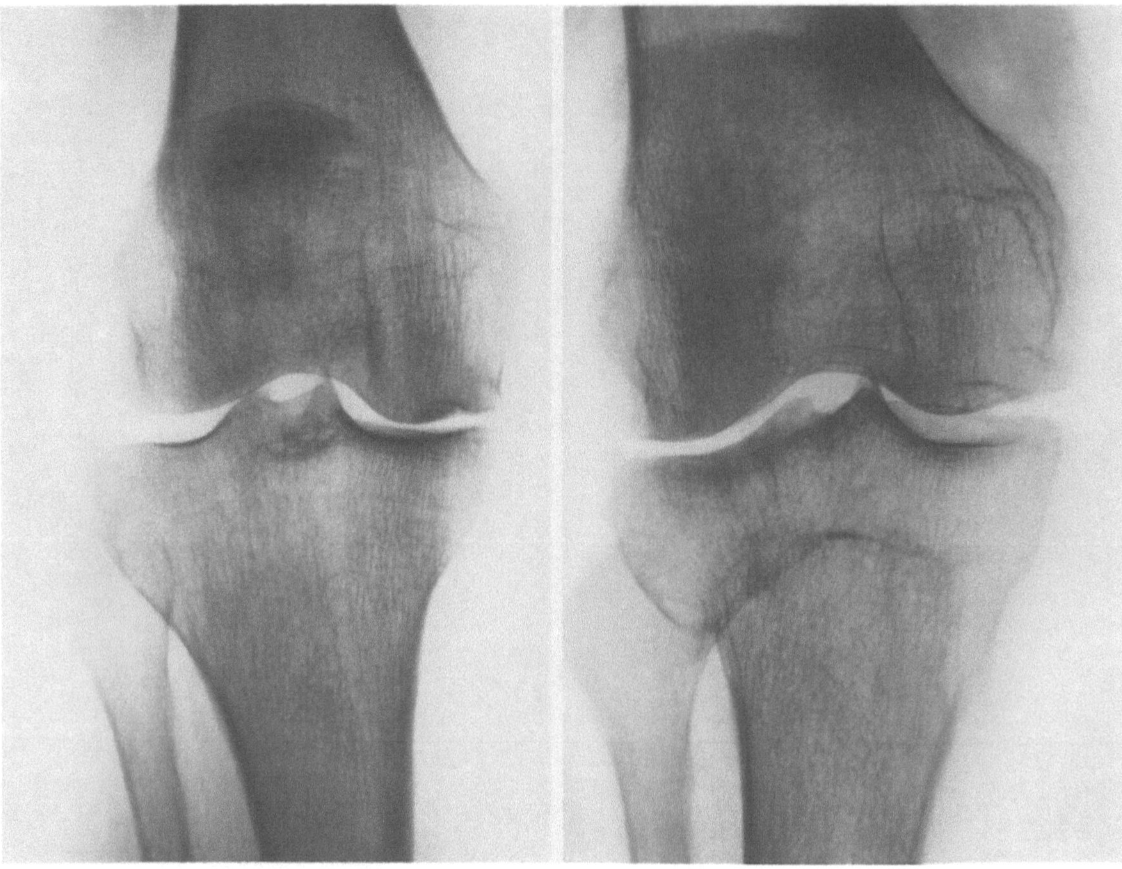

a b

Fig. 241 a and b. A 70-year-old female patient with osteonecrosis of the medial condyle of the femur (a). Seven years after a valgus osteotomy of the upper end of the tibia (b). Preoperative deformity (c). Planning of the barrel-vault osteotomy (d)

9. Osteonecrosis of the Medial Condyle of the Femur

In so-called idiopathic or spontaneous osteonecrosis of the medial femoral condyle, revascularization usually takes place, probably beginning before the condition appears in the X-rays. There exists a risk of collapse of the articular tissues before histological recovery has been completed. In order to minimize this risk, a valgus osteotomy of the upper end of the tibia is performed to produce a valgus deformity and thus bring the knee closer to the line of action of force P due to the body mass (Fig. 75). A smaller force L is required to counterbalance P. The resultant force R is reduced, as are the compressive stresses in the joint. The overall reduction of the articular pressure seems to allow the tissues to heal without collapsing.

A 70-year-old female patient (Fig. 241 a) complained of pain of sudden onset in her knee and started to limp. When she was seen some weeks later the affected knee presented signs of osteonecrosis. There was a varus deformity of 8° (Fig. 241 c). A 13° barrel-vault osteotomy was carried out (Fig. 241 d). At the last follow-up, seven years after the osteotomy, the knee is painfree, the range of movement is full and the patient does not limp. The X-ray picture shows an improvement of the condition (Fig. 241 b).

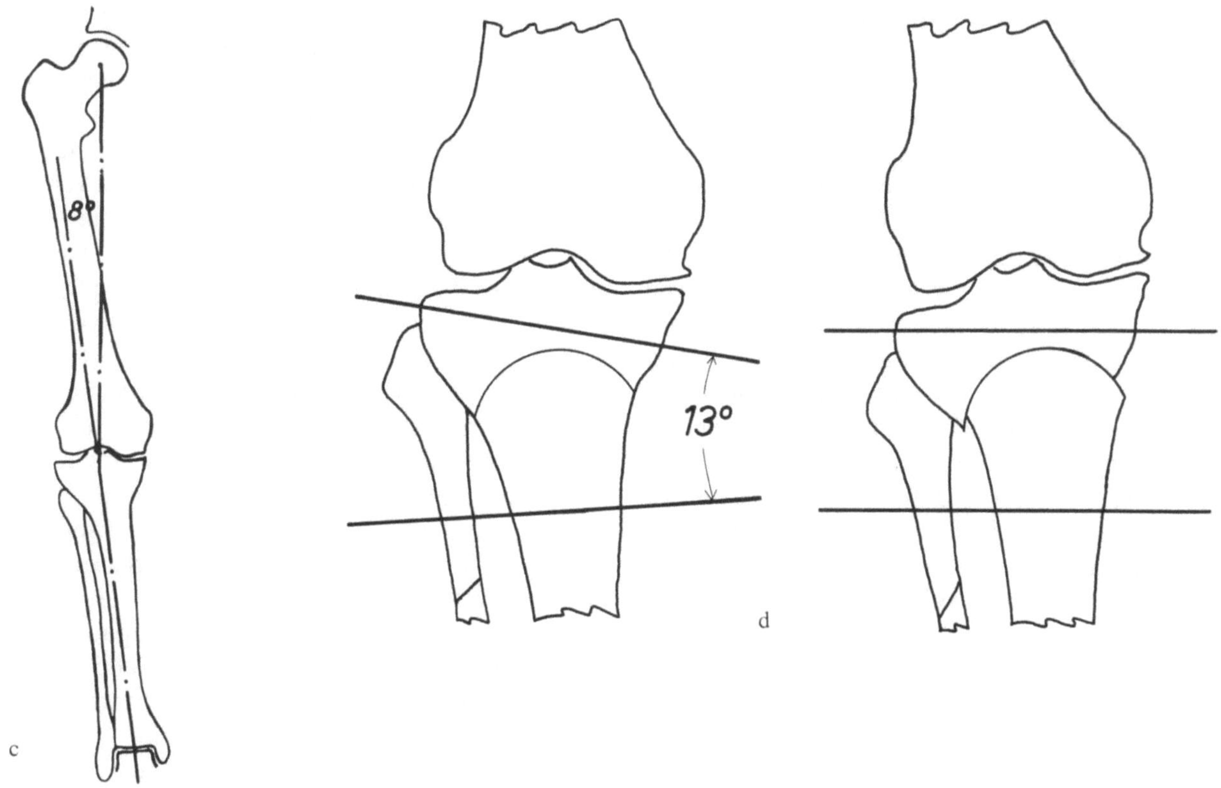

Fig. 241 c–d. Deformity of the leg (c) and planning of the barrel-vault osteotomy (d)

10. Widespread Osteoarthritis without Deformity

Osteoarthritis equally involving the whole joint may be due to a diminution of resistance of the tissues which are no longer able to sustain normal mechanical stress. Osteoarthritis symmetrically attacking both compartments of the knee indicates that the load R remains well centred. The only mechanical therapeutic possibility consists in diminishing this force R and the force exerted on the knee-cap. As seen before, anterior displacement of the patella tendon as far as possible reduces the force pressing the femur on the tibia and the force pressing the knee-cap against the femur. It is thus theoretically indicated in this sort of case. Our clinical experience with this has been limited so far but promising.

11. Histological Confirmation of the Regenerative Process

During the months following a barrel-vault osteotomy of the tibia or a supracondylar osteotomy of the femur carried out in order to redistribute the stresses in the joint appropriately, not only do the subchondral scleroses and the structure of the cancellous bone beneath the articular surfaces return to normal, but also a wide joint space reappears between the parts of the bones which were previously in contact. It must be remembered that all our A.-P. X-rays are taken with the patient standing and putting full weight on the knee. The reappearance of a joint space in the X-rays while standing implies the growth of tissue. It was of some interest to find out what kind of tissue regenerates in the joint subjected to tolerable pressure probably of physiological magnitude. To this end, Fujisawa et al. (1976)[15] carried out a series of arthroscopies of the knee just before surgery and at regular intervals afterwards. They also biopsied the load-transmitting surfaces of the joint in the area which was pre-operatively overstressed. Before surgery, the arthroscope showed eburnated bone on both femoral and tibial articular surfaces (Fig. 242a). In the cases in which the overcorrection of the deformity had been sufficient, white smooth tissue similar to cartilage covered the joint surfaces two years after surgery (Fig. 242b). Under the microscope, this tissue appears to be fibro-cartilage with a tendency of its deep layers to remodel into hyaline cartilage (Fig. 243).

Shiomi et al. (1978) carried out the same experiment on patients with patello-femoral osteoarthritis and after anterior displacement of the tibial tuberosity, as mentioned and illustrated on pages 156 and 157. They observed similar results when the anterior displacement had been sufficient.

This outcome demonstrates the capacity of the tissues of the skeleton for repair. When subjected to appropriate stressing, these tissues regenerate and differentiate into specific varieties such as cartilage, again confirming the concepts of Pauwels (1965, 1973).

[15] Among the co-authors, Matsumoto and Yamaguchi had each worked with the author for more than a year and had been surprised by the results. On their return to Japan they applied the theory and technical procedures described in this book

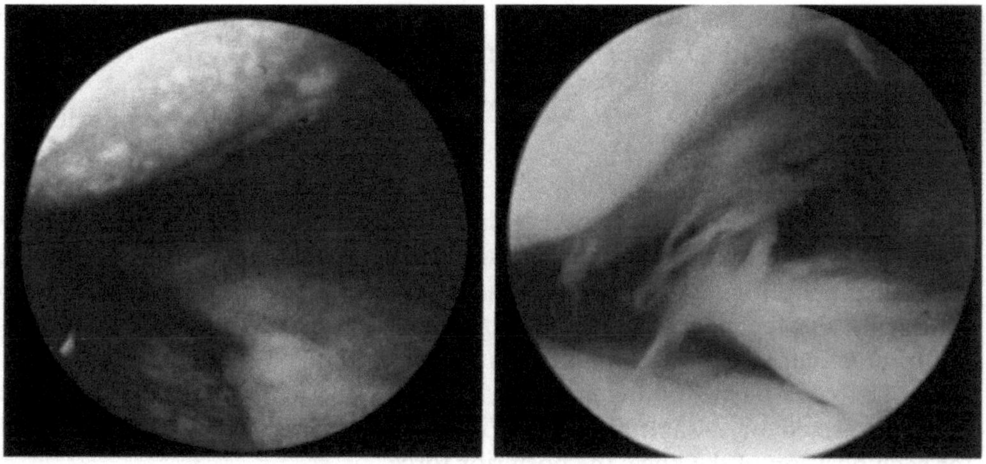
a b

Fig. 242a and b. Arthroscopy of the overstressed compartment of an osteoarthritic knee with a varus deformity (a). Arthroscopy of the same area 2 years and 3 months after an osteotomy of the upper end of the tibia overcorrecting the deformity (b): the meniscus appears between the two articular surfaces covered by smooth white tissue (from Fujisawa et al. 1976. Permission to use these pictures has been granted by the authors and by Igaku Schoin Ltd)

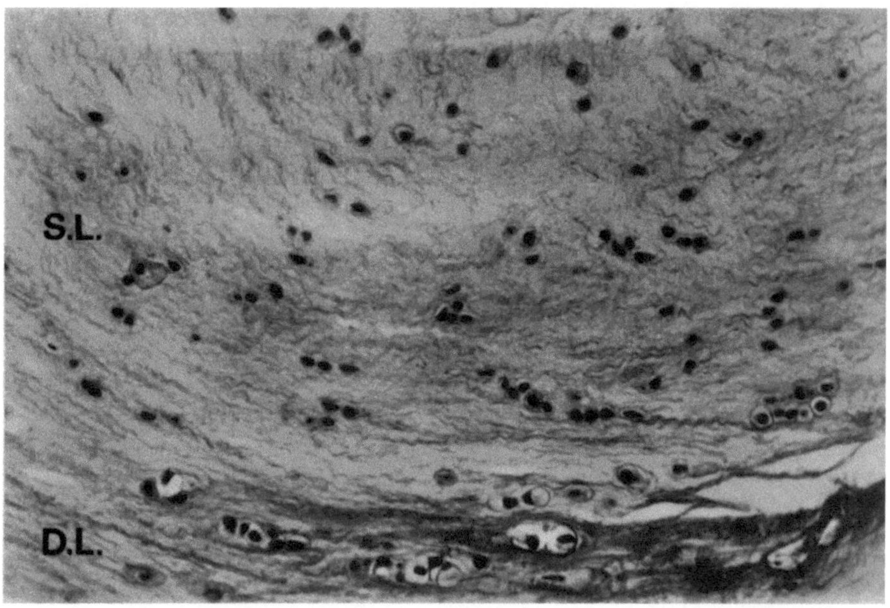

Fig. 243. Histological appearance of a postoperative biopsy from the tissue covering the articular surfaces of a preoperatively overstressed compartment. S.L.: superficial layer; D.L.: deep layer (from Fujisawa et al. 1976. Permission to use these pictures has been granted by the authors and by Igaku Schoin Ltd.)

D. Critical Analysis of Patellectomy and Other Procedures on the Patella

1. Standard Patellectomy

Benoist and Ramadier (1969) observe that "the risk of rupture of the extensor mechanism exists after all patellectomies". This risk actually results from the fact that patellectomy increases the force required from the quadriceps mechanism to carry out its work and maintain equilibrium. Projected on a sagittal plane (Fig. 244), for each position of the knee, the posterior force R_3 which tends to flex the joint is counterbalanced by an anterior force P_a, the patella tendon. The moment of each force (the product of the force and its lever arm) is what matters for equilibrium.

The equation of equilibrium is written:

$$R_3 \cdot e = P_a \cdot c.$$

For each given position, a shortening of the lever arm c must be compensated for by an increase of the force P_a. The femoro-tibial joint supports forces R_3 and P_a, which can be represented by their resultant R_4. The force R_4 is the vectorial sum of the forces R_3 and P_a and constitutes the load acting on the femoro-tibial joint. Its magnitude and direction depend on the magnitudes of the forces R_3 and P_a and on their lines of action. In a normal knee (Fig.

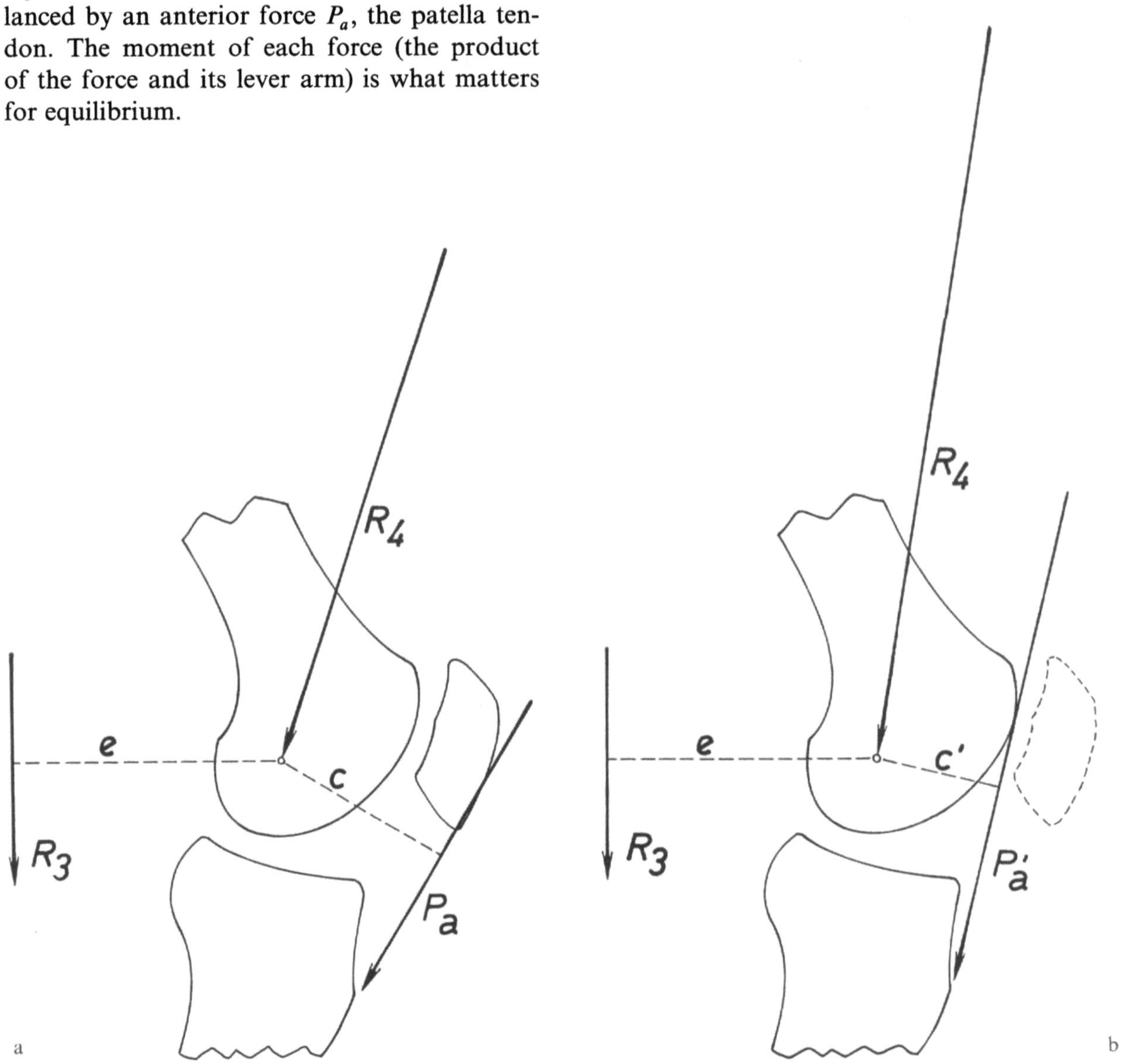

Fig. 244a and b. Effect of patellectomy. R_3: resultant of the flexion forces. e: lever arm of R_3. P_a: patella tendon. c: lever arm of P_a. R_4: resultant of forces R_3 and P_a. a: normal knee. b: after patellectomy

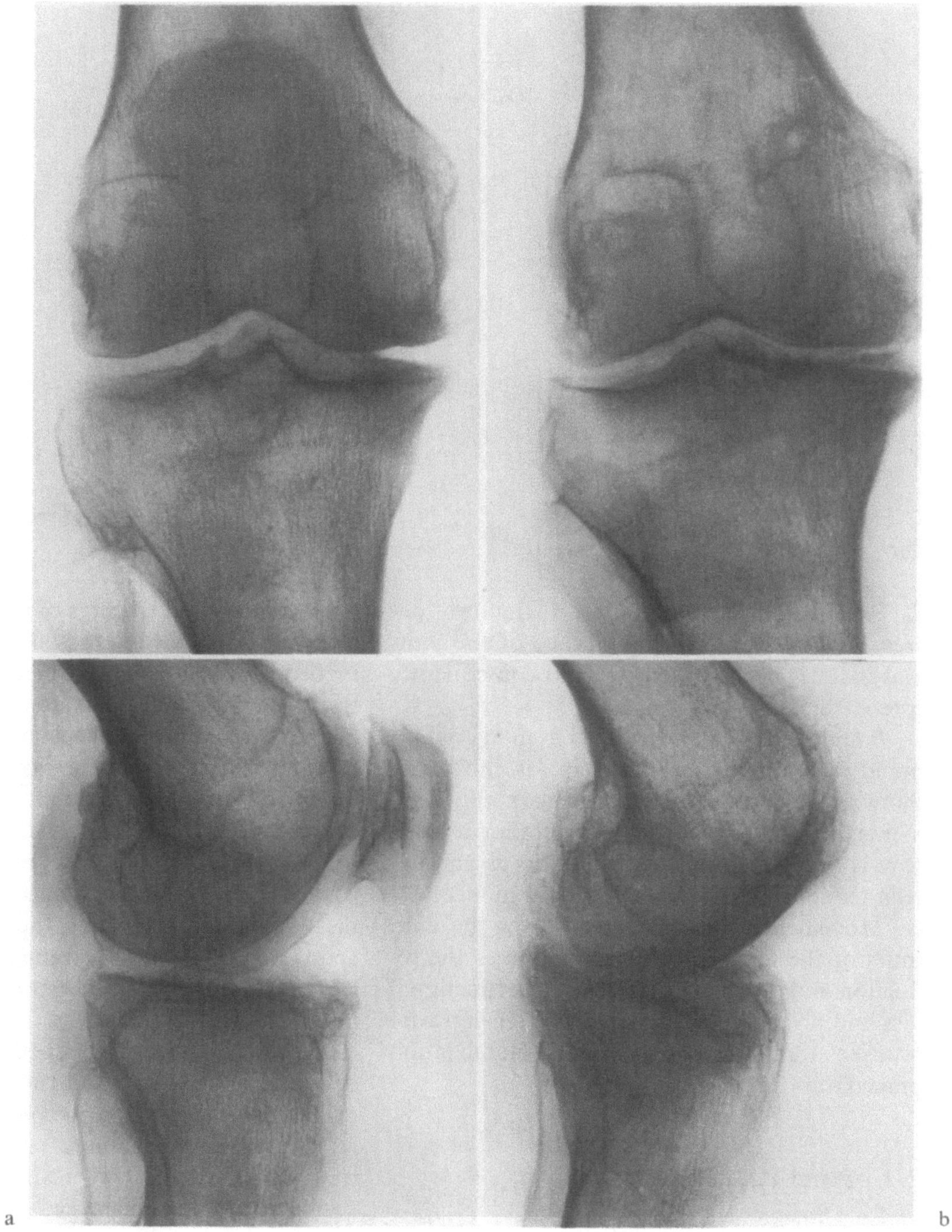

Fig. 245. (a) Before patellectomy. (b) Ten years after patellectomy

244a) the knee-cap ensures a certain length to the lever arm c of the patella tendon. Without a patella (Fig. 242b) the tendon P_a' falls into the intercondylar groove. This shortens its lever arm c'. Consequently, removing the patella considerably increases the force P_a and correspondingly the load R_4 exerted on the femoro-tibial joint.

Thus, performed on a non-osteoarthritic knee, patellectomy increases the risks of late osteoarthritis. It worsens the mechanical conditions in cases of osteoarthritis for which it has been carried out.

This is suggested by an instance of osteoarthritis of the knee (Fig. 245a) treated by patellectomy. Ten years later (Fig. 245b) the osteoarthritis is worse. The same happened to this patient's other knee.

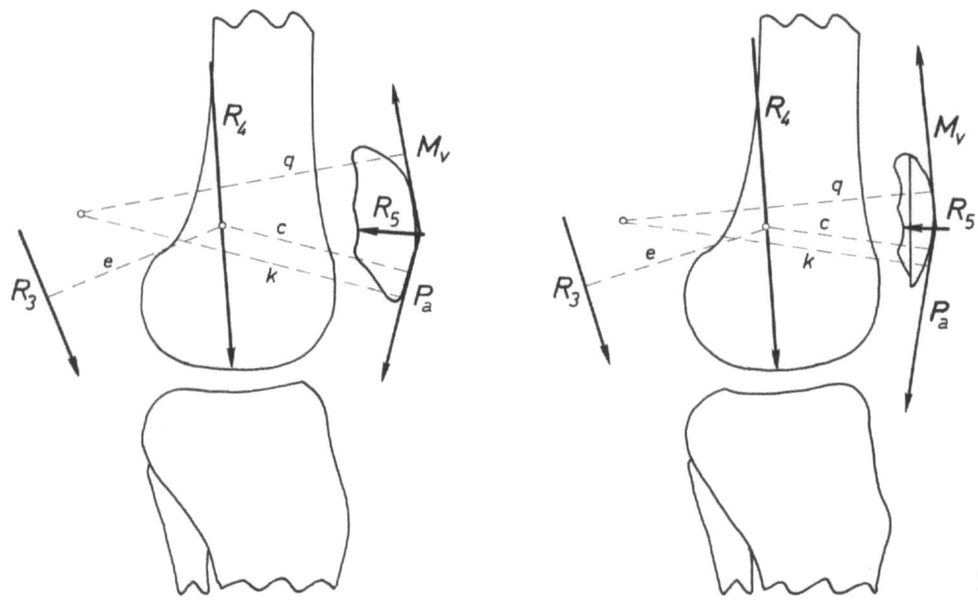

Fig. 246a

Partial patellectomy after a fracture decreases the weight-bearing surface of the joint and induces peaks of stresses and consequently osteoarthritis.

Patellectomy can be avoided even in comminuted fractures of the knee-cap, as shown by Pauwels (1965a). In such instances, the fragments are roughly reduced, an anterior tension wire is inserted through the quadriceps and patella tendons and the knee is moved immediately. Movement of the fragments over the contours of the femoral condyles improves the reduction and is obligatory. There is no functional loss and no muscular atrophy. The tension wire must be as anterior as possible and mobilization immediate.

2. Coronal Patellectomy

Removal of the posterior half or of a slice of the patella has been proposed in order to reduce the thickness of the bone. Shaving of the articular surface of the patella is unreliable (Bentley, 1970) and certainly destroys the congruence of the joint. But a longitudinal slice of the bone can be excised so as to decrease the thickness of the bone by about 1 cm. This shortens the lever arm c with which the patella tendon acts on the tibia, but opens the angle β formed by the lines of action of forces M_v, quadriceps, and P_a, patella tendon (Figs. 54 and 141).

We have used tracings of a knee at 4° hyperextension, 50° flexion and 90° flexion (Fig. 246) in order to evaluate the mechanical result of this procedure. We have assumed that in each of the three positions the same strength was required from the patella tendon, with the patella intact, to ensure equilibrium. The other forces are deduced from this assumption. We have graphically determined the geometrical changes provoked by removing a slice 1 cm thick from the patella and calculated the alteration of the acting forces. In all positions, the lever arm c with which the patella tendon moves the tibia about the femoral condyles is shortened, as are the lever arm k with which the patella tendon acts on the knee-cap and the lever arm q with which force M_v acts on the patella.

If equilibrium requires a given moment $P_a \cdot c$ to counterbalance the flexing moment $R_3 \cdot e$, then force P_a must increase by 21% at 4° hyperextension and at 50° flexion and by 25% at 90° flexion after thinning of the patella.

The augmentation of force P_a leads to a corresponding increase of force R_4, the load transmitted from the femur to the tibia (Figs. 244 and 246). Force R_4 is increased by 9% at 4° hyperextension, by 4% at 50° flexion and by 14% at 90° flexion. This increases somewhat the articular pressure in the femoro-tibial joint.

Despite the increase of forces P_a and M_v due to the shortening of their lever arms in all three analysed positions, the resultant force R_5 is a

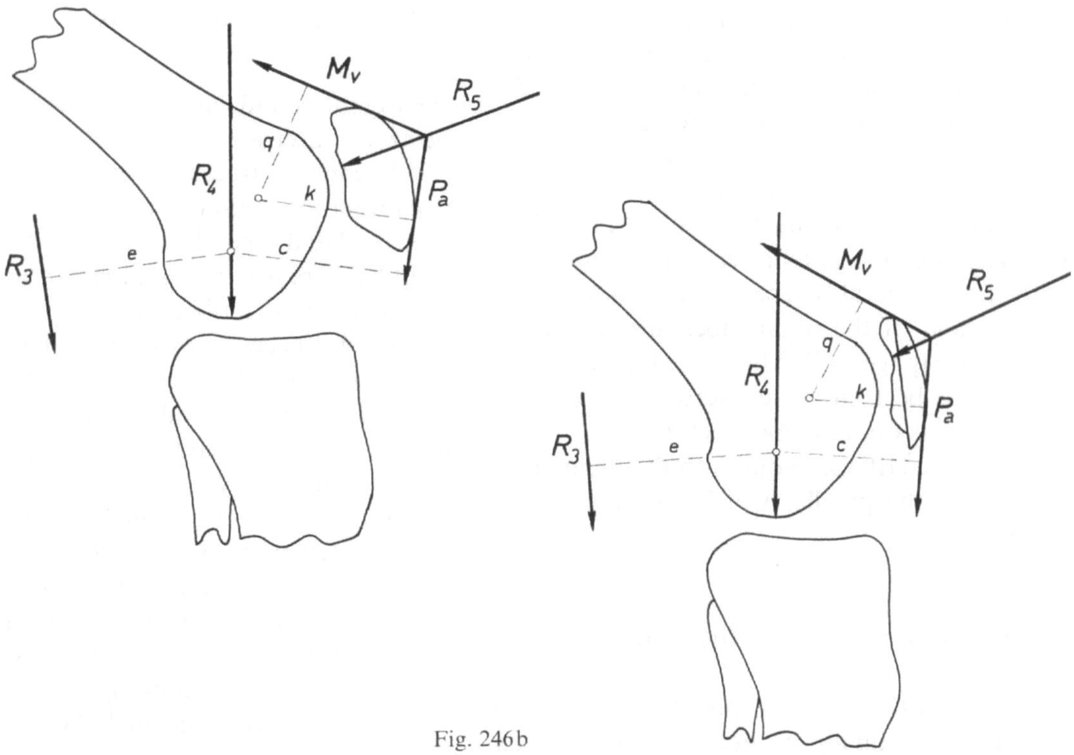

Fig. 246b

Fig. 246a–c. Thinning the patella by removing a 1 cm slice. (a) extension 4°, (b) flexion 50°, (c) flexion 90°. R_3: flexion force. e: lever arm of force R_3. P_a: force exerted by the patella tendon. c: lever arm of force P_a acting on the femoro-tibial joint. k: lever arm of force P_a acting on the patello-femoral joint. M_v: force exerted by the quadriceps tendon. q: lever arm of force M_v. R_4: resultant of forces R_3 and P_a. R_5: resultant of forces P_a and M_v

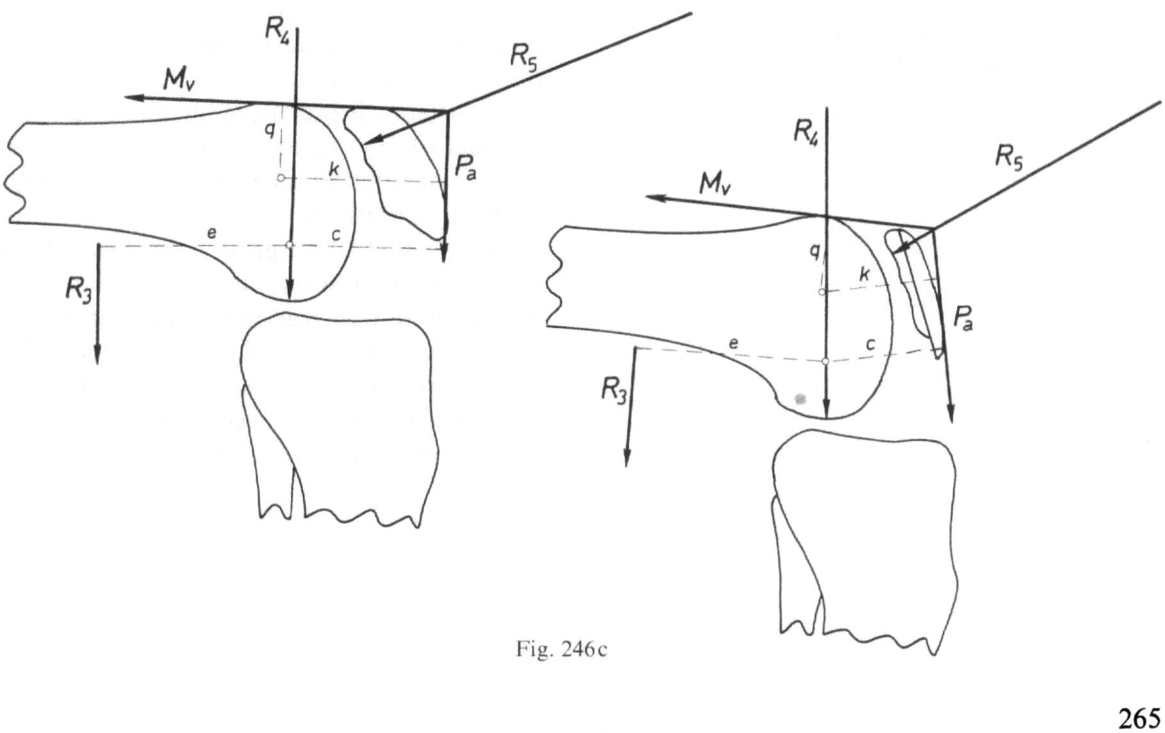

Fig. 246c

little reduced by thinning the patella, because the procedure opens the angle formed by the lines of action of forces M_v and P_a. Force R_5 pressing the patella against the femur is reduced by 32% at 4° hyperextension, by 1% at 50° flexion and by 10% at 90° flexion.

Reducing the patella to a thin "pancake" thus decreases the patello-femoral compressive force R_5 considerably less than a 2 cm anterior displacement of the tibial tuberosity, as advocated above (page 144). The procedure increases the load to be transmitted across the femoro-tibial joint, as does the patellectomy.

The same criticism applies to the "spongialisation" of the patella, which also diminishes its thickness.

3. Sagittal Osteotomy of the Patella

Morscher (1978) has proposed splitting the patella longitudinally at the junction between the medial and lateral facets. Opening the anterior aspect of the osteotomy, maintained by a graft, would ensure contact between the medial facet and condyle, thus enlarging the weight-bearing surface of the joint. Simultaneously, the osteotomy would decrease the blood pressure in the bone by ensuring a kind of drainage. We doubt that such an osteotomy can result in congruence of the articular surfaces of this complicated joint with irregular outlines, during movement. In any case this osteotomy must be considered as experimental until longer follow-ups are available.

E. Operative Indications

The operative indications can be deduced from the above discussion.

Flexion contractures, if not reducible by physiotherapy, require a posterior capsulotomy. The isolated patello-femoral osteoarthritis should be treated by as great an anterior displacement of the tibial tuberosity as possible, combined if necessary with a posterior capsulotomy. Osteoarthritis localized at the joint between the knee-cap and the lateral condyle requires an anterior and medial displacement of the tibial tuberosity.

Osteoarthritis involving the whole knee, without deformity in the coronal plane and without signs of locally concentrated overpressure, is a good indication for an anterior displacement of the tibial tuberosity combined with a posterior capsulotomy if needed.

Osteoarthritis of the medial compartment of the knee joint must be treated by a valgus proximal tibial osteotomy combined with an anterior displacement of the tibial tuberosity and, if necessary, a posterior capsulotomy.

A distal femoral osteotomy seems to be the most appropriate choice when dealing with osteoarthritis with a valgus deformity.

The surgical treatment of osteoarthritis either with a flexion deformity (as opposed to a flexion contracture) or with a recurvatum of the knee or with a malrotation of the leg must be chosen according to the mechanical conditions of each individual knee.

In osteoarthritis of the knee caused by a distant deformity correction of the deformity can be considered after a careful analysis of the mechanical conditions.

These several surgical procedures diminish the compressive forces acting on the osteoarthritic knee and enlarge the surfaces transmitting these forces. They enable us to decrease the articular compressive stresses significantly and to make them tolerable even for tissues with a lowered mechanical resistance.

Chapter VIII. Results

The examples presented in the preceding chapter support our theory of the pathogenesis of osteoarthritis of the knee. However, it remains to evaluate the outcome of the biomechanical approach to the problem statistically. It appeared that the long term results were predictable at the one-year follow-up. If they were good clinically and radiologically at one year, they usually remained good or even improved later. If they were poor at one year, they became worse later. Therefore, in order to collect a statistically significant series, we attempted to review all our patients with at least a one-year follow-up. This review of patients was also carried out in other Orthopaedic Departments and the overall results have been published (Maquet et al., 1982). However, such an analysis of knees operated on and reviewed by different surgeons using different techniques and perhaps basing their approach on different considerations, may be criticized for its lack of unity. Therefore, we shall deal here with our own cases only, referring to the common analysis when necessary.

We treated 376 osteoarthritic knees surgically. Three hundred forty could be reviewed. The 36 others belonged to patients who died in the meantime or were not available for follow-up for different reasons.

Some of the knees were reoperated on. They are accounted for twice. The first time, they are recorded as poor results and the follow-up ends at the date of the revision. The end result appears in the second record, at the last follow-up.

The 340 knees which were followed up for at least one year comprised:

1. 229 cases of femoro-tibial osteoarthritis:
   - 155 valgus tibial osteotomies
   - 22 varus tibial osteotomies
   - 37 varus femoral osteotomies
   - 9 valgus femoral osteotomies
   - 6 corrections of a deformity at a distance from the affected knee
2. 111 cases of patello-femoral osteoarthritis or chondromalacia:
   - 103 anterior displacements of the tibial tuberosity
   - 8 releases of the lateral retinaculum

For each knee operation we answered 198 questions concerning the history of the condition, the pre- and postoperative clinical and radiological examination, the surgical procedure, the assessment of the patient, our personal evaluation, etc.. All these data were computerized and analysed.

We used essentially the same approach as for the common analysis mentioned above (Maquet et al., 1982). Pain was rated from 0 corresponding to absence of pain, to 4 corresponding to permanent pain during day and night (Table 18). Similarly, the range of movement was rated from 0 for full range of movement to 4 for ankylosis. Extension or lack of extension are not mentioned in the table. However, they will be considered in the evaluation of the overall results. The gait was judged according to walking distance, rated from 0 for unrestricted walking to 5 for no walking at all, the use of walking aids and the presence or absence of limping, each rated from 0 to 4.

In the X-rays we considered the width of the joint space and the subchondral scleroses beneath the tibial plateaux or in the patella. These signs depend on articular pressure. Subluxation of the femur on the tibia, position of

Table 18

Rating	Pain	Range of movement	Gait		
			Walking capacity	Walking aid	Limping
0	No pain	Flexion to 130° or beyond	No limitation	None	None
1	Pain when starting or with fatigue	Flexion to 110° or beyond	More than 1 km	One stick or crutch outdoors	Intermittent
2	Pain after $^1/_4$ hour walking or sitting or managing steps, disappearing at rest or after changing position	Flexion to 90° or beyond	More than 100 m	One stick or crutch permanently	Slight
3	Considerable pain	Flexion to 70° or beyond	Possible outdoors	Two sticks or crutches	Severe
4	Pain day and night	Flexion less than 70°	Only indoors	No walking	No walking
5			No walking		

the patella, osteophytes and other signs were also considered. The presence or absence of osteophytes did not seem to influence the results. This information, therefore, has been discarded.

We arbitrarily chose clinical and radiological minimum requirements to classify the knees into excellent, good, fair and poor results (Tables 19 and 39). In planning the aforementioned common analysis, we arrived at these minimum requirements by trial and error (Maquet et al., 1982). We compared the results according to these requirements with the subjective assessment of the patient himself and that of the surgeon who reviewed the case (Fig. 247). We have adapted the level of the minimum objective requirements in order to achieve relative agreement with the subjective assessments. In order to be designated excellent, the knee had to meet all the prerequisites listed in the first horizontal column in Tables 19 or 39. If one or more of these prerequisites were not met, the result was considered less than excellent and classified further down in scale. In order to be called good the knee had to meet all the minimum requirements in the second horizontal column, etc. Thus the results were judged according to absolute clinical and radiological requirements rather than by comparison with the pre-operative state of the knee. Therefore, some patients were classified as having fair or even poor results although they had been considerably improved by the operation.

From Figure 247 it appears that we were more exacting than the patients and our opinion was more severe than the objective criteria.

The number of cases thus analysed was not exactly the same for each parameter. This resulted from an occasional lack of information in individual records. In order to avoid discrepancies and to facilitate comparison, the results are given in percentages, despite the small size of the series.

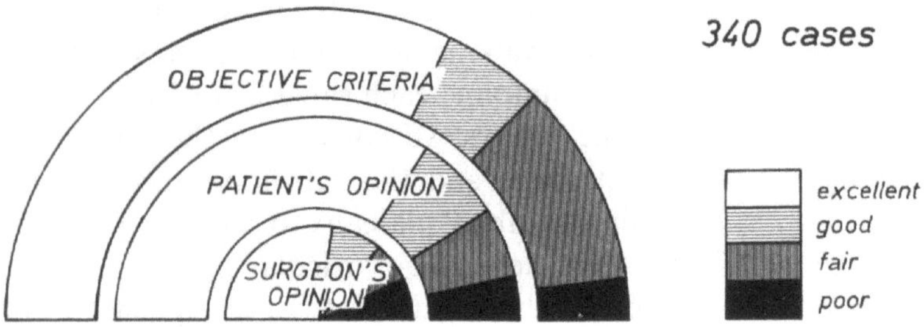

Fig. 247. Distribution of the results according to objective criteria, the patient's opinion and the surgeon's point of view

Table 19. Femoro-tibial osteoarthritis

Results	Clinical requirements			Radiological requirements	
	Pain	Range of movement	Walking capacity	Narrowing of joint space	Subchondral sclerosis
Excellent	0	Flexion ≧90° Extension 0° to 9°	>1 km	None to <50%	Normal
Good	0,1 or 2	Flexion ≧80° Extension −5° to 9°	>100 m	None to >50% but not complete	Normal or slightly increased
Fair	0, 1 or 2	Flexion ≧70° Extension −10° to 14°	Possible outdoors	None to >50% but not complete	Normal or slightly increased.
Poor	All the other cases				

We have analysed separately the results of treatment of femoro-tibial osteoarthritis and of patello-femoral osteoarthritis.

The complex programme which was used can entail up to 10% error. This results from the filling in of the questionnaires and from the manual perforating of the cards. Four cards each comprising 80 pieces of informations or parts of information were necessary for each knee thus analysed.

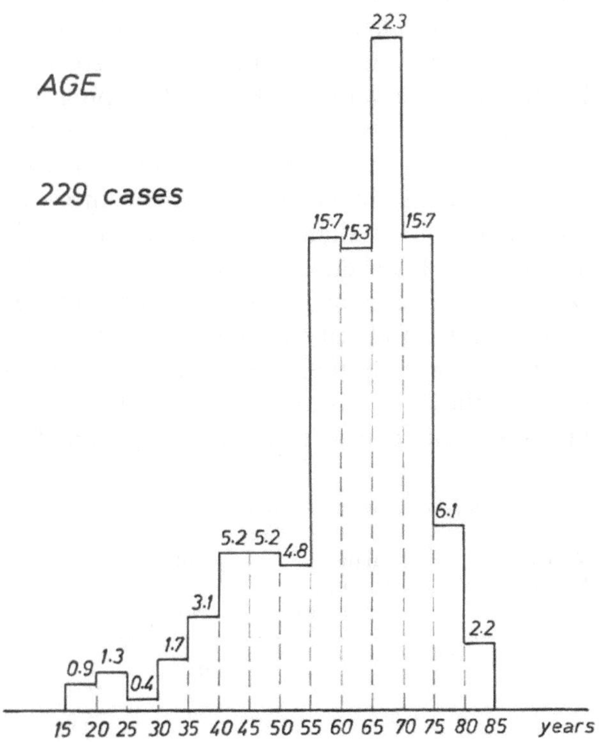

Fig. 248. Operations for femoro-tibial osteoarthritis

A. Femoro-Tibial Osteoarthritis

We consider that there is no physiological valgum of the knee (Fig. 120a). We speak of a varus knee when the centre of the knee lies laterally to the straight line drawn from the centre of the hip to the centre of the ankle (Fig. 120b), of a valgus knee when the centre of the knee lies medially to this line (Fig. 120c).

Most patients were female (76.9%). There were more left (55.5%) than right knees (44.5%). This difference is not significant statistically (n.s.). Age ranged from 17 to 81 years (Fig. 248) with a peak of frequency (37.6%) in the seventh decade. Follow-up (Fig. 249) extended to 17 years for one knee. It was three years or more in 57.1% of the knees.

Most of the patients were housewives (62.0%) but some of them were manual labourers (22.7%) (Table 20). After surgery, 89.5% of the patients returned to their previous occupations (Table 21).

Table 20. Occupations

	No.	%
Students	4	1.8
Retired	13	5.7
Housewives	142	62.0
Professionals	18	7.9
Manual labourers light	25	10.9
Manual labourers heavy	27	11.8

Table 21. Return to previous occupations

	No.	%
No	10	4.4
Yes	205	89.5
Changed because of knee	4	1.8
Changed for other reasons	1	0.4
Retired in meantime	9	3.9

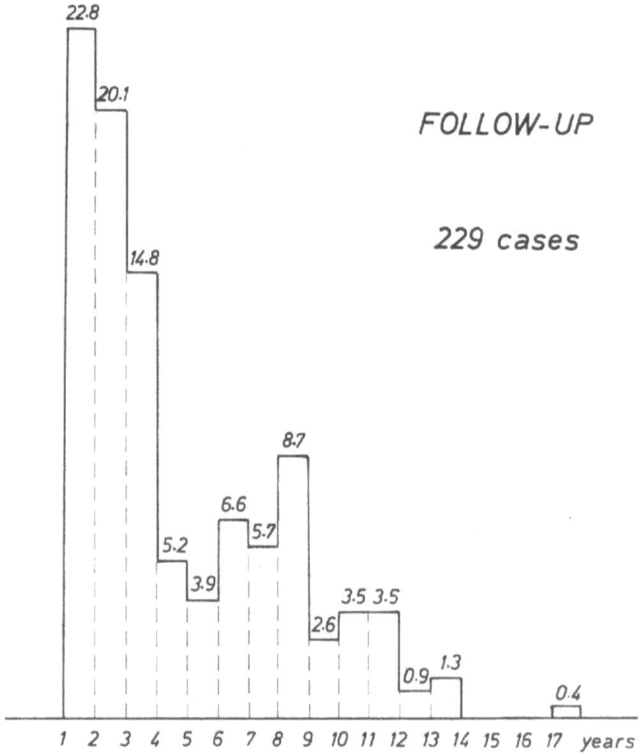

Fig. 249. Operations for femoro-tibial osteoarthritis

The patello-femoral compartment was involved in 79.4% of the knees with femoro-tibial osteoarthritis.

The surgical procedure consisted of a barrel-vault osteotomy fixed with compression clamps (page 165) in 155 (87.6%) of the 177 tibial osteotomies, of a supracondylar osteotomy with impaction of the fragments fixed with compression clamps (page 227) in 41 (89.1%) of the 46 femoral osteotomies. The other cases were operated on in the beginning of our series. We then used different techniques.

The correction of a varus or a valgus deformity by a tibial osteotomy nearly always was combined with an anterior displacement of the tibial tuberosity (page 170). Such an anterior displacement less often accompanied a femoral osteotomy (page 237) (Table 22).

An additional posterior capsulotomy was carried out whenever there existed a flexion contracture irreducible under anaesthesia (in 17.9% of the cases).

The use of a tourniquet was discarded after we had lost 6 patients from fatal pulmonary embolism. Thus 71.7% of the knees were operated on without tourniquet.

A plaster splint was used in 9.9% of our cases, in the beginning of our series, with the previous techniques.

Table 19 lists the minimum requirements according to which the results have been called excellent, good, fair or poor. Absence of pain certainly represents the most important achievement. For the range of movement, flexion and extension are taken into account. For gait, we have kept only the walking distance, disregarding the use of a walking stick and limping. On the one hand, because of their age, a fair number of the patients use a walking stick anyway, for reasons independent of their knee. On the other hand, a slight limp seems acceptable at this age. However, even a diminution of the walking distance may result from other causes than the knee and impair the evaluation.

X-rays provide the most objective parameters. All the A.-P. X-rays referred to have been taken with the patient standing and putting full

Table 22. Anterior displacement of the tibial tuberosity combined with a tibial or a femoral osteotomy

	Valgus Tibial Osteotomy	Varus Tibial Osteotomy	Varus Femoral Osteotomy	Valgus Femoral Osteotomy	Total	
					No.	%
Less than 6 mm	21	5	33	9	68	30.5
6 to 10 mm	72	6	1	0	79	35.4
11 to 15 mm	51	6	0	0	57	25.6
16 to 20 mm	11	1	0	0	12	5.4
More than 20 mm	0	4	3	0	7	3.1

weight on the studied leg. Therefore, the width of the joint space reveals the amount of tissue present between the femoral condyles and the tibial plateaux. The shape and magnitude of the subchondral sclerosis enable us to draw conclusions as to the magnitude and distribution of the stresses in the joint.

1. Osteoarthritis with a Varus Deformity

Most of the knees (155) underwent a valgus tibial osteotomy, some (9) a valgus femoral osteotomy.

a) Valgus Tibial Osteotomy

Early in our series we used previous classical procedures. Since 1968 all the patients underwent a barrel vault osteotomy as described above (page 165). The procedure combines an overcorrection of the varus deformity and an anterior displacement of the tibial tuberosity. In the presence of a flexion contracture, a posterior capsulotomy was carried out additionally.

Pain is completely relieved in 71.8% of the knees and almost completely in 14.8% (Fig. 250). The walking distance considerably increases; 60.8% of the patients walk without limping and 70.4% without walking stick (Fig. 251).

5.9% of the knees were competely unstable and 11.1% gave way frequently before operation. At the latest follow-up only 0.7% remain unstable and 4.6% give way frequently (Table 23).

Collateral laxity decreases significantly (Table 24).

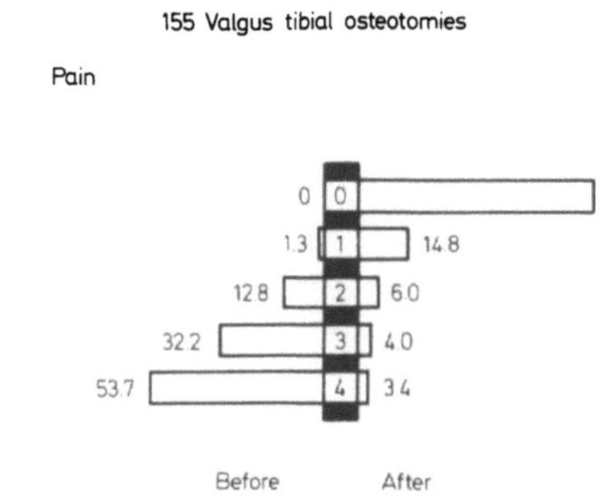

Fig. 250. See Table 18 for the meaning of the central figures

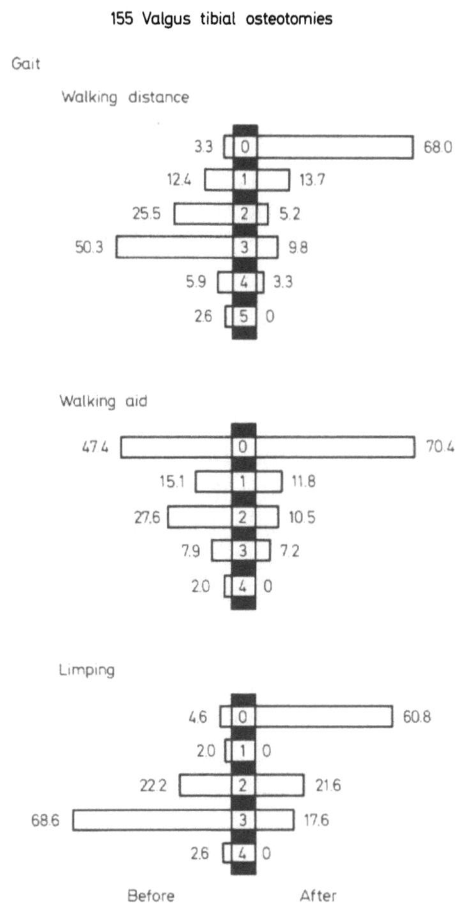

Fig. 251. See Table 18 for the meaning of the central figures

Table 23. Stability

Valgus tibial osteotomy	Normal	Giving way sometimes	Giving way frequently	Instability
Before	82.4	0.7	11.1	5.9
After	92.8	2.0	4.6	0.7

Table 24. Collateral laxity

Valgus tibial osteotomy	None	Slight	Significant
Before	80.9	8.6	10.5
After	90.8	5.3	3.9

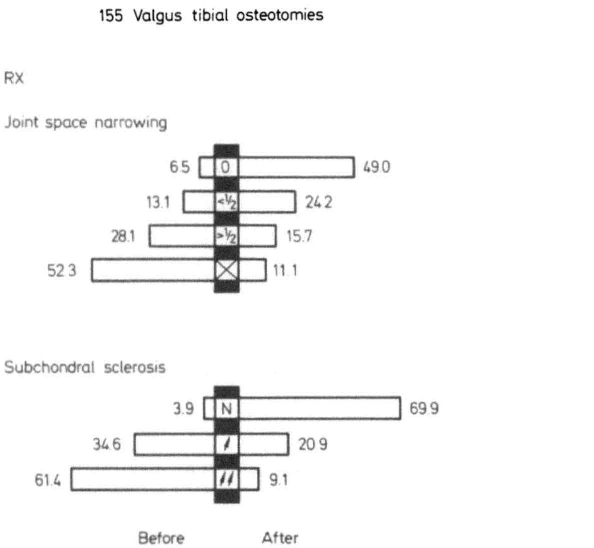

Fig. 252. 0 no narrowing; $<{}^{1}/_{2}$ narrowing less than half the normal space; $>{}^{1}/_{2}$ narrowing more than half the normal space; X no joint space visible; N normal subchondral sclerosis; ↗ subchondral sclerosis slightly increased; ↗ ↗ subchondral sclerosis significantly increased

In the X-ray, widening of the joint space and regression of the abnormal subchondral sclerosis (Fig. 252) are impressive.

47.3% of the knees presented some degree of femoro-tibial subluxation in the A.-P. view before operation. At the latest follow-up 89.2% look well aligned, only 10.8% remain with some residual subluxation (Table 25).

Table 25. Femoro-tibial subluxation

Valgus tibial osteotomy	None	Up to ${}^{1}/_{10}$	Up to ${}^{2}/_{10}$	Up to ${}^{3}/_{10}$	Up to ${}^{4}/_{10}$
		of tibial width			
Before	52.7	32.4	11.5	2.7	0.7
After	89.2	6.1	3.4	1.3	0

The patella is recentred by the operation in most knees in which it lay off-centre in the tangential view (Table 26). The patello-femoral joint space evolves concomitantly (Table 27).

Table 26. Position of the patella in the tangential view

Valgus tibial osteotomy	Centred	Centred but tilted laterally	Slight lateral subluxation	Significant lateral subluxation
Before	83.5	5.5	9.5	1.6
After	91.3	3.9	3.9	0.8

Table 27. Narrowing of the patello-femoral joint space

Valgus tibial osteotomy	None	Medial	Lateral	Overall
Before	78.2	0.7	15.2	6.0
After	91.4	0.7	5.7	2.0

According to our objective criteria (Table 19) there are 71.6% excellent and good results in this group of 155 valgus tibial osteotomies and only 3.2% poor results (Fig. 253). Only 53% of the varus knees were overcorrected at follow-up and 34.8% were barely corrected. 12.4% remained varus (Fig. 254). When the deformity has not been sufficiently overcorrected, osteoarthritis progresses and the varum tends to recur, resulting in obvious undercorrection.

Figure 254 shows that 25.8% of the knees in this group presented a pre-operative varus deformity of 20° or more. The severity of the initial deformity did not affect the end result.

There exists a correlation between the quality of the result and the amount of overcorrec-

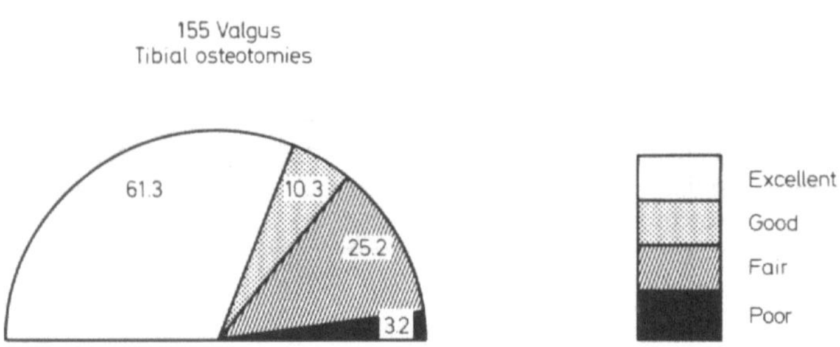

Fig. 253. Osteoarthritis with a varus deformity. Results according to objective criteria

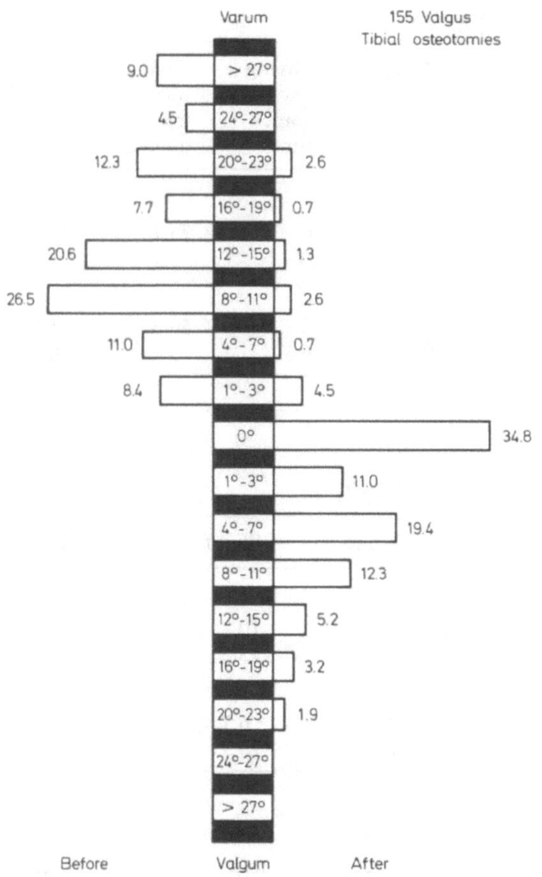

Fig. 254. Femoro-tibial angle before and after the valgus tibial osteotomies

tion of the varus deformity (Fig. 255). In order to assess this correlation we considered together the excellent and good results on one hand, the fair and poor on the other. The patients had to be subdivided into groups of sufficient numbers. Therefore, the range of femoro-tibial deformity in each group was relatively large. It appears that the knees with a postoperative valgum of 1° to 6° correspond to the highest percentage of satisfactory results. In the common analysis mentioned above (Maquet et al., 1982) we dealt with a greater number of knees (333 varus tibial osteotomies). Consequently, the range of femoro-tibial deformity in each group could be smaller. Then the best results were observed in the knees with a postoperative valgum of 4° to 9°.

At a first glance, severe laxity, complete disappearance of the joint space and significantly increased subchondral sclerosis seem unfavourably to influence the results. However, closer analysis shows that, in the presence of these factors, the results are as good as in their absence, as long as a sufficient overcorrection has been achieved (Maquet et al., 1982). The necessary overcorrection may be more difficult to assess when these factors are present.

b) Valgus Femoral Osteotomy

The nine varus knees in which a valgus femoral osteotomy was carried out either presented particular features (pages 209 and 212) or resulted from an exaggerated overcorrection of a valgus deformity (page 240). All of them did well and can be considered as excellent or good results.

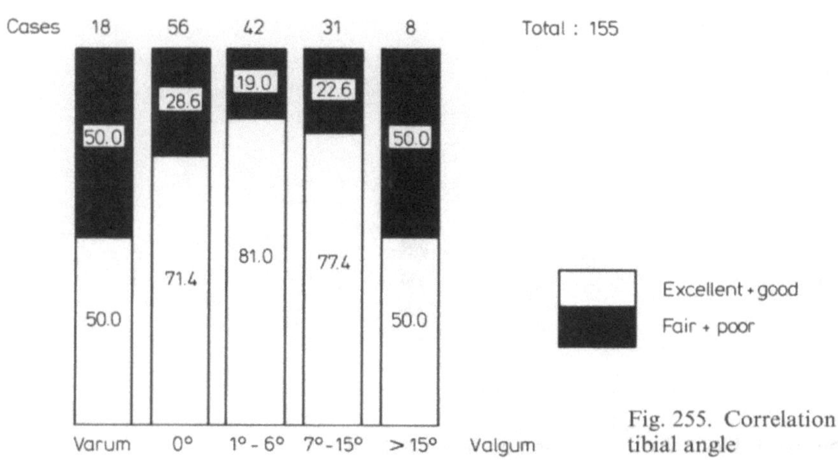

Fig. 255. Correlation between the result and the femoro-tibial angle

2. Osteoarthritis with a Valgus Deformity

Among the 22 varus tibial osteotomies, 3 were reoperations aiming at decreasing the valgum which resulted from an exaggerated overcorrection of a varus knee. These 3 revisions gave excellent results.

The other 19 varus tibial osteotomies were carried out early in our series, to deal with osteoarthritis in valgus knees. Unsatisfactory results in this group stimulated us to consider the problem again. From this study, a femoral osteotomy appeared as the method of choice to deal with most of these knees (page 218). Only the group of 19 varus tibial osteotomies carried out for lateral osteoarthritis will be compared with that of 37 varus femoral osteotomies.

Comparison of the results demonstrates a better outcome for the patients who underwent a femoral osteotomy than for those who were subjected to a tibial osteotomy. 86.1% of the patients were completely or nearly completely painfree at the latest follow-up after a varus femoral osteotomy as opposed to the 56.3% after a varus tibial osteotomy ($p<0.02$) (Fig. 256). The difference is less spectacular as far as gait is concerned (Fig. 257).

Giving way is no longer complained of in most cases of the two groups (Table 28). Collateral laxity also disappears in the majority of the knees of both groups (Table 29).

In the X-rays, widening of the joint space and regression of the abnormal subchondral sclerosis appear much more dramatic after a varus femoral osteotomy than after a varus tibial osteotomy (Fig. 258). This improvement in the X-rays suggests a more satisfactory future for the former group than for the latter.

The evolution of the femoro-tibial subluxation is strikingly different in the femoral and in the tibial osteotomy group (Table 30). The rate of subluxation decreases after a varus femo-

Table 28. Stability

	Normal	Giving way sometimes	Giving way frequently	Instability
Varus femoral osteotomy				
Before	54.1	0	35.1	10.8
After	94.6	0	2.7	2.7
Varus tibial osteotomy				
Before	18.8	0	56.3	25.0
After	93.8	6.3	0	0

Table 29. Collateral laxity

	None	Slight	Significant
Varus femoral osteotomy			
Before	54.1	10.8	35.1
After	78.4	13.5	8.1
Varus tibial osteotomy			
Before	26.7	6.7	66.7
After	73.3	26.7	0

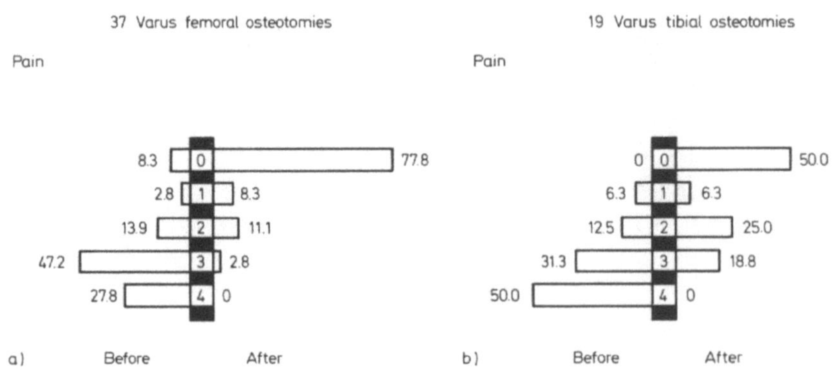

Fig. 256. See Table 18 for the meaning of the central figures

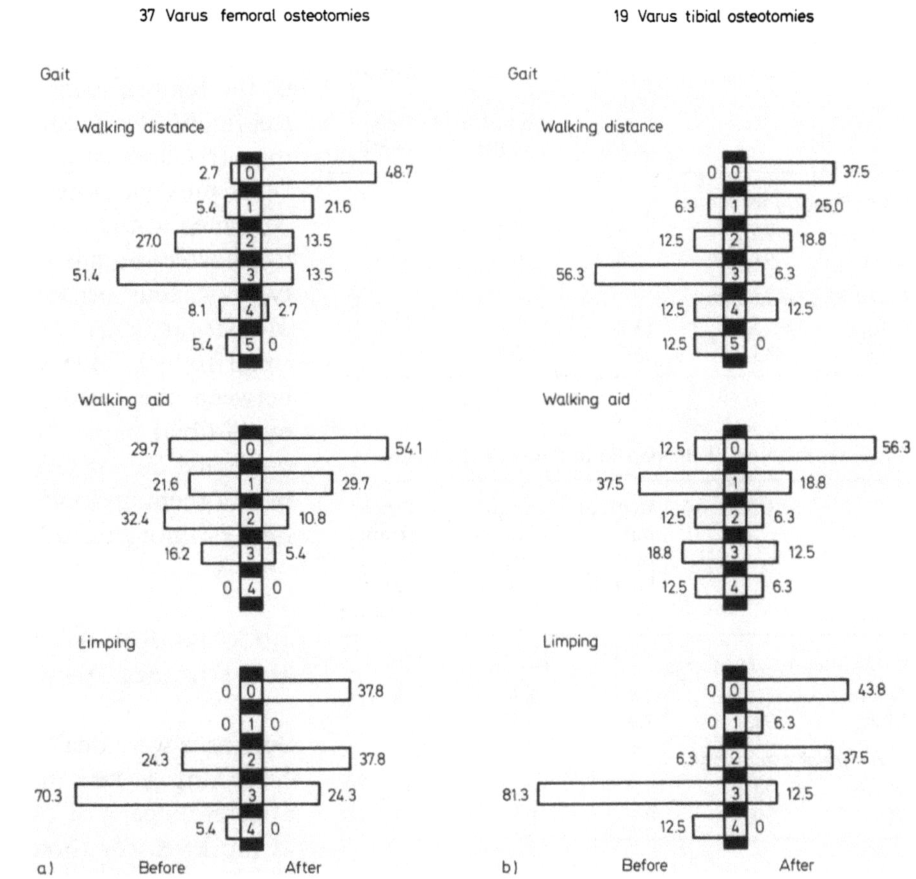

Fig. 257a and b. See Table 18 for the meaning of the central figures

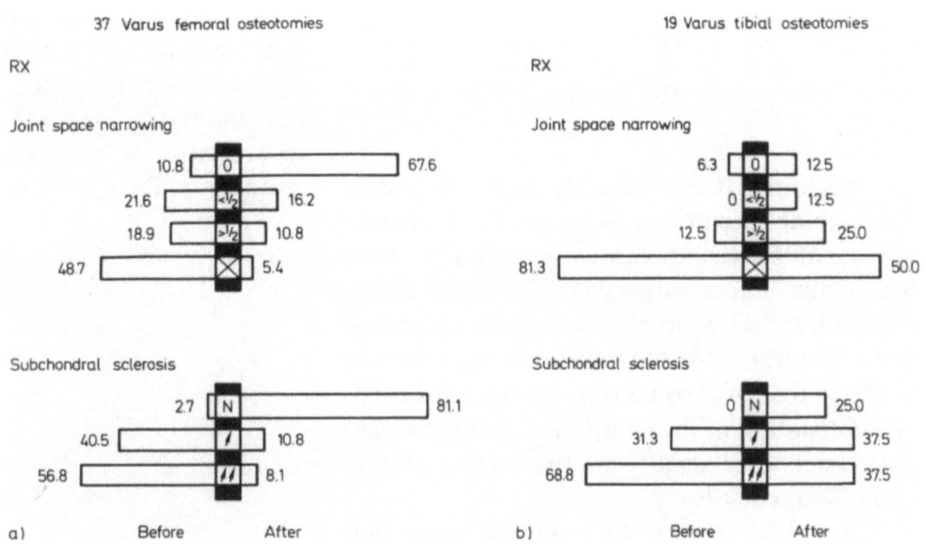

Fig. 258a and b. 0 no narrowing; $<{}^1/_2$ narrowing less than half the normal space; $>{}^1/_2$ narrowing more than half the normal space; X no joint space visible; N normal subchondral sclerosis; ↗ subchondral sclerosis slightly increased; ↗ ↗ subchondral sclerosis significantly increased

Table 30. Femoro-tibial subluxation

	None	Up to $^1/_{10}$ of tibial width	Up to $^2/_{10}$ of tibial width	Up to $^3/_{10}$ of tibial width
Varus femoral osteotomy				
Before	75.7	21.6	2.7	0
After	91.9	5.4	0	2.7
Varus tibial osteotomy				
Before	73.3	13.3	0	13.3
After	33.3	26.7	33.3	6.7

Table 31. Position of the patella in the tangential view

	Centred	Centred but tilted laterally	Slight lateral sub-luxation	Significant lateral sub-luxation
Varus femoral osteotomy				
Before	73.5	8.8	8.8	8.8
After	91.2	2.9	5.9	0
Varus tibial osteotomy				
Before	72.7	0	0	27.3
After	90.9	0	0	9.1

Table 32. Narrowing of the patello-femoral joint space

	None	Medial	Lateral	Overall
Varus femoral osteotomy				
Before	83.3	2.8	11.1	2.8
After	91.7	0	8.3	0
Varus tibial osteotomy				
Before	62.5	0	18.8	18.8
After	81.3	0	6.3	12.5

ral osteotomy. It considerably increases after a varus tibial osteotomy. Whereas the incidence of non subluxated knees was very similar in the two groups before surgery (75.7% and 73.3%), at the latest follow-up 91.9% are not subluxated in the femoral osteotomy group as against only 33.3% in the tibial osteotomy group. This residual subluxation of the femur on the tibia emphasizes the poorer quality of the results after a varus tibial osteotomy.

The patella tends to be recentred in the tangential view in both groups (Table 31). In both groups also the width of the patello-femoral joint space tends to improve (Table 32).

Comparison of the overall results in the two groups demonstrates the significant superiority of the femoral over the tibial osteotomy in the treatment of osteoarthritis with a valgus deformity ($p < 0.05$) (Fig. 259).

It must be noted that 22.1% remain undercorrected at the last follow-up (Fig. 260). When overcorrection has not been sufficient, deformity may recur because of the further evolution of osteoarthritis. This series, however, is too small to give clue to a significant correlation between the result and the postoperative femoro-tibial angle. In this group, 37.4% knees presented 20° or more initial valgus deformity, two of them up to 45°. The severity of the deformity did not preclude satisfactory end results.

3. Correction of a Deformity at a Distance from the Affected Knee

Six knees were dealt with by correcting or overcorrecting a deformity at a distance from the affected knee. For two knees the deformity lay at hip level, for three at midshaft of the femur and for one at mid lower leg. Correction of the hip deformity resulted in an improvement of both knees (page 250). Overcorrection of the femur gave two excellent results. In the third patient, the operation awakened a quiescent post-traumatic infection. This resulted in osteomyelitis. The patient was lost to follow-up. However, he is listed in the complications (Table 33). The overcorrection of a post-traumatic varus lower leg improved the knee of the patient but did not ultimately succeed in getting rid of osteoarthritis. A complementary barrel-vaul osteotomy of the tibia has been proposed.

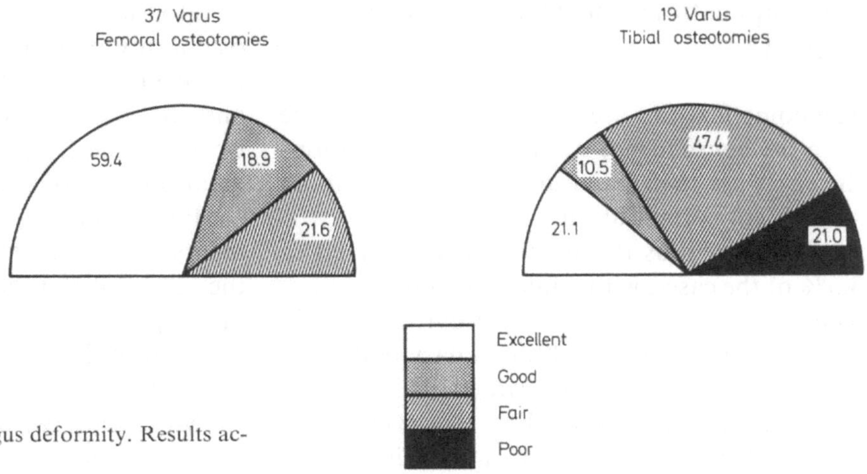

Fig. 259. Osteoarthritis with a valgus deformity. Results according to objective criteria

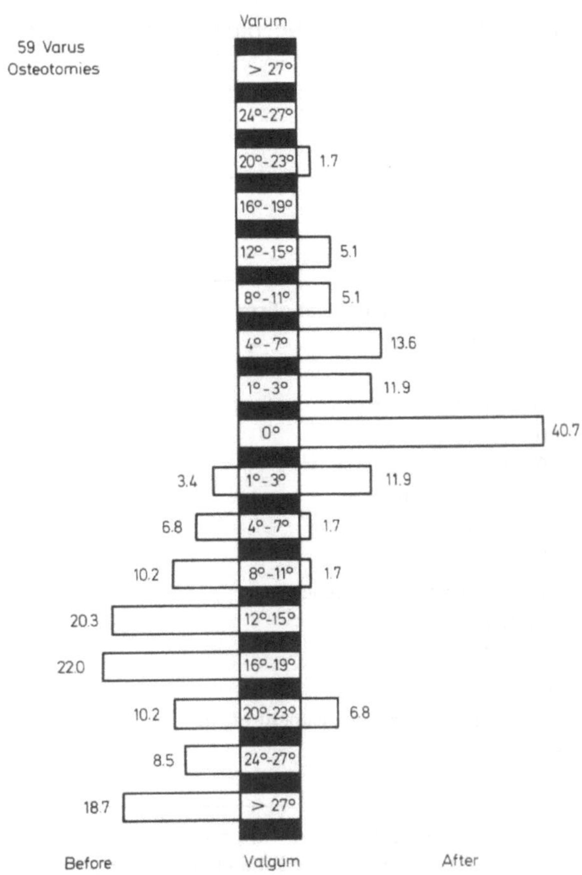

Fig. 260. Femoro-tibial angle before and after the varus tibial and femoral osteotomies

4. Complications and Unsatisfactory Results

The complications are listed in Table 33. Some were the consequence of another complication: the cases are then listed under the two or more subsequent complications.

Severe venous thrombosis was observed in 3.5% of the cases, with 6 fatal pulmonary embolism (2.6%). These deaths all occurred early in the series. No more fatal pulmonary embolism has occurred since we discarded the use of a tourniquet and a plaster splint. However, a correlation between the use of a tourniquet and the incidence of pulmonary embolism remains controversial.

Paresis of the extensors of the foot was recorded in 3.0% of the cases but persisted in 1.7% only. This complication occurred only in the group of valgus tibial osteotomies.

Paraesthesia of the dorsum of the foot was more frequent (9.4%) but persisted in 4.4% of the patients only.

Skin necrosis occurred in 3 cases and led to deep infection.

Deep infection was recorded in 3.9% of the cases. This includes one patient lost to follow-up.

Non-union occurred in 3 instances of tibial osteotomy early in our series when the Steinmann pins were retained for only three weeks and in 1 instance of femoral osteotomy when we fixed the fragments with two plates and a great number of screws.

Pin breakage or displacement of the fragments despite fixation could be traced either to the re-use of thin pins or to a too low insertion of the proximal pin after a tibial osteotomy (10 cases) or to the fixation by two plates and screws after a femoral osteotomy (4 knees).

Most (40.7%) of the unsatisfactory results seemed to be due to insufficient correction of the deformity (Table 34). Most of those due to a complication occurred in the beginning of our series.

Unsatisfactory results led to reoperation in 24 knees (Table 35). The revision consisted in a new osteotomy in 17 of these knees.

Table 33. Complications

	Tibial osteotomies (182 cases[a])	Femoral osteotomies (48 cases[a])	Total (230 cases[a])	
			No.	%
Venous thrombosis and embolism				
severe, non fatal	0	2	2	0.9
fatal	5	1	6	2.6
Paresis of the extensors of the foot				
transitory	3	0	3	1.3
persistent	4	0	4	1.7
Paraesthesia of the foot				
transitory	11	0	11	4.8
persistent	9	1	10	4.8
Skin necrosis and dehiscence	2	1	3	1.3
Superficial sepsis	2	0	2	0.9
Deep sepsis	5	4	9	3.9
Non union	3	1	4	1.7
Material failure or displacement of the fragments	10	4	14	6.1
Others	3	1	4	1.7

[a] The postoperative fatalities are recorded in the complications as well as an instance of deep sepsis which was lost to follow-up. This brings the total amount from 223 cases (with at least one year follow-up) to 230

Table 34. Presumed causes of the unsatisfactory results

	No.	%
Errors in the indications	15	6.7
Insufficient correction	35	15.7
Exaggerated overcorrection	8	3.6
Technical failure	8	3.6
Complication	18	8.1
Biological disturbance	1	0.4
Influence of other joints	1	0.4

Table 35. Reoperation of unsatisfactory results

Revision	No.	%
Osteotomy	17	7.6
Posterior capsulotomy	1	0.4
Anterior displacement of the tibial tuberosity	2	0.9
Others	4	1.8

5. Conclusions

Severe deformity (some of 45°!), significant laxity, complete disappearance of the joint space and increased subchondral sclerosis do not influence the results unfavourably as long as the deformity has been sufficiently overcorrected. In our series old age did not jeopardize the end result.

Tibial osteotomy constitutes the method of choice in order to deal with osteoarthritis with a varus deformity. Femoral osteotomy must be used in osteoarthritis with a valgus deformity rather than tibial osteotomy.

Analysis of the results emphasizes the necessity of overcorrecting the varus or the valgus deformity. Most of the fair or poor results are due to an undercorrection. In our series the highest percentage of satisfactory results was observed after an overcorrection of the varus deformity by 1° to 6°. In a larger series, the ideal overcorrection ranged between 4° and 9°. The overcorrection of a valgus deformity should be less, probably 1° to 2°. However, our series of valgus knees was too small to allow for an accurate assessment of the desirable overcorrection.

B. Patello-Femoral Osteoarthritis

One hundred and eleven knees underwent surgical treatment for so-called chondromalacia or for osteoarthritis of the patello-femoral joint. The anterior displacement of the tibial tuberosity resulting from our technique of upper tibial osteotomy or combined with a femoral osteotomy is not considered in this section.

Eight patients were subjected to a lateral retinaculum release and 103 to an anterior displacement of the tibial tuberosity.

Age ranged from 16 to 78 years (Fig. 261) with a maximum of patients between 50 and 60 years (22.5%). The follow-up records extended from 1 to 12 years (Fig. 262).

1. Division of the Lateral Retinaculum

We divided the lateral retinaculum and resected part of it as described by Ficat (1973) in 8 carefully selected knees: young patients with painful patellae with a tendency to subluxation, without signs of osteoarthritis in the X-rays.

In four cases, the result looks excellent or good at the follow-up of respectively 1 year, 4 years, 6 years and 6 years. In the other four knees, osteoarthritis developed with clinical aggravation and obvious radiological signs.

This means 50% satisfactory results only, despite the careful selection of the patients. However, this series is much too small to allow for any conclusion to be drawn.

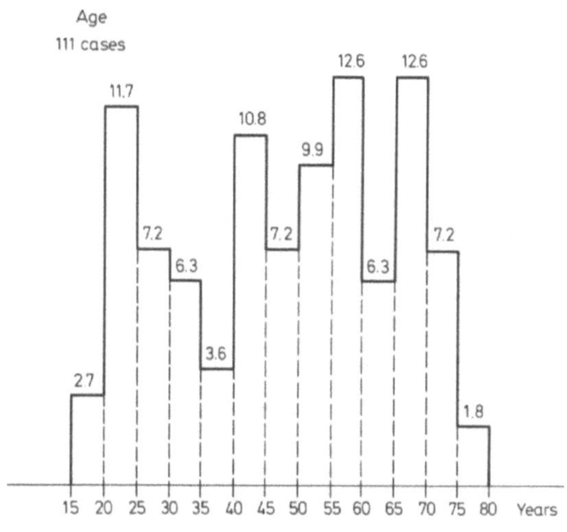

Fig. 261. Operations for patello-femoral osteoarthritis

2. Anterior Displacement of the Tibial Tuberosity

Under this title we have grouped the anterior as well as the anterior and medial displacements of the tibial tuberosity. The procedures were carried out for patello-femoral osteoarthritis (94 knees) or for so-called chondromalacia patellae (9 knees) which can be considered as the first stage of osteoarthritis. Anterior displacement of the tibial tuberosity was performed when the knee-cap was centred in the intercondylar groove, anterior and medial displacement of the tibial tuberosity when the knee-cap was subluxated laterally. In 85 knees the displacement was purely anterior, in 18 it was anterior and medial.

There were 51.5% right knees and 48.5 left knees.

78.6% of the patients were female. Here also more than half of the patients were housewives but a quarter were manual labourers (Table 36). Most of them returned to their previous occupations (Table 37).

The amount of anterior displacement of the tibial tuberosity was greater than 20 mm in 79.5% of the knees (Table 38). Smaller displacements either were carried out in the beginning of our series or resulted from complications.

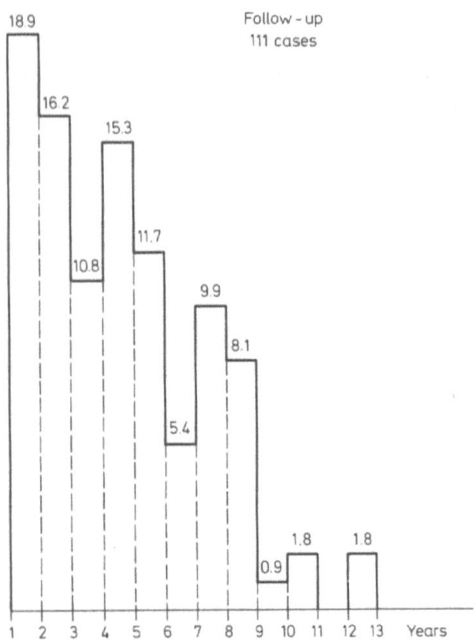

Fig. 262. Operations for patello-femoral osteoarthritis

Table 36. Occupations

	No.	%
Retired	9	8.7
Housewives	57	55.3
Professionals	11	10.7
Manual labourers light	13	12.6
Manual labourers heavy	13	12.6

Table 37. Return to previous occupations

	No.	%
No	2	92.2
Yes	95	1.9
Changed because of knee	1	1.0
Changed for other reasons	2	1.9
Retired in meantime	3	2.9

Table 38. Anterior displacement of the tibial tuberosity

	No.	%
Less than 11 mm	11	10.7
11 to 15 mm	5	4.9
16 to 20 mm	5	4.9
21 to 25 mm	47	45.6
26 to 30 mm	26	25.2
More than 30 mm	9	8.7

Table 39 lists the minimum requirements for the results to be considered as excellent, good, fair or poor. They differ from those related to femoro-tibial osteoarthritis only in what concerns the X-ray picture. Any narrowing of the joint space makes the result less than excellent.

The results were rated as excellent or good in 88.4% of the knees by the patient, in 78.7% of the knees by the surgeon.

Anterior displacement of the tibial tuberosity relieved pain completely in 86% of the knees and nearly completely in 9% (Fig. 263). It nearly always lengthened the walking distance. Most patients who used a walking stick before operation can dispense with it. The number of patients who limped diminished or their limp has become less pronounced (Fig. 264).

Table 39. Patello-femoral osteoarthritis

Results	Clinical requirements			Radiological requirements	
	Pain	Range of movement	Walking capacity	Narrowing of joint space	Subchondral sclerosis
Excellent	0	Flexion ≧90° Extension 0° to 9°	>1 km	None	Normal
Good	0, 1, 2	Flexion ≧80° Extension −5° to 9°	>100 m	None	Normal or slightly increased
Fair	0, 1, 2	Flexion ≧70° Extension −10° to 14°	Possible outdoors	None	Normal or slightly increased
Poor	All the other cases				

The Zohlen sign[16] became negative in most patients. It was recorded both before and after surgery in 48 charts. It was positive in all these patients before the anterior displacement of the tibial tuberosity. It became negative in 34 (70.8%).

The ability to kneel was recorded in 47 charts. More patients (63.0%) can kneel after the operation than before (52.2%). However, 17.4% patients who could kneel before can no longer kneel after the operation whereas 28.3% who could not kneel before can do so after surgery.

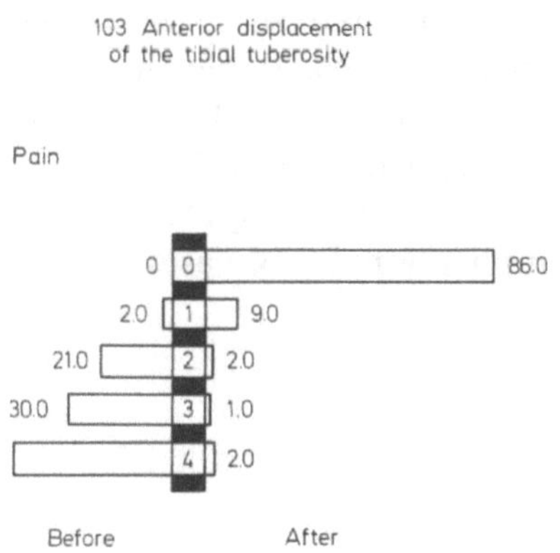

Fig. 263. See Table 18 for the meaning of the central figures

[16] The patient lying supine is asked to relax. The examiner pushes the patella distalward and holds it there with his two thumbs while the patient contracts the quadriceps. Pulling the patella over the proximal ridge of the articular surface of the condyles is very painful if some damage in the articular cartilage exists. This is the Zohlen sign

In the X-rays the patella which lay off-centre in 76.7% of the knees before lies in the centre of the intercondylar groove in 76.7% of the

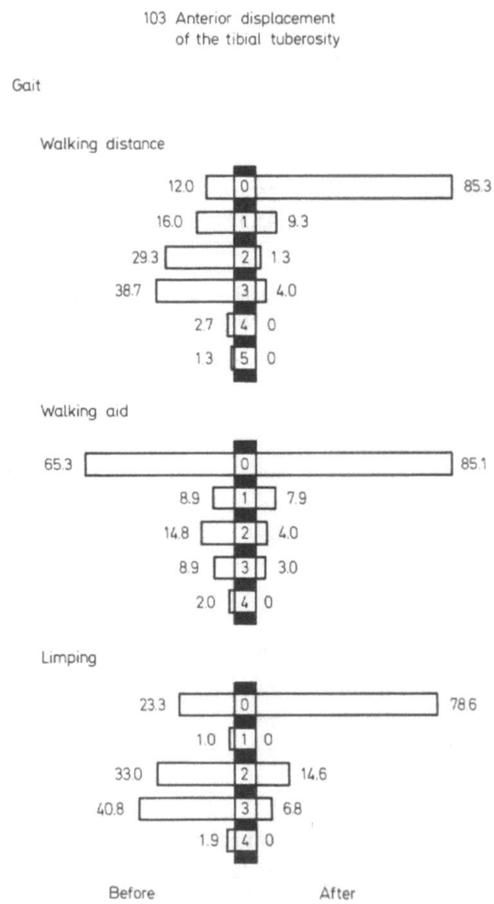

Fig. 264. See Table 18 for the meaning of the central figures

Table 40. Position of the patella in the tangential view

	Centred	Centred but tilted laterally	Slight lateral subluxation	Significant lateral subluxation	Lateral dislocation	Slight medial subluxation
Before	23.3	15.1	23.3	34.2	1.4	2.7
After	76.7	2.7	4.1	16.4	0	0

knees after the operation (Table 40). The proportion thus is reverted.

The patello-femoral joint space looks normal in 81.4% of the knees after the operation whereas it was normal only in 53.9% before (Fig. 265). The joint space actually has become normal in 66.0% of the knees in which it was narrowed before operation.

The subchondral sclerosis in the patella has decreased and returned to normal in most instances in which it was slightly or considerably increased before (Fig. 265).

Condensation of the anterior aspect of the patella was observed in 73.6% of the knees before the operation. It remains so in 54.7% after (Table 41).

Table 41. Condensation of the anterior aspect of the patella

	Absent	Present
Before	26.4%	73.6%
After	45.3%	54.7%

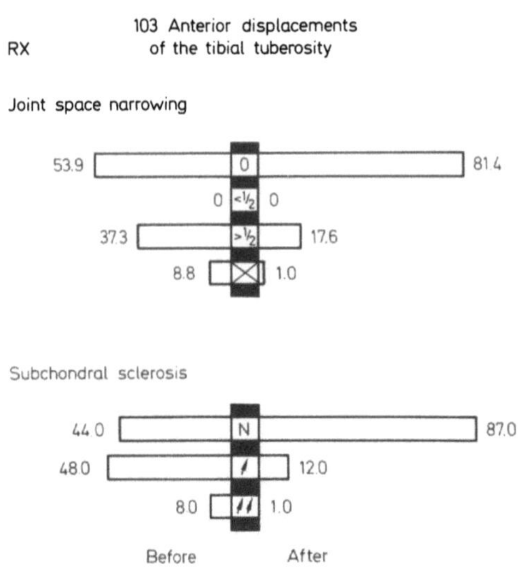

Fig. 265. 0 no narrowing; $<^1/_2$ narrowing less than half the normal space; $>^1/_2$ narrowing more than half the normal space; X no joint space visible; N normal subchondral sclerosis; ↗ subchondral sclerosis slightly increased; ↗ ↗ subchondral sclerosis significantly increased

The overall results according to our objective criteria (Table 39) were distributed into 86.4% excellent and good, 8.7% fair and 4.9% poor (Fig. 266).

The quality of the result depends on the magnitude of the anterior displacement as appears in Figure 267. In order to draw this figure we had to divide the knees in groups each of sufficient magnitude. It would be interesting to carry out the same analysis on a much larger series which would allow a distribution of the cases into groups of sufficient magnitude for anterior displacements incremented by 5 mm.

3. Complications of the Anterior Displacement of the Tibial Tuberosity (Table 42)

There was no death in this series.

Venous thrombosis occurred in 3.8% of the cases. It remained localized in 1.9% and was complicated by non-fatal pulmonary embolism in 1.9%.

We did not observe any paresis of the extensors of the foot in this series. Paraesthesia of the dorsum of the foot was recorded in 3 instances, transitory in 1, persistent in 2.

Skin necrosis and dehiscence appeared as the most frequent complication, in 10 cases. It can

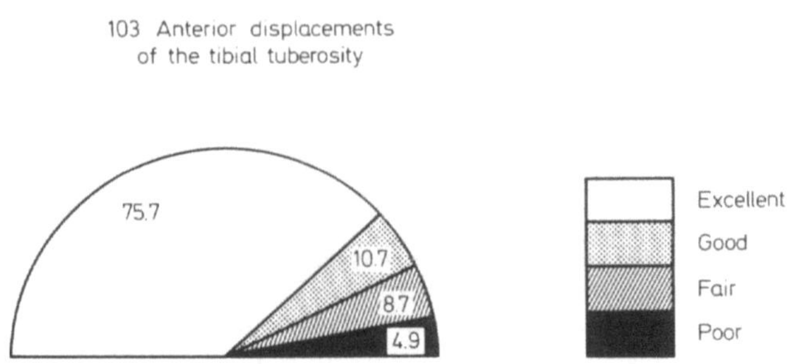

Fig. 266. Results according to objective criteria

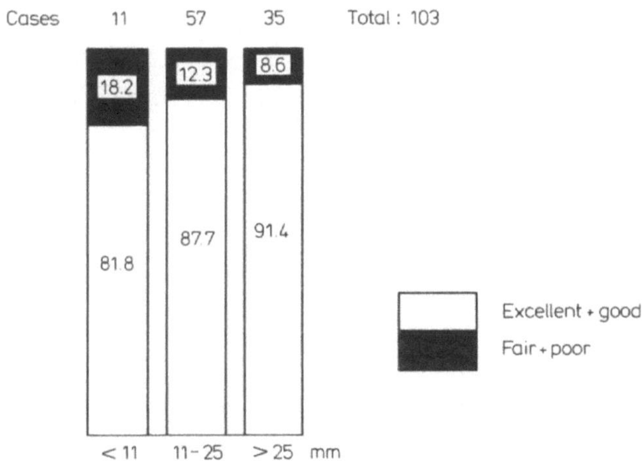

Fig. 267. Correlation between the results and the magnitude of anterior displacement of the tibial tuberosity

be avoided by carrying out relieving incisions when the skin is too thight over the tuberosity (page 148).

This complication led to a deep infection in 4 instances. In 2 of them the graft had to be removed. The others healed as did the 2 instances of superficial sepsis.

The grafts moved in 3 instances and had to be repositioned and fixed by one or two Kirschner wires.

Table 42. Anterior displacement of the tibial tuberosity (103 cases) complications

	No.	%
Venous thrombosis and embolism		
benign	2	1.9
severe	2	1.9
Paraesthesia of the foot		
transitory	1	1.0
persistent	2	1.9
Skin necrosis and dehiscence	10	9.7
Superficial sepsis	2	1.9
Deep sepsis	4	3.9
Displacement of the graft	3	2.9
Others	2	1.9

4. Conclusions

This analysis confirms the high percentage of satisfactory results from anterior displacement of the tibial tuberosity. The more the tuberosity is displaced anteriorly the better the results. This makes mechanical sense.

The satisfactory results counterbalance the disadvantages of considerable anterior displacement at least in severe osteoarthritis. The disadvantages are cosmetic and cutaneous. Necrosis of the skin over the tuberosity represents the most frequent complication. This complication can be avoided by relieving incision(s) if the smallest doubt about the viability of the skin exists.

Chapter IX. Conclusions

When standing on both feet, the load exerted on the knees is the weight of the supported part of the body, i.e. the weight of the body minus the weight of the lower legs and feet. But the knee is under much more stress during gait. In these conditions it eccentrically bears a heavier mass: the body minus only the supporting lower leg and foot. Furthermore, the displacements of the body segments through space provoke forces of inertia which are added to the body weight. If the weight of the body segments, their centre of gravity and their displacements are known it is possible to determine the successive positions of the centre of gravity of the part of the body supported by the knee and hence to calculate the forces of inertia due to accelerations of this part.

From these data, analytical geometry and trigonometry allow the magnitude and the line of action of the force exerted on the knee by the partial body mass to be deduced. This force acts eccentrically on the knee. It must be counterbalanced by muscular or ligamentous forces which can be calculated if the problem is formularized.

The muscular and ligamentous forces are vectorially added to the force exerted by the partial mass of the body. Their sum constitutes the load supported by the knee. During gait each knee alternately supports a load necessarily greater than in the standing position.

Such an approach, and the mathematical analysis it originates, can be applied to any walking individual. As an example, we have used them for the subject studied by Braune and Fischer in "Der Gang des Menschen." Therefore, the figures mentioned in Chapter IV are true for this subject and for his gait. But they give an order of magnitude of the forces which act on the knee in any normal individual. During gait (5.6 km/h) each knee alternately supports a load attaining 5–6 times body weight.

The forces acting on the knee joint in other individuals and in other conditions can be calculated in the same way by applying the formulae we have worked out, if we know the successive positions of the joints and of the centres of gravity of the different parts of the body projected on a system of three rectangular planes. These data can be obtained by cinematography or, better, by time lapse photography with a strobe light. Dots on the skin of the subject make it possible to analyse the displacements of the body segments.

The load acting on the knee, for reasons of equilibrium, must intersect the axis of the joint. It is transmitted from the femur to the tibia through a part of the articular surfaces. In order to know these weight-bearing surfaces we have relied on direct measurements of autopsy material. We have subjected knees of cadavers with their ligaments and capsule intact to a compression equivalent in order of magnitude to the force physiologically acting on the joint during gait. Indeed ligaments and capsule guide the movement as in the living joint. On the other hand, the load ensures the same contact between the articular surfaces as in the knee supporting the walking individual. The radio-opaque substance expelled by compression circumscribes the surfaces which transmit the load. We could establish that these weight-bearing surfaces are not limited to the contact between the femur and the tibia but extend onto the menisci. The involvement of the menisci in transmitting the compressive force had been deduced from their histological structure but it is now clearly confirmed by the pictures which we have obtained. The surface of the weight-bearing areas attains the order of 20 cm^2 in extension. It decreases to between 11 and 12 cm^2 at 90° of flexion.

Knowing the load acting on the knee and the weight-bearing surfaces, we can calculate the mean compressive stresses exerted on the joint.

They reach approximately 20 kg/cm^2 during the single support period of gait if they are evenly distributed over the weight-bearing areas.

To understand the distribution of the stresses we relied on a law formulated by Pauwels from the study of other articular surfaces and bone structures. This law states that at every place in the skeleton, within physiological limits, the quantity of bone tissue depends upon the stress exerted. Within these limits, increased stresses produce the apposition of bone, diminution of stresses the resorption of bone. To check this law at knee level requires proper X-rays making possible a comparison of the pictures either in several individuals or at different times in the same individual. It is observed that the bone densities underlying the tibial plateaux have a symmetrical outline if the forces acting on the knee have a normal magnitude and the articular surface presents no incongruence. We concluded that the stresses are evenly distributed in the femoro-tibial joint of a normal individual. Such a distribution is only possible when the compressive force is exerted in the centre of gravity of the weight-bearing areas. Consequently, force R, the resultant femoro-tibial compressive force, intersects the axis of flexion about its centre. This central position of load R results from an integration of its several successive positions. In fact, during flexion, force R can move medially or laterally at the axis of flexion between the centres of curvature O_1 and O_2 of the femoral condyles, its magnitude changing accordingly. R also pivots forward and backward about the axis of flexion and in so doing covers some surface of the femoral condyles and of the tibial plateaux in contact. This sweeping movement results from the fact that, like every joint between long bones, the knee is flexed by two groups of muscles inserted near the articulation. One consists of the hamstrings inserted near the proximal end of the tibia and fibula. The other is formed by the lower leg muscles inserted at the distal end of the femur. As shown by Pauwels (1963), such an anatomical disposition has as a rule the following consequences:

a) During flexion the force R moves in the central part of the joint surfaces.

b) The compressive stresses are evenly distributed over the weight-bearing areas.

c) There is no localized or permanent increase of the stresses.

In standing with symmetrical support on both feet, the load acting on each knee is 43% of the body weight. But during gait it reaches 5–6 times body weight. It is thus much greater than the compressive force supported by the hip in the same individual and in the same conditions. This force transmitted across the hip has been calculated by Pauwels and reaches about four times the body weight. The difference is easily explained if one considers the important distance between the knee and the line of action of the force exerted by the partial body mass S_7 at the beginning and at the end of the single support period of gait. In addition, the knee has to carry a heavier part of the body than the hip. For these two reasons, the moment of the force exerted eccentrically on the joint by mass S_7 is greater than the moment of the force exerted on the hip by a smaller mass S_5 (the body minus the whole loaded leg). But during gait and particularly at the end of the single support period the knee works in positions close to extension. In these positions it offers a larger weight-bearing area than the hip. Although supporting a greater compressive force it is subjected, thanks to these weight-bearing surfaces, to an average joint pressure of about 20 kg/cm^2. This is equivalent to the stresses exerted in the hip joint such as were predicted by Pauwels and calculated by Kummer. The order of magnitude of the stresses which we have calculated at knee level is in complete agreement with their data. This is true for the femoro-tibial joint as well as for the patello-femoral joint.

The load R acting on the knee is the resultant of forces due to the partial mass S_7 of the body and of muscular and ligamentous forces which counterbalance the former. The magnitude of force R and also its line of action can be modified if one or several of its components are disturbed.

Mathematical analysis makes it possible to determine the causes of an increase in and a permanent displacement of load R medially or laterally. The line of action of the compressive femoro-tibial force R can be shifted medially by a loosening of the lateral muscular tension band, by an increased body weight, by a varus deformity of the leg, or by a displacement of the centre of gravity of the body toward the opposite side. Inversely, R can be shifted laterally by an increased power of the lateral muscles made necessary for hip balance, by a valgus deformity of the leg or by a displacement of the centre of gravity of the body toward the side of the loaded knee. The compressive femoro-tibial and patello-femoral forces can be increased by a flexion contracture of the joint. A medial, lateral, posterior or anterior displacement of the femoro-tibial load, as well as an increase or a lateral displacement of the force pressing the patella against the femur, produces a localized and permanent augmentation of articular compressive stresses and causes the development and the aggravation of lesions generally known as degenerative osteoarthritis of the knee.

Progressively, the knee is deformed in varum or in valgum and can no longer extend completely. In many cases of osteoarthritis there is laxity of the ligaments with lateral instability aggravating the deformity. This deformity is now easy to understand as far as its origin is concerned.

Changes in the distribution of the joint stresses due to a displacement of the load acting on the knee can be demonstrated in photoelastic models. Localized increase of stress due to a displacement of the compressive force medially or laterally can be read directly in the X-rays. According to Pauwels' law concerning functional adaptation, the dense bone beneath the tibial plateaux has the same outline as the diagram of stresses. A more pronounced cup or a dense triangle underlying the medial plateau indicates the presence of abnormally high compressive stresses which can only be the consequence of a displacement of force R toward this plateau. They appear in cases of osteoarthritis with a varus deformity. An increased subchondral sclerosis beneath the lateral plateau results from a lateral displacement of force R and characterizes osteoarthritis of the knee with valgus deformity. A triangular subchondral sclerosis seen beneath the posterior aspect of the tibial plateaux is the consequence of a posterior displacement of force R (flexion deformity). Persisting anterior displacement of force R provokes a dense triangle beneath the anterior aspect of the plateaux (recurvatum). Increased sclerosis in the patella indicates an increase of patello-femoral compressive stresses.

All the described possibilities of displacement of load R find their confirmation in the X-ray patterns of subchondral bone density. Consequently, *osteoarthritis of the knee is characterized by a localized increase of the articular compressive stresses,* followed by destruction of the cartilage, narrowing of the joint space and remodelling of the bony tissue.

The treatment of mechanically induced lesions should logically be mechanical. In osteoarthritis of the knee, the joint pressure must be reduced. There are only two ways to achieve this. One consists of diminishing the load acting on the knee and the other of distributing this load evenly over the largest possible weight-bearing surfaces.

Spontaneous attempts empirically to reduce the joint pressure are regularly observed. Indeed, the patient with a painful knee limps and often uses a walking stick. Both decrease the joint pressure by diminishing the load exerted on the affected knee.

A rational treatment must attempt to achieve a better and longer lasting result than these more or less efficient empirical mechanisms. In order to reduce the joint pressure as much as possible, one must not only diminish the load supported by the knee but also distribute it evenly over the largest weight-bearing surfaces. In this way, correcting flexion contractures reduces the force transmitted from the femur to the tibia and from the patella to the femur. It also increases the femoro-tibial weight-bearing surfaces. Displacing the tibial tuberosity anteriorly significantly decreases the compressive forces on both femoro-tibial and patello-femoral joints partly by lengthening the lever arm of the patella tendon but mainly by opening the angle formed by the lines of action of the forces exerted by the quadriceps and the patella tendon. Proper recentring of the load makes a regular distribution of the compressive force possible over the largest weight-bearing surface. This is achieved in the patello-femoral joint by displacing the tibial tuberosity medially as well as anteriorly. In the femoro-tibial joint the load can be recentred by a posterior capsulotomy, by a proximal tibial osteotomy, by a distal femoral osteotomy or, in some instances, by the correction of a deformity at a distance from the knee.

When surgery changes the geometry of the leg sufficiently to diminish and recentre the compressive forces acting on the knee, the clinical and radiological signs of osteoarthritis regress. Pain is relieved. The range of movement improves and is often restored to normal. The knee becomes stable and the ligaments tighten spontaneously when the mechanical conditions are again made normal.

But the spectacular postoperative radiological evolution often best shows the changes in the distribution of the joint stresses. The excess of sclerosis underlying the previously overloaded tibial plateau disappears and is replaced by a cup-shaped normal density. The sclerotic cup beneath the opposite plateau becomes more pronounced and takes on a normal aspect. The underlying cancellous trabeculae are more clearly visible. A joint space often reappears in the standing X-ray of the loaded knee. It shows the regeneration of articular tissue, in most cases fibro-cartilage, functioning as hyaline cartilage.

These signs indicate an even distribution of the compressive stresses over the largest possible weight-bearing surfaces of both tibial plateaux. Such change in the distribution of the stresses can only be due to a recentring of the load favourably modifying the mechanical conditions in the knee.

The clinical and radiological results thus achieved persist for several years. This indicates the permanence of the new equilibrium created by surgery as between mechanical stresses and tissue resistance.

Nevertheless, the regression of the symptoms and signs of osteoarthritis could be attributed to a biological effect provoked by the osteotomy and affecting blood supply, nerves, local metabolism, etc.. Such a biological effect certainly exists but it is not essential, and indeed not every para-articular osteotomy causes a regression of the osteoarthritic changes. These changes persist and become even worse every time an osteotomy has not improved the mechanical conditions. For instance, when a varus osteotomy for osteoarthritis with valgus deformity does not succeed in distributing the load over both tibial plateaux the resultant force remains in the lateral part of the joint. Moreover, it is increased because of the lengthening of the lever arm of the body weight. That is why many patients having had a correction of their valgus deformity by a tibial osteotomy experience clinical and radiological deterioration. In cases of osteoarthritis with varus deformity, when the surgical overcorrection has not been sufficient or has been secondarily lost, failure is the rule.

In all these cases of lost or insufficient correction of varus or valgus deformity, the operation has certainly produced the non-specific biological effects due to any osteotomy. But these have not been sufficient for a regression of osteoarthritis. The described failures thus confirm the mechanical effect of surgery.

Some knees in which the deformity had not been properly overcorrected have been revised. After the second operation which achieved a sufficient overcorrection of the deformity, all these knees did well. The evolution of their subchondral scleroses confirms an optimal redistribution of the stresses in the joint. A large joint space has reappeared. Both osteotomies must have exerted a similar biological effect. However, the first did not restore proper mechanical conditions and was followed by failure. Only the revision which changed the mechanics significantly resulted in healing.

Still more convincing are the results achieved by an operation performed on a deformity at a distance from the knee. In such cases it is difficult to attribute the regression of osteoarthritis to local changes since the operation has not been carried out in the vicinity of the affected knee but at a distance from it and sometimes on the opposite leg. These results sufficiently demonstrate that the mechanical effects of the described surgical operations are essential for success.

In most patients an even distribution of the load over the largest possible weight-bearing surfaces can only be achieved by overcorrecting the pre-operative deformity. Indeed, restoring a normal anatomical form would only re-create the mechanical conditions from which osteoarthritis originated.

From a mechanical analysis of the forces acting on the normal and deformed knee, it appears that generally osteoarthritis with a varus deformity requires a valgus osteotomy of the upper end of the tibia, whereas osteoarthritis with a valgus deformity is best treated by a varus osteotomy of the lower end of the femur. Failures in cases dealt with in other ways and statistical study of the results confirm this theoretical conclusion.

To treat osteoarthritis of the knee one must forget the concepts of classical surgery based on a restitution of the anatomical forms. These concepts must be replaced by operations with biomechanical aims. Pauwels had already drawn this conclusion for the treatment of osteoarthritis of the hip. Applying such surgery of the stresses to the knee is more difficult than to the hip. In fact, for reasons of equilibrium, the line of action of the force acting on the hip joint must pass through the centre of rotation of the femoral head. When planning surgical operations it is thus always possible to locate this force. The problem consists of achieving a good articular congruence with the greatest possible articular surface. At knee level there is equilibrium when the compressive force R intersects the axis of flexion of the joint anywhere between the centre of curvature O_2 of the lateral condyle and the centre of curvature O_1 of the medial condyle or even, in many cases, the medial edge of the joint. In order to be distributed over the largest weight-bearing surfaces the load must be moved across the axis of flexion until it passes through the centre of the joint surfaces. To calculate the recentring accurately one should be able to measure precisely the forces developed by the muscular tension band during the single support period of gait. To determine the potential of the tensor fasciae latae and biceps femoris as suggested by Blaimont et al. (1971) is a first attempt but it is not enough. The potential of the other muscles participating in the lateral tension band should also be measured. Quantitative electromyography may attain this aim in the future. At the present time the overcorrection can only be evaluated by the surgeon empirically. He should take into account the patient's weight and muscular tone and the possible traumatic origin of osteoarthritis.

Despite these difficulties related to individual cases, it is often possible to diminish the compressive articular stresses sufficiently to make them tolerable to the tissues of the knee. Well planned and precisely performed surgery will accomplish this by reducing the load exerted on the joint and bringing it back to the centre of the articular weight-bearing surfaces. When the mechanical stress is made tolerable, the signs of osteoarthritis regress and the tissues regenerate. For the patient it is equivalent to healing.

Appendix. Remarks About the Accuracy of the Calculation of Forces and Stresses in the Knee Joint

A. Introduction

The magnitude of the force transmitted from the femur to the tibia in an average person during gait has been calculated. The mean stresses evoked by this compressive force acting on the weight-bearing surfaces have then been determined. The calculation has taken into account the forces of inertia and indications from X-rays and anatomical specimens.

For the determination of the forces of inertia one must know the masses of the different moving parts and the body weight minus the supporting lower leg. The laws of spaces, velocities and accelerations of the different parts of the body must also be known. Moreover, it is necessary to formularize the behaviour of the forces inside the body, i.e. in a bony skeleton supported, moved and stressed by considerable muscular forces. Finally, in order to determine the stresses σ in kg/cm^2, the weight-bearing surfaces of the joint must be measured. All the phases of the calculation depend on these four points – weights, laws, formularization and weight-bearing surfaces – which may influence the final result considerably.

One can ask what this result is worth. In other words one must look for the errors and variations which may change the magnitude of the forces and stresses attained.

This is the object of this appendix.

1. The Weights

First, it must be stated that the aim of the calculation is not only to provide a final figure but also better to show the transmission of the forces, the influence of the lever arms, and the result of surgery. That is why the formularization is important as well in a qualitative as in a quantitative sense. The latter is true only for the individual case. The figures are then applicable to the studied subject, selected as an example.

If, in another instance, there exists some asymmetry, if the relation of the masses – left and right for instance – is different, there follows a displacement of the centre of gravity S_6 sideways.

We assume the right upper limb of the subject standing on both feet to weigh 50 g more than the left and the right lower limb 100 g more than the left, the total weight being 58.7 kg. In these conditions

$$\Delta y_7 = 0.029 \text{ cm}.$$

This displacement of y_{S_6}, even for notable asymmetries, does not reach 0.03 cm. These influences are not very great and they are due more to differences of subject than to errors.

2. Formularization

A ball and socket joint can only transmit a force the line of action of which passes through its centre. If an eccentric force is applied there is movement. In such a case equilibrium can be restored only by adding another force.

Obviously the knee joint must transmit eccentric forces during movement. Equilibrium is made possible by muscular forces the magnitude of which may be great. These muscular forces are multiple. Consequently, the distribution of the forces is statically indeterminate and variable. When a musle is relaxed several others contract to maintain balance. It is impossible to analyse this complexity. It would turn the mind away from the essential. What is essential is to approach the problem by using a sound and rational formularization. The same problem arises in building when knowing the static behaviour of the whole is more important than that of the component parts. A sound formularization gives an understanding of the forces transmitted by the bones and of the stresses due to external circumstances: body weight, inertia, movement and relative positions of the masses.

3. The Laws

The laws of velocities and accelerations are deduced from that of the spaces in a purely mathematical way. The law of the spaces results from measurements made by Braune and Fischer between 1889 and 1904. Braune and Fischer wanted accurately to locate the centres of gravity of the different parts of the body during gait.

Because of the time when these measurements were made, one may doubt their accuracy. Nevertheless, it is essential to read the published work of the authors to appreciate their care and their rigid scientific approach. They used measuring instruments of remarkable accuracy. They received generous grants from the German army to improve their machinery. To study gait, they could choose from thousands of soldiers, subjects with a morphology similar to that of the cadavers which they first experimented on. They were not limited by time and published their results only after several years of research. These ideal conditions are not easily fulfilled today. Therefore, we think that we can rely on Braune and Fischer whose results have never been contradicted by more recent experiments. To repeat their work would be superfluous and would lack originality.

But we can try to find the influence on the final result of possible errors of Braune and Fischer, in other words of errors in the law of spaces. This will be done:

1. by analysing the influence of an error of time between two successive phases
2. by imagining that all the measurements are wrong by 10% without self adjustment (relative error)
3. by imagining that all the measurements of the spaces are wrong by 0.02 cm (absolute error) and by using the theory of cumulative errors.

4. Direct Personal Measurements

We have introduced into the calculation measurements which we have made on anatomical specimens: the weight-bearing surfaces and the distance r between the centre of the knee and the insertion of force F, the component of the muscular forces parallel to the tibia. We shall analyse the influence that errors in these measurements can exert on the final result.

B. Analysis of the Influence of the Variation of Time Between Two Successive Phases

One must first be reminded how Braune and Fischer measured this interval. The walking subject was equipped with Geissler tubes fixed on the different parts of the moving body. On the tubes the position of the projections of the centres of gravity was marked. Electrical flashes were produced by a Ruhmkorff coil. All the tubes were connected with the same secondary circuit of this inductor. To determine accurately the successive phases of gait with a constant interval, the interruptions of the primary circuit were regulated by musical vibrations of regular frequency. The frequency was measured with a seconds pendulum. The average of several measurements – all agreeing – gave a figure of 260.9 phases for 10 sec, a time of 0.03832886 sec between two successive phases. Since the authors lavished particular care on the measurement of time, we can admit a maximum error of 0.1 in the number of phases for 10 sec. This would change by 0.00001470 sec the interval between two successive phases.

The accelerations and the forces of inertia are inversely proportional to the square of the time. This means that the magnitude of the forces of inertia would be modified according to the equation:

$$\frac{(0.03832886)^2}{(0.03831418)^2}=1.000766724.$$

A force of inertia of 18.872 kg (D_x phase 14) would become 18.886 kg. It would be increased by 0.074%. A force of 1.555 kg (D_x phase 12) would become 1.156 kg. These changes are obviously too small significantly to modify the further calculation and alter the stress in the joint.

C. Influence of a Systematic Error of 10% in All the Measurements of Braune and Fischer

The upper curve of Figures 32, 33 and 34 shows the evolution of the co-ordinates x, y and z of S_7 during gait. The perfect continuity of these calculated results demonstrates the accuracy of the observations on which they are based. In such curves it is not possible to modify one of the measurements without disturbing the continuity of the geometrical process.

It can be imagined that all the measurements of Braune and Fischer are affected by a percentage of error. But this percentage must be common to all the measurements to result again in a smooth curve. This could happen for an individual whose partial weight P_7 would be different. For this reason the following analysis of the error gives no essential information.

We suppose that all the measurements of Braune and Fischer are affected by a considerable error of 10%, without any self-adjustment. The calculations are made for phases 14 and 21 taken as examples.

The values of D_x, D_y, and D_z (Table 2) must be multiplied by 1.1, as must be the values of x, y, z in Table 1 and x_G, y_G and z_G in Table 4.

The modified values of x'_g, y'_g, and z'_g are used in the equations on page 51.

For $P=54.560$ kg, we find:

Phases	x'_G	y'_G	z'_G	
14	18.095	10.769	52.349	
21	−15.686	7.854	54.043	
Phases	D_x	D_y	D_z	P_7+D_z
14	10.5919	5.9312	20.7592	75.3192
21	−5.9312	5.5077	11.4389	65.9989

Then the modified values of M_x, M_y, and M_z are:

Phases	M_x	M_y	M_z
14	−500.6201	808.4255	6.7391
21	−220.7028	−714.7189	39.8102

A, B, C, D, E, K are calculated by the equations on page 51.

Phases	A	B	C
14	14.258671	15.961534	73.227493
21	15.822796	9.636506	63.860307
Phases	D	E	K
14	−499.693148	805.236438	−78.219808
21	−195.813587	−706.203314	155.083062

The compressive force R of Table 7 is changed

14	from 247 to 262 kg
21	from 199 to 210 kg

In both cases force R is increased by 6% when the initial errors of 10% have been combined in order to increase their effect.

The results would be those of an individual of 62.200 kg instead of 58.700 kg.

D. Theory of Cumulated Errors, a Variation of 0.2 mm Being Assumed for All the Measurements

In order to know the tridimensional position of the partial centres of gravity for each phase of gait, the subject equipped with the Geissler tubes was photographed simultaneously from four directions. The information was read directly from the photographic glass plates with a specially built device to avoid deformation of the picture due to development on photographic paper. From their projections on the four photographs, the centres of gravity were introduced into a system of central projections and, in this way, located in space. For their calculations the authors corrected the slight deformations due to the lenses.

We assume $\delta = 0.02$ cm the maximal error on the co-ordinates of the centres of gravity and $\eta = 01'$ the maximal error on the orientation of the angles ψ and φ.

Calculation of the Cumulated Errors

1. The accelerations are given by formulae such as:

$$\mathrm{acc}\, z_p = \frac{z_{p+1} - 2z_p + z_{p-1}}{t^2}$$

and the constant error would be $\frac{4\varepsilon}{t^2}$.

2. The error on the forces of inertia is then

$$\varepsilon' = \frac{4\varepsilon}{t^2} \cdot \frac{P}{g} = 3.081259 \text{ kg.}$$

3. The cosines α_P, β_P and γ_P are given by the equations (page 37) $\cos\alpha_P = \frac{D_x}{P}$ and so on and

$$\Delta(\cos\alpha_P) = \frac{\varepsilon'}{P}\left(1 + \cos\alpha_P \frac{(D_x + D_y + D_z + P_7)}{P_7}\right).$$

The relations are similar for $\cos\beta_P$ and $\cos\gamma_P$.

Table 43 gives the results for phases 12 to 16.

Table 43

Phases	$\cos\alpha$	$\Delta(\cos\alpha)$	$\cos\beta$	$\Delta(\cos\beta)$	$\cos\gamma$	$\Delta(\cos\gamma)$
12	0.018898	0.051385	0	0.050414	0.999822	0.101763 ($\mathrm{Mx}\cos\gamma=1$)
13	0.244778	0.054312	0.026044	0.043011	0.969229	0.091743
14	0.129671	0.047904	0.072613	0.035615	0.98889	0.090373
15	0.111226	0.084060	0.101961	0.083235	0.988569	0.162239
16	0.058341	0.082796	0.048622	0.081962	0.997122	0.163421

Table 44

	12	13	14	15	16
a_u $\pm\Delta a_u$	0.993272 $\pm$(0.00657280)	0.99981342 $\pm$(0.000091896)	0.998809562 $\pm$(0.000131569)	0.997482209 $\pm$(0.000093518)	0.996143824 $\pm$(0.000086936)
c_u $\pm\Delta c_u$	0.115803934 $\pm$(0.000305593)	0.006108615 $\pm$(0.00033881)	0.048849937 $\pm$(0.000297674)	0.070450564 $\pm$(0.000764857)	0.087735492 $\pm$(0.000328040)
a_v $\pm\Delta a_v$	0.012889272 $\pm$(0.00066869)	−0.000760309 $\pm$(0.000038174)	−0.006593779 $\pm$(0.00054777)	−0.008665915 $\pm$(0.000115222)	−0.104015319 $\pm$(0.000329499)
b_v $\pm\Delta b_v$	0.993786569 $\pm$(0.000031821)	0.992223984 $\pm$(0.00037415)	0.9908448220 $\pm$(0.000077248)	0.992405798 $\pm$(0.000037053)	0.994533502 $\pm$(0.000031253)
c_v $\pm\Delta c_v$	0.110553656 $\pm$(0.001736134)	0.124462795 $\pm$(0.000307381)	0.134819612 $\pm$(0.000314369)	0.122697336 $\pm$(0.000310799)	0.104015319 $\pm$(0.000304323)

4. Calculation of $\cos\alpha_t$, β_t and γ_t. The formulae on page 49 give:

$$\cos\gamma_t=\frac{1}{\sqrt{1+\mathrm{tg}^2\psi_4+\mathrm{tg}^2\phi_4}},$$

$$\cos\alpha_t=\frac{\mathrm{tg}\,\phi_4}{\sqrt{1+\mathrm{tg}^2\psi_4+\mathrm{tg}^2\phi_4}},$$

$$\cos\beta_t=\frac{\mathrm{tg}\,\psi_4}{\sqrt{1+\mathrm{tg}^2\psi_4+\mathrm{tg}^2\phi_4}}.$$

In order accurately to know the maximum errors of these quantities we have successively calculated the $\sqrt{\cos\gamma_t}$ by adding 1′ to ψ_4 and φ_4, then by subtracting 1′ from ψ_4 and φ_4. In each case we kept the maximum error. The errors of $\cos\alpha_t$ and $\cos\beta_t$ have been determined by using the maximum value of $\mathrm{tg}\,\psi_4$ (φ_4) divided by the smallest value of $\sqrt{\cos\gamma_t}$ and inversely.

Every time we kept the maximum error. The errors of $\sin\beta$ and $\mathrm{tg}\,\beta$ could also be deduced.

5. Calculation of $\Delta(a_u, b_u, c_u, a_v, b_v, c_v)$. These magnitudes are given by the formulae on page 51.

The errors have been calculated by using:

$$(a_u)_{\mathrm{Mx}}=\frac{(\cos\gamma_t)_{\mathrm{Mx}}}{(\sin\delta_t)_{\min}}$$

or

$$(a_u)_{\min}=\frac{(\cos\gamma_t)_{\min}}{(\sin\beta_t)_{\mathrm{Mx}}}$$

and

$$\Delta a_u=|a_{u_{\mathrm{Mx}}}-a_u|$$

or

$$|a_u-a_{u\,\min}|.$$

Every time the maximum error was chosen. The result of this calculation appears in Table 44.

6. Calculation of M_x, M_y, M_z.

These magnitudes depend on the forces of inertia D_x, D_y, D_z which are affected by a constant error $\varepsilon'=3.081259$ kg and on the quantities:

$$x'=-x_7+x_G,$$
$$y'=-y_7+y_G,$$
$$z'=z_7-z_G$$

which are the differences between two co-ordinates of the centre of gravity. The latter are affected by a constant error $2\varepsilon = 0.04$ cm. The resulting differences would look like:

$$\varDelta M_x = 2\varepsilon(D_y + D_z + P_7) + \varepsilon'(z' + y').$$

7. The calculation of R and F requires knowledge of the quantities C, D, E, defined on page 51. The errors of those values are calculated by the equations

$$\begin{aligned}\varDelta C = {} & \varepsilon'(\cos\alpha_t + \cos\beta_t + \cos\gamma_t) \\ & + D_x \varDelta(\cos\gamma_t) + D_y \varDelta(\cos\beta_t) \\ & + (D_z + P_7)\, \varDelta(\cos\gamma_t),\end{aligned}$$

$$\varDelta D = a_u \varDelta M_x + c_u \varDelta M_z + \varDelta a_u M_x + \varDelta c_u M_z,$$

$$\begin{aligned}\varDelta E = {} & a_v\, \varDelta M_x + b_v \varDelta M_y + c_v \varDelta M_z + \varDelta a_v M_x \\ & + \varDelta b_v M_y + \varDelta c_v M_z.\end{aligned}$$

8. Calculation of R and F. F is given by the equation

$$F = \frac{\sqrt{D^2 + E^2}}{r}$$

and

$$\varDelta F = \frac{D \varDelta D + E \varDelta E}{r\sqrt{D^2 + E^2}},$$

$$R = C + F,$$

$$\varDelta R = \varDelta C + \varDelta F$$

and

$$\sigma = \frac{R}{S}, \quad \sigma = \frac{\varDelta R}{S}.$$

The result is given in Table 45.

E. Influence of a Variation of the Weight-Bearing Surfaces

The measurements are affected by an error due to parallax since the projected surfaces must be larger than the actual ones. But we do not use these measurements as such. They are reduced to a common ideal size, that of the knee of the subject of Braune and Fischer. Therefore the alterations due to parallax are irrelevant since they are automatically corrected.

But two objections can be raised as far as the contours of the weight-bearing surfaces are concerned. First, in some places the limit between light and dark areas is not properly defined. Second, the margin of the light area may represent a contact area but not a weight-bearing area. We estimate the width of the doubtful marginal strip on both sides of the contours converted to the ideal size of the subject of Braune and Fischer to be 0.1 cm.[17] The surfaces have been measured for four different contours. The results are given in Table 46.

The measured surfaces do not rigorously correspond to the phases of gait as defined by Braune and Fischer. Therefore, we have drawn a continuous curve of the surfaces from which we could determine the weight-bearing area for

[17] A layer of barium sulphate suspension 0.01 cm thick was shown to be of sufficient thickness to be radio-opaque, as determined by compressing a cylinder of methylmethacrylate on the bottom of a methylmethacrylate box layered with a suspension of barium sulphate. The area of clear space caused by the compression of the cylinder was measured in an X-ray and it showed that the barium could be seen when the thickness of the solution was 0.01 cm or more. If the articular surfaces of the femur and tibia form an angle of 6° at the point of contact and if these surfaces in the immediate vicinity are considered flat, the interval between the cartilage surfaces attains a thickness of 0.01 cm at 0.1 cm from the contact point. If the angle is 45°, the interval between the surfaces attains a thickness of 0.01 cm at 0.01 cm from the contact point

Table 45

F	$\varDelta F$	R	$\varDelta R$	$\sigma = \frac{R}{S}$	$\varDelta \sigma$	%
294.1556	±57.2303	354.6082	±60.681047	19.432	±3.325	17.11
189.2861	±56.8880	260.7491	±60.339147	15.135	±3.549	23.45
175.6290	±35.6571	247.1082	±39.2783	13.716	±2.170	15.82
55.3060	±33.7827	95.1271	±37.4323	5.1301	±1.980	38.60
54.7767	±38.4808	93.5045	±42.1274	5.699	±2.171	46.20

Table 46

Surf. init.	(Surf. + 1 mm)	Diff. in %	(Surf. − 1 mm)	Diff. in %
18.012 cm^2	20.55 cm^2	14%	15.39 cm^2	15%
20.178 cm^2	22.952 cm^2	14%	14.706 cm^2	12%
10.362 cm^2	11.76 cm^2	14%	8.67 cm^2	16%
10.656 cm^2	12.204 cm^2	15%	9.324 cm^2	13%

each phase of gait. From the quantities measured with a planimeter and given in Table 46, we calculated the mean possible error of the weight-bearing surfaces and found it to be 14%. The influence of this error on the maximal stress is also given in Table 47.

Table 47

Phases	σ	σ_{Mx}	%	σ_{min}	%
12	19.432	22.594	16.27	17.044	12.29
13	15.135	17.835	17.84	13.454	11.10
14	13.716	15.875	16.28	11.976	12.28
15	5.130	5.852	14.08	4.415	13.94
16	4.699	5.6044	19.27	4.228	10.03

F. Influence of an Error in Estimating *r*

We measured on anatomical cross sections the distance r between the centre of the knee and the point of application of force F. We adopted $r = 5$ cm, which is an average of our measurements on knees brought to the common ideal size of the subject of Braune and Fischer. But one can imagine that during gait there is not only one muscular force F acting, but two forces, one anterior, F_a, and one lateral, F_l. The radii ($r = 5$ cm) joining forces F_a and F_l and the centre G of the circle would form an angle of 90° (Fig. 52).

This extreme situation supposes a linear tendon in front of the knee and another one on the lateral aspect, acting simultaneously with no connection between them. This is a worse situation than that which actually exists. But to have an idea, we have calculated for some phases of gait the extreme values that R and σ would take in such conditions. Then the resultant F would intersect the middle of the cord BC joining two points orientated at 90° and r would become $\frac{r\sqrt{2}}{2}$.

The result of the calculation is given in Table 48.

From these five measurements the average error is 26.38%. This is a limit, practically nonexistent in reality. Consequently, if several muscles act at some phases of gait or if the point

Table 48

Phases	R	$R + \Delta R$	σ	$\sigma + \Delta\sigma$	%
12	254.6082	444.370	19.432	24.349	25.30
13	260.7491	333.768	16.135	19.663	29.72
14	247.1082	313.103	13.716	17.298	26.12
15	95.1271	119.552	5.130	6.325	23.30
16	93.5045	116.1938	4.699	5.9894	27.46

of insertion of the muscles is not 5 cm from the centre of the knee, the changes in the magnitude of R and σ would always be lower than 26%.

On the other hand, Table 7 shows the variations of R and F when force R does not intersect the axis of flexion at the centre G of the knee but at the point O_1, centre of the medial condyle, or at the point O_2, centre of the lateral condyle. Between these extreme positions, r varies from 2.4 cm to 7.6 cm.

G. Direct Measurements

No direct measurement could confirm the results of the calculations. Strain gauges in close proximity to the joint or a force plate on which the subject could walk have been mentioned. It should be possible to glue strain gauges to the tibia or the femur outside the articular capsule. This requires removal of sufficient periosteum with the consequent alteration of the periarticular tissues by scarring and by the permanent presence of electric wires. These changes and the presence of the wires would probably change the physiological function of the joint. On the other hand, the experiment should be done on a normal individual. We do not think it is opportune to subject a healthy individual to any surgical risk in order to get experimental information.

The information would only be approximate since it would come from points at some distance from the joint, on the cortex of the tibia or of the femur. But the deformations due to the acting forces occur in the cartilage and in the subchondral cancellous bone (Radin and Paul, 1970; Radin et al., 1970).

One can imagine inserting strain gauges directly under the tibial plateaux. Such devices and the damage necessary to introduce them into the bone would alter the architecture of the cancellous bone and the distribution of the stresses (Pauwels, 1973b; Radin and Paul, 1970; Radin et al., 1970). Moreover, each strain gauge would indicate the local pressure. The average pressure would still have to be calculated from these figures.

The force plate can only indicate the forces transmitted *to the ground* by the walking subject. It gives absolutely no direct information about the forces exerted *on the knee*. These forces attain six times the body weight. For the most part they are due to the muscles which compress the bones against each other and cannot be measured by the force plate.

They can be calculated from the force transmitted to the ground and from the masses and accelerations of the parts of the body (Paul, 1965, 1966, 1969; Morrison, 1968, 1970). We preferred to start the calculation at the other end, from the mass and accelerations of the part of the body supported by the knee.

H. Conclusion

The possible errors which we have calculated and the acceptable variations of the measured data do not essentially modify the order of magnitude of our results. Despite the impossibility of actual direct measurements in normal joints, the values we have calculated give a good idea of the forces and stresses in normal and in pathological knees.

References

Ahlbäck S (1968) Osteoarthrosis of the knee. Acta Radiol (Stockh) Suppl 127

Ahlbäck S, Bauer GCH, Bohne WH (1968) Spontaneous osteonecrosis of the knee. Arthritis Rheum 11:705

Amtmann E, Kummer B (1968) Die Beanspruchung des menschlichen Hüftgelenks. II. Größe und Richtung der Hüftgelenksresultierenden in der Frontalebene. Z Anat Entwickl-Gesch 127:286–314

Bandi W (1972) Chondromalacia patellae and femoro-patellare Arthrose. Helv Chir Acta Suppl II

Bandi W (1980) Les résultats de l'avancement de la tubérosité antérieure du tibia. Rev Chir Orthop 66:275–277

Basmajian JV (1962) Muscles alive. Williams and Wilkins, Baltimore

Bauer GCH (1970) Diagnosis and treatment of gonarthrosis (osteoarthritis of the knee). SICOT 11th congress, Mexico, 1969. Imprimerie des Sciences, pp 369–385

Beckers L (1982) Displacement osteotomy of the tibial tuberosity. Acta Orthop Belg 48:190–193

Benjamin A (1969) Double osteotomy of the painful knee in rheumatoid arthritis and osteoarthritis with an independent assessment of results by E.L. Trickey. J Bone Joint Surg B 51:694–699

Benoist JP, Ramadier JO (1969) Luxations et subluxations de la rotule (traumatiques exceptées). Rev Chir Orthop 55:89–109

Bentley G (1970) Chondromalacia patellae. J Bone Joint Surg 52A:221–232

Blaimont P (1970) The curviplane osteotomy in the treatment of the knee arthrosis. SICOT 11th congress, Mexico, Imprimerie des Sciences, Bruxelles, pp 443–446

Blaimont P (1982) L'ostéotomie curviplane dans le traitement de la gonarthrose. Acta Orthop Belg 48:97–109

Blaimont P, Burnotte J, Baillon JM, Duby P (1971) Contribution biomécanique à l'étude des conditions d'équilibre dans le genou normal et pathologique. Acta Orthop Belg 37:573–591

Bouillet R, Gaver P van (1961) L'arthrose du genou. Étude pathogénique et traitement. Acta Orthop Belg 27:1–188

Braune W, Fischer O (1889) Über den Schwerpunkt des menschlichen Körpers. Abhandl d Math-Phys Cl K Sächs Gesellsch Wissensch 15:561–589

Braune W, Fischer O (1891) Bewegungen des Kniegelenks nach einer neuen Methode an lebenden Menschen gemessen. Abh Math-Phys Kl K Sächs Ges Wiss 17:75–150

Braune W, Fischer O (1895) Der Gang des Menschen. I. Teil. Versuche am unbelasteten und belasteten Menschen. Abh Math-Phys Kl K Sächs Ges Wiss 21:153–322

Burke DL, Ahmed AM (1980) The effect of tibial tubercle elevation on patello-femoral loading. Orthop Res Soc, Atlanta

Casuccio C, Scapinelli R (1970) Surgical treatment of osteoarthritis of the knee. SICOT 11th congress, Mexico, 1969, Imprimerie des Sciences, Bruxelles, pp 467–475

Cauchoix J, Duparc J, Lemoine A, Deburge A (1968) L'ostéotomie dans les gonarthroses avec déviation angulaire dans le plan frontal. Résultats et indications thérapeutiques. Rev Chir Orthop 54:343–360

Coventry MB (1965) Osteotomy of the upper portion of the tibia for degenerative arthritis of the knee: a preliminary report. J Bone Joint Surg A47:984–990

Coventry MB (1970) Osteotomy for genuarthrosis. SICOT, 11th congress, Mexico, 1969, Imprimerie des Sciences, Bruxelles, pp 358–361

Coventry MB (1973) Osteotomy about the knee for degenerative and rheumatoid arthritis. Indications, operative technique and results. J Bone Joint Surg A55:23–48

Coventry MB, Bowman PW (1982) Long term results of upper tibial osteotomy for degenerative arthritis of the knee. Acta Orthop Belg 48:139–156

Debeyre J, Patte D (1962) Intérêt des ostéotomies de correction dans le traitement de certaines gonarthroses avec déviation axiale. Rev Rhum Mal Ostéo-artic 29:722–729

Debeyre J, Artigon JM (1972) Résultats à distance de 260 ostéotomies tibiales pour déviations frontales du genou. Rev Chir Orthop 58:335–339

Debrunner A, Seewald K (1964) Die Belastung des Kniegelenkes in der Frontalebene. Z Orthop 98:508–523

Despontin J, Thomas P (1978) Le scanner dans l'exploration de l'articulation fémoro-rotulienne. Acta Orthop Belg 44:84–88

Devas MB (1970) High tibial osteotomy. A method especially suitable for the elderly. SICOT 11th congress, Mexico, 1969. Imprimerie des Sciences, Bruxelles, pp 463–466

Deliss L (1977) Coronal patellar osteotomy. Preliminary report of use in chondromalacia patellae. Proc R Soc Med 70:257–259

Delouvrier JJ, Morvan G, Nahum H (1981) Mesure radiologique statique des déviations des membres inférieurs. J Radiol 62:147–156

Duparc J, Massare C (1967) Mesures radiologiques des déviations angulaires du genou sur le plan frontal. Ann Radiol 10:9–10

Elftman H (1940) Forces and energy changes in the leg during walking. Am J Physiol 129:672

Elftman H (1941) The action of muscles in the body. Biol Symp 3:191

Endler F (1972) Traitement biomécanique chirurgical de la nécrose avasculaire de la tête fémorale. Acta Orthop Belg 38:537

Engin AE, Korde MS (1974) Biomechanics of normal and abnormal knee joint. J Biomech 7:325–334

Ferguson AB, Brown TD, Fu FH, Rutkowsky R (1979) Relief of patello-femoral contact stress by anterior displacement of the tibial tubercle. J Bone Joint Surg A61:159–166

Ficat P (1970) Pathologie fémoro-patellaire. Masson, Paris

Fick R (1910) Handbuch der Anatomie und Mechanik der Gelenke, vol 3. Fischer, Jena, p 535

Fischer O (1899) Der gang des Menschen. II. Teil. Die Bewegung des Gesamtschwerpunktes und die äußeren Kräfte. Abhandl d Math-Phys Cl K Sächs Gesellsch Wissensch 25:1–163

Fischer O (1900) Der Gang des Menschen. III. Teil. Betrachtungen über die weiteren Ziele der Untersuchung und Überblick über die Bewegungen der unteren Extremitäten. Abhandl d Math-Phys Cl K Sächs Gesellsch Wissensch 26:87–185

Fischer O (1901) Der Gang des Menschen. IV. Teil. Über die Bewegung des Fußes und die auf denselben einwirkenden Kräfte. Abhandl d Math-Phys Cl K Sächs Gesellsch Wissensch 26:471–569

Fischer O (1903) Der Gang des Menschen. V. Teil. Die Kinematik des Beinschwingens. Abh Math-Phys Kl K Sächs Gesellsch Wissensch 28:321–428

Fischer O (1904) Der Gang des Menschen. VI. Teil. Über den Einfluß der Schwere und der Muskeln auf die Schwingungsbewegung des Beins. Abh Math-Phys Kl K Sächs Gesellsch Wissensch 28:533–623

Föppl L, Mönch E (1959) Praktische Spannungsoptik. Springer, Berlin Göttingen Heidelberg

Frankel VH, Burstein AH (1970) Orthopaedic biomechanics. Lea and Febiger, Philadelphia

Fujisawa Y (1981) The surgical treatment of the patellofemoral joint disorders. Anterior displacement of the tibial tuberosity. J Jpn Orthop Ass 55:891–894

Fujisawa Y (1982) The surgical treatment of osteoarthritis of the knee. An arthrosocopic study of 80 knee joints. Clin Orthop Surg (Jpn) 33:42–53

Fujisawa Y, Masuhara K, Matsumoto N, Mii N, Fujihara H, Yamaguchi T, Shiomi S (1976) The effect of high tibial osteotomy on osteoarthritis of the knee. An arthroscopic study of 26 knee joints. Clin Orthop Surg (Jpn) 11:576–590

Fujisawa Y, Masuhara K, Shiomi S (1979) The effect of high tibial osteotomy on osteoarthritis of the knee. An arthroscopic study of 54 knee joints. Orthop Clin North Am 10:585–608

Fürmaier A (1953a) Beitrag zur Ätiologie der Chondropathia Patellae. Arch Orthop Unfall-Chir 46:178

Fürmaier A (1953a) Beitrag zur Mechanik der Patella und des Kniegelenks. Arch Orthop Unfall-Chir 46:78

Gebhardt W (1911) Diskussion zum Vortrag. J. Schaffer: Trajektorielle Strukturen im Knorpel. Verh Anat Ges (Jena) 25:162–168

Goodfellow J, Hungerford D, Zindel M (1976) Patello-femoral joint mechanics and pathology. J Bone Joint Surg B58:287–290

Goronwy T (1962) Tibial osteotomy for osteoarthritis of the knee – Proceedings and reports of councils and associations. J Bone Joint Surg B44:956

Harrington IJ (1974) The effect of congenital and pathological conditions on the load action transmitted at the knee joint. Instrum Mech Eng

Henry AK (1959) Extensile exposure. Livingstone, Edinburgh

Herbert JJ, Bouillet R, Debeyre J, de Marchin P, Duparc J, Ficat P, Gariepy J, Glimet J, Judet J, Maquet P, Masse P, Ramadier JO, Simonet J, Trillat A (1967) Symposium sur les gonarthroses d'origine statique. Rev Chir Orthop 53:107–198

Hughes J, Paul JP, Kenedi RM (1970) Control and movements of the lower limbs: In Simpson DC (ed) Modern trends in biomechanics, vol I. Butterworths, London, pp 147–179

Husson A (1973a) The functional anatomy of the knee joint: the closed kinematic chain as a model of a knee joint. The knee joint. Excerpta Medica, Amsterdam, pp 163–168

Husson A (1973b) La chaine cinématique fermée. Bull Assoc Anat (Nancy) 57:887–894

Husson A (1974) Biomechanische Probleme des Kniegelenks. Orthopäde 3:119–126

Izadpanah M, Keönch-Fraknóy S (1977a) Entlastung des medialen oder lateralen Kniegelenkanteiles ohne Varisierungs- oder Valgisierungsosteotomie. Z Orthop 115:21–25

Izadpanah M, Keönch-Fraknóy S (1977b) Statische Auswirkung der Varisierungs- bzw. Valgisierungsosteotomie bei Genu valgum und varum. Z Orthop 115:100–105

Jackson JP, Waugh W (1961) Tibial osteotomy for osteoarthritis of the knee. J Bone Joint Surg B43:746–751

Jackson JP, Waugh W (1970) Tibial osteotomy for osteoarthritis of the knee. SICOT 11th congress, Mexico, 1969. Imprimerie des Sciences, Bruxelles, pp 386–392

Jackson JP, Waugh W (1982) Tibial osteotomy for osteoarthritis of the knee. Acta Orthop Belg 48:93–96

Johnston RC, Smidt GL (1969) Measurement of hip-joint motion during walking. J Bone Joint Surg A51:1083–1094

Judet J (1970) Traitement des gonarthroses. SICOT, 11th congress, Mexico, 1969. Imprimerie des Sciences, Bruxelles, pp 405–407

Kettelkamp DB, Chao EY (1972a) A method for quantitative analysis of medial and lateral compression forces at the knee during standing. Clin Orthop 83:202–213

Kettelkamp DB, Jacobs AW (1972b) Tibiofemoral contact area. Determination and implications. J Bone Joint Surg A 54:349–356

Knese KH (1955) Statik des Kniegelenkes. Z Anat Entwicklungsgesch 118:471–512

Kohn Trebner A, Orlando DM, Lamenza A (1970) La osteotomia de la tibia en la artrosis de rodilla. SICOT 11th congress, Mexico, 1969. Imprimerie des Sciences, Bruxelles, pp 476–485

Kostuik JP, Schmidt O, Harris WR, Woolbridge C (1975) A study of weight transmission through the knee joint with applied varus and valgus loads. Clin Orthop 108:95–98

Krempen JF, Silver RA (1982) Experience with the Maquet barrel-vault osteotomy. Clin Orthop 168:86–96

Kummer B (1956) Eine vereinfachte Methode zur Darstellung von Spannungstrajektorien, gleichzeitig ein Modellversuch für die Ausrichtung und Dichteverteilung der Spongiosa in den Gelenkenden der Röhrenknochen. Z Anat Entwicklunsgesch 119:223–234

Kummer B (1959) Bauprinzipien des Säugerskeletes. Thieme, Stuttgart

Kummer B (1962a) Gait and posture under normal conditions with special reference to the lower limbs. Clin Orthop 25:32–41

Kummer B (1962b) La Arquitectura funcional de los huesos largos de los miembros. Lecciones de anatomia humana. Marban, Madrid, pp 685–710

Kummer B (1963) Principles of the biomechanics of the human supporting and locomotor system. In: SICOT, 9th congress. Post-graduate course. Wiener Medizinische Akademie, Vienna, pp 60–80

Kummer B (1965) Die Biomechanik der aufrechten Haltung. Mitt Naturforsch Ges Bern N.F. 22, pp 239–259

Kummer B (1966) Photoelastic studies on the functional structure of bone. Folia Biotheoretica b:31–40

Kummer B (1968) Die Beanspruchung des menschlichen

Hüftgelenks. I. Allgemeine Problematik. Z Anat Entwicklungsgesch 127:277–285
Kummer B (1969) Die Beanspruchung der Gelenke, dargestellt am Beispiel des menschlichen Hüftgelenks. Verh dtsch Ges Orthop Traumat, 55th congress, Kassel, 1968. Enke, Stuttgart
Kummer B (1972) Biomechanics of bone: mechanical properties, functional structure, functional adaptation. In: Fung YC, Perrone N, Anliker M (eds) Biomechanics: its foundations and objectives. Prentice Hall, Englewood Cliffs, pp 237–271
Kummer B (1977) Biomechanischer Grundlager „beanspruchungsändernder" Osteotomien im Bereich des Kniegelenks. Z Orthop 19:923–928
Kummer B (1982) Cinématique du genou. Acta Orthop Belg 48:28–35
Lagier R (1970) Anatomo-pathologie de la gonarthrose. In: Nicod L (ed) Die Gonarthrose. Huber, Bern, pp 12–34
Langa GS (1963) Experimental observations and interpretations on the relationship between the morphology and function of the human knee joint. Acta Anat (Basel) 55:16–38
Lange M (1951) Orthopädisch-chirurgische Operationslehre, Bergmann, München, pp 660–664
Lemaire R (1982) Étude comparative de deux séries d'ostéotomies tibiales avec fixation par lame-plaque ou par cadre de compression. Acta Orthop Belg 48:157–171
Mac Intosh DL (1970) The surgical treatment of osteoarthritis of the knee. SICOT, 11th congress, Mexico, 1969. Imprimerie des Sciences, Bruxelles, pp 400–404
Magnuson PB (1946) Technique of debridement of the knee joint for arthritis. Surg Clin North Am 26:249
Maldague B, Malghem J (1976) Le faux profil rotulien ou profil vrai des facettes rotuliennes. Approche radiologique nouvelle de l'articulation fémoro-patellaire. Ann Radiol 19:573–581
Maldague B, Malghem J (1978) Chondromalacie de la rotule: apport de la radiologie. Acta Orthop Belg 44:21–40
Maquet P (1963) Considérations biomécaniques sur l'arthrose du genou. Un traitement biomécanique de l'arthrose fémoro-patellaire. L'avancement du tendon rotulien. Rev Rhum Mal Ostéo-artric 30:779–783
Maquet P (1966) Biomécanique des membres inférieurs. Acta Orthop Belg 32:705–725
Maquet P (1969a) Charge et sollicitation mécanique des os. Le principe du hauban. Rev Med Liege 24:115–132
Maquet P (1969b) Biomécanique du genou et gonarthrose. Rev Med Liège 24:170–195
Maquet P (1970) Biomechanics and osteoarthritis of the knee. SICOT, 11th congress, Mexico, 1969. Imprimerie des Sciences, Bruxelles, pp 317–357
Maquet P (1972) Biomécanique de la gonarthrose. Acta Orthop Belg 38 Suppl I:33–54
Maquet P (1974) Biomechanische Aspekte der Femur-Patella Beziehungen, Z Orthop 112:620–623
Maquet P (1976) Advancement of the tibial tuberosity. Clin Orthop 115:225–230
Maquet P (1980a) The biomechanics of the knee and surgical possibilities of healing osteoarthritic knee joints. Clin Orthop 146:102–110
Maquet P (1980b) Rappel biomécanique in "Déséquilibres et chondropathies de la rotule." Rev Chir Orthop 66:209–211
Maquet P (1980c) Osteotomy. In: Freeman MAR (ed) Arthritis of the knee. Springer, Berlin Heidelberg New York
Maquet P (1981) Les contraintes de compression patello-fémorales. Acta Orthop Belg 47:12–16
Maquet P (1982a) Pathogénie de la gonarthrose. Acta Orthop Belg 48:45–56
Maquet P (1982b) Traitement chirurgical de l'arthrose fémoro-tibiale. Acta Orthop Belg 48:172–189
Maquet P (1982c) Traitement chirurgical de l'arthrose patello-fémorale. Acta Orthop Belg 48:194–203
Maquet P, de Marchin P (1964) Biomécanique du genou. Rhumatologie 16:465–468
Maquet P, de Marchin P (1967) Guérison, par la chirurgie, des arthroses de la hanche et du genou. Méd Hyg (Genève) 25:1440–1441
Maquet P, Simonet J, de Marchin P (1966) Étude photoélastique du genou. Rev Chir Orthop 52:3–11
Maquet P, Simonet J, de Marchin P (1967a) Biomécanique du genou et gonarthrose. Rev Chir Orthop 53:111–138
Maquet P, de Marchin P, Simonet J (1967b) Biomécanique du genou et gonarthrose. Rhumatologie 19:51–70
Maquet P, Pelzer G, de Lamotte F (1975) La sollicitation mécanique du genou durant la marche, Acta Orthop Belg 41 Suppl II:119–132
Maquet P, Van de Berg A, Simonet J (1975) Femoro-tibial weight-bearing areas. J Bone Joint Surg A57:766–771
Maquet P, Watillon M, Burny F, Andrianne Y, Quintin J, Rasquin G, Donkerwolcke M (1982) Traitement chirurgical conservateur de l'arthrose du genou. Acta Orthop Belg 48:204–261
Marchin P de, Maquet P, Fontaine J (1963a) Quelques remarques sur la radiographie des genoux arthrosiques. Utilité des clichés «en charge». Rev Med Liège 18:148–152
Marchin P de, Maquet P, Simonet J (1963b) Considérations biomécaniques sur l'arthrose du genou. Quelques remarques sur les radiographies. Rev Rhum Mal Ostéo-artric 30:775–776
Menschik A (1974a) Mechanik des Kniegelenkes. I Teil. Z Orthop 112:481–495
Menschik A (1974b) Mechanik des Kniegelenkes. III Teil. Sailer, Wien
Menschik A (1975) Mechanik des Kniegelenkes. II Teil. Schlußrotation. Z Orthop 113:388–400
Milch H (1940) Photoelastic studies of bone forms. J Bone Joint Surg A 22:621
Mohing W (1966) Die Arthrosis deformans des Kniegelenkes. Springer, Berlin Heidelberg New York
Morrison J (1968) Bioengineering analysis of force actions transmitted by the knee joint. Biomed Eng 3:164–170
Morrison J (1970) The mechanics of the knee joint in relation to normal walking. J Biomech 3:51–61
Morscher E (1978) Osteotomy of the patella in chondromalacia. Arch Orthop Trauma Surg 92:139–147
Müller W (1929) Biologie der Gelenke. Barth, Leipzig
Murray MP, Drought AB, Kory RC (1964) Walking patterns of normal men. J Bone Joint Surg A46:335–360
Nicholson JT (1970) Patellectomy as an aid in management of pain in degenerative arthritis of the knee. SICOT, 11th congress, Mexico, 1969. Imprimerie des Sciences, Bruxelles, pp 413–415
Nicod L (1970) Die Gonarthrose. Huber, Bern
Ory M (1964) Des influences mécaniques dans l'apparition et le développement des manifestations dégénératives du genou. J Belg Méd Phys Rhum 19:103–120
Palazzi AS (1961) La greffe musculaire intra-osseuse comme traitement de l'arthrose du genou. Acta Orthop Belg 27:384–386
Paul JP (1965) Bioengineering studies of the forces transmit-

ted by joints. In: Kenedi RM (ed) Biomechanics and related bioengineering topics. Pergamon, Oxford, p 369
Paul JP (1966–67) Forces transmitted by joints in the human body. Proc Inst Mech Eng 181:8–15
Paul JP (1969) Magnitude of forces transmitted at hip and knee joints. In: Wright V (ed) Lubrication and wear in joints. Sector Publishing, London, pp 77–78
Pauwels F (1950) Über eine kausale Behandlung der Coxa valga luxans. Z Orthop Chir 79:305
Pauwels F (1951) Des affections de la hanche d'origine mécanique et de leur traitement par l'ostéotomie d'adduction. Rev Orthop 37:22
Pauwels F (1958) Neue Richtlinien für die chirurgische Behandlung der Coxarthrose. Langenbecks Arch Klin Chir 289:378
Pauwels F (1959) Directives nouvelles pour le traitement chirurgical de la coxarthrose. Rev Orthop 45:681
Pauwels F (1960) Neue Richtlinien für die operative Behandlung der Coxarthrose. Verh Dtsch Orthop Ges, 48th congress, p 332
Pauwels F (1963) Basis and results of an etiological therapy of osteoarthritis of the hip joint. SICOT 9th congress, postgraduate course, vol II. Vienna, pp 31–60
Pauwels F (1964) Directives nouvelles pour le traitement chirurgical de la coxarthrose. Acta Chir Belg 63:37–76
Pauwels F (1965a) Über die Bedeutung einer Zuggurtung für die Beanspruchung des Röhrenknochens und ihre Verwendung zur Druckosteosynthese. Verh Dtsch Orthop Ges, 52th congress, Stuttgart
Pauwels F (1965b) Gesammelte Abhandlungen zur funktionellen Anatomie des Bewegungsapparates. Springer, Berlin Heidelberg New York
Pauwels F (1968) Der Platz der Osteotomie in der operativen Behandlung der Coxarthrose. Triangel 8:196–210
Pauwels F (1973a) Atlas zur Biomechanik der gesunden und kranken Hüfte. Prinzipien, Technik und Resultate einer kausalen Therapie. Springer, Berlin Heidelberg New York
Pauwels F (1973b) Kurzer Überblick über die mechanische Beanspruchung des Knochens und ihre Bedeutung für die funktionelle Anpassung. Z Orthop 111:681–705
Pauwels F (1980) Biomechanics of the locomotor apparatus. Springer, Berlin Heidelberg New York
Pirard A (1960) Notes du cours de photoélasticité. Laboratoire de Photoélasticité, Univ, Liège
Pirard A, Sibille P (1954) Recherche des lignes de nulle pression. Le Génie Civil CXXI:4–8, 28–31, 45–48
Poulhes J, Chancholle AR, Lapage J, Ferret-Bouin P (1962) Etude photoélasticimétrique de l'articulation du genou. CR Ass Anat 47th Meeting, Naples, 1961. Nancy, pp 634–640
Poulhes J, Chancholle AR, Mourlan P (1962) Etude photoelasticimétrique de l'articulation du genou chez l'homme. CR Ass Anat 47th Meeting, Naples, 1961. Nancy, 1962, pp 1098–1105
Rabischong P, Courvoisier E, Bonnel F, Peruchon E, Devaud G (1970) Étude biomécanique de la répartition des forces au niveau des condyles fémoraux en charge statique. In: Nicod L (ed) Die Gonarthrose. Huber, Bern, pp 36–52
Radin E, Paul I (1970) Does cartilage compliance reduce skeletal impact load? Arthritis Rheum 13:139–144
Radin E, Paul I, Lowy M (1970) A comparison of the dynamic force transmitting properties of subchondral bone and articular cartilage. J Bone Joint Surg A52:444–456
Radin E, Paul I, Rose R (1972) Role of mechanical factors in pathogenesis of primary osteoarthritis. Lancet: 519–522
Ramadier JO (1965) Prévention et arrêt de l'arthrose du genou avec déviation transversale. Mém Acad Chir 24–25:815–825
Ramadier JO (1967) Étude radiologique des déviations dans la gonarthrose. Rev Chir Orthop 53:139–147
Ramadier JO, Buard JE, Lortat-Jacob A, Benoit J (1982) Mesure radiologique des déformations frontales du genou. Procédé du profil vrai radiologique. Rev Chir Orthop 68:75–78
Raynal L, Verheugen P (1961) Techniques opératoires dans la chirurgie de la gonarthrose. Acta Orthop Belg 27:394–408
Reilly DT, Martens M (1972) Experimental analysis of the quadriceps muscle force and patello-femoral joint reaction force for various activities. Acta Orthop Scand 43:126–137
Renard V, Renard Lefebvre AM (1970) La résection du tissu sous-chondral altéré, par voie extra-articulaire, dans le traitement des arthroses du genou. SICOT 11th congress, Mexico, 1969. Imprimerie des Sciences, Bruxelles, pp 425–431
Rubies Trias P (1970) Operacion de Magnusson en la artrosis de rodilla. SICOT, 11th congress, Mexico, 1969. Imprimerie des Sciences, Bruxelles, pp 421–424
Shinno N (1961) Statico dynamic analysis of movement of the knee. Tokushima J Exp Med 8:101–141
Shinno N (1962) Statico dynamic analysis of movement of the knee. Tokushima J Exp Med 8:189–202
Shinno N (1968a) Statico dynamic analysis of movement of the knee. Tokushima J Exp Med 15:53–57
Shinno N (1968b) Statico dynamic analysis of movement of the knee. In: Biomechanics I, Ist int seminar; Zürich, 1967. Karger, Basel, pp 228–237
Shoji H, Insall J (1973) High tibial osteotomy for osteoarthritis of the knee with valgus deformity. J Bone Joint Surg A55:963–973
Simonet J, Maquet P, de Marchin P (1963) Considérations biomécaniques sur l'arthrose du genou. Étude des forces. Rev Rhum Mal Ostéo-artric 30:777–778
Steindler A (1965) Kinesiology. Thomas, Springfield
Tansen HH (1976) A three-dimensional kinematic and force analysis of the human tibio-femoral joint during normal walking. Thesis, Purdue University Lafayette, Indiana
Tillmann B, Brade H (1980) Morphologische und biomechanische Untersuchungen an der Facies articularis patellae. Orthop Praxis 16:462–467
Torgerson WR (1965) Tibia osteotomy in the treatment of osteoarthritis of the knee. Surg Clin North Am 45:779–785
Townsend PR, Rose RM, Radin EL, Raux P (1977) The biomechanics of the human patella and its implications for chondromalacia. J Biomech 10:403–407
Tracy M, Hungerford DS, Hungerford MW, Jones L (1982) Biomechanical effect of tibial tubercle anteriorization. Orthop Res Soc, New Orleans
University of California (1947) Fundamental studies of human locomotion and other information relating to the designs of artificial limbs. National Research Committee on Artificial Limbs, final report no 12, Washington, vol 1–2
Van de Berg A (1976) La pneumo-arthro-tomographie et les lésions ligamentaires du genou. Acta Orthop Belg 42:193–209
Van de Berg A, Collard P, Quiriny M (1982) Gonarthrose et déviation angulaire du genou dans le plan frontal. Acta Orthop Belg 48:8–27

Wagner J, Bourgois R, Hermanne A (1982a) Comportement mécanique du cadre tibio-péronier dans les genoux varum et valgum. Acta Orthop Belg 48:57–93
Wagner J, Cheval P, Nelis JJ (1982b) Incidence du degré d'avancement du tendon rotulien sur les contraintes fémoro-patellaires. Acta Orthop Belg 48:639–650
Walker PS, Hajek JV (1972) The load bearing area in the knee joint. J Biomech 5:581–589
Walker PS, Erkman MJ (1975) The role of the menisci in force transmission across the knee. Clin Orthop 109:184–192
Weill D, Jacquemin MC (1982) L'ostéotomie cylindrique fémorale supracondylienne de varisation dans le traitement chirurgical de la gonarthrose. Acta Orthop Belg 48:110–130
Weill D, Schneider M (1982) Les ostéotomies du genou dans le traitement de la gonarthrose. A propos d'une expérience de douze ans et de plus de 500 interventions. Acta Orthop Belg 48:131–138
Wolff J (1892) Das Gesetz der Transformation der Knochen. Hirschwald, Berlin

Subject Index

MIX
Papier aus verantwortungsvollen Quellen
Paper from responsible sources
FSC® C105338

If you have any concerns about our products,
you can contact us on
ProductSafety@springernature.com

In case Publisher is established outside the EU,
the EU authorized representative is:
Springer Nature Customer Service Center GmbH
Europaplatz 3, 69115 Heidelberg, Germany

Printed by Libri Plureos GmbH
in Hamburg, Germany